Advances in Environmental Microbiology

Volume 1

Series Editor
Christon J. Hurst, Cincinnati, OH, USA
Universidad del Valle, Cali, Colombia

The frontispiece for this volume is titled "The Blue Marble" by the National Aeronautics and Space Administration. Globe west appears above and globe east below. Their accrediting is NASA Goddard Space Flight Center Image by Reto Stöckli (land surface, shallow water, clouds). Enhancements by Robert Simmon (ocean color, compositing, 3D globes, animation). Data and technical support: MODIS Land Group; MODIS Science Data Support Team; MODIS Atmosphere Group; MODIS Ocean Group Additional data: USGS EROS Data Center (topography); USGS Terrestrial Remote Sensing Flagstaff Field Center (Antarctica); Defense Meteorological Satellite Program (city lights)

More information about this series at http://www.springer.com/series/11961

Christon J. Hurst
Editor

Their World: A Diversity of Microbial Environments

 Springer

Editor
Christon J. Hurst
Cincinnati, OH
USA

Universidad del Valle
Cali, Colombia

ISSN 2366-3324   ISSN 2366-3332   (electronic)
Advances in Environmental Microbiology
ISBN 978-3-319-28069-1   ISBN 978-3-319-28071-4   (eBook)
DOI 10.1007/978-3-319-28071-4

Library of Congress Control Number: 2016935616

© Springer International Publishing Switzerland 2016
This work is subject to copyright. All rights are reserved by the Publisher, whether the whole or part of the material is concerned, specifically the rights of translation, reprinting, reuse of illustrations, recitation, broadcasting, reproduction on microfilms or in any other physical way, and transmission or information storage and retrieval, electronic adaptation, computer software, or by similar or dissimilar methodology now known or hereafter developed.
The use of general descriptive names, registered names, trademarks, service marks, etc. in this publication does not imply, even in the absence of a specific statement, that such names are exempt from the relevant protective laws and regulations and therefore free for general use.
The publisher, the authors and the editors are safe to assume that the advice and information in this book are believed to be true and accurate at the date of publication. Neither the publisher nor the authors or the editors give a warranty, express or implied, with respect to the material contained herein or for any errors or omissions that may have been made.

Printed on acid-free paper

This Springer imprint is published by Springer Nature
The registered company is Springer International Publishing AG Switzerland

Dedication

There are many people whom I should thank for their having encouraged me towards my career as a scientist. When I was 9 years old I was accepted a year early into the Junior Naturalist program instructed by Edith Blincoe at the Dayton Museum of Natural History in Dayton, Ohio. That was about the time when I received a chemistry set from my parents. I set up my laboratory on a folding aluminum picnic table and for years everyone around me suffered the strange and often sulfurous odors of experiments which I was restricted to doing outdoors. The only remains of that chemistry set are the two test tube racks which sit in my office, and a metal stand which I have used as a support for ceramic modeling. Both of my parents had been chemists, my mother worked at a drinking water treatment facility and my father studied air quality. Perhaps it was little surprise to them when eventually I too headed into environmental science. The microbiology aspect of my studies began a year later, after my family had moved to Cincinnati, Ohio. In fifth grade, when I was 10 years old, at Bond Hill Elementary School I first saw a microscope and was one of the few students whom the teacher trusted to use that amazing tool without her supervision. The teacher arranged for me to spend every morning for 2 weeks during the following summer, when I was still 10 years old, in the bacteriology laboratory of Cincinnati General Hospital in Cincinnati, Ohio. There, the bacteriologist taught me how to culture throat swabs, identify the cultured bacteria using multiple tube fermentation tests, plus stain and microscopically examine the cultured bacteria. My mother faithfully dropped me off at the hospital every morning, and then returned to take me back home at noon. My father eventually left his old college parasitology textbook in a place where I could find it, and my fate in science continued to unfold.

My efforts at learning microbiology continued when J. Robie Vestal became my undergraduate research advisor and subsequently also my post-doctoral advisor at the University of Cincinnati. Robie shared with me his enthusiasm about environmental microbiology. My enthusiasm about sailing eventually resulted in Robie buying his first sailboat, and I remember the day when I helped him learn how to ride across the waters surface by harnessing the wind. He became a very dear friend of mine and together we shared some heartfelt conversations. We even shared a common name, because the initial J represented the same name for him as it does for me, James. I am delighted to think about Robie and my fond memories of him. It has been nearly 25 years since Robie passed away and yet I almost can feel him beside me as I write this, he would be smiling and proudly say that I had become a big mucky-muck. It is with tremendous appreciation that I proudly dedicate this volume and my work on this book series in memory of Robie. By having accepted me into his laboratory, Robie also merits a commendation for bravery.

J. Robie Vestal (1942–1992)

A piece of artwork which has inspired me for many decades is titled Scientist, by Ben Shahn. The title of scientist is something towards which one optimistically strives during the course of their career. That is not a title which you can presume for yourself, but instead it must be awarded to you by others. Robie was one of the people in my life who clearly were scientists and they formatively encouraged me onward. Hopefully, someday I too could be considered to merit the title.

Dedication

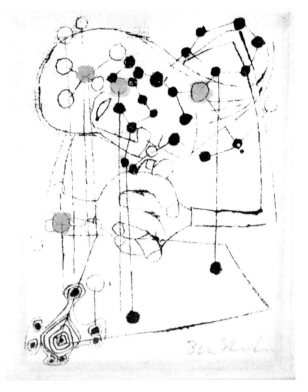

"Scientist" by Ben Shahn, used with permission

Series Preface

The light of natural philosophy illuminates many subject areas including an understanding that microorganisms represent the foundation stone of our biosphere by having been the origin of life on Earth. Microbes therefore comprise the basis of our biological legacy. Comprehending the role of microbes in this world which together all species must share, studying not only the survival of microorganisms but as well their involvement in environmental processes, and defining their role in the ecology of other species, does represent for many of us the Mount Everest of science. Research in this area of biology dates to the original discovery of microorganisms by Antonie van Leeuwenhoek, when in 1675 and 1676 he used a microscope of his own creation to view what he termed "animalcula," or the "little animals" which lived and replicated in environmental samples of rainwater, well water, seawater, and water from snow melt. van Leeuwenhoek maintained those environmental samples in his house and observed that the types and relative concentrations of organisms present in his samples changed and fluctuated with respect to time. During the intervening centuries we have expanded our collective knowledge of these subjects which we now term to be environmental microbiology, but easily still recognize that many of the individual topics we have come to better understand and characterize initially were described by van Leeuwenhoek. van Leeuwenhoek was a draper by profession and fortunately for us his academic interests as a hobbyist went far beyond his professional challenges.

It is the goal of this series to present a broadly encompassing perspective regarding the principles of environmental microbiology and general microbial ecology. I am not sure whether Antonie van Leeuwenhoek could have foreseen where his discoveries have led, to the diversity of environmental microbiology subjects that we now study and the wealth of knowledge that we have accumulated. However, just as I always have enjoyed reading his account of environmental microorganisms, I feel that he would enjoy our efforts through this series to summarize what we have learned. I wonder, too, what the microbiologists of still future centuries would think of our efforts in comparison with those now unimaginable discoveries which they will have achieved. While we study the many

Christon J. Hurst in Heidelberg

wonders of microbiology, we also further our recognition that the microbes are our biological critics, and in the end they undoubtedly will have the final word regarding life on this planet.

Indebted with gratitude, I wish to thank the numerous scientists whose collaborative efforts will be creating this series and those giants in microbiology upon whose shoulders we have stood, for we could not accomplish this goal without the advantage that those giants have afforded us. The confidence and very positive encouragement of the editorial staff at Springer DE has been appreciated tremendously and it is through their help that my colleagues and I are able to present this book series to you, our audience.

Cincinnati, OH   Christon J. Hurst

Foreword

The terms "Environmental Microbiology" and "Microbial Ecology" are often used interchangeably. In fact, a Wikipedia search for the former redirects to the latter. However, I always felt they were distinct disciplines.

To me, microbial ecology is the basic science whereas environmental microbiology focuses on specific environments, generally with a more "applied" focus. In fact, almost 50 years ago, when I published my book (*Principles of Microbial Ecology*, 1966, Prentice-Hall, Inc) I was motivated by the fact that microbial ecology had become fragmented into subfields such as soil, food, marine, aquatic, and medical microbiology. This fragmentation is now no longer the case, as the table of contents in the present series shows.

In *Principles* I wrote: "There are two groups of people who study ecological problems: those who are habitat-oriented and those who are organisms-oriented . . . . Microbial ecology embraces both approaches to ecology. Because microbes are so closely coupled to their environments, the habitat must always be taken into

Tom Brock in Yellowstone National Park, July 2002

account. And because of the peculiar experimental difficulties of microbiology, requiring the use of pure cultures for almost any study, the organism must always be reckoned with .... Thus I visualize microbial ecology not as a peripheral but as a central aspect of ecology .... Microbes play far more important roles in nature than their small sizes would suggest."

I also noted in *Principles* that microbial ecology was developing more slowly than the other branches of ecology, ... "because it ... has been out of fashion with microbiologists. Yet, microbial ecology has the potential of becoming the most sophisticated branch of ecology ..."

To a great extent, this prediction has been fulfilled, as this series illustrates and I am glad to see it published.

E.B. Fred Professor of Natural Sciences Emeritus,　　　　　　　　Thomas D. Brock
University of Wisconsin-Madison,
1227 Dartmouth Road,
Madison, WI 53705, USA

Volume Preface

This book introduces the series *Advances in Environmental Microbiology* by presenting a broad perspective regarding the diversity of microbial life that exists on our planet. Although we often identify Earth as the planet of humans, this planet does not belong to just ourselves, and the planets metabolic balance mostly is ruled by its microorganisms. And so, I use the word our in reference to all of the world's species. Humans are only one part of the biofilm that has developed on our wet, rocky world. The first two chapters of this book present theoretical perspectives that help to understand a species niche and habitat. Chapter three addresses the fossil record of microorganisms. The subsequent chapters in this volume then introduce microbial life that currently exists within various terrestrial and aquatic ecosystems.

I wish to thank the following people who very graciously served as reviewers for this volume: Teckla G. Akinyi, David A. Batigelli, Alexa J. Hojczyk, Karrisa M. Martino, and Lord Robert M. May of Oxford. I am tremendously grateful to Hanna Hensler-Fritton, Andrea Schlitzberger, and Isabel Ullmann at Springer DE, for their help and constant encouragement which has enabled myself and the other authors to achieve publication of this collaborative project.

Cincinnati, OH   Christon J. Hurst

Contents

1 Towards a Unified Understanding of Evolution, Habitat
   and Niche... 1
   Christon J. Hurst

2 Defining the Concept of a Species Physiological Boundaries
   and Barriers.. 35
   Christon J. Hurst

3 Microbes and the Fossil Record: Selected Topics in
   Paleomicrobiology... 69
   Alexandru M.F. Tomescu, Ashley A. Klymiuk, Kelly K.S. Matsunaga,
   Alexander C. Bippus, and Glenn W.K. Shelton

4 Endolithic Microorganisms and Their Habitats................ 171
   Christopher R. Omelon

5 The Snotty and the Stringy: Energy for Subsurface Life
   in Caves.. 203
   Daniel S. Jones and Jennifer L. Macalady

6 Microbiology of the Deep Continental Biosphere.............. 225
   Thomas L. Kieft

7 Microbiology of the Deep Subsurface Geosphere
   and Its Implications for Used Nuclear Fuel Repositories..... 251
   J.R. McKelvie, D.R. Korber, and G.M. Wolfaardt

8 Life in Hypersaline Environments............................ 301
   Aharon Oren

9 Microbes and the Arctic Ocean............................... 341
   Iain Dickinson, Giselle Walker, and David A. Pearce

Chapter 1
Towards a Unified Understanding of Evolution, Habitat and Niche

Christon J. Hurst

Abstract Evolution, the compartmentalization of habitats, and delineation of a species niche, function in a coordinated way and have unifying commonality in that each generally can be perceived as a balance of forces acting in opposition against one another. We intrinsically think of habitats as occupying a volume of space, but it also is possible to similarly imagine a niche in that way. This chapter suggests the depiction of a niche as having some theoretical volume, termed niche space. The concept of niche space can be assessed as pertaining either to an individual species or a larger taxonomic grouping as a whole, and can be envisioned as a multidimensional surface. Each biological group will attempt to occupy the greatest possible amount of niche space, and its competitive efforts will include speciation as either necessary or possible so as to achieve expansion into additional niche space while successfully defending that space which it already holds. Images of sculptures are used to help visualize the processes of how creating new niches and movement into existing but otherwise occupied niche space results from an outward evolutionary pressure acting against resistive exclusion. The niche determines the form of species which occupy it. If a niche closes then those species which occupy that niche will be doomed to extinction. Examples are presented which illustrate that if the same niche reopens at another time in either the same or a different location, then a new species with similar anatomical characteristics will evolve to occupy that reopened niche.

1.1 Introduction

The concepts of evolution, habitat, and niche, unite as three aspects of an interactive process. Evolution is the development of species through a process of sequential selections based upon the adequacy and comparative fitness of individuals metabolic and physiologic traits, driven by the search to find and utilize potential energy resources. The ecology of a species has two components, a habitat that is physically

C.J. Hurst (✉)
Cincinnati, OH 45255, USA

Universidad del Valle, Cali, Colombia
e-mail: christonjhurst@fuse.net

defined, and a niche that is biologically defined. If I could give you a phrase to ponder as we begin our journey together in an effort to mutually understanding these concepts, it would be that "a species inhabits its niche while dwelling within its habitat." More basically worded, the habitat is where members of a species live, and the niche is what members of the species do within that habitat. Without a potentially habitable location, there can be no evolution to develop and occupy the niches which exist within that habitat. If given the existence of a habitat, then evolution will produce species capable of using the energy resources available in that habitat, and utilization of those resources is done through interaction with other species which occupy connecting niches. Accomplishing that usage of resources will change the habitat and may include competition between species. Those changes and competitions can in turn help to drive additional evolution. Trying to comprehend how these three concepts of evolution, habitat and niche interrelate thus requires a unified understanding.

The potential habitat of a species may be very broad, and that potential habitat will have been defined by the evolutionary path which produced the species. However, the species will be restricted to residing within a more limited operational habitat which is defined by abiotic and biotic considerations. Those restrictions most notably are imposed by availability of resources and competition from other species. The potential niche of a species also will have been defined by evolution and that potential niche may be very broad, although the species will be restricted to a more limited operational niche as similarly defined by biotic and abiotic considerations. It is possible to understand the concepts of habitat and niche, both at the potential and operational levels, on a larger scale pertinent to the biological lineage of those taxonomic groupings which represent the evolutionary origin of a given species. Thus, each biological phylum, class, order, and family generally can be understood to have its collective habitat and niche.

Having prefaced this chapter with those statements, then it is important to add mention that once a species has evolved to occupy some given niche which exists within a specified habitat, the species becomes constrained by that evolutionary outcome. Those constraints include the necessity of surviving within a definable combination and range of environmental conditions that can be described mathematically in the form of vital boundaries (see Chap. 2, Hurst 2016) which are operationally defined by the physiological and metabolic limitations of a species.

1.2 Evolution

Evolution is a process of custom tailoring and its specificity is evidenced by the characteristic nature of the communities that are present within individual habitats. Thusly, many of the bacteria, fungi, and protozoa present in surface waters are not characteristic of soils. The types of autotrophic organisms, those which derive their operating energy either from photosynthesis (photoautotrophic) or chemosynthesis (chemoautotrophic) that are found in soils are often quite different from organisms found on the surfaces of leaves. These contrast with the often equally specific heterotrophs, organisms which derive their operating energy from organic compounds found in their surrounding environment. The forces of selection often are based upon nonbiological environmental characteristics that require a community able to cope with restraints or limitations such as a low pH, exposure to high radiation intensity, and absence of available oxygen. Such abiotic (not biological in origin) factors are often reasonably easy to demonstrate. However, more difficult to establish and often scientifically more interesting are the biotic (of biological origin) stresses such as the types and concentrations of readily available carbon sources that are available in environments. Biotic stresses can dominate the composition of communities in which major abiotic stresses do not make those determinations.

1.3 Habitat

Habitat is a term which could be defined as the place where everything must occur, and that definition requires understanding a complex combination of characteristics including not just the physical location but also the suitability of environmental conditions. Both habitat and niche exist not only in space but also in time. A location which represents a comfortable habitat, and accommodates the niche requirements for a species, may change to conditions that are less suitable and those changes may be cyclic. The result can be some requirement for a species either to enter hibernation until more favorable conditions hopefully resume, or for that species to move from one place to another. Examples of such movement include migrations of various distances, and we require an understanding that endosymbiont transference between hosts also represents migration.

Habitats are defined by gradients representable as mathematical variables with the physical and chemical conditions, abiotic and as well as biotic, combining to create locations that can be detailed either broadly or narrowly. Habitats can be considered from the perspective of zones, a concept that Grinnell (1917) termed as being "life zones." We can define such habitational zones in many ways, they are generally representative physical areas which include the more specified habitats occupied by individual species. Perhaps the most obvious approach would be to organize habitational zones by climate classification. That concept includes physical traits among which are the temperature of a particular location, whether that location is terrestrial versus aquatic, and altitude versus depth relative to oceanic surface level. One climate classification approach which includes both altitude and aquatic depth information yields the following list: alpine, subalpine, highland, lowland, riverine, estuarine, littoral, continental shelf, continental slope, and abyssal. Another approach is classification using latitude, which divides the earths geographical zones into frigid versus either temperate or torrid.

A large terrestrial surface area, for example, can be divided into zones on many different size scales by using increasingly selective criteria. A few of those criteria could be grouped into categories with a required understanding that the categories strongly overlap and interrelate. The category of climate is abiotic and comprised of such factors as insolation, temperature including the way in which temperature is impacted by altitude, plus atmospheric humidity which is related both to temperature and level of precipitation. Chemistry of the ground or surface matrix will be another abiotic component, in addition to whether that matrix is of either consolidated rock, loosely deposited material, or a developed soil. The physical aspects of valleys and either hilltops or mountain tops versus their slopes is a major abiotic consideration, along with a consideration of which parts of that terrestrial area are cultivated versus those which are not. Cultivation, including the application of agricultural chemicals and irrigation, will of course impact both the abiotic chemistry and biotic factors pertinent to a location. The consideration of ground surface is accompanied by additional abiotic factors including its porosity, which is a measure of how much of that matrix is open space, and its permeability which in this case would be a measure of the ease with which water as a vital fluid can move through those pores. The root zone and deep subsurface will have their own sets of abiotic and biotic factors and of these, at least the root zone will be chemically impacted by the plants and animals residing there. The living components of an ecosystem are, of course, its biotic factors. Typically the living components are sorted into two groups with regard to whether they are either autotrophs or heterotrophs as mentioned above. Autotrophs can satisfy their nutritional requirements from substances in their surroundings by using either radiant energy or chemical energy to obtain usable carbon from inorganic sources. Heterotrophs get their reduced carbon from other organisms. Biotic considerations include competitive exclusions, plus an understanding of which other predator or prey species may be present, and the possibility of beneficial symbioses and commensal interactions. When considering the biotic aspects affecting microorganisms that are endosymbiotic, we would need to address as distinguishing environmental characteristics a variety of factors relating to the hosts' individual cells, tissues and organs. In the microbial world, a life zone might be only a few millimeters thick and scarcely larger than that in circumference.

Elton (1927) did a good job of distinguishing potential terrestrial habitats in the state of California, United States of America, based upon the plants which lived in a particular area. For uncultivated land, he considered such distinct main zones as grassland, bracken with scattered trees potentially forming a sort of bracken savannah, and woodland. If considering only the woodland, then the biotic factors would include the types of trees. Zonations in the tree-tops would include leaves versus twigs and branches, plus the under bark and rotten wood of branches. For tree trunks, we would need to consider the upper part with lichens as being somewhat drier, versus the lower part with mosses and liverworts as being damper owing to run-off from the trunk. The undergrowth includes a litter containing dead leaves, underlain in some places by a moss carpet, beneath which is the surface soil, and finally the underground.

Ricketts and Calvin (1948) described that the general area between high and low tides on the ocean shore can be divided into zones based upon the periodicity of immersion, depth within the deposited shore materials, presence of other biotic materials such as plants or invertebrates, and abiotic conditions such as the presence of either rocks or anthropogenic structures. The invertebrates present, and correspondingly of course the microbial populations present, would be different in each zone. An overlapping range of abiotic and biotic factors would similarly divide aquatic environments into zones, with each zone definably having its own suitability for the species found there (Hutchinson 1957). Each of these three authors, (Elton 1927; Ricketts and Calvin 1948; Hutchinson 1957) addressed their subject from the perspective of studying macroorganisms, and their concepts of zones often were defined from the relative perspective of which macroorganism species are found in any particular place. However, their insights directly apply to the study of microorganisms.

Each habitat contains its own characteristic set of niches along with the species which correspond to those niches. The groups of species, which are termed communities, and the location of their individual component species down to the level of each individual member of a species within a habitat, relates to abiotic factors as well as biotic factors. This seems like a circular path of understanding, and indeed it truly is! The biotic considerations and to some extent the abiotic considerations, including the important way in which both relate to the availability of nutrients, vary as a result of non-equilibrium conditions as explained by Hutchinson (1961). The result of an increasingly detailed zonation is subdivision of a physical habitat into increasingly smaller spaces. Presence of the individual species found within any given place relates both to the suitability of that place, as expressed in terms of whether the corresponding needs for the niche of a particular species are satisfied, along with the possible existence of exclusionary pressure exerted by other species which are competing for similar space and resources. The questions of suitability and exclusion produce minute distinctions resulting in a three dimensional patchwork of niche spaces within a habitat and this corresponds to an arrangement of how various species are located within a biotope (in German, *biotop*) as illustrated by Hutchinson (1957).

The habitat of a species can be defined on two levels. Its potential habitat would consist of all locations where the abiotic and biotic conditions are suitable for permanent settlement. The operational habitat will be smaller, reflecting a truth that the location in which members of any one species can live often will be limited by competitive pressure from other species. Figure 1.1 is a presentation of how mutually exclusive forms appear in a three dimensional space, and is being used at this point in the chapter to visualize the subdivision of a habitat. In this case, each color represents a different species. The species represented by orange presumably could occupy this entire volume as its potential habitat, but instead this species is being restricted to occupying a smaller volume as its operational habitat due to competitive pressure from its neighboring species. This same illustration also could serve as a visualization of competition between species for niche space, and to that end I again will refer to this same figure later in the chapter.

The parts of a species' habitat either may be contiguous or separated by barriers, with some examples of barriers being listed in Table 1.1. Many barriers can be

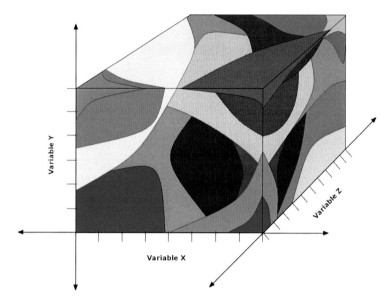

Fig. 1.1 Mutually exclusive forms in a three dimensional space. This figure is being used to represent the result of competition between different species, depicted here as different colors, for living space described as habitat within a three-dimensionally defined location. This figure also could be used more basically as a representation of competition for niche space, although that type of usage needs to be accompanied by an understanding that truly depicting niche space would require incorporating a far greater number of dimensions as variables. The factor of time has not been included here for the sake of simplicity. However, the importance of time cannot be considered negligible because the competition for niche space and its accompanying partitioning of habitat space results in three-dimensionally defined spatial arrangements or patterns of species appearance that consequently change with time. The term biotope often is used to describe the location pattern which results when species partition available living space

physically defined and are determinable by physical measurements. Such barriers also may be tangible. For example, a species' oxygen requirements may turn geographical features into barriers. Barriers are not fixed with regard to either location or time. Barriers can appear, disappear, and move, with examples being the fact that mountain ranges rise and fall, glaciers advance and retreat, continents shift, and competing species evolve or become extinct. Species tend to evolve survival mechanisms that allow their individual members to successfully move as the conditions of a habitat change and as their pertinent barriers move. Cyclical migrations are a particularly noticeable example of species movement, in which the continuing survival of a species depends upon periodic relocation between parts of their habitat. For example, seasonal migrations occur among many species of bats, birds, butterflies, caribou, salmon, and whales. Even trees migrate, albeit on a longer time scale, as evidenced by their having advanced and retreated in correspondence with cyclical patterns of glaciation (Pitelka and Plant Migration Workshop Group 1997). In the cases of monarch butterflies and maple trees, no individual member of the species completes the full migration. The populations

Table 1.1 Barriers to species movement as generally considered from the perspective of preventing infectious diseases

Categories of Chemical Barriers
Ionic (includes pH and salinity)
Surfactant
Oxidant
Alkylant
Desiccant
Denaturant
Categories of Physical Barriers
Thermal
Acoustic (usually ultrasonic)
Pressure
Barometric
Hydrostatic
Osmotic
Radiation
Electronic
Neutronic
Photonic
Protonic
Impaction (includes gravitational)
Adhesion (adsorption)
Electrostatic
van der Waals
Filtration (size exclusion)
Geographic features
Atmospheric factors
(Includes such meteorological aspects as humidity, precipitation, and prevailing winds)
Categories of Biological Barriers
Immunological
(Includes specific as well as nonspecific)
Naturally induced
(Intrinsic response)
Naturally transferred
(Lacteal, transovarian, transplacental, etc.)
Artificially induced
(Includes cytokine injection and vaccination)
Artificially transferred
(Includes injection with antiserum and tissue transfers such as transfusion and grafting)
Biomolecular resistance
(Not immune-related)
Lack of receptor molecules
Molecular attack mechanisms

(continued)

Table 1.1 (continued)

(Includes nucleotide-based restrictions)	
Antibiotic compounds	
(Metabolic inhibitors, either intrinsic or artificially supplied)	
Competitive	
(Other species in ecological competition with either the microbe, its vectors, or its hosts)	

of microbial species whose lives are either dependent upon or interdependent with migrating macrobial hosts must successfully comigrate in conjunction with their hosts, because a failure of successful comigration could well result in extinction of the microbial species. Many other species choose metabolic hibernation rather than migration as a way of surviving periodically unfavorably habitat conditions, and similarly their microbial symbionts must survive that period of hibernation.

1.4 Niche

Niche is a term which could be defined as the way in which a species biologically fits into its habitat, including its total collective activities and interactions both with its surrounding physical environment and also with other species encountered in that habitat. The term niche similarly can apply to the collective actions of even larger taxonomic groups within their respective habitats.

Each species evolves to occupy a single specific niche. And, it is important to remember that when we examine the morphological characteristics and behavior of a species we actually are viewing the biological representation of its niche. Where evolution occurs so to does homeostasis, which is a collective biological force generally perceived as preferring constancy and resisting change. Homeostasis is a stabilizing mechanism that represents the interlocking biological activities of those species which occupy connecting niches. Not everything represents happiness within that interlocking community, because the activities include competition and predation in addition to cooperation. It is important to note that both biological invasions including those attributable to pathogens, as well as natural phenomenon such as cyclical fluctuations and unidirectional changes in climate, periodically throw a figurative wrench into this mixture of activities. Homeostasis is powerful but not infallible, and conceivably it needs a measure of flexibility in order to allow the community some long term potential for change in membership and activities. Unless there were at least some capability for change, having a rigid sense of homeostatic inflexibility could result in an entire community collapsing should a part of the community fail consequent to the appearance of that figurative wrench.

Extinction is the consequence of a species failure to successfully thrive in this milieu. These three forces or evolution, homeostasis and extinction metaphorically could be envisioned as the Hindu trimurti, depicted in Fig. 1.2. The trimurti consists

1 Towards a Unified Understanding of Evolution, Habitat and Niche

Fig. 1.2 The trimurti of Hindu Gods. This image titled "Trimurti ellora" is by Redtigerxyz and used under the Creative Commons Attribution 2.0 Generic license. It shows a sculpture depicting *from left to right* Brahma, Vishnu, and Shiva at Ellora Caves. It is being presented here with Brahma the creator philosophically representative of evolution, Vishnu the preserver of balance philosophically representing homeostasis, and Shiva the destroyer philosophically representing extinction

of three gods with Brahma representing creation, Vishnu representing the existence of a balancing force, and Shiva representing destruction.

1.4.1 Evolution and Niche

The availability of an energy resource enables evolution and for this reason we cannot completely separate the two concepts of evolution and niche. Both the characteristics of the available energy resource as well as the physical and chemical characteristics of the available habitat in which that resource is located represent some of the pressures which define a niche. In turn, the requirements of a niche determine the form, physiology and metabolic characteristics of those species which evolve to occupy that niche.

As is the case with a species habitat, the niche of a species can be defined on two levels. The species potential niche would consist of all possible conditions and interactions for which a species has become suited as a consequence of its evolutionary development. The operational niche of that species will be smaller, reflecting a truth that the conditions under which members of any one species can live often will be limited by such factors as competitive interactions with other species existing within the same habitat area. At this point I refer you again to Fig. 1.1, which can be perceived as representing how the opportunities of one species are operationally restricted by competition from neighboring species. In this case, we may envision Fig. 1.1 as representing a competition for energy resources, and thus relate Fig. 1.1 to interspecies competition for niche space rather than imagining this same figure as earlier used to depict competition for habitat space.

Each individual member of a species strives to survive, as does its species, collectively trying to cling onto their niche within their habitat. However, the other member species of that ecosystem generally will not be dependent upon this one particular species. Instead, those other species selfishly are concerned only for themselves, and concerned that the niches which in turn connect to their own will remain open and be occupied successfully. Interspecies support seldom will be altruistic! If a species is displaced from its niche, then the newly occupant species may physically resemble the old one in many key aspects because form is determined by function. The displaced species may be doomed to extinction because, although the displaced species had the correct form required for that niche, the displaced species lost a contest of fitness.

If the requirements of a niche change, then the species which occupies that niche may also need to change. And, just as the opening of a niche drives evolution, so to will the closing of a niche result in extinction. As if a chain reaction, the extinction of one species often dooms to extinction many of the other species which occupy connecting niches and consequently the entire ecosystem may experience a collapse. Changing opportunities within an ecosystem, both in terms of habitat and niche, produce the result that each biological species, genus, family, order and class, has a cycle of evolution and extinction. An example would be the fact that during the Triassic and Jurassic periods, the terrestrial browsing animals were dinosaurs. Mass extinctions, such as the one which caused loss of the dinosaurs, often will be the consequence of simultaneous habitat destructions accompanied by the resulting collapse of niches. When the same habitats were restored, the niches for terrestrial browsing animals subsequently reopened and were occupied by mammals. We find that if any particular niche reopens at another time in either the same location or someplace else, then the new species which evolves to occupy that reopened niche often will have a physical appearance similar to the extinct previous occupant species because the niche determined the form of species which occupied it.

The subjects of both displacement from a niche, and the cyclical nature of niche openings and closings, as evidenced by the physical similarity of species will receive further examination later in this chapter.

1.4.1.1 Evolution and Community Cooperation

Community involvement is a consequence of evolution, with the result being that microbial communities function as interacting, coevolved assemblages. Our increasing understanding of this is based upon earlier philosophical considerations of how food chains are integrated into trophic webs, notably the writings of G. Evelyn Hutchinson (1959).

Helpful examples for understanding the coordination between evolution and community cooperation would include the following publications:

Bosak et al. (2009), Brune (2014), Chen et al. (2014), Costello and Lidstrom (1999), Fernandez et al. (2000), González-Toril et al. (2003), Gray et al. (1999), Joneson and Lutzoni (2009), Takai et al. (2001), Tomescu et al. (2008), Voolapalli and Stuckey (1999).

1.4.1.2 Morphometrics and Morphospace: Studying Evolution by Measuring a Physical Form

Morphometrics is a term which refers to the statistical description of a biological form. As an example, the comparative anatomical aspects of wing configuration in birds can be measured and statistically assessed. When those measurements are plotted in a multidimensional space with each axis representing a measurable characteristic as its variable, and each plotted point representing an individual, then collectively those plotted points define the morphospace occupied by that measured group relative to those particular variables. This concept of morphospace also is used as one of the theoretical considerations that underlies an adaptive landscape. The early description of morphometrics, which eventually developed into that of morphospace, was created by measuring the dimensions of coiled invertebrate shells (Raup 1966). David Raup (1966) described in that publication a cube with three measured variables as its axes, and he presumed any space within the cube which was not occupied by measured individuals represented shapes that were theoretically possible but not naturally realized. Bernard Tursch (1997) suggested a need to consider that restrictions upon the morphospace occupied by members of a biological group may arise not from limitations inherent to the groups evolutionary morphological capacity, but rather those restrictions may be imposed due to the fact of this group being confined to that morphospace through exclusion from other space. Figure 1.3 illustrates the concept proposed in Raup's 1966 publication along with the limiting exclusionary force concept proposed by Tursch (1997). Additional references for understanding the concepts of morphometrics and morphospace are those by Mitteroecker and Huttegger (2009), and Rohlf and Bookstein (1990). It needs to be understood that the concept of morphospace should not be limited to only three variables, but rather could represent a multidimensional space with an almost indefinite number of variables.

While the subjects of morphometrics and morphospace were developed from physically measuring macrobes, those same concepts would apply to studying the characteristics of microbes. Furthermore, physiological characteristics could be considered in addition to just physical dimensions. It also is important to understand that the concept of morphospace lacks a dimension of fitness, and as such, the fact of some species having a functional form which is both allowed and appropriate may not guarantee long term survival! Some biological groups have shown a very quick radiative evolution, which possibly suggests those groups were newcomers

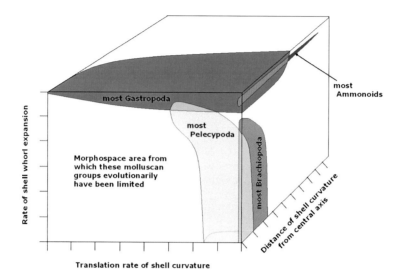

Fig. 1.3 Molluscan morphospace in three dimensions. This figure has been redrawn from Raup (1966), with incorporation of the limiting exclusionary force concept proposed by Tursch (1997) which considers that the morphospace occupied by members of a biological group may not be determined by the groups evolutionary morphological capacity but rather by limitations imposed due to the fact of that group being restricted from any other morphospace

moving into a preexisting morphospace that might already have been occupied but populated differently by other species (Shen et al. 2008), and the newcomers defeated those other species during a test of fitness. A similarly rapid expansion into existing but unoccupied morphospace very likely can occur after mass extinctions.

1.4.2 Are There Definable Rules?

Definable, yes, but perhaps they also are deniable!

1.4.2.1 First Rule: Nothing Lasts Forever

It is important to understand that the opening and closing of niches often drives the cycle of evolution and extinction. If a niche which has closed subsequently reopens in either the same place or somewhere else, then a similar species will evolve to occupy it. But, new is not necessarily better! When a change does occur in occupancy of a niche, that replacement does not automatically mean the resulting situation will be an improvement for the community. Thus, biologically selective extinctions need to be understood as a force in evolution along with the important fact that they may not be constructive (Raup 1986).

1.4.2.2 Second Rule: Evolution Is Driven by Either Necessity or Opportunity

It is true that when a new opportunity arises, something will adapt and evolve to maximally benefit from that opportunity. Alex Fraser, who was the first person to try instructing me in the subject of evolution, told me that the best way to win the game of evolution is successful evolutionary cheating rather than attempting a fair competition. A good example of evolutionary cheating is learning how to eat your competition! Or, if you are an anaerobe competing with other anaerobes that produce oxygen, then learning how to use oxygen would be a way to cheat.

1.4.2.3 Third Rule: Evolution Leads to Conformity with the Requirements of Function

A friend, Aaron Margolin, once asked me what I though about the pressures which created the form of the members of a species. He thought that the pressures of the niche determine the form of what occupies that niche. In time, I have come to agree with his conclusion. Because a species evolves to inhabit its niche, the pressures exerted by a niche determine not only the physical form but also the physiological characteristics of what biologically occupies that niche. The giant size of a whale reflects the niche which it occupies. The small size of bacteria and virus similarly represent the correct form for the niches that they occupy. We often find examples of species that have similar body forms suggesting similar ecological function, but which vary tremendously in size and evolutionary heritage. A comparative example would be the microorganism *Amoeba proteus* and the macroorganism known as the common octopus *Octopus vulgaris*, compared in Fig. 1.4.

1.4.3 Evolutionary Pressure and How It Works Along with the Concept of Niche Space

I suggest it is possible to perceive of a niche as being represented by a theoretical volume, and I will term that volume to be "niche space." This concept of niche space can be assessed as pertaining either to an individual species or a larger taxonomic group as a whole. The entire assemblage of species which comprises a community can be imagined as occupying a sum amount of niche space. Biological groups will attempt to occupy the greatest possible amount of niche space with that action including speciation as either necessary or possible, and consequently

Fig. 1.4 *Amoeba proteus* and *Octopus vulgaris*. The *upper image* titled "Amoeba proteus" is credited to Cymothoa exigua and used under the Creative Commons Attribution-Share Alike 3.0 Unported license. The *lower image* is of *Octopus vulgaris* and titled "Octopus2," it is by Albert Kok and used under the Creative Commons Attribution-Share Alike 3.0 Unported, 2.5 Generic, 2.0 Generic and 1.0 Generic license. These images have not been corrected to relative size

biological groups compete with each other both to achieve successful evolutionary expansion and to defend the niche space which they already hold. Creation of new niches and movement into existing but otherwise occupied niche space occurs through a combination of two forces, an outward evolutionary pressure and a resistive exclusion, acting in opposition against one another.

Evolutionary pressure represents the compulsive evolution of biological groups as driven by a pressure to occupy additional niche space. Factors which contribute to evolutionary pressure will include overcrowding and the necessity of competition for available resources within the niche space presently occupied by that biological group. Evolutionary pressure is facilitated by opportunity. Facilitation includes the desired possibility through adaptive radiation to gain additional resources by either occupying unused niche spaces, creating new niches, or perhaps successfully competing for coveted niches occupied by some other group.

Evolutionary pressure will be restricted by a force of exclusion which occurs when something else already occupies the desired additional niche space, and that something else does not wish to leave. The exclusionary force actually is a sum of individual competitive forces from multiple species, and collectively that force represents part of the homeostatic mechanism which resists the introduction of new species. Figure 1.1 can serve to represent how competition for niche space might appear in three dimensions, with the individual colors in that figure each representing a different species. Indeed, this is how the three-dimensional allocation of niche spaces, the respective volumes of those niche spaces, and the surfaces of those individual allocations, might appear if we tried to judge an ecological situation by looking for and identifying the presence of individual microbial species within something like a complex biofilm.

1.4.3.1 Using Sculpture to Envision the Dimensionality of an Individual Species Niche Space

Figure 1.1 is inadequate in some ways because it restricts our perspective by requiring us to imagine and depict the definition of niche volumes and their surfaces in terms of only the three variables used as axes for that illustration. In actuality, the niche of a species will be delineated by a huge number of variables acting in combination, and consequently the niche will have a very complex dimensionality. It is possible to consider some measure of higher dimensionality by envisioning regular dodecahedrons and icosahedrons, allowing their greater numbers of axes to represent a correspondingly greater number of variables. Figure 1.5 depicts a regular dodecahedron and regular icosahedron, both of which still have limitations with respect to the number of axes and hence variables which can be envisioned. Figure 1.6 shows models of viral capsid structures in which the regular icosahedral surfaces are extensively subdivided, potentially allowing us to metaphorically better imagine the complex geometry which would define the volume of a niche

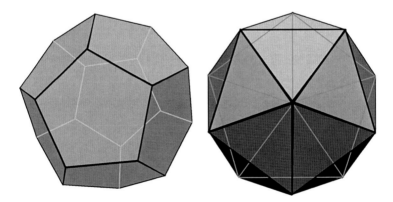

Fig. 1.5 Regular dodecahedron and regular icosahedron. These figures are being used to visualize a niche as a volume with a definable surface. The axes of these regular polyhedrons could represent a corresponding number of those variables which would define a niche. The dodecahedron is on the *left*

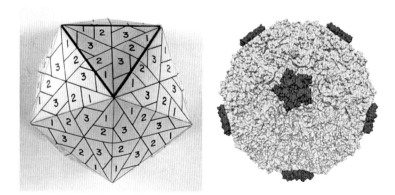

Fig. 1.6 Three dimensional picornavirus models. The *left image* is of a paper sculpture titled "Three dimensional model of human rhinovirus type 14" and appears with permission of the author (Hurst et al. 1987). That paper sculpture depicts the relative positions of the three major capsid proteins, respectively designated viral peptides 1, 2, and 3, which comprise the majority of a picornavirus protein shell. The *right image* titled "Rhinovirus," was computer generated by an anonymous author and is used under the Creative Commons Attribution-Share Alike 3.0 Unported license. These two images are being presented to illustrate that by subdividing its edges and faces, the basic structure of a regular icosahedron with its given number of axes can be depicted as a volume whose surface structure has greater complexity. The resulting subdivision potentially allows us to visually imagine the volume of these polyhedrons, depicted by their surfaces, as representing a greater number of variables than might the comparatively simpler mathematical constructions shown in Fig. 1.5

as visualized by its surface. Still, if we tried to visualize a niche with its true multidimensional character, then a single species niche might resemble something more like the rhombic dodecahedron sculpture which Vladimir Bulatov created using a three dimensional printer. Continuing with this metaphorical representation, the voids in Bulatov's sculpture would be occupied by niche space belonging to other species (Fig. 1.7) whose activities interconnect with the single depicted species.

1.4.3.2 And, Using Sculpture to Envision the Total Niche Space of a Biological Group

The creation of new niches, and movement of biological groups into preexisting but otherwise occupied niche space, conceptually occurs through a combination of evolutionary pressure and exclusion acting as forces in opposition. The exclusion force represents in total those species which already occupy other, preexisting niches. Description of this concept can be made easier with the right illustrations, and for that purpose I again am using sculptures.

Figure 1.8, which is an aluminum casting made of a fire ant nest (genus *Solenopsis*), can be perceived as metaphorically representing how a biological group occupies niche space. This figure can be considered an illustrative example

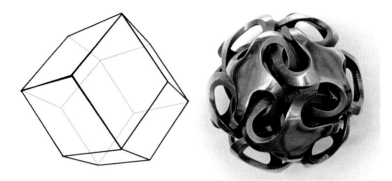

Fig. 1.7 Rhombic Dodecahedron as a line drawing and a sculpture. The *left image* shows the edges that define a basic rhombic dodecahedron. The *right image* of a printed metal sculpture is titled "Rhombic Dodecahedron I" and appears courtesy of Vladimir Bulatov. This sculpture by Bulatov is being used here to suggestively depict a species niche as a hypothetical volume that mathematically represents those numerous variables and their parameter ranges which define the niche, with that mathematical volume visualized by its surface complexity

on many different levels. We can view the total nest as occupying a generally conical volume of space as defined by three variables, an approach conceptually similar to the illustration used in Fig. 1.1. The total volume of the nest's tunnel system figuratively could represent for us the total amount of niche space occupied by some particular taxonomic group, much as we visually can perceive the total amount of environmental space that had been occupied by this nest. Appropriately, we might imagine arthropods as being the group whose niche space is represented by this casting.

The taxonomic group metaphorically depicted in Fig. 1.8 will have attempted to achieve an evolutionarily successful existence by adapting and integrating its way into an interconnecting network which included niches belonging to other taxonomic groups, with all of the groups together producing an ecosystem. By examining in greater detail the total volume of niche space occupied by this taxonomic group, we can perceive that the volume is not simply conical but does instead have a geometry which is very complex both in terms of the occupied volume of space and its delineated surface. Each of the extending outward branches is defined by its own set of axes and occupies its own definable space.

Biologists often tend to envision the path of evolution as only a series of lineage lines which connect sequentially evolved species, as if we simply were connecting sequentially numbered dots on a page. However, the full understanding of evolution is not just a matter of drawing connecting lines. We do instead need to comprehend that each series of lineage lines represents a process through which individual species developed over the course of time as an evolutionary response, through sequential modification and adaptation of preexisting biological forms, to the pressures of those niches which each species would come to occupy. Thus, by drawing lineage lines that connect sequentially evolved species, we actually are

Fig. 1.8 Aluminum casting of a fire ant nest, genus *Solenopsis*. This casting symbolically is being used to represent an aspect of evolution. Its purpose is illustrating the concept that creation of new niches and movement into existing but otherwise occupied niche space occurs through a combination of two forces, evolutionary pressure and resistive exclusion, acting against one another as forces in opposition. Biological groups attempt to occupy the greatest possible amount, depicted here as a volume, of niche space with that action including speciation as either necessary or possible and consequently biological groups compete with each other to achieve successful evolutionary expansion. In this example, initial formation of the nest volume and its structure by the ant colony can be perceived as representing an outward expansion force acting as a pressure applied against resistance from the surrounding soil and that expansion pressure was accompanied by removal of the surrounding soil. If the colony was alive when the casting was made, then the molten aluminum could be perceived as having represented a successful displacement of existing species (the ants) from their niches by overwhelming evolutionary pressure arising from some other biological group (represented here by the molten aluminum). Following mass extinctions, the surviving biological groups often expand their total niche space by movement into already available morphospace that had been populated differently. If this nest was vacant when the casting was made, then flow of molten aluminum into the existing but vacated space could represent the relatively rapid evolutionary expansion into unoccupied niche space which is effected by biological groups as they quickly undergo radiative evolution following a mass extinction. This casting weighs 25.3 lbs and is 21.5 in. high. The image courteously was provided by David Gatlin of Anthill Art

1 Towards a Unified Understanding of Evolution, Habitat and Niche

following how the niche of that biological group has changed over time. Changes which may occur in the requirements of a niche occupied by some given species then oblige evolution to subsequently produce an appropriate new species which is adapted to the new set of requirements. By examining the sequential development of species we can derive an understanding of what the niche characteristics and requirements were at any given time point both collectively for some biological group and individually for its member species, and we also can understand how those niches changed through time.

The individual nest tunnels whose branches are shown in Fig. 1.8 could each represent a different pathway taken during the evolution of species and, as with speciation, many of the pathways interconnect. The branchings and relative volumes observable for each section of this ant nest thus could represent how the total amount of niche space occupied by a biological group both historically and presently has been divided among its subgroups. As we follow a single tunnel outward from the center of the nest, we could imagine that we are following the history of one evolutionary branch of this biological group, and the terminal end of each tunnel could represent the niche of a single species.

The next time that either you or I squash an insect in our home that has violated the "size and proximity rule," with larger bugs being tolerated if they are further from where we happen to be sitting inside of our home, we could think of that species as being the end of one ant tunnel extension and the entire nest representing either all arthropods in general or just some subgrouping of arthropods. The same conceptual construction must hold true each time that we look at any plant, animal, or microorgansim. The ants which created the nest were an original radiative pressure acting against the resistance of the soil. If the ant nest had been abandoned prior to the casting, then pouring in the aluminum could be perceived to represent evolutionary expansion into vacant niches following a massive extinction. If the ants still were occupying the site when the aluminum was poured, then the aluminum could be perceived as a new radiative evolutionary pressure which occupied those niches by overpowering and destructively displacing the previous occupants in a challenge of adaptive fitness.

There are many different ways for a group to evolutionarily adapt and resultingly occupy niche space. As such, not all radiative expansions would need to take the same form, with some expansions branching more extensively than do others. Examples of this variable expansion concept can be imagined by comparing Fig. 1.8, which shows the nest type made by genus *Solenopsis*, versus Fig. 1.9 which shows the less branching types of tunnel systems typically made by *Aphaenogaster treatae*, and by the carpenter ant *Camponotus castaneus*. The weights of these two nest castings shown in Fig. 1.9 are closely similar to one another, and thus metaphorically we could imagine them as representing how two different biological groups might occupy similar volumes of niche space. The far greater weight of the nest casting shown in Fig. 1.8 metaphorically would represent a biological group that occupies a greater volume of niche space.

Fig. 1.9 Aluminum castings of *Aphaenogaster* and *Camponotus* nests. These two aluminum castings of ant nests are: *top*, *Aphaenogaster treatae* (cast weight 2 lbs, 6.5 in. high); and *bottom*, carpenter ants *Camponotus castaneus* (cast weight 2.5 lbs, 22 in. high). These images are being used to symbolically represent the concept of there being many different ways to occupy an approximately similar volume of niche space. These images kindly were provided by David Gatlin of Anthill Art

1.4.4 Does Persistence of a Niche Equate to Long Term Evolutionary Success for Its Occupying Species?

It clearly does in some cases, but success is relative and may be temporary.

1.4.4.1 Persistence of a Niche may Lead to Success of the Occupant Group

Stromatolites are biofilm structures produced by growing microorganisms, especially cyanobacteria, with a concomitant accretion of inorganic materials. In the fossil record, stromatolite structures have been found which might date back to 3.5 billion years ago. Stromatolites still are produced in certain areas of the world. Although we do not know whether the cyanobacterial species which actively form todays stromatolites are the same species which existed that long ago, we at least do know that the niches have persistence and that this microbial group has maintained its place in those niches. The crocodilians, members of the order Crocodylia, are large semiaquatic predatory reptiles which first appeared roughly 83.5 million years ago as evolutionary descendents of an even older group. The continued presence of crocodilians and their ecological prominence indicate not only persistence of their niches but also the ability of that order to maintain its place in those niches.

1.4.4.2 But, Keeping Ones Place in a Niche Is Not Guaranteed and Your Defenders may Be Few!

Developing the correct form for functioning within a niche may not confer upon a species the certainty of success, even if its niche has long term stability, because there always is competition for a viable niche and all niches are valuable. Every individual member of a species strives to survive in this constant contest of fitness, as does its species which tries to cling onto the niche that it is occupying, because displacement from ones niche can lead to extinction. We need to remember a key truth, which is that other species around you will care only about what you do and not about who you are! In the long term, the many other species which occupy connecting niches in an ecosystem are not dependent upon the survival of a specific neighboring species. Instead, the only concern of those other species is that the niches upon which they themselves depend are occupied successfully.

I will use trilobites as a possible example of how displacement from a niche can occur on a large scale. The trilobites (Levi-Setti 1993), members of the class Trilobita, occupied the largest possible grouping under the arthropods and successfully existed in the oceans for more than 270 million years with about 17,000 species having been identified. Appearance of the class Trilobita started in the Early Cambrian about 521 million years ago and its members flourished for a long time. The trilobites obviously had a general body form which was appropriate for their niches. It is necessary to make a cautious guess when thinking that two species with similar physical forms, one extinct and the other living, are occupants of a similar niche and that guess could be quite wrong! Given that caveat, I will provide some examples suggesting that although the trilobites are long gone, their niches may have continued to exist until the present time. Success of the trilobite body form unfortunately did not guarantee the group's survival even while their niches continued to exist. Eventually, by the end of the Devonian period, only one order of trilobites existed and that remaining order disappeared about 252 million years ago during the mass extinction at the close of the Permian.

The earths climate has changed considerably since the time when trilobites were a prominent biological group. First evolution of trilobites occurred in the Cambrian period during which it is believed the mean atmospheric O_2 level was approximately 12.5 % (percent) by volume, mean atmospheric CO_2 content was 4500 ppm (parts per million) which roughly is equal to 0.45 %, with a peak atmospheric CO_2 level of 7000 ppm, and mean surface temperature was 21 °C. Decline of the trilobites occurred in the Devonian period during which the mean atmospheric O_2 level was approximately 15 % by volume, the mean atmospheric CO_2 content was 2200 ppm, and the mean surface temperature was 20 °C. Extinction of the trilobites occurred at the end of the Permian period. It is believed that during the Permian period the mean atmospheric O_2 level was approximately 23 % by volume, the mean atmospheric CO_2 content was 900 ppm, and the mean surface temperature was 16 °C. The contrasting environmental data for our current timepoint are: mean atmospheric O_2 level approximately 21 % by volume, mean atmospheric CO_2 content 400 ppm (about 0.04 %), and mean surface temperature of about 14.6 °C. Even if we could try to discount a role for temperature change, these modifications in atmospheric chemistry would have produced severe changes in ocean chemistry and may have influenced the competitive fitness of the trilobites against their eventual marine successors.

When looking at trilobite fossils, we can identify by similarity of body form that many of the niches which trilobites once occupied seem to again be occupied by groups of arthropods. Most of the arthropod groups which now seem to occupy the previously trilobite niches had a temporal coexistence during the gradual demise phase of the trilobites, leading me to speculatively suggest that many of the current niche occupants may be descendents of species that successfully outcompeted the trilobites. The demise of trilobites might also have been affected by long term environmental changes. If environmental changes could have proven less deleterious to other arthropod groups than to the trilobites, then environmental change may have affected species fitness and contributed to displacement of trilobites from their niches.

One example which I can suggest of an extant aquatic biological group that may exist in niches once held by trilobites, is the arthropod group known as horseshoe crabs (order Xiphosura which evolved during the Silurian period, about 443 to 419 million years ago) as shown in Fig. 1.10. Other aquatic inhabitants of niches potentially similar to those once occupied by trilobites could include the

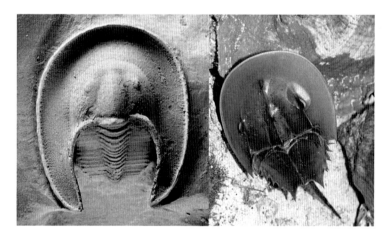

Fig. 1.10 *Harpes macrocephalus* and *Limulus polyphemus*. The *left image* is of *Harpes macrocephalus* and appears courtesy of Riccardo Levi-Setti. The *right image* of a horseshoe crab titled "Limulus polyphemus (aq.)" is by Hans Hillewaert and used under the Creative Commons Attribution-Share Alike 4.0 International license. This is a young *Limulus*, and so the two animals depicted would have been similar in size

Fig. 1.11 *Drotops armatus* and *Acanthaster planci*. The *left image* titled "Drotops armatus, Middle Devonian, Bou DOb Formation, Jbel Issoumour & Jbel Mrakib, MaOder Region, Morocco - Houston Museum of Natural Science - DSC01626" is by Daderot and used under the Creative Commons CC0 1.0 Universal Public Domain Dedication. The *right image* titled "Crown of Thorns-jonhanson" is of the starfish *Acanthaster planci*, by jon hanson and used under the Creative Commons Attribution-Share Alike 2.0 Generic license. These images have not been corrected to relative size

Fig. 1.12 *Dalmanites limuloides* and *Arnoglossus laterna*. The *left image* is of *Dalmanites limuloides* and appears courtesy of Riccardo Levi-Setti. The *right image* titled "Arnoglossus laterna 2" is of a scaldfish, by Hans Hillewaert and used under the Creative Commons Attribution-Share Alike 4.0 International license. These images have not been corrected to relative size

echinoderms known as starfish (class Asteroidea which evolved about 485 to 443 million years ago), see Fig. 1.11. Niches that once belonged to the trilobite genera *Dalmanites* perhaps now are occupied by vertebrates with an example being flatfish belonging to the order Pleuronectiformes (see Fig. 1.12). The flatfish are relative newcomers by having evolved 66 to 56 million years ago, which was long after demise of the trilobites.

Fig. 1.13 *Isotelus maximus* and *Bathynomus giganteus*. The *left image* is of the trilobite *Isotelus maximus* and appears courtesy of Riccardo Levi-Setti. The *upper right image* titled "Bathynomus giganteus" is by Borgx and used under the Creative Commons Attribution-Share Alike 3.0 Unported license. The *lower right image* titled "Giant isopod" by NOAA exists in the public domain because it contains materials that originally came from the U.S. National Oceanic and Atmospheric Administration, either taken or made as part of an employee's official duties. These two species have a similar mature size

All of the trilobites were marine and yet we can identify many existing terrestrial species whose physical forms are similar to those of trilobite groups. This similarity of physical forms leads to a suggestion that these existing terrestrial species may be occupying niches similar to those which once existed in marine environments and were occupied by trilobites. Examples which I can suggest of arthropod groups that span the range of aquatic and terrestrial environments and may occupy niches similar to those of trilobites are isopods (order Isopoda which evolved 300 million years ago) and amphipods (order Amphipoda which evolved about 38 to 33 million years ago), see Figs. 1.13 and 1.14 respectively. Neither the isopods nor amphipods as groups are old enough to have competed directly against trilobites. However, the isopods and amphipods belong to the class Malacostraca which evolved roughly 541 to 485 million years ago, a time period when the trilobites did exist. Examples which I can suggest of arthropod groups that now exist in terrestrial niches similar to those of the trilobites include: centipedes (class Chilopoda which evolved 418 million years ago), Fig. 1.15; millipedes (class Diplopoda which evolved about 428 million years ago), Fig. 1.16; and springtails (class Collembola which evolved perhaps 400 million years ago), Fig. 1.17.

1 Towards a Unified Understanding of Evolution, Habitat and Niche

Fig. 1.14 *Triarthrus eatoni*, *Hyperia macrocephala*, and *Eurydice pulchra*. The *top image* is an x-ray photograph of the trilobite *Triarthrus eatoni* and appears courtesy of Riccardo Levi-Setti. The *lower left image* of *Hyperia macrocephala* titled "Hyperia" is by Uwe Kils and used under the Creative Commons Attribution-ShareAlike 3.0 license. The *lower right image* titled "Eurydice pulchra" is by Hans Hillewaert and used under the Creative Commons Attribution-Share Alike 4.0 International license. These images have not been corrected to relative size

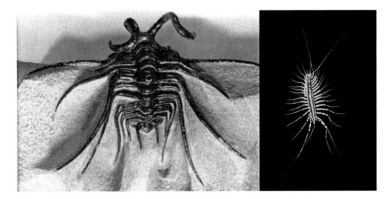

Fig. 1.15 *Dicranurus monstrosus* and *Scutigera coleoptrata*. The *left image* is of *Dicranurus monstrosus* and appears courtesy of Riccardo Levi-Setti. The *right image* titled "Scutigera coleoptrata MHNT" is of a house centipede, by Didier Descouens and used under the Creative Commons Attribution-Share Alike 4.0 International license. These images have not been corrected to relative size

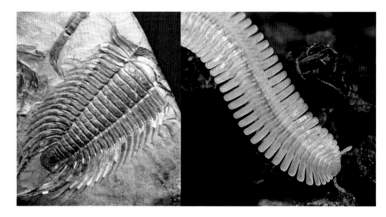

Fig. 1.16 *Paradoxides paradoxissimus* and *Brachycybe lecontii*. The *left image* is of *Paradoxides paradoxissimus* and appears courtesy of Riccardo Levi-Setti. The *right image* titled "Brachycybe lecontii (Platydesmida) millipede (3680001399)" is by Marshal Hedin and used under the Creative Commons Attribution 2.0 Generic license. These images have not been corrected to relative size

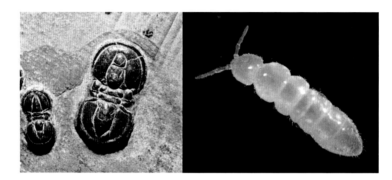

Fig. 1.17 *Itagnostus interstrictus* and *Folsomia candida*. The *left image* titled "Itagnostus interstrictus - Wheeler Shale, Utah, USA - Cambrian period ($\approx$ −507 MA) - 39.25°N 113.33°W" is by Parent Géry and used under the Creative Commons Attribution-Share Alike 3.0 Unported license. The *right image* titled "Folsomia candida - Soil Fauna Diversity" is of a springtail, by Cristina Menta and used under the Creative Commons Attribution 3.0 Unported license. These images have not been corrected to relative size

There is no certainty as to why other arthropod groups may have outcompeted the trilobites so severely. One interesting possibility is that eye structure was involved. The eye lenses in trilobites were made of calcite (Schwab 2002). Today, the only existing group of animals known to use calcite eye lenses are the brittle stars (Aizenberg et al. 2001). With the current global climate change patterns and increasing level of atmospheric carbon dioxide, our planet may be headed back to a climate similar to that of the Silurian period (443.4 to 419.2 million years ago) during which time the trilobites still were thriving, when the mean atmospheric carbon dioxide content over the period's duration is believed to have been approximately 4500 ppm, equal to 16 times the pre-industrial level. The Silurian mean atmospheric molecular oxygen content over the period's duration was approximately 14 % by volume (70 % of modern level), the mean surface temperature was approximately 17 °C (2.5 °C above modern level), and sea level was approximately 180 m above present day with short-term negative excursions. It is impossible to know what will be occupying the niches once held by the trilobites if our planets climate reverts to conditions of the Silurian period, but it will not be a matter of the trilobites enjoying a return to those "good old days."

1.4.4.3 Using Vertebrates as a Reference Point for Studying the Cyclical Nature of Niche Openings and Closings

Just as the opening of a niche drives evolution, so to does the closing of a niche result in extinction. Each biological species, genus, family, order and class, may have a cycle of evolution and extinction. In Hindu, the cosmic functions of creation, maintenance, and destruction are personified as a trinity. As I alluded to earlier in this chapter, perhaps from the standpoint of speciation we could consider Brahma the creator to represent the fact that opening of a niche drives evolution, Vishnu the preserver could represent homeostasis, and Shiva the destroyer might then represent the fact that closing of a niche produces extinction (Fig. 1.2).

We generally can conclude that most of the same niches which we know to exist today also existed during at least some point in prehistoric times. However, the result of niche similarity presumably producing similar forms through evolution allows us to understand that many niches which previously existed are not open now. I would give as one example the extinct fungal genus *Prototaxites*, see Fig. 1.18, which produced a body structure that was the size of a modern tree at a time when the existing trees were comparatively minute (Hobbie and Boyce 2010). That same niche seems never to have reopened even though the general form of

Fig. 1.18 Fossilized *Prototaxites*. Photograph by Charles R. Meissner, Jr. who formerly was associated with the United States Geological Survey. This image is being used with permission from the Smithsonian Institution

Prototaxites still is suitable and used by the saguaro cactus *Carnegiea gigantea*, which has a slightly similar size and shape. Perhaps the saguaro could be said to occupy a somewhat similar niche if one presumes that *Prototaxites* had photosynthetic symbionts.

The opening and closing of niches can be a process that cyclically repeats itself many times. When a niche opens, something will evolve to fill that niche. And yet, if the niche closes then the species which occupied that niche will become extinct. Should that same niche reopen, either in the same place or somewhere else, then a similar and perhaps nearly identical species will evolve to occupy it. Subsequent reclosure of the niche will doom its new occupying species to extinction. This chapter is, of course, about microbiology and microbes must follow that same rule! The fossil record increasingly provides us with valuable knowledge about the evolutionary past of microorganisms (Bosak et al. 2009; Tomescu et al. 2008) some of which were indeed quite large (Hobbie and Boyce 2010). And yet, because of the small size and relatively few distinguishing physical features of most microorganisms, historically we have looked at the fossil record of macroorganisms to help illustrate the cyclical process of niche existence and corresponding similarity of the occupying species.

If we narrowed our focus to only vertebrates, from the beginning of the Jurassic period until the end of the Cretaceous period, the large browsing terrestrial vertebrate animals were dinosaurs. During the Cretaceous–Paleogene extinction event, when many habitats were destroyed, those niches collapsed. When the habitats were restored, the niches subsequently reopened and were occupied by mammals. There were numerous stout, horned quadrupedal species of the herbivorous dinosaur family Ceratopsidae which evolved in North America and Asia during the Late Cretaceous period. The cerotopsids existed from perhaps 77 million years ago until the time of that extinction event when their niches closed as their habitats collapsed 66 million years ago. Perhaps the most famous of those large, horned ceratopsids was the genus *Triceratops* which existed during the end of that family's time period, from about 68 to 66 million years ago. The mammalian family Rhinocerotidae, whose body characteristics bear some resemblance to the Ceratopsidae, eventually evolved perhaps 38 million years ago after those earlier niches again had opened. Members of the group rhinoceros still exist in at least some parts of the world. Although, in many instances hunting by humans rather than niche closure has contributed to extinction of the later rhinocerotids.

If for a moment we chose to consider just mammals, we could turn our attention to the sabre-toothed cats and cat-like mammals as obvious examples for the cyclic principle of niches repeatedly opening and closing. These animals were carnivores with exceptionally long canine teeth and they have evolved and gone extinct a number of times as their corresponding niches opened and closed in different locations, see Fig. 1.19. Niches appropriate for sabre-toothed cat-like animals opened in Africa, Europe, and North America about 37.2 million years ago. That opening produced the various species which belonged to the genus *Eusmilus* of the placental family Nimravidae. When those niches closed about 28.4 million years ago, *Eusmilus* went extinct. The reopening of those niches, similarly in North

1 Towards a Unified Understanding of Evolution, Habitat and Niche

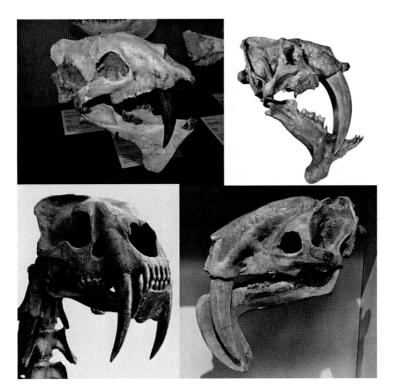

Fig. 1.19 Skulls of extinct sabre-toothed cat-like mammals. *Clockwise beginning in the upper left* these appear in chronological sequence based upon the order of their species evolutionary appearance. The *upper left image* is a painted cast of a *Eusmilus* skull and appears courtesy of Kesler A. Randall, San Diego Natural History Museum. The *upper right image* is a skull of *Barbourofelis* and appears courtesy of Ross Secord, Nebraska State Museum. The *lower right image* titled "Thylacosmilus sp skull BMNH" is by Alexei Kouprianov and used under the Creative Commons Attribution 3.0 Unported license. The *lower left image* titled "Smilodon head" is by Wallace63 and used under the Creative Commons Attribution-Share Alike 3.0 Unported license. These images have not been corrected to relative size

America, Eurasia and Africa, about 16 million years ago produced the various species which belonged to the genus *Barbourofelis* of the placental family Barbourofelidae. Those niches gradually closed, with the last ones remaining open in North America until around 5.3 million years ago and consequently *Barbourofelis* became doomed to extinction. The niches for sabre-toothed cat-like animals were open in South America from approximately 10 to 3 million years ago, producing the genus *Thylacosmilus* of the marsupial family Thylacosmilidae and then dooming *Thylacosmilus* to extinction. Eventually, about 2.5 million years ago, the appropriate niches did open once more in the Americas at which point the placental genus *Smilodon* of the family Felidae (true cats) evolved and existed until those niches closed roughly 10,000 years ago. Drawing conclusions about possible correlations between physical traits and their functionality can be deceptive. An

Fig. 1.20 *Tiarajudens eccentricus*. This image titled "Tiarajudens eccentricus skull" is by Juan Cisneros and used under the Creative Commons Attribution 3.0 Unported license. *Tiarajudens eccentricus* was a herbivorous species of therapsid reptiles that had sabre-like teeth and lived about 260 Ma (million years ago)

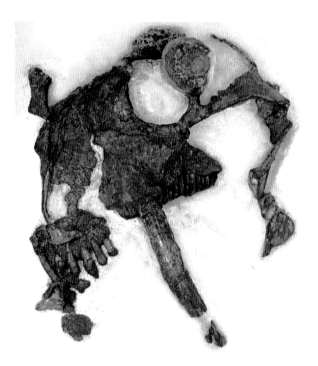

example of this would be the guess that such pronounced sabre-like upper canine teeth are used for killing animal prey. That very well may have been the usage employed by *Eusmilus*, *Barbourofelis*, *Thylacosmilus*, and *Smilodon*, but evolutionarily need not have been the purpose for such teeth. Similarly prominent sabre-like teeth also evolved in herbivorous reptiles such as *Tiarajudens eccentricus*, see Fig. 1.20, a therapsid that lived about 260 Ma (million years ago). The history of herbivorous mammals with sabre-like teeth includes both the extinct pantodont *Titanoides primaevus*, whose temporal range was approximately 59 to 56 Ma, and the extant Siberian musk deer *Moschus moschiferus*, see Fig. 1.21. We thus need to consider that those sabre-like teeth may have served more for display activities than predation, and the reasons for extinction of their species may have been entirely unrelated to possessing such teeth.

1.4.4.4 The Only Species Which Can Be Said to Represent End-Products of Evolution Are Those Species Which Are Extinct

The trilobites, the four groups of sabre-toothed cats and cat-like mammals, the cerotopsids, pantodonts, *Tiarajudens eccentricus*, and *Prototaxites*, all represented end products of evolution. It is likely that their fateful demise also doomed to extinction numerous species of symbiotic microorganisms. Coextinction

1 Towards a Unified Understanding of Evolution, Habitat and Niche

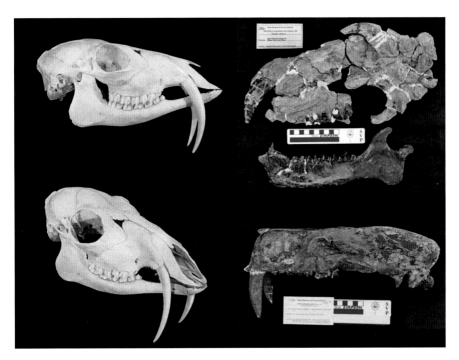

Fig. 1.21 Skulls of mammalian herbivores with sabre-like teeth. The *images on the left* are of the extant Siberian musk deer *Moschus moschiferus*. The *images on the right* are of the extinct pantodont *Titanoides primaevus*, with the temporal range of *Titanoides* having been approximately 59 to 56 Ma. Both the *upper left image* titled "Porte musc Profil 2," and the *lower left image* titled "Porte musc perspective 5," are by Didier Descouens (2011) and being used under the Creative Commons Attribution-Share Alike 4.0 International license. Both the *upper right image* "P 15520," and the *lower right image* "P 15523" are courtesy of William F. Simpson, Field Museum of Natural History. These images have not been corrected to relative size

particularly is imaginable as having occurred with the loss of such a large taxonomic group as the trilobites.

The crocodilian species existing today are not end products of evolution, although we can say that the crocodilians as a group successfully have demonstrated residential stability in their niche space. Stability of the crocodilians may have provided their microbial symbionts with similar stability. Why would I suggest only a maybe? Each microbial symbiont of the crocodilians would have been challenged to many tests of fitness during the 83.5 million years which have passed since members of the order Crocodylia first appeared, and those challenges may have resulted in numerous of the microbial symbiont species having been displaced from their niches. The crocodilians would not have been bothered by those displacements. Instead, the crocodilians would have continued onward by accepting the new symbionts in those niches. We also must consider the reverse perspective, which is that success of the crocodilians may in part be due both to the success of their microbial symbionts, and to success of the interactions between the crocodilians and their microbial symbionts.

If we somehow could presume that the microbial species which formed stromatolites 3.5 billion years ago are the same species as currently occupying those continuing niches, a presumption that may not be very likely, then the stromatolite builders would be extremely good examples of evolutionary success and stability in a niche although those species also would not represent end-products of evolution.

Acknowledgements I long have appreciated my conversations with Alex Fraser (1923–2002), which started me on the path of understanding evolution, and my conversation with Aaron Margolin which helped me to understand how the requirements of a niche direct evolution. During my consequent journey towards biological enlightenment I have learned to identify some of the sign posts along this path, although I still do not completely understand all of the processes which create the path. My journey continues towards the goal of achieving a better and more unified understanding of evolution, habitat and niche, and I have been grateful for the company of my traveling companions.

I wish to thank Vladimir Bulatov, David Gatlin, Carol L. Hotton, Riccardo Levi-Setti, Kesler A. Randall, Ross Secord, and William F. Simpson for generously provided me with images to use in this chapter. I further wish to thank Ross Secord and William F. Simpson for helpfully corresponding with me regarding the ecology of mammals with sabre-like teeth.

References

Aizenberg J, Tkachenko A, Weiner S et al (2001) Calcitic microlenses as part of the photoreceptor system in brittlestars. Nature 412:819–822
Bosak T, Liang B, Sim MS et al (2009) Morphological record of oxygenic photosynthesis in conical stromatolites. Proc Natl Acad Sci USA 106(27):10939–10943
Brune A (2014) Symbiotic digestion of lignocellulose in termite guts. Nat Rev Microbiol 12:168–180
Chen L-X, Hu M, Huang L-N et al (2014) Comparative metagenomic and metatranscriptomic analyses of microbial communities in acid mine drainage. Int Soc Microb Ecol J 2014:1–14
Costello AM, Lidstrom ME (1999) Molecular characterization of functional and phylogenetic genes from natural populations of methanotrophs in lake sediments. Appl Environ Microbiol 65:5066–5074
Elton C (1927) Animal ecology. Macmillan, New York
Fernandez AS, Hashsham SA, Dollhopf SL et al (2000) Flexible community structure correlates with stable community function in methanogenic bioreactor communities perturbed by glucose. Appl Environ Microbiol 66:4058–4067
González-Toril E, Llobet-Brossa E, Casamayor EO et al (2003) Microbial ecology of an extreme acidic environment, the Tinto River. Appl Environ Microbiol 69:4853–4865
Gray ND, Howarth R, Rowan A et al (1999) Natural communities of *Achromatium oxaliferum* comprise genetically, morphologically, and ecologically distinct subpopulations. Appl Environ Microbiol 65:5089–5099
Grinnell J (1917) The niche-relationships of the California thrasher. Auk 34:427–433
Hobbie EA, Boyce CK (2010) Carbon sources for the Palaeozoic giant fungus *Prototaxites* inferred from modern analogues. Proc R Soc Biol Sci 277:2149–2156
Hurst CJ (2016) Defining the concept of a species physiological boundaries and barriers. In Hurst CJ (ed) Their world: a diversity of microbial environments. Advances in environmental microbiology, vol 1. Springer, Heidelberg, pp 35–67

Hurst CJ, Benton WH, Enneking JM (1987) Three-dimensional model of human rhinovirus type 14. Trends Biochem Sci 12:460

Hutchinson GE (1957) Concluding remarks. Cold Spring Harb Symp Quant Biol 22:415–427

Hutchinson GE (1959) Homage to Santa Rosalia or why are there so many kinds of animals? Am Nat 93(870):145–159

Hutchinson GE (1961) The paradox of the plankton. Am Nat 95(882):137–145

Joneson S, Lutzoni F (2009) Compatibility and thigmotropism in the lichen symbiosis: a reappraisal. Symbiosis 47:109–115

Levi-Setti R (1993) Trilobites, 2nd edn. University of Chicago, Chicago

Mitteroecker P, Huttegger SM (2009) The concept of morphospaces in evolutionary and developmental biology: mathematics and metaphors. Biol Theory 4(1):54–67

Pitelka LF, Plant Migration Workshop Group (1997) Plant migration and climate change. Am Sci 85:464–473

Raup DM (1966) Geometric analysis of shell coiling: general problems. J Paleontol 40:1178–1190

Raup DM (1986) Biological extinction in earth history. Science 231:1528–1533

Ricketts EF, Calvin J (1948) Between pacific tides an account of the habits and habitats of some five hundred of the common, conspicuous seashore invertebrates of the Pacific Coast between Sitka, Alaska, and Northern Mexico, rev edn. Stanford University Press, Stanford

Rohlf FJ, Bookstein FL (eds) (1990) Proceedings of the Michigan morphometrics workshop. Museum of Zoology, University of Michigan, Ann Arbor, MI

Schwab IR (2002) The eyes have it. Br J Opthalmol 86:372

Shen B, Dong L, Xiao S et al (2008) The avalon explosion: evolution of ediacara morphospace. Science 319:81–84

Takai K, Komatsu T, Inagaki F et al (2001) Distribution of archaea in a black smoker chimney structure. Appl Environ Microbiol 67:3618–3629

Tomescu AMF, Honegger R, Rothwell GW (2008) Earliest fossil record of bacterial–cyanobacterial mat consortia: the early Silurian Passage Creek biota (440 Ma, Virginia, USA). Geobiology 6:120–124

Tursch B (1997) Spiral growth: the 'museum of all shells' revisited. J Mollusc Stud 63:547–554

Voolapalli RK, Stuckey DC (1999) Relative importance of trophic group concentrations during anaerobic degradation of volatile fatty acids. Appl Environ Microbiol 65:5009–5016

Chapter 2
Defining the Concept of a Species Physiological Boundaries and Barriers

Christon J. Hurst

Abstract This chapter presents the concept that physiological boundaries can be described for each species based upon the evolutionarily established capabilities and metabolic needs of that species. This concept is envisioned as a vital boundary consisting of a center point enveloped by two concentric theoretical closed surfaces. The boundary center represents optimum environmental conditions for that species, enabling members of the species to achieve their normal longevity. Suitability of the environmental conditions lessens with outward distance from the boundary center. The species inner vital boundary surface defines the minimum limit of environmental conditions which would allow a sufficient longevity for achieving numerical replacement. Physical locations where the environmental conditions meet the requirements either for inclusion within the inner boundary region which is encompassed by the inner vital boundary, or equate the inner vital boundary itself, would represent potential habitat areas for that species. The outer vital boundary represents combinations of environmental conditions which allow members of the species a survival time of 1 min. Physical locations meeting the environmental requirements for the interboundary region, which lies between the inner and outer vital boundary surfaces, represent areas where that species could survive only temporarily. Physical locations where the environmental conditions are beyond the outer vital boundary represent barriers to movement by that species. Two species could interact in nature only if their vital boundaries overlap. These boundaries theoretically could be depicted mathematically. Calculations estimating conditions just inside the outer vital boundary would have application for ascertaining short term survival under extreme conditions.

C.J. Hurst (✉)
Cincinnati, OH 45255, USA

Universidad del Valle, Cali, Colombia
e-mail: christonjhurst@fuse.net

2.1 Introduction

Species differ with regard to the habitat locations in which they live, and in large part such ecological differences result from the ways in which those species have evolutionarily adapted to survive in accessible environmental locations (Grinnell 1917, Hutchinson 1957). This chapter offers one perspective for studying those differences by presenting the concept that each species can be described as having physiological boundaries defined by the evolutionarily established physiological capabilities and metabolic needs of that species. This concept is envisioned as a vital boundary consisting of two concentric theoretical closed surfaces. For the purpose of illustration I have represented this concept in Fig. 2.1 with those surfaces depicted as two concentric polyhedrons. The surface of the inner polyhedron represents the species inner vital boundary. The surface of the outer polyhedron represents the species outer vital boundary. There would be a centerpoint for the inner polyhedron, which would represent the center of the species boundary, but for lack of artistic skill I have not depicted that centerpoint in Fig. 2.1. The favorability of environmental conditions is presumed to be optimum for a given species at the center point of their inner vital boundary.

Biologists and environmental microbiologists often become accustomed to thinking of a species habitat in terms of geospatial coordinates, and thus we mentally pin species to a permanent map location just as museum specimens get pinned into a fixed location within an exhibiton case. When we pin species to a map location we do not account for the fact that environmental conditions at the identified location will change over the course of time, and species must be allowed to move when those conditions change. Otherwise, our geospatially pinned species would reach the same fate as do museum specimens, existing only as evidence of what once had life. One of my goals in presenting this habitat definition concept is removing those geospatial pins by offering an alternative, which is to understand the criterion that define a species choice of where it resides.

According to this habitat definition concept, members of a species would have a predictable average population longevity time (L_t) for each combination of environmental conditions to which they were exposed. At the center of the species vital boundary the favorability of environmental conditions would be optimum, and correspondingly the expected longevity time for members of the species would be greatest. When living under the conditions at the center of a species boundary, members of that species potentially could achieve their average normal longevity (L_n). The favorability of environmental conditions decreases as the mathematically defined distance increases from the center of the boundary outward. The surface of the inner polyhedron shown in Fig. 2.1 represents the species inner vital boundary, where the environmental conditions are minimally adequate for members of that species to experience the average longevity they would require from birth in order to achieve numerical replacement (L_r). Although the maximum radius from the centerpoint outward theoretically would be infinite, in a practical sense we would need to presume that population survival time has some functional lower limit. The

2 Defining the Concept of a Species Physiological Boundaries and Barriers

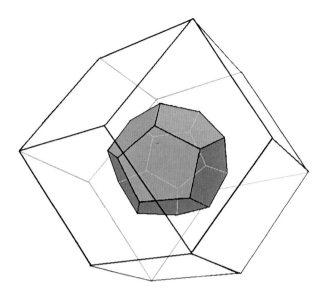

Fig. 2.1 Inner boundary depicted within an outer boundary. This illustration depicts the concept of a species having an inner vital boundary concentrically located within an outer vital boundary. The boundaries would be theoretical closed mathematical surfaces with the axes of these polyhedrons potentially representing some of the environmental variables that would be used in defining those boundary surfaces. The boundary surfaces would be less symmetrical than depicted here. Also, the mathematical distance between the inner and outer vital boundary surfaces would not be uniform

outer polyhedron outlined in Fig. 2.1 would represent the surface of the species outer vital boundary, where the evironmental conditions are adequate for members of that species to experience an average longevity of one minute and that survival duration has been chosen as the functional lower limit for longevity. Thus, L_t equals L_n at the center of a species boundary, L_t equals L_r at the surface of the species inner vital boundary, and L_t equals 1 min at the surface of the species outer vital boundary. By presenting each L_t value as an average, I allow each of those values to have a statistical deviation range which can account for variation between individual members of a species and also can account for differences between subpopulations of a single species.

The members of a species are obliged to live within their species vital boundaries. This concept of boundaries is not something marked out as if a territory on a map, it is instead a statement of the functional requirements and limitations of that species. From the perspective of a given species, any geographically or otherwise identifiable physical locations which satisfy either requirements of the inner boundary region which is encompassed by the surface of the inner vital boundary, or at minimum the environmental conditions of the inner vital boundary itself, would represent areas where members of the species potentially could have a permanent residence. Those physical locations would be potential habitat areas for the species because the environmental conditions in those locations could allow members of the species to have sufficient longevity for completing their normal replicative cycle. Many species have evolutionarily established inner vital boundary conditions that allow a very broad range of suitable habitat locations. Contrastingly, in the case of some symbiotic microorganisms, the species inner vital boundary conditions may limit the members of that symbiont species to permanent residence only within a few specific tissues of certain host species. A species continually must find at least one habitat location in which its members successfully can reside within the requirements of their inner vital boundary, because otherwise that species would become extinct.

From the perspective of a species, any physical locations which satisfy either the interboundary region which spans between the inner and outer vital boundaries, or at minimum the environmental requirements defined by their outer vital boundary, would represent places where members of the species could survive but only temporarily rather than permanently. Those places have only temporary suitability for the indicated species because a population of that species could not survive to complete its life cycle under such relatively inadequate conditions and thus the species population eventually would be fated to die off unless the members of that species could transition to a more suitable location.

Physical locations where the environmental conditions are beyond the outer vital boundary of a species represent barriers to movement by that species. While a particular species cannot survive in a metabolically active condition for very long beyond its own outer vital boundary, its habitat locations may be sufficiently diverse that while moving within their habitat the members of this species may cross the boundaries of many other species. This overlap of boundaries allows biological interactions between the members of different species.

From the perspective of this habitat definition concept, the mathematically representable boundaries of a species will have been established by evolution and those boundaries remain fixed. The physical locations which are adequate for habitation by a species will, however, shift with changing environmental conditions. Species do tend to evolve survival mechanisms that allow their members to successfully shift their physical location as changing environmental conditions shift the physical locations of their pertinent barriers so that a species will not become trapped beyond its outer vital boundary. The only way for members of a species to survive the conditions beyond their outer vital boundary is by having developed a metabolically inert survival structure. Examples of such structures are bacterial endospores and viral capsids. The vital boundaries of a species can be moved only by that species undergoing further evolution. Many species indeed have moved their boundaries through evolution and thereby successfully crossed existing barriers.

2.2 Further Understanding the Nature and Limitations Imposed by Boundaries and Barriers

The fact that a particular physical location meets the environmental conditions required for permanent residence by some given species does not mean that the suitability is unchangeable. Environmental changes can be either permanent or cyclical, and those changes often result in the need for species to develop migrational capabilities. Examples of seasonal migrations are those made by species of bats, birds, caribou, and whales. In the cases of migrations by monarch butterflies and maple trees (Pitelka and Plant Migration Workshop Group 1997), no individual member of the species completes a full migration cycle. The capability of a species' members to migrate also facilitates colonization of new areas. Cyclical migrations could be viewed as highly evolved forms of dispersal. And, as the macroorganisms migrate, so too must there be a corresponding migration of those microbial species whose lives are either dependent upon or interdependent with the migrating macrobes. Successful migration requires an absence of physiologically defined barriers that could block the migrational routes.

Suitable residential locations for a given species may be either contiguous or separated by barriers and what represents a barrier to one species many not represent a barrier to some other species. Figure 2.2 presents one aspect of the concept that barriers often separate those locations which could represent potentially suitable habitat locations for a species. The outer vital boundary for the big cats extends some distance from the shoreline, provided that either the water depth is sufficiently shallow that the animals could walk or the distance is not greater than the cats capacity for swimming. An expanse of water which exceeds the travel limitations of the big cats will represent a barrier for those cats. The shoreline represents a barrier from the perspective of sharks because they have no terrestrial capabilities. For some aquatic reptiles such as sea turtles, and fish such as the grunion (genus *Leuresthes*), their species inner vital boundary spans the shoreline as evidenced by the fact that those aquatic species lay their eggs on the land. The fact that Grunion rely upon aquatic respiration makes them susceptible to quickly suffocating if they strand when spawning on the land surface, a vulnerability which suggests the mathematical distance between their inner and outer vital boundaries may be relatively small with regard to this aspect of their environmental requirements.

2.2.1 Barriers Are Not Fixed in Location and Time

Barriers determined by the physiological capabilities and metabolic requirements of a species often can be physically defined and determinable by physical measurements. An example of the tangible nature of many barriers is the fact that a species' oxygen (molecular oxygen) requirements may turn geographical features such as

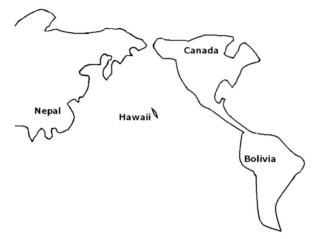

Fig. 2.2 Big cats and sharks. This illustration has been given a humorous title, but represents the concept that for each species there potentially might be many different places that could serve as suitable locations for residence. Unfortunately, it may not be possible for a single species to reach all of those locations. The depiction here is of mountain dwelling cats which could move by land connection between the Canadian Rockies and South American Andes. An insurmountable aquatic barrier currently prevents large cats from moving between the Americas and the Nepalese Himalayas. Mountain dwelling cats have not crossed the aquatic barrier which blocks them from reaching the Hawaiian mountains. Sharks can reach all of the land masses but the shoreline represents a barrier which they cannot cross

either high mountains or shorelines into barriers. It is important to remember that the physical locations of barriers are not permanently fixed. Some barriers may shift and others even disappear as with the rise and fall of mountain ranges and water surface levels. Some types of barriers have a relatively short term cyclical occurrence resulting from either tidal, daily, seasonal, or annual climatic fluctuations. Examples of short term cyclical barriers would include those produced by weather patterns including temperature and changes in precipitation; the presence, levels and flowrates of surface and subsurface water including tidal patterns; thermoclines and haloclines. Other cyclical changes have longer periodicity such as those which involve glaciation cycles and plate tectonic movements.

We humans have shown an exceptional capability for cultural evolution which has allowed us to survive in physical locations that would otherwise be too inhospitable. Controlled combustion (Berna et al. 2012), insulating clothing (Toups et al. 2011), and the use of containers as means for storing water have allowed us to establish permanent residence in geographical locations where we otherwise could reside only temporarily. These same technological achievements plus additional developments including the invention of boats as a mode of transportation, the ability to store and carry breathable atmosphere, climbing equipment, and pressurized suits have helped us to travel beyond our species natural geographical barriers and survive in locations where the ambiental environmental conditions are well beyond our outer vital boundary. It is important to understand that an astronaut in deep space who is wearing a pressure suit hasn't really changed his vital boundaries, he is just ensuring that those vital boundary conditions are satisfied within his suit.

The capacity of our species for accomplishing cultural evolution has expanded our abilities in many regards, enabling us to interact with other species whose vital boundaries do not overlap with our own outer vital boundary. However, if inability to cross a physiologically defined barrier precludes movement of a particular species from an unsatisfactory location to some suitable location, and presence of the barrier does not change, then that species must hope to move its vital boundaries. The presumptions are that a species vital boundaries are fixed by biological evolution and that movement of a species vital boundaries can be achieved only through further biological evolution.

2.2.2 Understanding the Factors that Define a Species Vital Boundaries

The concept of a species vital boundaries includes a complicated mixture of factors and sometimes those factors are very highly species specific. Table 2.1 lists some examples of environmental factors that are useful in understanding the vital boundaries of a species. Such environmental factors visually could be represented by the axes of the polyhedrons depicted in Fig. 2.1 and those factors could be employed as variables for calculating population survival time as described later in this chapter. Thus, Fig. 2.1 helps with understanding the conditions of the boundary center, inner boundary region, inner vital boundary, interboundary region, and outer vital boundary. Figure 2.3 is intended to help with understanding how an overall optimum combination of environmental conditions defines the center of a species vital boundary. When examining Table 2.1 and Fig. 2.3 it is important to consider how individual environmental factors, which can be considered and represented as mathematical variables, affect the survival conditions for an individual species.

It is important to notice that micronutrient minerals are included in Table 2.1 because they would be considered natural environmental factors. Other micronutrients such as vitamin C (World Health Organization 1999), which are organically generated by food species and acquired by ingesting those food species, are of key importance but I have not considered those food-generated micronutrients as being environmental factors for the purpose of this proposed concept. Similarly, neither the availability of food species, nor presence of predators, nor infectious disease are included in the estimation of L_t values because they also are not being considered as environmental variables for the purpose of this proposed concept. Thus, there are limitations when using this proposed concept for stating whether or not the combination of environmental factors which describes a physical location would suggest that location to represent an adequate habitat for allowing some species to achieve either their L_n or L_r. The environmental

Table 2.1 Examples of environmental factors representing variables for use in understanding vital boundaries

Factors with potential applicability to all species
Ambiental temperature
Ambiental external body pressure
Ambiental level of ionizing radiation
Distance to suitable resting surface
Inclination angle of the surface
Potential for adherence to the surface
Potential toxicity of the surface
Heavy metals including those representing micronutrient minerals
Natural and synthetic toxins (includes antibiological compounds)
Photoperiod
Level of specifically required wavelengths for photosynthesis
Factors which could apply to species using terrestrial respiration
Atmospheric gases
Carbon dioxide
Carbon monoxide
Chlorine
Oxygen
Ozone
Sulfur dioxide
Distance to available drinking water
Flow velocity of the surrounding atmosphere
Humidity
Precipitation
Factors which could apply to species using aquatic respiration (includes microbes living in liquid medium)
Dissolved gasses
Carbon dioxide
Carbon monoxide
Oxygen
Ozone
Sulfur dioxide
Dissolved halogens
Chlorine
Iodine
Flow velocity of the surrounding water
pH
Salinity

conditions in that location might be entirely suitable, but the species might not thrive in that location for other reasons including absence of food species, predation, and also competition against other species that have similar niche requirements.

2 Defining the Concept of a Species Physiological Boundaries and Barriers

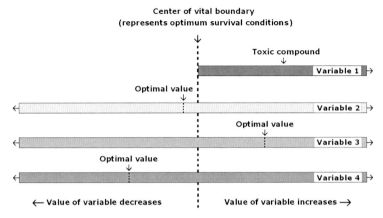

Fig. 2.3 Center of vital boundary relative to optimal parameter values. This figure gives a hypothetical representation of how the parameter values of different environmental variables relate to the expected longevity for members of an example species. Most of the individual environmental variables which are important in defining the boundaries for a given species will have a survivable parameter range and optimal value. Some of the variables could be considered unidirectional, signifying that the deleteriousness of variance from the optimal value is assessed in only one direction as represented by Variable 1. Toxic compounds such as pesticides and antimicrobial compounds would be representative examples of variables that qualify for unidirectional assessment because their optimal parameter value may be zero and any presence of the compound acts to decrease survival of the affected species. Most variables will be assessed bidirectionally, signifying that the deleteriousness of variance from the optimal value would be assessed in two directions. Variables 2 through 4 depicted in this figure would be assessed bidirectionally. Conceptually the center of a species vital boundary represents its optimum environmental conditions and allows members of that species the possibility of surviving to fully realize their normal longevity

2.2.2.1 Defining Those Factors as Mathematical Variables

If we reflect upon the suitability of an available resting surface (yes, pun intended) as one example of the factors listed in the Table 2.1, many aquatic species seemingly need no resting surface. This same example factor can have only a plus or minus influence for numerous other species that we might consider, such as many microbial species which survive best in a biofilm (Huq et al. 1983, Kiørboe et al. 2003, van Schie and Fletcher 1999), and some of the sessile aquatic invertebrates which physically attach to their resting surfaces, of which both groups prefer having the presence of a suitable surface but may seem unaffected by the inclination angle of the surface onto which they have become attached. However, the inclination angle of the resting surface does affect numerous species in a unidirectional sense meaning that the optimal may be zero, any deviation from zero is detrimental, and the deviation effectively can be measured in only one direction. A consideration would be species for which a surface angle of zero inclination may be optimally required and increasing the deviation from that optimal inclination angle will make the surface less suitable although assessing that angular deviation as positive versus negative is unimportant because falling off to one side of a physical surface is equally deleterious as falling off to the other side. As one general example, many aquatic species such as water striders of the family Gerridae reside by resting at the water surface and unsteady water surface angles due to water turbulence may make the surface unsuitable for meeting the requirements of that species.

Interactions between species occur in locations where the environmental requirements of the involved organisms, and thus their vital boundaries, overlap. Species often differ with regard to the variables which define their environmental requirements. Determining the likelihood of boundary overlap and thus potential for interaction between the members of two species requires the use of common variables along with an understanding of the parameter value ranges required by the two species. Even when the parameter value ranges of most environmental variables generally would suggest that a physical location either meets or exceeds the requirements for being encompassed by a species inner vital boundary, a single key factor can make a huge difference between the ability for two species to interact. I once viewed a television program which showed a fox chasing a goat onto a rocky hillside. The goat species represented customary prey for the fox species. But, when the goat climbed onto a steep rock outcropping upon which the fox could not stand, the fox abandoned the pursuit. Clearly, in that one aspect the species otherwise similar vital boundary requirements did not overlap because the surface conditions of the rock outcropping were beyond the outer vital boundary of the fox, and thus the inclination angle of that surface represented a barrier which blocked interaction between the goat and fox. Ionizing radiation (United States Nuclear Regulatory Commission 2015) and also many toxic compounds (Health and Safety Executive 2013) seem to affect species in a unidirectional manner, meaning that the ideal level of exposure to that factor is zero, and there are no measurable parameter value ranges below zero. There are still other environmental factors that may either be essential for a species survival or generally considered benign, but which seemingly can become unidirectionally toxic beyond a given parametric value range particularly as a combination effect. An example of that category would be nitrogen (molecular nitrogen), which to our species is innocuous in a normal concentration at typical atmospheric pressures and we generally seem unaffected by either low atmospheric levels or partial pressures of nitrogen, but nitrogen becomes toxic to us at higher environmental pressures (Edmonds et al. 2013).

Most environmental variables, however, affect species in a bidirectional sense meaning that an optimal value for the variable can be determined and that optimal value is likely to be near the center of the species vital boundary. For a bidirectionally affecting variable, the suitability of environmental conditions for a given species progressively lessens as the parameter value for that variable either increases or decreases beyond the optimal value, eventually reaching either an upper or lower parameter value which only minimally allows the species to have permanent residence (the species inner vital boundary), and potentially reaching either an upper or lower parameter value beyond which the species functionally cannot survive (the species outer vital boundary).

There are species for which surface inclination angle might either be bidirectional or only apparently bidirectional. As an example, some plants prefer to grow in sloping soil for reasons of soil drainage. In those cases, the soil surface inclination angle might only appear to be bidirectional because soil water retention has several definable component variables including porosity (Mohanty and Mousli 2000) that are bidirectionally affective. The contact angle rather than the inclination angle of a surface would represent a bidirectional component that is useful for assessing the possibility of microbes colonizing some types of surfaces, including medical devices (Chandra et al. 2005). Atmospheric oxygen (molecular oxygen) content is another example of an environmental variable which acts bidirectionally. Suggestions for the use of modeling equations to describe how environmental variables affect species survival are presented later in this chapter.

2.2.2.2 Level of Available Atmospheric Molecular Oxygen as an Example Variable

The level of available oxygen, upon which a large percentage of species depend for respiration, can be used as an example variable which has bidirectional effects. Some species use terrestrial respiration, meaning that their oxygen needs must be met by the surrounding atmosphere, while other species use aquatic respiration (Raven and Johnson 2001) which means that they depend upon oxygen dissolved in water. Many microbial species can obtain oxygen from either a surrounding atmosphere or a liquid medium. Macroorganisms tend to be far more physiologically specialized and thus ecologically limited in this regard. There are a few aquatic vertebrate species which normally use terrestrial respiration and successfully also can utilize aquatic respiration (Root 1949). Humans, of course, have evolved to depend upon terrestrial respiration.

Presumably the optimal atmospheric molecular oxygen concentration for our species is 20.9 % at 1 atmosphere of pressure, an oxygen concentration and total pressure which is equal to average sea level atmosphere (molecular oxygen 159 mm Hg; total pressure 760 mm Hg). The safe breathing range for humans is an oxygen concentration of 19.5–23.5 % by volume (Occupational Safety and Health Administration 2015).

Oxygen deficiency results when the level of inhaled oxygen is too low, and oxygen toxicity results when the level of inhaled oxygen is too high. It is difficult to estimate the atmospheric oxygen levels which would represent points on our species inner vital boundary because the lifetime effects of oxygen levels have received little study. But, it is known that humans begin experiencing ill effects at oxygen concentrations of 16 % and below (McManus 2009), which means that an oxygen concentration of 16 % would be beyond our inner vital boundary. For humans, death occurs within minutes at oxygen concentrations of less than 6 % (McManus 2009) which means that an oxygen concentration of 6 % would be just inside our outer vital boundary. Human death occurs within seconds at oxygen concentrations of 4 % and below (McManus 2009), which means that an oxygen concentration of 4 % would be beyond our outer vital boundary. I don't like considering the experiments which would have been performed to produce those results! Humans experience oxygen toxicity effects beginning with oxygen levels of about 53 % (molecular oxygen 400 mm Hg, total pressure 760 mm Hg), with those effects starting as respiratory irritation but progressively increasing in severity at even higher oxygen pressures (McManus 2009), clearly indicating that such high oxygen levels represent environmental conditions that are beyond our inner vital boundary but within our interboundary region.

It is important to understand how the different environmental factors, examples of which are listed in Table 2.1, interrelate. For example, breathing atmospheric air containing oxygen at a partial pressure level which would be normal for us at sea level, will exceed our outer vital boundary at either altitudes where we experience hypobaric conditions (Raven and Johnson 2001) or aquatic depths where we experience hyperbaric conditions (Edmonds et al. 2013) that by definition are too extreme for our species. Physical locations having either those extreme altitudes or depths therefore represent physiologically definable barriers for our species in part because of the way in which our bodies respond to oxygen exposure. The oxygen partial pressure which represents an optimal level for humans at a total pressure equal to normal atmospheric sea level, does instead prove beyond our inner vital boundary at the elevated pressure levels that are used for emergency medical treatments (Patel et al. 2003). The physiological limitations due to pressure related toxicities of molecular oxygen and molecular nitrogen which severely affect the ability of humans to freely venture deep beneath the ocean surface have far less effect upon aquatic mammal species such as Cuvier's beaked whale (*Ziphius cavirostris*) whose vital boundary limits are quite different from ours in many respects. That whale species can dive to 2992 m depth and hold its breath for 137.5 min (Schorr et al. 2014), evolutionarily acquired capabilities which many humans including myself would envy. The critical issue for oxygen dependent species is the oxygen pressure within body organs and tissues (Saglio et al. 1984). Interestingly, we have learned medical usage for the knowledge that tissue oxygen demand decreases with hypothermic conditions (Luscombe and Andrzejowski 2006), representing another way in which different environmental factors interact.

2.2.3 Organisms Which Utilize Other Species as Biological Vectors

The concept of boundaries and barriers as presented here applies to the external environmental conditions which are being faced by the members of a species. As a practical example, anaerobic microorganisms can exist within and interact with aerobic hosts if the internal environment of the host contains zones that are suitably anaerobic. Crossing a physiological barrier to reach a more suitable location, as defined by this theorectical concept, means that a species must survive some

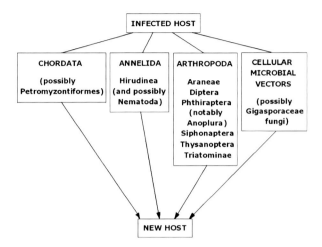

Fig. 2.4 Vectors of infectious disease. Many of the microbial species which are dependent upon biological hosts survive very poorly in the open environment, and thus the open environment represents a barrier which blocks interhost transfer of those dependent species. The evolutionary solution often has been for the dependent species to utilize still other species as vectors for facilitating transmission between hosts. Virus often use their hosts as biological vectors, an interaction that particularly is evident for those virus groups that are capable of residing genomically in near quiescence, as either proviruses or prophage, within a host cell

transition across its outer vital boundary. Figure 2.4 represents the way in which this concept of successfully crossing environmental barriers applies to some microbial species that are dependent upon close physical interactions with biological host species. Of course, not all microbes are welcome guests! Infectious disease is a type of interaction in which a microorganism acts as a parasitic predator and in such interactions the microorganism is referred to as being a pathogen. Both host accessibility and potential disease hazards can change when members of either the potential host species, a potential pathogen species, or a potential vector species crosses a barrier of some other species.

2.2.4 Understanding the Use of Vital Boundaries When Treating Infectious Disease

We intentionally use a variety of approaches that are termed barrier concepts to create obstacles which can lessen the probability of acquisition and transmission of infectious diseases (Hurst 1996, Hurst 2007, Hurst and Murphy 1996). Those types of barriers are classified by their nature as either chemical, physical, or biological, with examples listed in Table 2.2 as described by Hurst (2011). The categories of chemical barriers that are listed in Table 2.2 represent utilization of environmental conditions which exceed the survival limits of the infectious agents. Those

Table 2.2 Barriers to species movement as generally considered from the perspective of preventing infectious diseases

Categories of Chemical Barriers
Ionic (includes pH and salinity)
Surfactant
Oxidant
Alkylant
Desiccant
Denaturant
Categories of Physical Barriers
Thermal
Acoustic (usually ultrasonic)
Pressure
Barometric
Hydrostatic
Osmotic
Radiation
Electronic
Neutronic
Photonic
Protonic
Impaction (includes gravitational)
Adhesion (adsorption)
Electrostatic
van der Waals
Filtration (size exclusion)
Geographic features
Atmospheric factors (includes such meterological aspects as humidity, precipitation, and prevailing winds)
Categories of Biological Barriers
Immunological (includes specific as well as nonspecific)
Naturally induced (intrinsic response)
Naturally transferred (lacteal, transovarian, transplacental, etc.)
Artificially induced (includes cytokine injection and vaccination)
Artificially transferred (includes injection with antiserum and tissue transfers such as transfusion and grafting)
Biomolecular resistance (not immune-related)
Lack of receptor molecules
Molecular attack mechanisms (includes nucleotide-based restrictions)
Antibiotic compounds (metabolic inhibitors, either intrinsic or artificially supplied)
Competitive (other species in ecological competition with either the microbe, its vectors, or its hosts)

chemical barriers thus also represent barriers as defined in this proposed concept by creating environmental conditions that are beyond the pathogens outer vital boundaries. Many of the physical barriers listed in Table 2.2 similarly represent the use of environmental conditions which are beyond the pathogens outer vital boundaries and thus match with this proposed concept of defining barriers on the basis of a species physiological capabilities and metabolic requirements. The use of size exclusion barriers such as filtration, as listed in Table 2.2, represents physical obstacles which are not related to the pathogens vital boundaries even though size exclusion often effectively prevents the movement of species between suitable habitat locations. The biological barriers listed in Table 2.2 likewise are not related to the pathogens vital boundaries and thus not related to this proposed concept of boundaries and barriers.

The goal in treating infectious diseases is to help terminate the infection by decreasing the favorability of the host environment in which the invading microbe is surviving. We often attempt to accomplish that goal by successfully changing the local environmental conditions found either on the surface or within the body of the host. Because such treatments as the use of hyperbaric oxygen, wound disinfectants, and the selective toxins termed antibiotic compounds, represent attempts to achieve local environmental conditions representing either the pathogens interboundary region, or even better the achievement of conditions beyond the pathogens outer vital boundary, then those treatments are encompassed by this proposed concept of boundaries and barriers.

2.2.5 How This Concept of Vital Boundaries Applies to Virus

From the perspective of virus, a host cell can be described as either permissive, semipermissive, or nonpermissive, for a particular virus strain depending upon the extent to which that virus can replicate within the indicated host cell (McClintock et al. 1986). The internal environment of a host cell that is permissive, which signifies the virus can fully replicate and produce progeny virus particles within that cell, would represent the virus' inner boundary region and also the surface of the virus' inner vital boundary. A semi-permissive cell, one in which the virus can replicate partially but not produce progeny virus particles, would represent the interboundary conditions for that virus species. From the perspective of the virus, the cell wall or cytoplasmic membrane of a host cell would represent the outer vital boundary for a viral species because a virus cannot sustain its metabolic activity beyond the intracellular environment. A nonpermissive cell, one in which not even partial replication of a particular viral species is possible, also would have to be considered as representing a location beyond the outer vital boundary of that virus.

2.3 Example Species

The average longevity that can be expected for a particular species under a specified combination of environmental conditions is defined as the species L_t value for that combination of conditions. L_n is the average natural longevity for members of that species under the optimum conditions presumed to exist at the center of its species vital boundary. The goal of this section is comparing the L_t values for each species relative to that species estimate of L_n. If L_t at birth under some combination of environmental conditions equals L_r, which is the longevity required to achieve numerical replacement, then that combination of conditions qualifies as a point on the species inner vital boundary. Any combination of environmental conditions for which a species L_t is one minute qualifies as a point on the species outer vital boundary. Relative survival time is presented here as logarithmically transformed longevity ratio values, given as $\log_{10}L_t/L_n$, and those ratio values describe the expected longevity value L_t for a particular species under a given combination of environmental conditions relative to the normally expected longevity, L_n, for that species. This approach is analogous but not identical to the use of logarithmically transformed survival ratio values in which $\log_{10}N_t/N_o$ (N_o represents the number of individuals alive at time 0 which is the outset of an observation period, N_t represents the number of individuals remaining alive at elapsed time t) is used for describing exponential population decay rates (Hurst et al. 1980) as briefly explained later in this chapter.

I have chosen to give information representing five species as examples, and my selection was of species that similarly are vertebrates and which have at least some degree of overlap in their inner vital boundaries. The example species human, dog, and house mouse, have such strongly overlapping inner vital boundaries that these three species have developed successful commensal relationships and can complete their life cycles within the same room of a dwelling. The Atlantic ridley turtle and red-tail hawk have been chosen to expand the presentation beyond terrestrial mammals. Table 2.3 summarizes the longevity values for these five example species, expressed as $\log_{10}L_t/L_n$, determined for different mathematical distance points from the center of each species vital boundary outward to its outer vital boundary. The following section explains how the values in Table 2.3 were derived.

2.3.1 Human (Homo sapiens)

The value that I have chosen to use for average global human life expectancy at birth is 67.1 years (United Nations 2013), and that represents my choice of an L_n (average normal longevity) value for humans. The United Nations report (United Nations 2013) indicated that a fertility level of 2.1 children per woman represented our species replacement level. The median of national averages for age at first birth as reported by the Central Intelligence Agency (Central Intelligence Agency 2015)

Table 2.3 Longevity values expressed as $\log_{10} L_t/L_n$ for different mathematical distance points from the center of a species vital boundary outward to its outer vital boundary

Relative outward distance	Example species				
	Human	Atlantic ridley	Dog	Red-tailed hawk	House mouse
Center of boundary, estimated survival time equals normal species longevity ($L_t = L_n$)	0.0	0.0	0.0	0.0	0.0
Inner vital boundary, estimated survival time equals that required to achieve numerical replacement ($L_t = L_r$)	-3.6×10^{-1}	-5.5×10^{-1}	-6.9×10^{-1}	-5.8×10^{-1}	-9.2×10^{-1}
Survival time ($L_t = 1$ year)	-1.8	-1.6	-1.1	-9.3×10^{-1}	-3.0×10^{-1}
Survival time ($L_t = 1$ month)	-2.9	-2.7	-2.2	-2.0	-1.4
Survival time ($L_t = 1$ day)	-4.4	-4.2	-3.7	-3.5	-2.9
Survival time ($L_t = 1$ h)	-5.8	-5.5	-5.1	-4.9	-4.2
Outer vital boundary, species survival essentially impossible beyond this point ($L_t = 1$ min)	-7.5	-7.3	-6.8	-6.7	-6.0

L_t represents the expected longevity time, which presumably can be estimated for members of a given species under any definable combination of environmental conditions. L_n represents the estimated normal longevity under optimum conditions, which identifies the center of the species boundary. The L_n values used in this representation are: human, 67.1 years; Atlantic ridley, 40.0 years; dog, 12.67 years; red-tailed hawk, 8.5 years; and house mouse, 2.0 years. L_r represents the estimated longevity time required at birth for the members of a species to achieve numerical replacement. The L_r values used in this representation are: human, 29.0 years; Atlantic ridley, 11.15 years; dog, 2.6 years; red-tailed hawk, 2.25 years; and house mouse, 0.24 years

is 22.9 years. Human births are singlets in most instances and thus an average of 1.1 additional births per woman would be required to achieve numerical replacement. I have estimated an average of 3.0 years between human births by the same woman. Tsutaya and Yoneda (2013) have estimated that 2.8 years is required for human weaning, and I presume all offspring would be capable of feeding independently at that time point following birth of the last child. By combining these values relative to human births I have estimated that the longevity humans typically would require from birth in order to achieve numerical replacement (L_r) is 29.0 years, and that has been derived as follows: 22.9 years as median age at first birth, plus 3.3 [equals

3.0 × 1.1 additional births per woman to reach a replacement number of 2.1], plus 2.8 years for weaning. The human value for L_r/L_n is 0.432 and the value of $[1-(L_r/L_n)]$ for this species is 0.568, which means that on average 57 % of the normal human lifetime remains at replacement age.

The human life expectancy values used in this chapter are from the United Nations report (United Nations 2013), and they represent the average life expectancy at birth for the period of 2000–2005. It is important to avoid making broad mathematical assumptions based upon statistical values that pertain to only subpopulations of a species, and for that reason I have tried to use estimates of both species longevity and also numerical replacement age that are as broadly representative as might be possible. The global estimate for average human life expectancy at birth is 67.1 years and that number has been used in this study as the normal longevity, L_n, value for humans. The greatest national life expectancy listed in that report was for Japan, which had an estimate of 81.8 years. The least national life expectancy was for Sierra Leone, which had an estimate of 40.1 years. My choice of using the estimated global human life expectancy at birth, which is 67.1 years, as the general estimate for human L_t at birth under optimum conditions means that 67.1 years is the L_n for humans. Resultingly, by definition the estimated relative survival time for humans expressed as $\log_{10}L_t/L_n$ is zero when 67.1 years is inserted as the L_t value. If instead I were to use the L_t values at birth for individual countries and the global L_n value, then the $\log_{10}L_t/L_n$ value at birth would be 0.086 for those humans born in Japan, where longevity is estimated at 81.8 years. The corresponding $\log_{10}L_t/L_n$ value at birth would be -0.223 for those humans born in Sierra Leone, where the longevity is estimated at 40.1 years. Our conclusion from those individual nation longevity values might be that environmental conditions are more favorable for humans born in Japan, but in fact other variables such as nutrition and availability of health care may be more important in understanding the human longevity differences between these two countries. In either case, for both Japan and Sierra Leone the human L_t at birth is greater than the L_r for humans, indicating that environmental conditions in both countries should be suitable for permanent residence by humans. That presumption of suitability could be distorted if we considered only the human fertility estimates for individual nations rather than their national L_t at birth. The United Nations report (United Nations 2013) indicates that world wide human fertility (average number of children per woman) for the 2005–2010 time period was 2.53. The fertility numbers for individual nations were 1.34 for Japan and 5.16 for Sierra Leone. Using only those fertility numbers for individual countries rather than the national L_t values as an assessment of environmental suitability incorrectly would produce a conclusion that the environmental conditions in Japan are inadequate as habitat for humans, because humans do not achieve numerical replacement in Japan. That same incorrect conclusion also could result if age at first birth rather than L_t was used as the indication of environmental suitability. Information for human age at first birth from a Central Intelligence Agency report (2015) indicated that the 2012 estimated age at first birth in Japan was 30.3 years while the 2013 estimated age at first birth in Sierra Leone was 19.2 years, and we incorrectly might think that environmental conditions in Japan correspondingly are less conducive for human fertility. Social pressure rather than environmental suitability factors is more likely to be the controlling reason for these differences in age a first birth between Japan versus Sierra Leone.

2.3.2 Atlantic Ridley (Lepidochelys kempii)

The value which I have used as the estimated average normal longevity, L_n, for this marine turtle species is 40 years and that is the mean of the 30–50 year range estimate (Texas Parks and Wildlife Department 2015). The estimates for age at maturity for this species range from 7 to 15 years (Texas Parks and Wildlife Department 2015, Turtle Expert Working Group 2000) and I have presumed the middle of that range which is 11 years. The number of eggs laid in the first clutch (Texas Parks and Wildlife Department 2015, Turtle Expert Working Group 2000) should be sufficient to satisfy numerical achievement. The eggs hatch within 55 days (Texas Parks and Wildlife Department 2015) and there is no postnatal care in this species. Thus, my estimated L_r for this species is 11.15 years (11 years plus 55 days). The value of L_r/L_n for this species is 0.279 and the value of $[1-(L_r/L_n)]$ for this species is 0.721, which means that potentially 72 % of this species normal lifetime remains at replacement age.

2.3.3 Dog (Canis lupus familiaris)

The value that I have used as my estimated average normal longevity, L_n, for this species is 12.67 years, which is equal to 12 years plus 8 months as reported by Michell (1999). It is presumed that the typical age at which females commence breeding is 2 years (value for dingo, Corbett 2004). The first litter produces enough offspring to satisfy numerical replacement presuming there are no losses due to infanticide (Corbett 2004). My estimate of the gestation period is 0.18 years, equal to 65 days, which is the middle point of the 61–69 day range stated by Corbett (2004). Denning typically ends 0.42 years after pups are born, which is equal to 5 months as reported by Boitani and Ciucci (1995), beyond which time the offspring presumably can feed independently. Thus, my estimated L_r for this species is 2.6 years (2 years plus 65 days plus 5 months). The value of L_r/L_n for this species is 0.205 and the value of $[1-(L_r/L_n)]$ for this species is 0.795, which means that typically 79 % of this species normal lifetime remains at replacement age.

2.3.4 Red-Tailed Hawk (Buteo jamaicensis)

The value that I have used as my estimated average normal longevity, L_n, for this species, is 8.5 years (de Magalhães 2015). Females reach sexual maturity at 730 days (de Magalhães 2015) which is 2.0 years of age. It is presumed that numerical replacement could be achieved with the first clutch of eggs (de Magalhães 2015). The estimated incubation period is 31 days (de Magalhães 2015). The young leave the nest at about 6–7 weeks after hatching but are not capable of strong flight for at least another 2 weeks (National Audubon Society 2015). I have used 45.5 days, equal to 6.5 weeks, as my estimate of the time period required for leaving the nest plus added 2 weeks for the strengthening of flight muscles. Based upon that information, my estimated L_r for this species is 2.25 years (2 years plus 31 days plus 6.5 weeks plus 2 weeks). The value of L_r/L_n for this species is 0.265 and the value of $[1-(L_r/L_n)]$ for this species is 0.735, which means that typically 74 % of this species normal lifetime remains at replacement age.

2.3.5 House Mouse (Mus musculus)

The value that I have used as my estimated average normal longevity, L_n, for this species is 2 years (Berry 1970). The females reach sexual maturity at 6 weeks (Berry 1970). I have used 5 days as my estimate for their delayed implantation period (Berry 1970). It is presumed that the first litter of offspring satisfies numerical replacement. Gestation typically lasts 3 weeks and is followed by a nursing period of 3 weeks (Berry 1970), after which the offspring can feed independently. Based on that information, my estimated L_r for this species is 0.24 years (12 weeks plus 5 days). The value of L_r/L_n for this species is 0.12 and the value of $[1-(L_r/L_n)]$ is 0.88, which means that potentially 88 % of this species normal lifetime remains at replacement age. The fact that the ratio of L_r to L_n is so dramatically different for this species, as compared to the other example species, is possibly suggestive of the high predation rate typically suffered by this species.

2.4 The Possiblilty of Mathematically Estimating Vital Boundaries

Theoretically, both the inner and outer vital boundaries of a species could be depicted mathematically and calculations approximating conditions just inside the outer vital boundary would have a special application for ascertaining and potentially predicting short term survival for species members under extremely adverse environmental conditions. Importantly, those numerous environmental factors which could serve as variables when defining the potential longevity for members of a species would differ in the extent to which they influence various sections of the inner and outer vital boundary surfaces.

Trying to incorporate all potential variables into such calculations would be as difficult as the challenge faced by the fictional character Hari Seldon, professor of mathematics in Isaac Asimov's Foundation series. Fortunately, the number of variables required to functionally estimate the outer vital boundary and L_t values just inside the outer vital boundary would be relatively fewer than required to estimate the inner vital boundary, because fewer variables can impact life so drastically over extremely short intervals of time. For example, levels of available dietary micronutrients are unimportant if you are dying from hypothermal exposure. Examples of variables that do have importance at the outer vital boundary are ambiental temperature, gases such as oxygen, carbon dioxide, carbon monoxide and chlorine (molecular chlorine), ionizing radiation, certain natural and synthetic toxins including those which have been used for military purposes, and ambiental pressure. In is not necessary for all possible variables to be included in a model. The choice of how many variables need to be included in a given calculation, and which variables those need to be, will depend upon the species and situation being considered.

Figure 2.1 depicts the two concentric theoretical closed surfaces which represent a species inner and outer vital boundaries as polyhedrons. Figure 2.5 delves further into the proposed concept by illustrating the center of a species vital boundary, the species inner vital boundary, and the species outer vital boundary, as points that can be graphed by a plot of relative survival time versus relative distance from the center of the boundary. This distance from the center of the boundary is not a physical measurement, but instead this distance is determined by the parameter values of pertinent environmental variables. The area marked in Fig. 2.5 as 'permanent residence possible' represents the inner boundary region. The area marked in Fig. 2.5 as 'temporary survival possible' represents the interboundary region. The area in Fig. 2.5 marked as 'survival essentially impossible' represents environmental conditions that constitute barriers for the species.

I have represented relative survival time as $\log_{10} L_t/L_n$ for this proposed concept because doing so allows the use of modeling techniques analogous to those procedures developed for modeling microbial population survival versus environmental variables as an exponential decay rate function (Hurst et al. 1980, Hurst et al. 1992). The concepts presented by Hurst et al. (1980, 1992) used ratio values expressed as $\log_{10} N_t/N_o$, in which N_t represented the surviving number of individuals at elapsed time t relative to the number of individuals which had been alive at time 0, with time 0 having been the outset of the observation period, and regressed those ratio values as a dependent variable against time and environmental factors as independent variables. Although analogous, those techniques developed by Hurst et al (1980, 1992) would need to be modified for use with this proposed concept because this proposed concept includes time as part of the dependent variable, relative survival time. In both types of analyses, determining population survival time using ratio values calculated either as $\log_{10} L_t/L_n$ or $\log_{10} N_t/N_o$ requires stating that, during the time period for which the survival is either being observed or

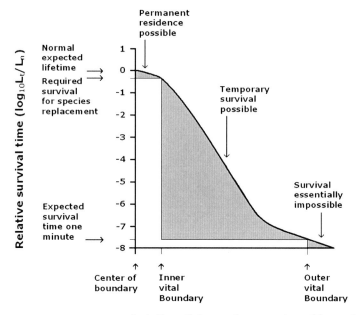

Fig. 2.5 Human relative survival time and boundaries. This figure represents the concept that the relative survival time for some given population of a species, expressed as $\log_{10} L_t/L_n$, will depend upon the parameter values for those environmental variables which are important to the species survival. The favorability of environmental conditions, and thus the expected longevity time (L_t) for members of a species, decreases with outward distance from the center of its vital boundary. By definition, the L_t value which members of a species potentially can achieve at the center of their vital boundary is equal to their normal expected longevity time (L_n) and is their normal expected lifetime under optimum conditions. The inner vital boundary is defined by all combinations of conditions under which expected longevity time would equal the longevity required from birth for members of that species to achieve numerical replacement (L_r). Physical locations where the environmental conditions could allow an expected longevity either equal to or greater than L_r potentially would represent permanent residence locations for that species. The outer vital boundary is defined by all combinations of environmental conditions under which expected longevity time would be one minute. Physical locations where the environmental conditions could allow an expected longevity greater than one minute but less than L_r would allow only temporary survival, thus representing only temporary residence locations, as the conditions in those locations would not allow the species to numerically sustain its population. A species would find its survival essentially to be impossible under any combination of environmental conditions which is beyond that species outer vital boundary, due to the extremely low expected L_t values beyond its outer boundary. Thus, locations where a species L_t would be less than one minute will represent barriers which severely restrict the movement of that species. The curve shown here was not calculated for humans but generally is illustrative of population survival curves

estimated, the environmental conditions remain within statistical limits of the stated parameter values for those environmental factors that are being used as mathematical variables. Thus, for any given species, each possible combination of parameter values for the considered environmental variables would generate a single estimated population survival time, the L_t value for that set of conditions. Only one combination of environmental conditions, the optimum for that species, would yield an L_t value equal to L_n. The surface of the inner vital boundary for that species would be comprised of all combinations of environmental conditions for which the L_t value equals L_r. The surface of the outer vital boundary for that species would be comprised of all combinations of environmental conditions for which the L_t value equals 1 min.

2.4.1 Linear Regression Model

Historically, a presumption has been made that population survival satisfactorily is estimated as a log-linear function (Fig. 2.6) for a given set of environmental conditions. This section of the chapter presents that type of linear modeling approach using a technique developed by Hurst et al. (1980). Their linear regression analyses technique involves two steps, as represented in Fig. 2.7, and was used to examine and model how the survival of viruses in soil was affected by multiple soil characteristics. The general linear model equation is Eq. (2.1).

$$Y = B_0 + B_1 X \tag{2.1}$$

2.4.1.1 Step One of the Linear Regression Technique

The first step in the linear regression technique was done by using Eq. (2.2) to determine the rate of population change, calculated as B_t, for each combination of organism and environmental conditions that had been studied.

$$\log_{10} \frac{N_t}{N_o} = B_0 + B_t t \tag{2.2}$$

2.4.1.2 Step Two of the Linear Regression Technique

The B_t values developed in step one were termed survival slope values, and subsequently used as dependent variable Y for the second step of the analysis. Environmental variables were used as independent variables in the second step of the analysis. Insight was gained by using scatterplots and simple linear regression analysis to examine the relationships between the survival slope values and individual environmental variables using Eq. (2.1). Development of models during the second step of the process was achieved by performing multivariate linear

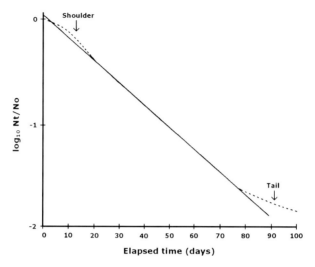

Fig. 2.6 Population survival plot. This is a visual presentation of Eq. (2.2). It illustrates exponential population decay with Log_{10} transformed titer ratio values used as the dependent variable; these values are regressed linearly with respect to time (t) as the independent variable. The *solid line* is the slope, B_t, which represents the rate of population change, or population decay, expressed as $[\log_{10}(N_t/N_o)]/t$. N_o is the number of population members, or titer if considering a population of microorganisms, existing at time 0 which is the outset of the observation period. N_t is the number of population members, or titer, existing at elapsed time t. B_o is the point where the solid line intercepts the y axis. The dashed lines demonstrate deviations from log linearity that are termed shouldering and tailing. Those deviations reflect the fact that more accurately and precisely modeling population survival requires time to be allowed a coefficient in the form of an exponent as described by Eq. (2.6), which was the basis of the Multiplicative error II equation format developed by Hurst et al (1992)

regression to analyze the survival slope values against a number of different environmental variables using Eq. (2.3).

$$Y = B_0 + B_1X_1 + B_2X_2 + \bullet\bullet\bullet + B_nX_n \qquad (2.3)$$

The independent variables, depicted either as X in Eq. (2.1) or as X_1 through X_n in Eq. (2.3), were physical and chemical soil characteristics because that study was designed to assess how soil characteristics determined the survival time of virus populations. Using a backwards elimination regression technique to select which independent variables were to be included in the model equation proved to be the best approach for simplifying the modeling equation, by eliminating as variables those soil characteristics which had least influence upon survival. Once a set of four soil characteristics had been selected as key independent variables, the backwards elimination regression analysis process was repeated several times with the difference being that for each additional trial the regression was run with a single one of the key variables excluded from consideration for incorporation into the multivariate regression equation. Those trials in which variables individually were excluded

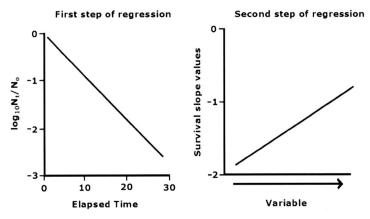

Fig. 2.7 Two step linear regression technique. The first step depicts a survival slope determination which represents Eq. (2.2), with the survival slope values expressed as the rate of $\log_{10}$ numerical change in the population ($\log_{10}N_t/N_o$), termed titer change if considering the survival of microorganisms, per unit time. Those slope values were expressed as [($\log_{10}N_t/N_o$)/time], and the slope values were negative which indicated that the surviving fraction of the population decreased with time. The generated slope values then can be used as the dependent variable in a second regression step, using environmental factors as independent variables to determine the statistical relationship between survival and those environmental factors. This statistical approach initially was developed for studying the survival of viruses with time measured in days, and with the independent variables being temperature and either water or soil characteristics. Equation (2.1) would be used for the second regression step if the choice were to perform simple linear regression, with [($\log_{10}N_t/N_o$)/time] as the dependent variable and a single environmental factor as the independent variable. The basic equation used by Hurst et al. (1980) for that second regression step was Eq. (2.3), with [($\log_{10}N_t/N_o$)/time] incorporated as the dependent variable in a multivariate linear regression which simultaneously examined survival as a function of numerous environmental factors as independent variables. The example shown here for the second regression step represents a positive correlation between the variable and survival, with slope values decreasing as the value of the variable increases. Alternatively, the signs of both the slope values and the y-axis intercept from the first step of the regression can be inverted, and then termed inactivation rate values rather than slope values, before regressing those slope values as dependent variable against environmental factors as the independent variables. If this type of mathematical approach were used for modeling survival, i.e., L_t, as the dependent variable against environmental factors as independent variables, then only the second step of the regression technique would be used with $\log_{10}L_t/L_n$ values substituted as dependent variable in place of $\log_{10}N_t/N_o$ values, and consequently the modeling equations could not use time as an independent variable since longevity expressed as either L_t, L_n, or L_r already incorporates time. This figure has been redrawn from Hurst et al. 1980

helped with understanding the interrelationships between the four key characteristics. The final model equation incorporated all four key characteristics as independent variables, and we better understood the role of each characteristic.

2.4.1.3 Adapting the Two Step Linear Regression Technique for Use with Longevity Values

Using the backwards elimination approach (Hurst et al. 1980) for evaluating the relative importance and interactions between environmental factors as pertaining to vital boundary conditions would require eliminating their first step which utilized Eq. (2.2), and instead directly proceeding to Eq. (2.3) with the values of $\log_{10} L_t/L_n$ used as Y. The reason for modeling vital boundary conditions by using just the second step of the Hurst et al. (1980) procedure, is that production of the slope values during the first step of their analysis utilized change in titer represented by $\log_{10} N_t/N_o$ for the dependent variable with time incorporated as an independent variable. The slope values generated from Eq. (2.2) thus represented change per unit time as [($\log_{10} N_t/N_o$)/time], and usage of those slope values as dependent variable during the second step of the regression technique then placed time on the dependent side in Eq. (2.3). The values of $\log_{10} L_t/L_n$ already incorporate time, and since time should not be used on both sides of an equation, it would be appropriate to use only the second step of their two step analysis technique. Equation (2.4) will be the result from using $\log_{10} L_t/L_n$ as the dependent variable for this type of modeling approach.

$$\log_{10} \frac{L_t}{L_n} = B_0 + B_1 X_1 + B_2 X_2 + \bullet \bullet \bullet + B_n X_n \qquad (2.4)$$

2.4.2 Non-Linear Regression Model

The study by Hurst et al. (1992) compared how well eight different equation formats served for developing regression models to understand the survival of viruses in environmental water. Their approach was to regress common sets of experimentally determined N_t/N_o survival ratio values as dependent variable versus a predetermined set of water characteristics as independent variables. The models generated using those different equation formats, along with their respectively calculated coefficients, then were used to predict what the outcome from that study should have been. Predictions were made both for the same parameter value ranges that had been used to generate the models, and also for parameter value ranges beyond those that had been used to generating the models. Then, the sets of values predicted by each of the different modeling equation formats were compared by simple regression against the actual experimentally observed values. The best equation format proved to be one based upon the general non-linear Eq. (2.5):

$$Y = B_0 X_1^{B_1} X_2^{B_2} \bullet \bullet \bullet X_n^{B_n} \qquad (2.5)$$

with N_t/N_o used as the dependent variable Y. Time was used as one of the independent variables. Following linearization the result was Eq. (2.6), which the authors termed their Multiplicative error II equation format.

$$\log_{10}\frac{N_t}{N_o} = \log_{10}B_0 + B_1\log_{10}X_1 + B_2\log_{10}X_2 \\ + \bullet \bullet \bullet + B_n\log_{10}X_n + B_t\log_{10}t \qquad (2.6)$$

Importantly, the Multiplicative error II equation format demonstrated the greatest ability to accurately predict survival under parameter value ranges beyond those for which the coefficients had been developed. It also showed that time requires a coefficient in the form of an exponent.

2.4.3 My Best Suggestion for Mathematically Estimating Vital Boundaries

Either the linear or non-linear regression approach could be used for developing modeling equations to analyze and predict vital boundary conditions. My best suggestion would be to use Eq. (2.7) as the starting point, which would be analogous to the Multiplicative Error II equation format but with relative survival time substituted as the dependent variable and time deleted from the independent side of the equation.

$$\log_{10}\frac{L_t}{L_n} = \log_{10}B_0 + B_1\log_{10}X_1 + B_2\log_{10}X_2 \\ + \bullet \bullet \bullet + B_n\log_{10}X_n \qquad (2.7)$$

2.5 Relating This Concept to the Process of Evolution and the Niche of a Species

Evolution is a process which attempts to maximize usage of available energy resources by the development of species. Each evolved species has an ecology which we can define as including two main aspects. One of these aspects is the collective set of actions and functions performed by that species, by which we define the species niche. Another aspect is the established range of environmental conditions under which this species can survive. Physical locations which meet the requirements for that range of environmental conditions define the potential habitat areas within which this species can reside.

Which came first, the habitat or the niche? Posing this question leads us to a cyclical process of understanding. My personal guess would be the habitat, because no species can function without a location capable of supporting life. And yet, the opportunities and requirements of a niche help to establish the range of environmental conditions in which the occupying species will reside. Evolution therefore is a reactionary process that must establish both the niche and environmental requirements of a species simultaneously. By studying the ways in which the habitat of an individual species overlaps, and its niche interconnects, with those of other species, we can gain an understanding of the evolutionary path that consequently produced the observed species. That path will have led the species to an eventual success which may be either long term or simply a quick demise.

Once a species has evolved in response to selection pressures, the species becomes constrained in ways that we can define both with regard to the set of environmental conditions under which the species successfully can reside, which is the purpose of the habitat definition concept presented in this chapter, and also with regard to its niche (Hurst 2016). My hope in presenting this habitat definition concept is that it may prove to be a helpful tool, but that success depends upon the possibility of someone recognizing this concept as being useful. Unless it finds utility, the fate of this concept will be just that of an unused wrench which can serve as little more than either a paperweight or doorstop. But then, I own a set of wrenches that seldom get used.

2.6 Some Questions and Answers Regarding This Concept

Why the choice of 1 min for the outer vital boundary? The mathematical distance from the center of the boundary outward would be infinite, and correspondingly the value of $\log_{10}L_t/L_n$ approaches its minimum value as an asymptote. It therefore is necessary to pick some L_t value as representing a practical minimum duration of survival. The choice of 1 min was not entirely random, but did seem to represent a practical choice.

Are there more than just two boundaries? There actually would be an enormous number of theoretical closed surfaces layered as concentric shells between the center of a species boundary and its outer vital boundary, with each possible L_t value which is less than L_n representing a different shell. L_n itself represents a single point. As an example, my L_n value for humans is 67.1 years and if L_t were estimated in units of minutes, then for humans there would be 35,291,915 concentric shells [(67.1 years × 365.25 days per year × 24 h per day × 60 min per hour) − 1]. That minus 1 represents the center point, L_n. Six of those shells are identified in Table 2.3, and they are: $L_t = L_r$, $L_t = 1$ year, $L_t = 1$ month, $L_t = 1$ day, $L_t = 1$ h, and $L_t = 1$ min. I have designated two of those shells as perhaps representing the more significant demarcations from the perspective of a species survival, and termed those two shells as the inner and outer vital boundaries.

Where would L_{max} fit into this concept? Each L_t value would consist of a designated average value and include a statistical deviation range. As an example, I have used 67.1 years as my L_n for humans which is the reported world value for life expectancy at birth (United Nations 2013). L_{max}, which would be the maximum normal lifetime for members of a species, marks the upper limit of the species L_n range. The L_{max} for humans is approximately 114 years.

Are there other usages for the concept of L_t/L_n ratio values? It would be possible to develop L_t/L_n data for subgroups of a species and valuable comparative information could be gained in that way. When we consider either animals or plants, those groupings might be based upon subspecies with an example being that different breeds of dogs vary in their life expectancy. Subgroupings for humans might be based upon their location of residence, such as by country, and valuable demographic information could be gained in that way. These other usages of L_t/L_n ratio values are a departure from the concept of vital boundaries. As mentioned earlier, although it is not possible to determine a value of L_t/L_n that represents L_r for Japan alone since the human population in that country currently does not achieve numerical replacement, we can assess a value of L_t/L_n at the time of first human birth for Japan and the value would be 0.37 (30.3 years divided by 81.8 years). This value of 0.37 indicates that on average the first childbirth in Japan occurs when 37 % of a womans lifetime has passed, and she then has 63 % of her lifetime remaining. The corresponding number would be 0.48 for Sierra Leone (19.2 years divided by 40.1 years) indicating that on average the first human birth in Sierra Leone occurs when 48 % of a womans lifetime has passed, and then 52 % of her lifetime remains. Comparing those calculations for Japan versus Sierra Leone tells us that although the first human birth for women in Japan comes at a later age of 30.3 years rather than 19.2 years, that first human birth for Japanese women comes at a relatively earlier point when considered from the perspective of a womans average life expectancy.

How can we understanding the vital boundaries of a virus? Virus are one of the special cases in biology because of their nature as obligately intracellular parasites that lack metabolic activity outside the confines of a host cell. And for many, the issue of "Is a virus even alive?" further complicates the discussion. So, how could we really define the vital boundaries for a virus, as its boundaries would seem to change depending upon whether the virus is inside a host cell or resting dormant on a surface somewhere? We know that the vital boundaries of a virus would seem to be heavily defined by its host because a virus exhibits metabolic "life" only within either a permissive or semipermissive host cell. We even know that retroviral DNA apparently can be incorporated permanently into the host genome, at which point the virus and its host have the same biological agenda and symbolically have joined to create a single species (Hurst 2011). The hypovirulence elements of the fungi which cause Chestnut blight disease are another clue, these elements apparently evolved from a virus and have achieved symbiosis with their fungal host to such an extent that often the fungus seems unable to have permanent survival in nature without its viral-derived symbiont. These hypovirulence elements sustain their existence by reducing virulence of their host fungi so that the host fungus does not kill the tree upon which the fungus feeds, enabling survival of the hypovirulence element, fungus and tree. It thus would seem that the vital boundaries of the virus evolutionarily have changed by conforming to the vital boundaries of their infected hosts, and this is particularly true both for those virus species which have become symbiotic and also for many of the viral-derived elements. The vital boundaries of a virus species therefore would be represented by the vital boundaries of its infected host. The inner vital boundary of the virus would be the outer vital boundary of a permissive host. The outer vital boundary of the virus would be the outer vital boundary of a semipermissive host. Mathematically modeling the survival of a species population, such as animals, under defined environmental conditions actually represents modeling the rate at which members of that population die by becoming metabolically inactive and is the function of L_t values and L_t/L_n values presented in this chapter. Virus particles are by themselves metabolically inactive, and mathematically modeling the survival of a virus population using N_t/N_o values actually represents determining not the rate at which virus die, but rather the rate at which virus lose their ability to be revived.

How to model the fate of bacterial endospores? Bacterial endospores, like virus particles, are metabolically inert survival structures. Nonenveloped virus particles have a notable environmental robustness and are capable of reactivating after decades of appropriate storage. Bacterial endospores have an even more amazing capability for environmental stability that extends for perhaps 25 million years (Cano and Borucki 1995). Both virus particles and bacterial endospores function by protectively allowing genetic material to survive in a revivable form under environmental conditions which are too extreme for allowing their species to demonstrate metabolic activity. Both bacterial endospores and virus particles also aid the dispersal of their species in addition to aiding their species survival. There certainly are notable morphological differences between these two types of metabolically inert survival structures. It also is important to remember that production of virus particles is a reproductive strategy which generates numerous virus particles per host cell, in contrast with the fact that endospore production is not reproductive and instead yields a ratio of only one spore per vegetative cell. Equations (2.4) and (2.7) could be used for modeling the vital boundaries of bacterial species when their cells are in a metabolically active vegetative state, in just the same manner as those equations would be used for modeling the vital boundaries of the animals which have been presented as example species for this publication. It would be appropriate to model the persistence of bacterial endospores by using the concept of $\log_{10} N_t/N_o$ ratio values and the analysis techniques presented by Hurst et al. (1980, 1992) which included Equations (2.2) and (2.6), and thereby assess the rate at which the spores lose their capacity for revitalization just as done for modeling population decay rates of virus particles. It also would be appropriate to model the persistence of either lyophilized or cryopreserved vegetative cells using the concept of $\log_{10} N_t/N_o$ ratio values and the analysis techniques presented by Hurst et al. (1980, 1992) which included Eqs. (2.2) and (2.6), because both lyophilized as well as cryopreserved cells are being held in a metabolically inactive state under environmental conditions beyond the outer vital boundary of their species.

Acknowledgements I dedicate my work on this idea to my memories of Mehdi Shirazi, whom in my childhood I knew as Uncle Mike. Mehdi was a mathematician who modeled arterial and cardiac flow. He rented rooms in my grandmothers house and eventually was adopted into the family. My grandmother made certain that Mehdi occasionally had a good home cooked meal, and even after he was able to afford residing in his own home Mehdi often would be present at our family gatherings. He helped me to learn that mathematics could be used to describe and understand the science with which I was fascinated. I cannot even begin to understand the mathematical models which Mehdi wrote, but optimistically it is my hope that he would be proud of those comparatively simple models which I have been able to generate during the science career I eventually pursued.

I wish to thank Teckla G. Akinyi, David A. Batigelli, Alexa J. Hojczyk, Rachel S. Hurst, Karrisa M. Martino, Lord Robert M. May of Oxford, Aharon Oren, and David W. Uhlman for having provided helpful review suggestions as I was writing this chapter. My daughter Rachel is the only fashion designer whom I could imagine capable of being entrusted with identifying the flaws in a mathematical modeling equation, and I am tremendously proud of her abilities.

References

Berna F, Goldberg P, Horwitz LK et al (2012) Microstratigraphic evidence of in situ fire in the Acheulean strata of Wonderwerk Cave, Northern Cape province, South Africa. Proc Natl Acad Sci USA 109:1215–20. doi:10.1073/pnas.1117620109, Accessed 03 April 2012

Berry RJ (1970) The natural history of the house mouse. Fld Stud 3:219–262

Boitani L, Ciucci P (1995) Comparative social ecology of feral dogs and wolves. Ethol Ecol Evol 7:49–72

Cano RJ, Borucki MK (1995) Revival and identification of bacterial spores in 25- to 40-million-year-old Dominican amber. Science 268:1060–1064

Central Intelligence Agency (2015) Mothers mean age at first birth. In: The world factbook. Central Intelligence Agency. https://www.cia.gov/library/publications/the-world-factbook/fields/2256.html#download. Accessed 11 June 2015

Chandra J, Patel JD, Li J et al (2005) Modification of surface properties of biomaterials influences the ability of *Candida albicans* to form biofilms. Appl Environ Microbiol 71:8795–8801

Corbett LK (2004) Australia and oceania (Australasian). In: Sillero-Zubiri C, Hoffmann M, Macdonald DW (eds) Canids: foxes, wolves, jackals and dogs. Status survey and conservation action plan. International union for conservation of nature and natural resources, Gland, pp 223–230

de Magalhães JP (2015) AnAge entry for *Buteo jamaicensis* classification HAGRID (01028). In: AnAge the animal ageing and longevity database. Human ageing genomic resources. http://genomics.senescence.info/species/entry.php?species=Buteo_jamaicensis. Accessed 16 June 2015

Edmonds C, Thomas B, McKenzie B et al (2013) Diving medicine for scuba divers, 5th edn. Edmonds, Manly, http://www.divingmedicine.info. Accessed 14 August 2015

Grinnell J (1917) The niche-relationships of the California thrasher. Auk 34:427–433

Health and Safety Executive (2013) EH40/2005 Workplace exposure limits. HSE books, Merseyside

Huq A, Small EB, West PA et al (1983) Ecological relationships between *Vibrio cholerae* and planktonic crustacean copepods. Appl Environ Microbiol 45:275–83

Hurst CJ (1996) Preventing foodborne infectious disease. In: Hurst CJ (ed) Modeling disease transmission and its prevention by disinfection. Cambridge, Cambridge, pp 193–212

Hurst CJ (2007) Estimating the risk of infectious disease associated with pathogens in drinking water. In: Hurst CJ, Crawford RL, Garland JL et al (eds) Manual of environmental microbiology, 3rd edn. ASM, Washington, pp 365–377

Hurst CJ (2011) Defining the ecology of viruses. In: Hurst CJ (ed) Microbial and botanical host systems. Studies in viral ecology, vol 1. Wiley, Hoboken pp 3–40

Hurst CJ (2016) Towards a unified understanding of evolution, habitat and niche. In Hurst CJ (ed) Their world a diversity of microbial environments. Advances in envronmental microbiology, vol 1. Springer, Heidelberg, pp 1–33

Hurst CJ, Gerba CP, Cech I (1980) Effects of environmental variables and soil characteristics on virus survival in soil. Appl Environ Microbiol 40:1067–1079

Hurst CJ, Murphy PA (1996) The transmission and prevention of infectious disease. In: Hurst CJ (ed) Modeling disease transmission and its prevention by disinfection. Cambridge, Cambridge, pp 3–54

Hurst CJ, Wild DK, Clark RM (1992) Comparing the accuracy of equation formats for modeling microbial population decay rates. In: Hurst CJ (ed) Modeling the metabolic and physiologic activities of microorganisms. Wiley, New York, pp 149–175

Hutchinson GE (1957) Concluding remarks. Cold Spring Harb Symp Quant Biol 22:415–427

Kiørboe T, Tang K, Grossart H-P et al (2003) Dynamics of microbial communities on marine snow aggregates: colonization, growth, detachment, and grazing mortality of attached bacteria. Appl Environ Microbiol 69:3036–3047

Luscombe M, Andrzejowski JC (2006) Clinical applications of induced hypothermia. Contin Educ Anaesth Crit Care Pain 6(1):23–27

McClintock JT, Dougherty EM, Weiner RM (1986) Semipermissive replication of a nuclear polyhedrosis virus of *Autographa californica* in a gypsy moth cell line. J Virol 57(1):197–204

McManus N (2009) Oxygen: health effects and regulatory limits, Part I: Physiological and toxicological effects of oxygen deficiency and enrichment. NorthWest Occupational Health & Safety, North Vancouver, nwohs@mdi.ca www.nwohs.com. Accessed 13 August 2015

Michell AR (1999) Longevity of British breeds of dog and its relationships with sex, size, cardiovascular variables and disease. Vet Rec 145(22):625–629

Mohanty BP, Mousli Z (2000) Saturated hydraulic conductivity and soil water retention properties across a soil-slope transition. Water Resour Res 36:3311–3324

National Audubon Society (2015) Red-tailed Hawk (*Buteo jamaicensis*). In: Guide to North American birds. National Audubon Society. https://www.audubon.org/field-guide/bird/red-tailed-hawk. Accessed 16 June 2015

Occupational Safety and Health Administration (2015) Appendix A, OSHA's Permit-required confined spaces standard, 29 CFR 1910.146. Occupational Safety and Health Administration, Washington. https://www.osha.gov/dep/etools/eprcs/prcsappendices.pdf. Accessed 13 June 2015

Patel DN, Goel A, Agarwal SB et al (2003) Oxygen toxicity. J Indian Acad Clin Med 4(3):234–237

Pitelka LF, Plant Migration Workshop Group (1997) Plant migration and climate change. Am Scientist 85:464–473

Raven PH, Johnson GB (2001) Biology, 6th edn. McGraw-Hill, New York

Root RW (1949) Aquatic respiration in the musk turtle. Physiol Zool 22(2):172–178

Saglio PH, Rancillac M, Bruzan F et al (1984) Critical oxygen pressure for growth and respiration of excised and intact roots. Plant Physiol 76:151–154

Schorr GS, Falcone EA, Moretti DJ et al (2014) First long-term behavioral records from Cuvier's beaked whales (*Ziphius cavirostris*) reveal record-breaking dives. Plos One 9(3), e92633

Texas Parks and Wildlife Department (2015) Kemp's ridley sea turtle (*Lepidochelys kempii*). Texas Parks and Wildlife Department. http://tpwd.texas.gov/huntwild/wild/species/ridley/. Accessed 16 June 2015

Toups MA, Kitchen A, Light JE et al (2011) Origin of clothing lice indicates early clothing use by anatomically modern humans in Africa. Mol Biol Evol 28(1):29–32

Tsutaya T, Yoneda M (2013) Quantitative reconstruction of weaning ages in archaeological human populations using bone collagen nitrogen isotope ratios and approximate bayesian computation. Plos One 8(8), e72327

Turtle Expert Working Group (2000) Assessment update for the Kemp's ridley and loggerhead sea turtle populations in the Western North Atlantic, NOAA Technical Memorandum NMFS-SEFSC-444. U.S. Department of Commerce, Miami

United Nations (2013) World population prospects: The 2012 Revision, highlights and advance tables. Working paper No. ESA/P/WP.228

United States Nuclear Regulatory Commission (2015) Biological effects of radiation. United States Nuclear Regulatory Commission, Rockville, http://www.nrc.gov/reading-rm/basic-ref/teachers/09.pdf. Accessed 13 June 2015

van Schie PM, Fletcher M (1999) Adhesion of biodegradative anaerobic bacteria to solid surfaces. Appl Environ Microbiol 65:5082–5088

World Health Organization (1999) Scurvy and its prevention and control in major emergencies. WHO/NHD/99.11 World Health Organization, Geneva

Chapter 3
Microbes and the Fossil Record: Selected Topics in Paleomicrobiology

Alexandru M.F. Tomescu, Ashley A. Klymiuk, Kelly K.S. Matsunaga, Alexander C. Bippus, and Glenn W.K. Shelton

Abstract The study of microbial fossils involves a broad array of disciplines and covers a vast diversity of topics, of which we review a select few, summarizing the state of the art. Microbes are found as body fossils preserved in different modes and have also produced recognizable structures in the rock record (microbialites, microborings). Study of the microbial fossil record and controversies arising from it have provided the impetus for the assembly and refining of powerful sets of criteria for recognition of bona fide microbial fossils. Different types of fossil evidence concur in demonstrating that microbial life was present in the Archean, close to 3.5 billion years ago. Early eukaryotes also fall within the microbial realm and criteria developed for their recognition date the oldest unequivocal evidence close to 2.0 billion years ago (Paleoproterozoic), but Archean microfossils >3 billion years old are strong contenders for earliest eukaryotes. In another dimension of their contribution to the fossil record, microbes play ubiquitous roles in fossil preservation, from facilitating authigenic mineralization to replicating soft tissue with extracellular polymeric substances, forming biofilms that inhibit decay of biological material, or stabilizing sediment interfaces. Finally, studies of the microbial fossil record are relevant to profound, perennial questions that have puzzled humanity and science—they provide the only direct window onto the beginnings and early evolution of life; and the methods and criteria developed for recognizing ancient, inconspicuous traces of life have yielded an approach directly applicable to the search for traces of life on other worlds.

A.M.F. Tomescu (✉) • K.K.S. Matsunaga • A.C. Bippus • G.W.K. Shelton
Department of Biological Sciences, Humboldt State University, Arcata, CA 95521, USA
e-mail: Alexandru.Tomescu@humboldt.edu

A.A. Klymiuk
Department of Ecology and Evolutionary Biology, University of Kansas, Lawrence, KS 66047, USA

3.1 Microbial Fossils: A Vast Field of Study

Knowledge of the microbial fossil record has expanded tremendously in more than 50 years since early discoveries (e.g., Tyler and Barghoorn 1954; Barghoorn and Tyler 1965), both in depth—geologic time—and breadth—types of organisms, modes of preservation, and types of fossil evidence. Along with the new discoveries of fossil microbes and microbially induced structures, and keeping pace with technological advances in analytical tools, the paleontological community developed and expanded the set of methods used to study these fossils and refined the types of questions addressed, as well as the criteria applied to them. At the same time, the community of scientists itself broadened its scope and expanded its ranks to include paleobiology, geobiology, geochemistry, taphonomy, and other areas of research in its sphere of investigation. As a result of this explosive growth, paleomicrobiology is today just as vast an area of science as its "neo" counterpart and could itself be the subject of a multivolume book. That is why for this chapter, we had to select only some of the topics of major interest in paleomicrobiology which we review to summarize the current state of the art.

One of the topics is the recognition of microbial fossils and the criteria used for it (Sect. 3.3). These provide the foundation of all work involving microbial body fossils and are especially relevant to the search for the earliest traces of life. The development of these criteria over time, by discovery and critical scrutiny of increasingly older Precambrian microbial fossils, provides a telling example of the workings of science, in general, and paleobiology, in particular, as an objective empirical approach to questions about nature. As a logical follow-up on the criteria for recognition of microbial body fossils, we discuss microbially induced sedimentary structures and other traces of microbial activity (microbially induced structures), their classification, and criteria of recognition. These provide a powerful complement to the study of the microbial fossil record, even in the absence of body fossils, and are active and growing fields of inquiry. This section is prefaced by a review of microbial fossil preservation (Sect. 3.2), which provides a broader context for the different aspects comprising the recognition of the fossils. The next topic involves the earliest records of life and a review of the Archean fossil record (Sect. 3.4). Aside from pushing back in deep time the history of life on Earth, these fossil discoveries and the controversies they engendered were crucial in shaping both the methods and the theoretical bases for the study of microbial fossils. As a part of this topic, we summarize a few of the now-classic debates which animated (or are still animating) the scientific community and provided much of the impetus for the development of a powerful set of criteria for microbial fossil recognition. Another topic covers the rise of early eukaryotes as reflected by the microbial fossil record, with a discussion of the criteria used to recognize them and a survey of the earliest types (Sect. 3.5). Next, we review the role of microbially mediated processes in the various fossilization pathways of other organisms—microbial-associated mineralization, plant, animal, and trace fossil preservation (Sect. 3.6)—and the fossil record of symbioses that involve microbial participants

(Sect. 3.7). The chapter ends with a discussion of future directions of investigation in the study of microbial fossils and of the role of paleomicrobiology in the study of life.

Throughout the chapter, we focus mainly on the record of body fossils, with some detours into geochemistry and sedimentology for discussions of biogenicity, microbially induced structures, and fossilization processes. The survey of the prokaryote and eukaryotic fossil record is limited to early occurrences—Archean for the former (4.0–2.5 Ga = billion years) and Paleoproterozoic and Mesoproterozoic (2.5–1.0 Ga) for the latter. However, the discussions of the roles of microbes in fossilization and of the fossil record of symbioses draw on examples from throughout the geologic time scale.

3.2 Microbial Fossil Preservation

Traces of microbial life occur as (1) body fossils, which can be preserved in several modes, (2) structures (micro- and macroscopic) generated by microbial presence and activities, and (3) chemical compounds present in the rock record as a result of microbial metabolism (chemical biosignatures) (Fig. 3.1). This chapter deals mostly with body fossils and, to a somewhat lesser extent, with microbially induced structures. While chemical biosignatures can offer very useful insights into early life on Earth and investigations of biogenicity of candidate microfossils (Brasier and Wacey 2012), in this chapter, the impressive body of work produced by geochemists [e.g., Knoll et al. (2012) and references therein] is touched upon only lightly.

3.2.1 Body Fossils

The modes of preservation of microbial body fossils parallel those described for plant fossils (e.g. Schopf 1975; Stewart and Rothwell 1993) and include *permineralization* (also known as petrifaction), *coalified compression, authigenic* or *duripartic preservation*, and *cellular replacement* with minerals. In permineralization, minerals (usually calcium carbonate, silica, iron sulfide) precipitate from solutions inside and around cells, so the organisms end up incorporated in a mineral matrix and preserved three dimensionally, sometimes in exquisite detail. The quality of cellular preservation depends on the extent of decomposition of the organisms preceding the permineralization phase. Many early prokaryotes are preserved as permineralizations. In filamentous types, such as cyanobacteria, permineralized specimens often preserve mainly the external cellular envelopes (sheaths), whereas cells and their contents are altered to various degrees or not preserved at all (e.g., *Eoschizothrix*; Seong-Joo and Golubic 1998) (Fig. 3.1a, b). This is consistent with the results of chemical, structural (Helm et al. 2000), and

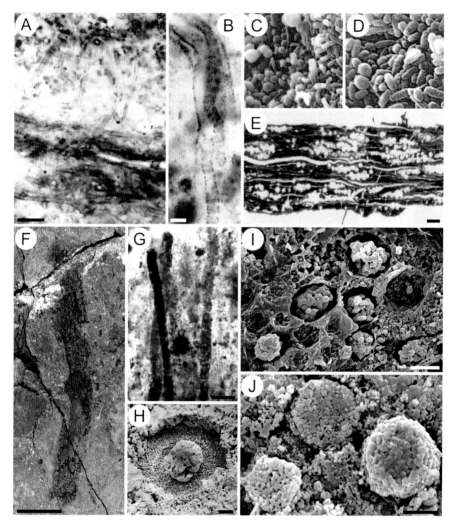

Fig. 3.1 Modes of microbial fossil preservation. (**a**) Permineralization, mats of cyanobacterial filaments (*Eoschizothrix*) preserved in silicified stromatolites of the Mesoproterozoic Gaoyuzhuang Formation, China. (**b**) Permineralization, multiple filaments of *Eoschizothrix* in a common extracellular polysaccharide sheath; Gaoyuzhuang Formation. (**c**) Cellular replacement, phosphatized fossil bacteria preserved in the eye of a fish from Tertiary oil shales in Germany. (**d**) Cellular replacement, calcified fossil bacteria preserved in the eye of a fish from tertiary oil shales in Germany. (**e**) Coalified compression and cellular replacement; cross section through a cyanobacterial colony (*Prattella*) showing brown coalified material representing extracellular polysaccharide sheath material and molds left by dissolution of mineral replaced (pyritized cells), Early Silurian Massanutten Sandstone, Virginia, USA. (**f**) Coalified compression, the coaly material formed by fossilization of *Prattella* extracellular polysaccharide sheath material, Massanutten Sandstone. (**g**) Cellular replacement, fragment of *Prattella* colony where the coalified extracellular polysaccharide matrix was cleared using oxidizing agents to expose filaments consisting of pyrite-replaced cells, Massanutten Sandstone. (**h**) Authigenic preservation, cross section of a possible cyanobacterial filament preserved in carbonate deposits, with the extracellular

experimental taphonomic (Bartley 1996) studies, which have stressed the higher resistance to degradation of extracellular polymeric substances (i.e., sheath and slime) in contrast that of the cell contents.

Coalified compressions are formed when the layers of sediment that incorporate the organisms are subjected to lithostatic pressure during rock forming processes (diagenesis). The pressure and temperature associated with burial in the Earth's crust induce changes in the geometry and chemistry of cells along a gradient of coalification of the carbonaceous material. For unicellular microfossils, a high degree of coalification can lead to complete obliteration of diagnostic features, to the point of rendering them unrecognizable as biogenic objects. However, lesser degrees of coalification can preserve diagnostic features even down to the ultrastructure level, as in the case of some early unicellular eukaryotes (Javaux et al. 2001, 2004) (Figs. 3.9e, f, 3.10e, 3.11b), while also increasing the preservation potential of fossils by rendering their organic compounds more chemically inert. Sometimes, microbial colonies can form compressions, as in the case of cyanobacterial colonies whose copious extracellular sheath material is coalified (Tomescu et al. 2006, 2008) (Fig. 3.1e, f).

Authigenic preservation refers to removal of the organic material previously enclosed in rock (by oxidation, decomposition) and its replacement with secondary material (precipitated minerals or sediment) that forms casts, whereas duripartic preservation involves the precipitation of minerals due to metabolic processes of the organisms that are fossilized. Duripartic mineral precipitation can occur in the cell walls, in the extracellular sheaths of colonies, or around the organisms, forming molds that can preserve cell-level structural details. Various minerals are known to form on microbial cell surfaces as a consequence of interactions between the microbial metabolism and the chemistry of its environment (reviewed by Southam and Donald 1999). Many fossil cyanobacteria are preserved as calcium carbonate (micrite) rinds that coated the organisms (e.g., *Girvanella*; Golubic and Knoll 1993) (Fig. 3.1h), which corresponds to a combination of authigenic and duripartic preservation. Cellular replacement is a cell-to-cell process in which diagenetic minerals precipitate inside individual cells, replacing their content. Different by its discrete nature from both authigenic and duripartic preservation which involve wholesale processes, cellular replacement is nevertheless more akin to authigenic preservation. Pyrite is widespread in cellular replacement (e.g., Munnecke et al. 2001; Kremer and Kazmierczak 2005) (Fig. 3.1i, j), but the list of minerals

⬅

Fig. 3.1 (continued) sheath preserved as micrite and the filament lumen filled with sparry calcite. (**i**) *Prattella* SEM of framboidal pyrite aggregates replacing individual cyanobacterial cells of *Prattella* and occupying molds in the coalified extracellular polysaccharide matrix, Massanutten Sandstone. (**j**) Cellular replacement, coccoid cells of a cyanobacterial mat replaced by framboidal pyrite aggregates, Silurian, Poland. Scale bars: (**a**) 50 μm, (**b**) 10 μm, (**c**) the bacterial cells are 0.5–1.5 μm, (**d**) the bacterial cells are 0.5–1 μm, (**e**) 20 μm, (**f**) 1 cm, (**g**) 50 μm, (**h**) 10 μm, (**i**) 5 μm, (**j**) 5 μm. Credits—images used with permission from (**a**) John Wiley & Sons (Seong-Joo and Golubic 1998). (**c**), (**d**) Blackwell Science Ltd. (Liebig 2001). (**h**) Blackwell Science Ltd. (Golubic and Knoll 1993). (**j**) Society for Sedimentary Geology (Kremer and Kazmierczak 2005)

is broad and includes apatite, calcite, siderite, hematite, and silica (Liebig 2001; Noffke et al. 2013b) (Fig. 3.1c, d).

In some instances, modes of preservation are combined, as in the case of the cyanobacterium *Prattella* (Tomescu et al. 2006, 2009) in which the macroscopic colonies of multitrichomous filaments are enveloped in high amounts of extracellular slime which forms coalified compressions, but within these compressions individual cyanobacterial cells are preserved by cellular replacement with framboidal pyrite aggregates (Fig. 3.1e, f, g, i).

3.2.2 Microbially Induced Structures

Microbially induced structures are structures produced by interactions between microbes and the sediment or rock. The different types of microbially induced structures encompass a wide spectrum of morphologies and range in size from the submillimeter scale up several orders of magnitude; some can form associations and layers with sizes ranging up to the kilometer scale. Microbially induced structures are known in large numbers from Proterozoic and Phanerozoic rocks, but their Archean record is less extensive (Awramik and Grey 2005; Noffke et al. 2003, 2006, 2013a). Two broad categories of microbially induced structures can be distinguished: *microbialites* and *microborings*.

3.2.2.1 Microbialites

Microbialites (Burne and Moore 1987) are constructional organosedimentary structures formed as a result of interactions between microbial mats and sediment. Depending on whether their formation involves the precipitation of minerals or not, microbialites are further divided into *stromatolites* (Kalkowsky 1908) and *microbially induced sedimentary structures* (MISS), respectively (Noffke and Awramik 2013).

Microbially Induced Sedimentary Structures

Because they do not involve mineral precipitation, MISS do not form thick or tall buildups like the stromatolites; instead, they have a more flattened, two-dimensional form. Microbial mats form at the interface between their substrate (usually the sediment surface) and water or air, and MISS formed as a result of interactions between the microbial mat and sediment (stabilizing, trapping) mark the transient location of such surfaces. Modern analogues of MISS found in tidal siliciclastic and terrestrial environments, which host a wide range of types with different morphologies (Noffke et al. 1996; Beraldi-Campesi and Garcia-Pichel

2011), allow us to understand the processes that lead to the formation of MISS encountered in the fossil record (Schieber et al. 2007).

The types of interactions between mats and sediment that lead to both MISS and stromatolite formation are binding and biostabilization, baffling and trapping, and growth (Noffke and Awramik 2013). Binding of microbial populations by cooperative secretion of extracellular polymeric substances leads to formation of a microbial mat. Once established, the mat alters the structure and physical properties of the sediment, stabilizing it (biostabilization) and, thus, reducing the effects of erosion (e.g., Fang et al. 2014). Sudden changes of fluid dynamic patterns at the interface between the sediment and water (or air) that increase the risk of bacterial mat erosion trigger increased or renewed biostabilizing activity in the microbial mat. Baffling and trapping occur when microbial mats baffle the water current around them, causing suspended particles to sediment and become trapped in the mat. Growth is the lateral and vertical expansion of the microbial mat by production of more extracellular polymeric substances or more cells, and sediment may become trapped within the mat while growth and associated binding are taking place.

MISS are ultimately formed when the surfaces hosting microbial mats are buried in sediment and lithified; then, the finer-grained layer trapped and bound by the microbial mat serves to separate two sedimentary beds and assist the preservation of any surface structure (Noffke 2009; Noffke and Awramik 2013). Thus, in the sedimentary record, MISS are recovered as structures on bedding planes, associated with recognizable microscale sedimentary patterns immediately beneath the bedding plane (laminae of organic matter concentration, grain size sorting, heavy mineral concentrations, etc., microtextures—Noffke 2009) (Figs. 3.2a–c, 3.3a, 3.11f). The wide morphological variety of MISS is the result of different sedimentary environments, substrates (sediment), types of interaction, and taphonomic processes. Because these interactions and processes have not changed significantly over 3.5 billion years, the morphologies of MISS in the fossil record are directly comparable with microbial mat-induced sedimentary structures documented in modern environments (Noffke et al. 2013b). Although identified as such relatively recently, MISS have received a lot of attention as traces of early life and a considerable body of literature on their genesis, recognition, classification, and geologic record has accumulated (e.g., Noffke 2009, 2010; Noffke and Awramik 2013; Noffke et al. 2013b).

Stromatolites

We use the term stromatolites here to designate any sedimentary structure formed as a result of microbial mats baffling, biostabilizing, and trapping sediment, in association with mineral precipitation. In other words, stromatolites can be regarded as MISS cemented by mineral precipitation. On the centimeter-to-meter scale, these structures exhibit varied morphologies, including wavy laminations, domes, and branched or unbranched columns (Awramik and Grey 2005) (Figs. 3.3b–d, 3.6b, 3.7h, 3.8f, g), some of which have received specific names—

e.g., dendrolites and thrombolites (Aitken 1967; Kennard and James 1986; Shapiro 2000). However, here we group all these different morphologies under the umbrella term stromatolite as they all share a laminated organization (the precipitated mineral is often calcium carbonate) and origin involving the activity of microbial mats (Shapiro 2000; Awramik and Grey 2005). Reitner et al. (2011) provide a comprehensive review of stromatolite research.

The most widely accepted mechanism for stromatolite formation involves sediments that are trapped and bound within microbial mats by the same basic mechanisms that form MISS and carbonate or some other type of mineral precipitation that cements the layer thus formed (Noffke and Awramik 2013). Over time, the microbial mat grows above the cemented portion forming a new layer which is, in turn, cemented, leading to upward growth of the stromatolite. The repetition of recolonization of the upper stromatolite surface by the microbial mat and subsequent cementation with precipitated minerals produces the characteristic laminated structure of stromatolites. Laboratory studies suggest that the extracellular polymeric substances produced by the microbial mats play an important role in carbonate precipitation (Altermann et al. 2006; Dupraz et al. 2009) and that heterotrophic members of microbial mat communities may initiate mineral precipitation (Noffke and Awramik 2013). If the precipitating minerals permineralize the microbial cells that form the mats, body fossils can be associated with the stromatolites. Although such occurrences tend to be rare (Wacey 2009), they have been documented both in the fossil record (e.g., Schopf and Blacic 1971; Schopf and Sovietov 1976; Knoll and Golubic 1979) and in modern stromatolites (Kremer et al. 2012a).

Structures described from shallow, hypersaline marine environments, such as Shark Bay in Western Australia (e.g., Awramik and Riding 1988), are often presented as good modern analogues for ancient stromatolites (Fig. 3.2f). This is because these structures are formed by microbial mats that trap and bind sediment, producing morphologies similar to those of stromatolites from the geologic record (Awramik and Grey 2005; Awramik 2006). However, most modern structures proposed as stromatolites are primarily formed by binding or trapping of sediments, without significant precipitation, whereas mineral (usually carbonate) precipitation is a prominent feature of ancient stromatolites (Kazmierczak and Kempe 2006; Noffke and Awramik 2013). To date, the only modern stromatolite analogues formed by both microbial mat activity and carbonate precipitation are those described from caldera lakes in Tonga (south Pacific) by Kazmierczak and Kempe (2006) and Kremer et al. (2012a) (Fig. 3.2d, e). Kazmierczak and Kempe (2006) propose that the highly alkaline chemistry of the caldera lakes is similar to the chemistry of Precambrian seas that hosted the wealth of stromatolites documented in the geologic record and that the rarity of this particular type of conditions in modern marine environments should account for the infrequency of modern carbonate-precipitating stromatolites and the presence of only comparable but nonprecipitating structures in modern benthic environments.

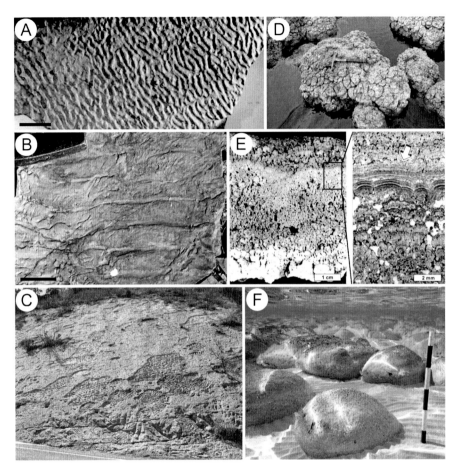

Fig. 3.2 Microbially induced structures. (**a**) Wrinkle structures representing buried microbial mats, Archean Mozaan Group, Pongola Supergroup, South Africa. (**b**) Wrinkle structures representing in situ preserved, thin microbial mats, Neoproterozoic Nama Group, Namibia. (**c**) Landscape-scale preservation of MISS on a hill side; tidal flat morphology resulting from partial erosion of a mat-stabilized sedimentary surface—the raised flat-topped areas are ancient microbial mats and the ripple marked depressions represent areas where the mats were eroded, Cretaceous Dakota Sandstone, Colorado, USA. (**d**) Stromatolites from caldera lakes on Niuafo'ou Island (Tonga Archipelago, south Pacific), recently recognized as the closest modern analogues of Precambrian Stromatolites. (**e**) Vertical sections through Niuafo'ou Island stromatolites showing the variety of internal structures. (**f**) Stromatolites such as these, from Shark Bay (Western Australia), were recognized early as modern analogues of Precambrian stromatolites. Scale bars: (**a**) 2 cm, (**b**) 5 cm, (**d**) hammer for scale 28 cm, (**f**) measuring pole painted in 10 cm intervals. Credits—images used with permission from (**a**) Geological Society of America (Noffke et al. 2003). (**b**) Elsevier Science Publishers (Noffke 2009). (**c**) Society for Sedimentary Geology (Noffke and Chafetz 2012). (**d**) and (**e**) Springer-Verlag (Kazmierczak and Kempe 2006). (**f**) Geological Society of America (Noffke and Awramik 2013)

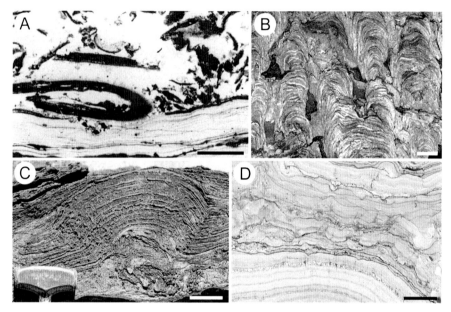

Fig. 3.3 Microbially induced structures. (**a**) Fine carbonaceous laminations with one layer folded over itself and overlain by a deposit of microbial mat-like fragments, Archean Kromberg Formation, Onverwacht Group, South Africa. (**b**) Vertical section through stromatolite deposit, Neoproterozoic Shisanlitai Formation, China. (**c**) Stromatolite in the Archean Tumbiana Formation, Fortescue Group, Western Australia. (**d**) Complex lamination at several scales in a thin section through Tumbiana Formation stromatolite. Scale bars: (**a**) 500 μm, (**b**) 2 cm, (**c**) 5 cm, (**d**) 5 mm. Credits—images used with permission from (**a**) Elsevier Science Publishers (Walsh 1992). (**b**) Geological Society of America (Noffke and Awramik 2013). (**c**) Elsevier Science Publishers (Lepot et al. 2009b). (**d**) Springer Science + Business Media B.V. (Wacey 2009)

3.2.2.2 Microborings

Microborings are microscopic, tubular, usually branched cavities that record the activity of euendolithic microbes. Although the differences are sometimes blurred, only some of the rock-inhabiting (endolithic) microbes, the euendoliths, actively bore into rock, whereas others occupy preexisting fissures and pore spaces of the rock (chasmoendoliths and cryptoendoliths, respectively) (Golubic et al. 1981; McLoughlin et al. 2007). In the rock record, microborings are found in both sedimentary (Campbell 1982; McLoughlin et al. 2007) (Fig. 3.8h) and volcanic rocks (Furnes et al. 2004, 2007; McLoughlin et al. 2012) (Figs. 3.4d–f, 3.7d, g, 3.8i). Living analogues have also been discovered in a wide range of environments, including near-surface sedimentary rocks (Knoll et al. 1986) and volcanic glass (Thorseth et al. 1991; Fisk et al. 1998).

Modern euendoliths employ several metabolic strategies, including photoautotrophy in near-surface environments and chemolithoautotrophy in deeper endolithic environments (McLoughlin et al. 2007). In carbonate rocks, endoliths

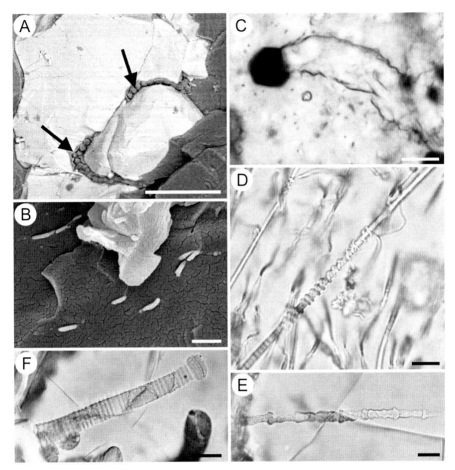

Fig. 3.4 (a) Chains of non-syngenetic endolithic fossilized coccoidal cells (*arrows*) penetrating a crack around a magnetite grain in Archean rocks of the Isua Greenstone Belt, Greenland. (b) Indigenous and syngenetic heterotrophic bacteria preserved in fossil extracellular polysaccharide sheaths of a cyanobacterial colony (*Prattella*); the bacterial cells are exposed in fresh breaks of the carbonaceous material (as seen here in SEM) which demonstrate that they did not penetrate through fissures at a later time; the bacterial cells also exhibit plastic deformation characteristic of soft, organic bodies (toward *bottom right*), corroborating hypotheses of biogenicity, Early Silurian Massanutten Sandstone, Virginia, USA. (c) Ambient inclusion trail with terminal pyrite grain and jagged tube edges, Archean Apex chert, Warrawoona Group, Western Australia. (d), (e), and (f) Microbial bioerosion structures (microborings) in volcanic glass of the Troodos ophiolite (Cretaceous, Cyprus), (d) *Tubulohyalichnus spiralis*, (e) *Tubulohyalichnus annularis* with unevenly spaced annulations, (f) *Tubulohyalichnus annularis* with uniformly spaced annulations and a terminal swelling. Scale bars: (a) 50 μm, (b) 1 μm, (c) 10 μm, (d) 10 μm, (e) 20 μm, (f) 20 μm. Credits—images used with permission from (a) Elsevier Science Publishers (Westall and Folk 2003). (c) Geological Society of London (Wacey et al. 2008). (d), (e), and (f) Geological Society of London (McLoughlin et al. 2009

have been shown to dissolve the host rock by producing organic acids or bioalkalization (McLoughlin et al. 2007) and exant euendoliths that live within volcanic glass or other siliceous rocks dissolve the host rock using similar pH-altering mechanisms (Callot et al. 1987; Thorseth et al. 1995; Staudigel et al. 1998, 2008; Büdel et al. 2004).

3.3 Recognizing Microbial Fossils

As their biogenicity receives support from both chemical and morphological lines of evidence, microbially induced structures represent more reliable indicators of prehistoric life than exclusively chemical biosignatures. Nevertheless, the most robust line of evidence in documenting the presence of microbial life, especially in the search for the earliest records of it in the Archean, is represented by body fossils (Ueno et al. 2001a). Yet when objects are recognized as candidate microbial fossils in the rock record, their identification as actual fossilized microbes (biogenicity) can be hampered by a series of factors (Buick 1990; Schopf and Walter 1983). The small size of microbes renders them easily degradable during diagenesis and, hence, unrecognizable. A wide variety of non-biogenic objects are known that mimic biogenic morphologies (mineral dendrites, crystallites, spheroids, filaments; Schopf and Walter 1983; Westall 1999; Brasier et al. 2006). Because of their simple morphology, fossil microbes are difficult to tell apart, unequivocally, from abiogenic microbial-looking objects, and we are missing a lot of the data needed to predict what kinds of such abiogenic objects may have been produced by diagenesis in the host rocks (Buick 1990). Furthermore, it has been argued that early microbial life may have looked and lived differently than modern microbes (Buick 1990), but exactly because we are looking for the earliest forms of life, we have no reference base, so we don't know what types of fossil to expect (Schopf and Walter 1983); for example, Archean microbes may not be directly comparable to modern counterparts whose morphologies and metabolisms may have been shaped by adaptation to living in a world filled with complex eukaryotes which were absent in the Archean (Brasier and Wacey 2012). Because of all these reasons, the literature on microfossils includes various terms conveying different degrees of certainty about the biogenicity (or absence thereof) of fossil-like objects: dubiofossils (fossil-like objects of uncertain origin; Hofmann 1972), pseudofossils (fossil-like objects undoubtedly produced by abiogenic processes; Hofmann 1972), bacteriomorphs (abiotic structures morphologically similar to bacteria; Westall 1999), biomorphs (abiogenic structures that mimic biological structures; Lepot et al. 2009a).

3.3.1 Recognizing Microbial Body Fossils

The need to recognize bona fide microbial fossils and distinguish them from abiogenic fossil-like objects on empirical bases was fueled by discoveries of putative microbial fossils in Precambrian rocks. Questions on the biogenicity of such fossils coming from progressively older rocks have been approached from two epistemologically opposite directions. One of these relies on inductive lines of reasoning focusing on demonstration of biogenicity by application of a set of criteria, whereas the other emphasizes falsification of non-biogenicity in a suite of contexts that range from geologic to metabolic. The former approach (*traditional approach* hereafter) was perfected in time by trial and error, whereas the latter (*contextual approach* hereafter) is a more recent development that stems from work on some of the oldest putative traces of life for which the simple application of the traditional set of criteria does not produce sufficient resolution and unequivocal conclusions (Brasier et al. 2006; Brasier and Wacey 2012). However, in theory, if they are applied rigorously and given enough relevant data, the two approaches to demonstrating biogenicity should ultimately lead to similar conclusions.

3.3.1.1 The Traditional Approach

The now-classic set of criteria for biogenicity used in the traditional approach was distilled over many years (e.g., Cloud 1973; Cloud and Hagen 1965; Knoll and Barghoorn 1977; Cloud and Morrison 1979; Schopf and Walter 1983; Buick 1990; Walsh 1992; Golubic and Knoll 1993; Horodyski and Knauth 1994; Morris et al. 1999; Schopf 1999; Southam and Donald 1999; Westall 1999; Schopf et al. 2010) and broadened based on the accumulation of knowledge brought about by successive discoveries of putative fossil microbiota. Each new discovery presented scientists with its own type of putative fossils, set of geologic conditions leading to fossilization (taphonomy), and modes of preservation. Each claim for the oldest record of fossils in a given category at a given moment was thoroughly scrutinized by the community (Brasier and Wacey 2012) which resulted in rejection or acceptance of the new record—see, for example, Barghoorn and Tyler's (1965) reevaluation of the initial inferences of Tyler and Barghoorn (1954) or Knoll and Barghoorn's (1975) rejection of the presence of eukaryotes in the 800 Ma (million years) Bitter Springs Formation (Australia); numerous other examples are summarized by Schopf and Walter (1983), and some are discussed below for the Akilia, Isua, Apex Chert, and Martian meteorite ALH84001 controversies and debates. Of these grew an increasingly more comprehensive, objective, and stringent set of criteria which is in use today (with some differences between authors). In general, application of these criteria involves addressing two fundamental types of questions: (1) Is the putative fossil indigenous to, and formed at the same time with, the host rock (*indigenousness* and *syngenicity*), as opposed to a modern contaminant or material introduced in the rock at a later time after rock formation? (2) Is the nature of the putative fossil demonstrably biological (*biogenicity*) (Table 3.1)? Only if it passes these two tests is a candidate fossil confirmed as a bona fide microbial body fossil.

Indigenousness

To demonstrate indigenousness of the putative fossils, one has to demonstrate that they are embedded in the prehistoric rock matrix. Contamination by modern biota within the rock can arise by percolation through cracks and microfissures or during sampling, but it can also be comprised of modern endolithic organisms inhabiting pores and fissures beneath the rock surface (e.g., recent endoliths inhabiting the cracks and fissures of the 3.7 Ga rocks at Isua, Greenland, along with carbonaceous remains washed into cracks by rainwater; Westall and Folk 2003) (Fig. 3.4a). Because of these, many authors recommend use of fresh samples from beneath the weathering front of outcrops and use of petrographic thin sections to ascertain microscopically that the putative fossils do not occur on fissures (Schopf and Walter 1983; Buick 1990; Morris et al. 1999). Scanning electron microscopy can also provide evidence for indigenousness when fresh breaks in the rock are analyzed and they reveal breaking of the putative fossils in the same plane, which also indicates syngenicity (Fig. 3.4b); alternatively, putative fossils found exclusively on surfaces with dissolution features are suspect of representing contamination (Morris et al. 1999). If the microfossils are extracted by dissolution of the rock, special care must be taken to avoid any modern contaminants in the facilities and on equipment (Buick 1990) and to exclude from analysis the outermost layers of rock samples that may introduce contaminants acquired during sampling (Redecker et al. 2000).

Syngenicity

Demonstrating the syngenicity (also referred to as *syngeneity*) of candidate fossils involves proving that they were placed in the rock matrix upon its formation and not later (Schopf and Walter 1983; Buick 1990). For this, the age of the rock and the processes which led to its formation need to be well understood. Fossils need to be fully enclosed in the host rock as identifiable in petrographic thin sections, and when broken, they should fracture in a manner consistent with the way the groundmass of the host rock breaks around them (Morris et al. 1999). If the candidate fossils form only very localized assemblages or are consistently associated with discontinuities in the rock structures, their syngenicity is questionable as they may have been transported and emplaced along veins or other secondary diagenetic structures. Additionally, one expects to see overall consistency between the chemistry of syngenetic fossils and host rock, therefore presence in the candidate fossils of elements or compounds that are absent in the groundmass of the host rock supports non-syngenicity (Morris et al. 1999). In one example, Javaux

Table 3.1 Criteria for recognition of bona fide microbial fossils; based primarily on Schopf and Walter (1983), Buick (1990), Morris et al. (1999), Westall (1999), Brasier and Wacey (2012)

Traditional approach		Contextual approach	
Age	• Age and stratigraphy of rocks is resolved	Geologic context: ancient and viable for life and fossil preservation	• Age of rocks is resolved
Indigenousness	• Candidate fossils are indigenous to the host rock		• Candidate fossils are indigenous to the host rock • Lithology and stratigraphy denote an environment that could have harbored life
Syngenicity	• Petrography indicates emplacement of candidate fossils in host rock at the time of its deposition • Occurrence at several locations in rock • Chemistry consistent with that of host rock		• Petrology reflects diagenetic history favorable to preservation of traces of life • Petrography indicates emplacement of candidate fossils contemporaneous with formation of rock unit
Biogenicity	Direct: morphology and chemistry of candidate fossils • Morphology is consistent with that of living and fossil organisms (cellular organization) • Range of morphological variation within the population of candidate fossils is consistent with that seen in living and fossil organisms (non-uniformity) • Chemistry roughly matching that of host rock; carbonaceous or formed by biologically mediated mineral precipitation or mineral replacement • Dissimilar from potentially coexisting abiological organic objects	Morphological context: candidate fossils fit within the morphospace of cellular organization	• Morphology is consistent with cellular organization • Range of morphological variation within the population of candidate fossils is consistent with that seen in living organisms • Distinct from abiogenic mimics expected within the same kind of setting • All plausible explanations for abiogenic origin of like morphologies can be falsified
	Indirect: patterns of association, geologic, chemical, and evolutionary context • Abundance of occurrence • Associated in a multi-component assemblage of such objects (non-monospecific assemblage) • Geologic context is consistent with an environment that could have harbored life and conditions favorable to fossil preservation • Chemistry of candidate fossils/host rock consistent with that of microbial metabolic processes • Age and level of organization consistent with overall context of the evolution of life	Behavioral context: candidate fossils associate in patterns consistent with biological behavior	• Associated with microbially induced structures (e.g. biofilms, cements) • Associated with structures of the same kind in clusters or mats • Position within rock denotes biologically mediated substrate preferences
		Metabolic context: chemistry of candidate fossils and surrounding rock is consistent with living organisms and their metabolic products	• Cell walls, when present, have chemical composition consistent with a metabolic pathway • Host rock around candidate fossils bears chemical signatures of metabolic extracellular effusions and their effects on the mineral environment • Host rock bears chemical signatures of interlinked metabolic pathways (and their spatial zonation) characteristic of functioning ecosystems

et al. (2010) demonstrated syngenicity of 3.2 Ga microfossils from the Moodies Group (South Africa) (Fig. 3.10a) by showing that the organic matter comprising the microfossils had undergone the same degree of metamorphism as dispersed organic matter in the host rock, based on Raman spectrometry. Recently, Olcott Marshall et al. (2014) demonstrated that the carbonaceous material in the 3.46 Ga Apex Chert (Warrawoona Group, Australia) represents four generations of material with different thermal alteration histories and associated with different episodes of matrix formation, indicating that at least some of the four generations (if not all) are not syngenetic.

Biogenicity

The biogenicity of candidate fossils is demonstrated both by direct assessment of the objects themselves—morphology and chemistry—and indirectly, based on their broader taphonomic, geologic, chemical, and evolutionary-biostratigraphic context. The shapes and sizes of candidate fossils have to be consistent with those of known fossil and living organisms (Schopf and Walter 1983). Morphologies indicative of biogenicity include cells exhibiting phases of division (Knoll and Barghoorn 1977) (Fig. 3.7c, f) or plastic deformation characteristic of soft, organic bodies (Tomescu et al. 2008) (Fig. 3.4b). Morphological requirements for confirmation of biogenicity include sizes within the range of known microbes (>0.01 μm^3) and hollow objects (coated in carbonaceous material), i.e., walls or sheaths of cells or cell colonies (filaments) with or without internal divisions (Buick 1990).

Ideally, the candidate fossils show cellular elaboration, but this criterion is the source of much debate (Buick 1990) as abiogenic objects can mimic some features of cellular organization. Several authors recommend special caution in the interpretation of spheroids comparable to coccoid prokaryotes, even when these exhibit morphologies comparable to dividing cells, as such morphologies can be formed by abiogenic processes (Westall 1999; Brasier and Wacey 2012). In such cases, independent lines of evidence are required to corroborate biogenicity. Furthermore, even more complex filamentous morphologies comparable to Precambrian microfossils can be generated abiotically, as shown by Garcia-Ruiz et al.'s (2003) experiments on precipitates formed by metallic salts in silica gels; however, the structures thus formed are not hollow. Abiogenic structures mimicking microbial filaments are also formed when local dissolution of the rock matrix allows for displacement of crystals representing mineral inclusions which leave trails (Knoll and Barghoorn 1974) (Fig. 3.4c); when carbonaceous inclusions from the rock are also included in the trails, these can be easily mistaken for microbial filaments (Lepot et al. 2009a). Only careful study of the microstructure and distribution of carbonaceous matter, along with the fact that the "filaments" have mineral crystals at their ends, reveals the abiogenic nature of such biomorphs (Brasier et al. 2002; Lepot et al. 2009a).

The chemistry of candidate fossils can help in assessment of their biogenicity, which is supported by the presence of cell walls or internal structures consisting of kerogen (geologically transformed organic matter; see Sect. 3.6.2.1) and by chemical compositions that roughly match that of the rock groundmass but show elevated carbon content (Buick 1990; Morris et al. 1999). Stable carbon isotope ratios ($\delta^{13}C$) of carbonaceous material in candidate fossils have been used extensively in discussions of biogenicity (Westall 1999), and ^{13}C-depleted values are thought to indicate biological fractionation of carbon and, thus, biogenicity (e.g., Ueno et al. 2001b). However, a survey of the modern biota reveals that biogenic $\delta^{13}C$ values can vary at least as broadly as −41‰ to −3‰ PDB (Pee Dee Belemnite, a standard used for reporting carbon isotopic compositions and based on the Cretaceous marine fossil cephalopod *Belemnitella americana*), overlapping toward the top of this range with inorganic carbon (Buick 2001; Schidlowski 2000; Fletcher et al. 2004). Therefore, caution should be applied in drawing generalizations based on $\delta^{13}C$ values (Buick 2001), which should at best be used to support biogenicity only in conjunction with other independent sources of evidence.

The requirement for presence of organic carbon compounds (kerogen) excludes most traces of microbial life comprised exclusively of inorganic material, such as some microbially induced structures for which biogenicity criteria are discussed below. A particular case of inorganic structures of biogenic origin are the magnetosomes, ferromagnetic magnetite particles that are biomineralization products of magnetotactic bacteria. Magnetosomes are produced inside the bacterial cells, and when the latter are degraded, their magnetosomes form chains that mark the location of former filaments, but very similar magnetite grains can also have a fully abiogenic origin. The equivocal nature of such magnetite grains fuelled a significant part of the Martian meteorite ALH84001 debate (McKay et al. 1996; Thomas-Keprta et al. 2001; Golden et al. 2004; see below Sect. 3.4.4). More recently, Gehring et al. (2011) were able to identify dispersed magnetite particles in Holocene lake sediments as magnetosomes using two-frequency ferromagnetic resonance spectroscopy, thus opening the way to detection of this group of bacteria based on acellular but biogenic body fossils.

Whether organic carbon is present or not, another set of morphological criteria address biogenicity in terms of assemblage-level features. Candidate fossils co-occurring with more clearly discernable microfossils (e.g., spheroids co-occurring with rod-shaped fossils or a fossilized biofilm) are more likely to have a biogenic origin (Westall 1999). While some morphological variation is to be expected in assemblages of bona fide microbial fossils, the fossils have to be consistent in morphology and size throughout the assemblages (Figs. 3.7c, f and 3.8a, b, d), and significant disparities in the size of morphologically similar objects within an assemblage indicate abiogenic origin (Buick 1990; Westall 1999). Also at the scale of the entire candidate fossil assemblage, occurrence in abundance throughout the rock volume (a criterion for indigenousness and syngenicity as well) and the presence of multiple morphological types, thus non-monospecific assemblages (Fig. 3.7a), support biogenicity (Schopf and Walter 1983).

In a broader perspective, beyond the realm of morphology, the geology of the host rock has to reflect both genesis in an environment favorable to the presence of life and a subsequent geologic history conducive to fossil preservation (Schopf and Walter 1983; Buick 1990). Microbial body fossils are not thought to preserve in metamorphic rocks formed beyond low-grade metamorphism conditions. Microbe-like objects found in medium- to high-grade metamorphic rocks or in igneous rocks are either abiogenic, or if they are bona fide fossils, they represent nonindigenous microbes (contaminants) or non-syngenetic microbial fossils. The chemistry of the host rock can also be used to support inferences of biogenicity of candidate fossils when it roughly matches that of the putative fossils, and it is characterized by significant levels of elements and minerals formed by the direct or indirect activities of microbes (e.g., pyrite produced by sulfate-reducing bacteria or magnetite, as discussed above) (Morris et al. 1999; Westall 1999). Finally, the level of biological complexity and organization of candidate fossils has to be consistent with the age of the host rock and the overall context of the known history of life on Earth (Schopf and Walter 1983). In this context, candidate fossils that appear out of context are likely to be abiogenic, nonindigenous, or non-syngenetic.

3.3.1.2 The Contextual Approach

The search for the oldest traces of life adds another set of challenges (as discussed by Brasier et al. 2006) to those encountered in documenting the microbial fossil record in younger rocks. First, whereas Proterozoic (<2500 Ma old) rocks have yielded a rich microfossil record, older rocks (especially pre-Neoarchean; >2800 Ma old) have produced very rare candidate fossils that the traditional approach to biogenicity can resolve unequivocally. Second, the environments of early Earth were very different from those we are familiar with or that we can even imagine today, and the potential life forms they hosted were very likely more similar to those of modern environments we are just exploring today (e.g., deep intraterrestrial endoliths, hyperthermophiles, anaerobes) than to anything else. Third, there is currently increasing recognition that a variety of abiogenic self-organizing structures generated by natural processes can mimic the morphological complexity of bona fide traces of life.

In most cases, the situations generated by these constraints reside beyond the sphere of resolution of the traditional inductive approach to biogenicity. Furthermore, the initial recognition of putative microfossils is based on intuition and experience; however, intuition and experience are double-edged swords, as they can easily lead one down the path of simply seeking evidence in support of a preferred interpretation, without consideration of alternative explanations. Such considerations, along with the challenges of identifying traces of life deeper and deeper in the rock record, have led some workers (e.g., Brasier et al. 2002, 2006) to reject the traditional approach and adopt a falsificationist approach wherein microbial structures are not accepted as biogenic until the alternative null hypothesis of abiogenicity is falsified. This approach emphasizes the integrative use of a comprehensive set of methods and an outlook based on asking open-ended questions about types of processes (biogenic and abiogenic) and geologic settings, and whether they could have produced the candidate fossil structures, rather than on comparisons with known fossil or modern organisms and structures (Brasier et al. 2006). In other words, instead of proving the biogenicity of structures, this approach strives to demonstrate that they cannot be abiogenic. For this, questions

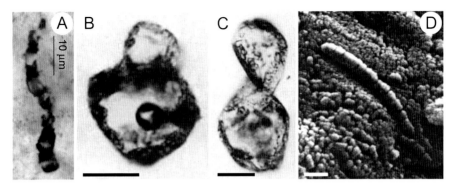

Fig. 3.5 Debates and controversies. (**a**) Putative microbial filament (*Primaevifilum amoenum*) from the Archean Apex Chert (Warrawoona Group, Western Australia). (**b**) and (**c**) Limonite-stained inclusions or cavities in Archean rocks of the Isua Greenstone Belt (Greenland), initially interpreted as microbial fossils (*Isuasphaera isua*). (**d**) Structures from the ALH 84001 Martian meteorite initially interpreted as bacterial magnetosomes. *Scale bars*: (**a**), (**b**), and (**c**) 10 µm; (**d**) 10 nm. Credits—images used with permission from (**a**) Elsevier Science Publishers (Schopf et al. 2007). (**b**) and (**c**) Springer-Verlag (Pflug 1978b). (**d**) Wikimedia Commons; file: ALH84001_structures.jpg; author: NASA

are asked to assess the level of support for biogenicity in a hierarchy of contexts: (1) geologic, (2) morphological, (3) behavioral-taphonomic, and (4) metabolic (Table 3.1).

The method of the contextual approach has been formalized by Brasier et al. (2006), Wacey (2009), and Brasier and Wacey (2012). Not surprisingly, some of the criteria applied in the contextual approach necessarily overlap with those of the traditional approach. However, due to the degree of generality of questions asked in applying it, the applicability of this approach extends beyond body fossils, to microbially induced structures. The approach has been used to reject the biogenicity of putative prokaryote fossils of the 3.46 Ga old Apex Chert in Australia (Brasier et al. 2002) (Fig. 3.5a) and to demonstrate the biogenicity of prokaryote fossils in the 3.4 Ga old Strelley Pool Formation in Australia (Wacey et al. 2011a) (Fig. 3.8a, b) and in an extensive critical analysis of all claims for early Archean life (Wacey 2009). Below we summarize the elements of the contextual approach as set forth by Brasier and Wacey (2012)—see also Table 3.1.

1. In a *geologic context*, questions are aimed at establishing the age of the rock hosting the candidate fossils, as well as the indigenousness and syngenicity of the latter, much in the same way that these are addressed traditionally. Furthermore, regional stratigraphy and petrology are mapped and sampled at the kilometer -to-meter scale in order to gain an understanding of whether the past local environments reflected by the rock record could have harbored life and whether the rock sequence reflects a diagenetic and post-diagenetic history favorable to fossil preservation. Detailed mapping of petrography and geochemistry at the cm-to-nm scale are then used to document spatial and temporal

relationships between the candidate fossils and their emplacement in the rock, on the one hand, and the host rock and its history, on the other hand. These allow for more in-depth assessment of the suitability of the host rocks for fossil preservation and for reconstruction of a detailed time line of the events of putative fossil formation and rock genesis, which allows for assessment of indigenousness and syngenicity. Inability to document in detail all of these aspects of the geologic context leaves the door open for alternative untestable (because unknown) hypotheses of nonindigenousness, non-syngenicity, or non-biogenicity.

2. The *morphological context* provides information important in assessing the biogenicity of candidate fossils regarded as members of a population of similar objects. The level of support for biogenicity is tested by documenting the morphospace occupied by the population (ranges of morphological and size variation) and asking whether it fits within the morphospace of cellular organization in terms of both shape and size and range of variation. For example, ranges that are too broad may indicate abiogenic structures whose variability is not constrained by genetics or habitat. Importantly, biogenicity is assessed at the same time in terms of the potentiality for the documented morphospace to be occupied by abiogenic objects expected within the geological context under consideration—for this the morphology of candidate fossils has to be considered in concert with their chemistry. All plausible explanations for abiogenic origin of objects with morphologies similar to that of the candidate fossils have to be falsified to demonstrate biogenicity. In this context, Brasier and Wacey (2012) emphasize the need for greater emphasis on improved mathematical modeling of morphospace occupation by different populations of objects, in order to produce more powerful tests for distinguishing biological populations from amalgamations of abiogenic structures (e.g., Boal and Ng 2010).

3. Because living organisms have behaviors which may be reflected in the taphonomy of their fossils, support for biogenicity has to be tested in a *behavioral-taphonomic context*. Patterns of association documented within assemblages of candidate fossils are assessed for the presence of features characteristic of biological behaviors. These include populations of candidate fossils assembled in clusters or mats reflecting colonial associations (Figs. 3.6a, 3.7a, b, 3.8d), or populations associated with microbially induced structures and textures (biofilm-like textures, biogenic or organomineral cements), as well as assemblages of candidate fossils positioned in response to substrate preferences (Fig. 3.8b). Just like with the morphological context, care must be taken to falsify abiogenic explanations for these types of association patterns: abiotic mineral growth can also form clusters, sometimes at the contact between contrasting lithologies which may be interpreted as a substrate surface colonized by a microbial mat; and inferences of biogenicity need to be corroborated by chemical data.

4. The metabolism of living organisms influences their environment and this may be reflected in the chemistry of candidate fossils and their host rock, which offers a *metabolic context* for testing hypotheses of biogenicity. Candidate fossils that comprise a carbonaceous fraction are tested for ^{12}C-enriched stable carbon

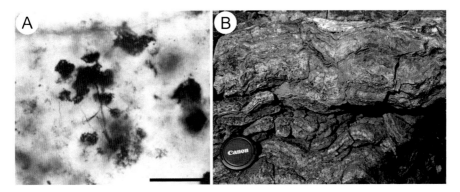

Fig. 3.6 Archean Dresser Formation (Warrawoona Group, Western Australia). (**a**) Putative filamentous microfossils; some exhibit helical geometries, others are interwoven, branched, or radiate from kerogen clots. (**b**) Putative stromatolite. Scale bars: (**a**) 50 µm, (**b**) lens cap diameter ca. 6 cm. Credits—images used with permission from (**a**) and (**b**) Springer Science + Business Media B.V. (Wacey 2009)

isotope ratios characteristic of biogenic carbonaceous material (kerogen). An enrichment in chemical elements comprising major building blocks of living matter (hydrogen, oxygen, nitrogen, sulfur, phosphorus) relative to the host rock matrix is also to be expected in biogenic objects. Due to the small size of microfossils, their chemistry is compared to that of surrounding mineral grains. Additionally, bona fide body fossils will be associated with chemical signatures generated by their metabolic extracellular effusions in the host rock (discussed in some detail below—see Sect. 3.6.2 Authigenic mineralization). In such cases, if microbial communities are preserved in situ, the host rock can record the chemical signatures of interlinked metabolic pathways (e.g., carbon fixation and carbon respiration; Brasier and Wacey 2012) and their spatial zonation characteristic of functioning microbial ecosystems.

3.3.2 Recognizing Microbially Induced Structures

The recognition of diverse structures in the geologic record as traces of microbial life follows the same paradigms as that of microbial body fossils. Some authors apply sets of criteria in a traditional inductive approach, while others favor a context-based falsificationist approach, and the criteria used vary somewhat among authors. The methods used for assessing the criteria also vary somewhat for different types of microbially induced structures (MISS, stromatolites, microborings) because of differences in the types of microorganisms that generated the structures and their mode of formation. However, the fundamental requirements for recognizing diverse structures in the geologic record as traces of microbial life are the same as for microbial body fossils. Biogenic-like morphologies are what initially recommends them as candidate microbially induced structures, and the

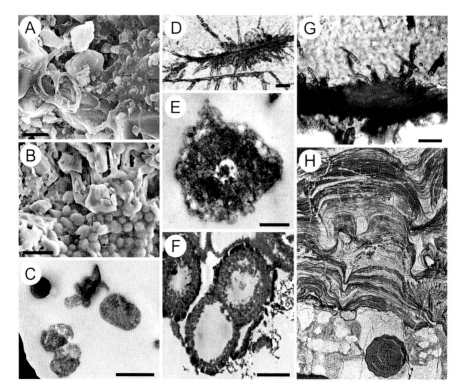

Fig. 3.7 (**a**) and (**b**) Microfossils from the Archean Kitty's Gap Chert (Warrawoona Group, Western Australia), filamentous, rod-shaped, and coccoid microorganisms. (**c**)–(**g**) Microbial fossils of the Archean Hooggenoeg Formation (Onverwacht Group, South Africa), (**c**) cell-like bodies showing possible stages of division, (**d**) candidate microborings—titanite (*brown*) in volcanic glass (*green*), (**e**) granular-textured body showing porosity and central cavity, (**f**) aggregate of cell-like bodies showing central cavities, (**g**) candidate microborings. (**h**) Stromatolites of the Archean Onverwacht Group. Scale bars: (**a**) and (**b**) 2 µm, (**c**) 0.5 µm, (**d**) 50 µm, (**e**) 0.2 µm, (**f**) scale 2 µm, (**g**) scale 50 µm, (**h**) coin ca. 20 mm diameter. Credits—images used with permission from (**a**) and (**b**) Geological Society of America (Westall et al. 2006). (**c**), (**e**), and (**f**) Elsevier Science Publishers (Glikson et al. 2008). (**d**) Geological Society of America (McLoughlin et al. 2012). (**g**) Elsevier Science Publishers (Furnes et al. 2007). (**h**) Blackwell Science Ltd. (Golubic and Knoll 1993)

geologic context needs to reflect conditions suitable for life and syngenicity (except for microborings which can be emplaced subsequent to the formation of their host rock) (Schopf and Walter 1983; Buick 1990; McLoughlin et al. 2007). Aside from morphology, geochemical analyses are called upon in the assessment of biogenicity, to test for evidence for metabolic activity (Brasier et al. 2006; McLoughlin et al. 2007; Brasier and Wacey 2012). Furthermore, in a contextual approach (as outlined above), the null hypothesis of abiotic origin must be rejected for any candidate microbially induced structure (Brasier et al. 2006; Brasier and

3 Microbes and the Fossil Record: Selected Topics in Paleomicrobiology

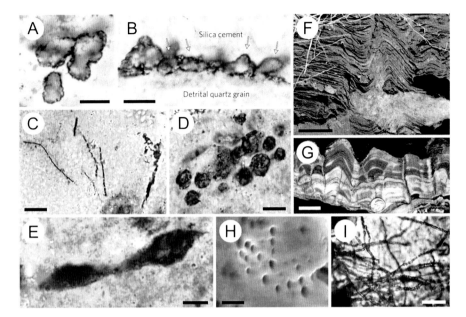

Fig. 3.8 Microbial fossils of the Archean Kelly Group (Western Australia); (**a**)–(**h**) Strelley Pool Formation, (**i**) Euro Basalt. (**a**) Cluster of cells, some showing cell wall folding and invagination. (**b**) Cells attached to quartz grain exhibiting preferred alignment parallel to the surface of the quartz grain. (**c**) Carbonaceous threads of rodlike objects representing putative microbial filaments. (**d**) Colony of loosely clustered hollow spheroidal microfossils. (**e**) Pair of linearly arranged lenticular carbonaceous microfossils. (**f**) and (**g**) Stromatolites. (**h**) Microbial etch pits (microborings) in pyrite. (**i**) Segmented microborings in volcanic glass. *Scale bars*: (**a**) and (**b**) 20 µm, (**c**) 20 µm, (**d**) 20 µm, (**e**) 10 µm, (**f**) 4 cm, (**g**) 3 cm, (**h**) 3 µm, (**i**) 25 µm. Credits—images used with permission from (**a**) and (**b**) Nature Publishing Group (Wacey et al. 2011a). (**c**), (**d**), and (**e**) Elsevier Science Publishers (Sugitani et al. 2013). (**f**) and (**g**) Springer Science + Business Media B.V. (Wacey 2009). (**h**) Elsevier Science Publishers (Wacey et al. 2011b). (**i**) Elsevier Science Publishers (Furnes et al. 2007)

Wacey 2012). Several authors have proposed different recognition criteria and ways in which these may be satisfied for specific types of structures.

3.3.2.1 Microbialites

Most microbialites lack microfossils and determining their biogenicity requires attention to the geologic context, overall morphology, microstructure, and chemistry. Modern analogues of stromatolites are very rare (Kazmierczak and Kempe 2006) (Fig. 3.2d, f) and little is known about their formation from direct observations. As a result, definitions of stromatolites and the criteria used to recognize them as biogenic structures are widely different between authors and are still debated. In contrast to stromatolites, which have been recognized and studied for at least a century (Awramik and Grey 2005), the study of microbially induced sedimentary structures as fossil traces of life, and especially in the search for the earliest traces of life, is a relatively recent development. In part because of its recent rise, but also due to the wealth of modern sedimentary environments that host microbial mats providing as many modern analogues, the study of MISS is firmly grounded in an empirical comparative approach that was pioneered by Noffke and Krumbein (Noffke et al. 1996; Noffke et al. 2001).

Microbially Induced Sedimentary Structures

Noffke (2009) provided the most recent treatment of the lines of evidence used for recognizing MISS, summarizing them in a set of six criteria of biogenicity which we discuss below. These criteria are based on numerous studies of both modern and fossil microbial mats and MISS, but emphasize MISS formed by photoautotrophic mat-building microbes in aqueous environments.

1. *Broader geologic context*: MISS occur in sedimentary rocks that have experienced, at most, low-grade metamorphism (lower greenschist facies). Higher-grade metamorphism is unfavorable not only to the preservation of body fossils but also to MISS preservation—the often complicated diagenesis of metamorphic rocks can render syngenicity difficult to assess, metamorphic textures and structures can overprint the sedimentary structures, and, overall, biosignatures are altered and difficult (or impossible) to recognize.
2. *Sequence stratigraphy*: MISS are associated with regression-transgression turning points (the term transgression indicates a relative rise in sea level, while regression indicates a relative drop in sea level). More specifically, it is the transgressive phases that succeed those turning points that witness expansion of tidal flat and shallow shelf areas favorable to microbial mat formation along the passive continental margins, due to sea level rise.
3. *Depositional environment*: MISS occur in rocks formed in environments favorable to the establishment and preservation of microbial mats. Although modern photoautrophic mat-building microbes are often found in environments characterized by fine-grained sand substrates and low current velocities (10–25 cm/s), it is conceivable that chemoautotrophic microbes could build mats beyond the photic zone and on substrates suitable to their metabolic needs.
4. *Hydraulic regime*: the types of MISS and their spatial and stratigraphic distribution are consistent with the hydraulic regime implied by sedimentology, as specific types of modern MISS have been shown to characterize environments with different hydraulic regimes (e.g., shallow shelf, lagoon, different tidal zones).
5. *Morphology*: constrained by the same biology throughout the ages, microbial mats have maintained the same functional morphology traits, therefore geometries and dimensions of fossil MISS match those of modern MISS.
6. *Microtexture*: microscopic textures and structures related to, caused by, or representing microbial mats are found associated with the macroscopic MISS.

These include (often poorly) fossilized or mineral-replaced microbial filaments, wavy laminae concentrating organic matter and associated with sharp microstratigraphic geochemical gradients (Noffke et al. 2013b), finer-grained clasts or heavy minerals, upward-fining microsequences, and minute pores left by gases accumulating under the microbial mat.

Stromatolites

The realization, early in the history of stromatolite studies, that some laminated precipitation structures represent microbially induced structures led to a period of somewhat undiscerning application of the label of biogenicity to any finely laminated structure in the rock record, which resulted in a plethora of reports of stromatolites (and therefore microbial fossils) from rocks of all ages [e.g., the 3.4 Ga Buck Reef Chert of South Africa, listed as a stromatolite by Schopf (2006) but not treated as one by Tice and Lowe (2004)]. The subsequent realization that not all laminated structures are biogenic (e.g., Lowe 1994) led to disagreement over what defines a stromatolite (Hofmann et al. 1999; Awramik and Grey 2005). Because of that, distilling a set of criteria of biogenicity for stromatolites is much more difficult than for MISS (e.g., Riding 2011). Whereas some authors define stromatolites as laminated structures that are formed by microbial activities and mineral precipitation (Awramik and Margulis 1974; Buick et al. 1981), others use the term stromatolite for laminated, lithified sedimentary structures, regardless of the involvement of microbes in their formation (e.g., Semikhatov et al. 1979; Antcliffe and McLoughlin 2009). Consequently, the disagreement also extends over whether such laminated structures could represent good evidence for ancient life, and a number of different sets of criteria for stromatolite biogenicity have been proposed.

Buick et al. (1981) assembled probably the most stringent set of criteria that focuses on morphological lines of evidence:

1. The candidate structures (termed stromatoloids; Buick et al. 1981) occur in sedimentary or metamorphosed sedimentary rocks;
2. The structures are synsedimentary with the host rock; because most stromatolites are built by photosynthetic microbial communities that tend to be thicker in positions receiving the most light and therefore raised with respect to the rest of the substrate;
3. Most structures in a stromatolite have a convex-upward morphology (e.g., Figs. 3.2d–f, 3.3b–d, 3.6b, 3.7h, 3.8f, g) and
4. Individual laminae are thicker over the upward-facing convex surfaces (e.g., Figs. 3.3b, c, 3.7h, 2.8g);
5. Laminations are wavy and wrinkled or have several orders of curvature (e.g., Figs. 3.2e, 3.3c, d, 3.7 h) because bedding irregularities are amplified by the phototropic tendencies of the microbial communities listed above;
6. Microbial body or trace fossils are present in the structures;

7. Changes in the microfossil assemblages are associated with changes in the morphology of sedimentary structures, indicating a relationship between the microbial communities and the formation of the structure; and
8. Microfossils are preserved in situ and in positions that indicate accretion activities—binding, trapping, or precipitation.

It is worth noting that, based on strict morphological criteria and including requirements that microfossils be present and proven to contribute to sedimentary accretion, this set of criteria diagnoses as abiogenic most structures accepted as stromatolites by other systems of evaluation.

In his criteria for stromatolite biogenicity, Walter (1983) was concerned with syngenicity—orientation of laminated structures with respect to bedding planes indicating formation at the same time with adjacent layers and occurrence in rocks where the laminated structures can only be explained as primary sedimentary features—and biogenicity as reflected in morphology, microbial origin, and chemistry. Hofmann et al. (1999) used careful studies of morphology, sedimentology, and local microstratigraphic relationships in their assessment of the biogenicity of 3.45 Ga conical laminated structures from the Warrawoona Group (Australia). Their approach was aimed at rejecting hypotheses of abiogenic origin and their arguments were later regarded as a set of criteria for biogenicity (Awramik and Grey 2005). Hofmann et al. (1999) rejected regional deformational processes as a potential explanation, based on the geographically broad extent of the structures, combined with their narrowly constrained stratigraphic and lithologic circumstances. They also used the geometry and relative position of laminae to reject sideways compression (folding), downright directed slumping, and strictly chemical precipitation as explanations for different morphological aspects of the structures, supporting their formation by upward accretion. Furthermore, these authors used the steep angles of the structures to reject simple sedimentation as a formation mechanism, supporting the presence biological binding. Finally, Hofmann et al. (1999) invoked the wide acceptance of other independent lines of evidence (at the time; some of these were later contested), such as body fossils and chemical biosignatures, for the presence of life in coeval rocks, as making biogenic causes even more plausible.

The contextual approach (see above; Brasier et al. 2002; Brasier and Wacey 2012) is applicable, with some changes, to MISS and stromatolites (Wacey 2009). The fact that stromatolite-like morphologies have been proven to form by abiogenic processes (Grotzinger and Rothman 1996; McLoughlin et al. 2008) implies that morphology alone is not enough to determine biogenicity and that microstructural and geochemical analyses are necessary to reject the null hypothesis of abiogenic formation for any given stromatolite. It is interesting to note that Awramik and Grey (2005) have expressed doubt about whether any abiogenically formed structures can convincingly resemble stromatolites. On a more general level, these authors criticized the contextual approach for setting lofty standards for paleontology. They point out that alternative modes of genesis are thoroughly considered by workers adopting a traditional inductive approach to stromatolite biogenicity, even if the associated deliberations are not included in the final publication. Instead of what they call the binary character of the contextual approach, Awramik and Grey (2005) advocate an approach that emphasizes morphology (without excluding other types of data) aimed at finding evidence for the contribution of biogenic processes to the formation of candidate structures and placement of that evidence on a scale of credibility (unequivocal-compelling-presumptive-permissive-suggestive evidence).

Most of the disputes on stromatolite biogenicity and the criteria to assess it are certainly rooted in the absence of microbial body fossils from many candidate stromatolites (Hofmann et al. 1999). Addressing this topic, Kremer et al. (2012a) studied calcification and silicification processes in cyanobacterial mats that form the best modern analogues of fossil stromatolites, in Tonga (south Pacific; Kazmierczak and Kempe 2006) (Fig. 3.2d, e). They showed that morphological preservation of cyanobacteria by primary mineralization depends on two main factors: the type of mineral phase and the time of mineralization. Variations in the two factors produce a wide spectrum of modes of morphological preservation which encompass several stages of degradation that were documented for both filamentous and coccoidal cyanobacteria. These observations led Kremer et al. (2012a) to suggest that Archean life may have been more abundant than previously thought based on meager findings, but difficult to recognize because of the alteration of original microbial body fossils or sedimentary structures due to recrystallization and mineral replacement.

3.3.2.2 Microborings

In contrast to stromatolites, microborings have abundant modern counterparts. However, the abundance and ubiquity of modern microborings (included by some authors in the broader category of *bioalteration textures*; Furnes et al. 2007) is only starting to become appreciated and studied. At a general level, the criteria used to establish that microborings are ancient and biogenic are the same as those applied to other microbially induced structures and to body fossils. However, due to their relatively simple morphology and unique mode of formation of microborings, the specific application of these criteria is somewhat different than for other fossil biosignatures. In their treatment of microborings, McLoughlin et al. (2007) and Furnes et al. (2007) propose an approach that includes evaluation in three contexts—geologic, morphological, and geochemical-metabolic—aimed at determining the age of structures and at falsifying an abiogenic origin.

1. *Age*: since microbes producing microborings are at work today just as they were in the Proterozoic, it is imperative that microborings reported from ancient rocks and presumably of ancient age be demonstrated to have originated early during diagenesis of the host rocks. Age is established based on the relationships of the microboring with surrounding features of the rock. Early diagenetic microborings will crosscut early stage fractures in the host rock and will be crosscut by later (younger) metamorphic mineral growths (McLoughlin

et al. 2007). Furthermore, ancient microborings will show the same level of metamorphosis as their host rock (Furnes et al. 2007), and in the case of microborings in volcanic rock filled with titanite (Fig. 3.7d, g), the absolute age of the titanite can be directly determined by U-Pb dating (e.g., the microborings of the 3.35 Ga Euro Basalt, Pilbara Craton, Australia; Banerjee et al. 2007) (Fig. 3.8i).

2. *Geologic context*: the regional and local scale context of the microborings has to be consistent with the presence of life. This is becoming an increasingly less stringent requirement as modern and fossil biogenic microborings are being discovered in (unforeseen) settings that broaden our ideas about the range of environments where we can expect to find traces of life (e.g., subseafloor oceanic crust; Furnes et al. 2001; McLoughlin et al. 2012). The geologic setting of microborings is expected to be consistent with behavioral aspects of microbial biology—microborings should demonstrate substrate preferences in the host rock for metabolically useful compounds, or if their origin is cyanobacterial, the stratigraphy and sedimentology of the host rock should be consistent with a shallow, photic depositional environment (McLoughlin et al. 2007).

3. *Morphological context*: the morphology of microborings has to be consistent with biogenic origin. Since microborings have a relatively simple morphology, consisting only of branched tubes, it is difficult to establish a set of morphological criteria that applies to all microborings and excludes all abiogenic mimics. Nevertheless, a list of features that would lend support to a biogenic origin of such tubes would include μm-scale size and branching or changes in direction as they encounter other such structures (Furnes et al. 2007). These features, considered singly, cannot provide compelling evidence for biogenicity, and morphology alone can rarely be used to determine the biogenicity of putative microborings, without supporting geochemical evidence (McLoughlin et al. 2007).

4. *Geochemical-metabolic context*: geochemical evidence provides the strongest support for the biogenicity of most microborings. Euendoliths producing microborings are chemolithoautotrophic prokaryotes, so microborings preferentially occupying areas of the host rock rich in compounds that are metabolically important for bacteria, such as metal inclusions, provide evidence for biological processing (Brasier et al. 2006). Finer scale geochemical evidence from both the microtubes and their filling can provide even stronger support for biogenicity. For example, depletion of Mg, Ca, Fe, Na in the rock matrix around putative microborings indicates metabolic processing by euendoliths (Alt and Mata 2000). Fine (<1 μm thick) linings of C, N, and P in both recent and prehistoric microtubes have been interpreted as cellular remains (Giovannoni et al. 1996; Furnes and Muehlenbachs 2003), so similar linings of biologically important elements may be used as evidence for a biogenic origin of putative microborings. Additionally, analyses of in situ carbon within the microborings can also provide insights into their potential biogenicity (Furnes et al. 2004; McLoughlin et al. 2012).

5. *Null hypothesis of abiogenicity*: in most cases, the alternative explanation of micron-sized tubular structures resides in *ambient inclusion trails* (AIT) (Lepot et al. 2009a); therefore it is essential that putative microborings are shown to not belong in this category of structures. Although the exact conditions that lead to formation of AIT are not well understood (McLoughlin et al. 2007), they have been suggested to form by the movement of a crystal (e.g., pyrite, garnet) through the crystalline silica matrix of the rock. The movement is thought to be the result of pressure-solution processes and potentially driven by the thermal decomposition of organics into gases (Tyler and Barghoorn 1963; Knoll and Barghoorn 1974; Wacey et al. 2008; Lepot et al. 2009a).

While relatively few morphological characters can be used in support of the biogenicity of putative microborings [although see McLoughlin et al.'s (2009) formal taxonomy of microborings as trace fossils (Fig. 3.4d–f)], many more features can be used to identify tubular structures as AIT of abiogenic origin. McLoughlin et al. (2007) provided a list of characters present in AIT that can be used to distinguish them from bona fide microborings: (1) presence of a mineral grain at the end of the tube (Fig. 3.4c); (2) longitudinal striations caused by the facets of the mineral grain as it was driven through the rock; (3) angular cross-sectional geometry and twisted paths, particularly toward the end of the tubes (due to the increasing resistance of the host rock); and (4) a tendency to crosscut other tubes or branch, with sudden changes in diameter at branching points—branches with different diameters are caused when the mineral grain forming the AIT splits and the resulting fragments continue to form one AIT each.

Despite the abiogenic mode of AIT formation, an interesting aspect of these structures that can be relevant to early microbial life is the nature of the organic material that drives the movement of the trail-forming crystals. Wacey et al.'s (2008) study of the 3.4 Ga Strelley Pool sandstone (Australia) provides a good example of biogenicity assessment for tubular trace fossils and AIT recognition using the criteria outlined above. Furthermore, using nanoSIMS (secondary ion mass spectrometry) analyses, these authors were able to confirm the role of decomposing organic matter in AIT formation (proposed by Knoll and Barghoorn 1974) and to demonstrate features consistent with biogenic origin for the organic material associated with the trails.

3.4 Earth's Oldest Fossils and the Archean Fossil Record

3.4.1 An Archean-Proterozoic Disparity

Archean evidence for life includes all types of fossils discussed thus far: microbial body fossils (microfossils), microbially induced structures (stromatolites, MISS), and chemical biosignatures. This section focuses primarily on microfossils, with microbially induced structures addressed only in the summary of the oldest evidence for life (Sect. 3.4.5). A sweeping glance at the Precambrian microfossil record reveals a marked change in the quality and number of fossils from the Proterozoic to the Archean (Schopf et al. 2010). The Proterozoic rock record hosts numerous localities with well-preserved, unequivocal microfossils (Sergeev 2009). The high diversity of microfossil types and morphologies seen among these localities suggests that prokaryotic life arose and started diversifying much earlier, in the Archean (Altermann and Schopf 1995; Sergeev 2009). However, despite decades of research aimed at finding signals of life in the Archean, microfossils from this time period are comparatively rare and often highly controversial (Schopf and Walter 1983; Altermann and Schopf 1995; Wacey 2009). A historical focus on the Proterozoic by researchers may have been at the origin of this disparity in the early stages of the discipline. Nevertheless, continued studies are now indicating that this discrepancy in the quality and abundance of microfossils between the Archean and the Proterozoic is predominantly an artifact of the geologic record and not the result of lack of study or differences in the early biosphere (Schopf and Walter 1983; Schopf et al. 2010; Wacey 2012). This can be attributed to a series of factors that include (1) relatively low levels of cratonic sedimentation (i.e., on continental landmasses) during the Archean; (2) comparatively intense metamorphism and deformation of Archean sedimentary rocks, which are predominantly found today in metamorphic greenstone belts (Schopf and Walter 1983; Buick 1990) that consist of metamorphosed basalt with minor sedimentary rock interlayers; and (3) the fact that the silica forming most Archean cherts that preserve candidate fossils precipitated from relatively hot, acidic, and concentrated hydrothermal fluids, whereas Proterozoic cherts formed from cooler, more neutral, and dilute surficial fluids (Buick 1990). Such processes degrade the remains of organisms and biological compounds, complicating efforts to identify them in Archean rocks. Thus, although Archean rocks can be found in many areas, those that are well exposed, have undergone relatively little metamorphism, are of clear sedimentary origin, and have yielded bona fide body fossils thus far are found in only two places on Earth: the Kaapvaal Craton in South Africa and the Pilbara Craton of Western Australia (Wacey 2009; Hickman and Van Kranendonk 2012), which may have been part of the same landmass in the Archean (Zegers et al. 1998) and possibly in close vicinity to each other. It is from these two regions that nearly all Archean microfossils known to date originate, including the very earliest evidence of life.

3.4.2 Brief History of Discovery

The first bona fide Precambrian microfossils were reported in 1907 from ca. 1 Ga old Torridonian sedimentary phosphates in Scotland (Peach et al. 1907; Wacey 2009), but it was not until Tyler and Barghoorn (1954) first published descriptions of microfossils from the Gunflint iron formation (Canada) that sustained work went into documenting microbial diversity in the Precambrian (Schopf and Walter 1983). Dated to 1.9 Ga, the Gunflint iron formation microfossils provided the first solid evidence of Proterozoic life older than 1 billion years. Much older microfossils were reported from Archean rocks of South Africa during the 1960s and from Australia as early as the late 1970s (Knoll and Barghoorn 1977; Dunlop et al. 1978; Awramik et al. 1983). However, nearly all of these claims were questioned or contested in some capacity (Nisbet 1980; Schopf and Packer 1987; Buick 1984). It was not until microfossils from the 3.46 Ga Apex Basalt (Australia) (Fig. 3.5a) were described that an assemblage of early Archean (Paleoarchean) microfossils was widely accepted as evidence of early Archean life (Schopf and Packer 1987; Schopf 1993). For over a decade, these fossils were embraced as the best and earliest evidence for life on Earth (Brasier et al. 2004). In 2002 Brasier et al. presented evidence refuting all the lines of evidence on which the biogenicity claims for the Apex microfossils were based, sparking a debate that continues to the present day (e.g., Schopf and Kudryavtsev 2012; Pinti et al. 2013). Despite their later contentiousness, the initial wide acceptance of these fossils established an enduring paradigm that shapes our conception of the timing of the emergence of life. This idea of an early Archean origin for life has been dubbed the "Early Eden Hypothesis" (Brasier et al. 2004). Regardless of the biogenicity of the Apex microfossils, a look at the Archean microfossil record known to date reveals numerous lines of evidence which, even in the absence of a smoking gun, point strongly toward the emergence of life around 3.5 billion years ago (see below).

3.4.3 Salient Patterns in the Archean Fossil Record

Several major patterns emerge from a review of the Archean record of life. In terms of the host rock, Archean microfossils occur in sedimentary rocks, most commonly cherts and sandstones, formed in marine environments and metamorphosed to greenschist facies. Most of these fossils are found in cherts (Table 3.2) and a small proportion originate from sandstones such as those from the Strelley Pool Formation in Australia and Moodies Group in South Africa (Noffke et al. 2006; Wacey 2009). Despite their rarity in the Archean microfossil record, sandstones record valuable data that are difficult to extract from other types of rock, including successive depositional and diagenetic stages or microbially induced structures (e.g., Noffke et al. 2006, 2013a; Wacey et al. 2011b). Sandstones may also provide some protection to the fossil structures from mechanical stress and strain during metamorphism (Wacey 2009). Although early studies focused on rocks originating from environments thought to be most conducive to early life at the time, such as shallow marine environments, more recent work has focused on searching for biosignatures in what were previously considered unlikely contexts, such as in hydrothermal deposits and pillow basalts (Rasmussen 2000; Furnes et al. 2004; Duck et al. 2007; Furnes et al. 2007; Wacey 2009; McLoughlin et al. 2012).

Geographically, as previously mentioned, Archean microfossils occur exclusively in rocks of the Kaapvaal Craton in South Africa and in the Pilbara Craton of Western Australia. However, within each of these regions, fossil localities have

Table 3.2 Archean microfossil record, with an emphasis on rock units older than 3 Ga and microbial body fossils, *MISS* microbially induced sedimentary structures

Age (Ma)	Rock unit	Location	Fossil evidence	References
3480	Dresser Formation (Warrawoona Group, Pilbara Supergroup)	Australia	Microfossils[a]—filaments spiral, tubular branched and unbranched, some septate; spheroids MISS ?Stromatolites	Buick et al. (1981), Ueno et al. (2001a, b), Awramik and Grey (2005), Van Kranendonk (2006), Schopf (2006), Wacey (2009), Noffke et al. (2013a)
?3470[b]	?Mount Ada Basalt (Warrawoona Group, Pilbara Supergroup)	Australia	Microfossils—tubular sheaths; filaments unbranched, some septate	Awramik et al. (1983)
3466	Kitty's Gap Chert in Panorama Formation (Warrawoona Group, Pilbara Supergroup)	Australia	Microfossils—colonies of primarily coccoid cells, some chain-like; some filaments and rare rod-shaped cells	Westall et al. (2006)
3465	Apex Chert in Apex Basalt (Warrawoona Group, Pilbara Supergroup)	Australia	Microfossils[c]—filaments unbranched septate	Schopf and Packer (1987), Schopf (1993, 2006)
3450	Hooggenoeg Formation (Onverwacht Group, Swaziland Supergroup)	South Africa	Microfossils—narrow filaments; clusters of spherical, subspherical structures, cell walls; rod-shaped cells Microborings in pillow lavas	Walsh (1992), Walsh and Westall (2003), Westall et al. (2001, 2006), Furnes et al. (2004), Glikson et al. (2008), Wacey (2009), Fliegel et al. (2010), McLoughlin et al. (2012)
3426–3350	Strelley Pool Formation (Kelly Group, Pilbara Supergroup)	Australia	Microfossils—threadlike and hollow tubular filaments, filmlike structures, hollow spheroids, lenticular microfossils (ellipsoids) Stromatolites Microborings	Hofmann et al. (1999), Brasier et al. (2006), Allwood et al. (2007), Wacey et al. (2006, 2011a, b), Sugitani et al. (2010, 2013)

(continued)

Table 3.2 (continued)

Age (Ma)	Rock unit	Location	Fossil evidence	References
3416–3334	Kromberg Formation (Onverwacht Group, Swaziland Supergroup)	South Africa	Microfossils—spheroids, ellipsoids, rod-shaped cells, spindle-shaped microfossils Microborings in pillow lavas	Walsh (1992), Furnes et al. (2007), Wacey (2009)
3416	Buck Reef Chert (Onverwacht Group, Swaziland Supergroup)	South Africa	Microfossils—occasional filaments in laminations MISS	Tice and Lowe (2004), Wacey (2009)
3350	Euro Basalt (Kelly Group, Pilbara Supergroup)	Australia	Microborings	Banerjee et al (2007), Wacey (2009)
3260	Swartkoppie Formation (Onverwacht Group, Swaziland Supergroup)	South Africa	Microfossils—coccoid cells	Schopf (2006)
3245	Sheba Formation (Fig Tree Group, Swaziland Supergroup)	South Africa	Microfossils—coccoid cells	Schopf (2006)
3240–3235	Kangaroo Caves Formation (Sulphur Springs Group, Pilbara Supergroup)	Australia	Microfossils—unbranched filaments and bundles of tubes, spheroids (small, 50–100 nm diameter)	Rasmussen (2000), Duck et al. (2007), Wacey (2009)
3200	Dixon Island Formation (West Pilbara Superterrane)	Australia	Microfossils—filaments, spheroids, microbial mat remnants	Kiyokawa et al. (2006), Schopf et al. (2006), Wacey (2009)
3200	Moodies Group (Swaziland Supergroup)	South Africa	Microfossils—spheroids MISS	Noffke et al. (2006), Wacey (2009), Javaux et al. (2010)
3000	Cleaverville Formation (Gorge Creek Group, De Grey Supergroup)	Australia	Possible microfossils—spheroids	Wacey (2009)
3190–2970	Farrell Quartzite (Gorge Creek Group, De Grey Supergroup)	Australia	Microfossils—threadlike, filmlike, spheroidal, lenticular, or spindle-like	Sugitani et al. (2007), Wacey (2009)
2750	Hardey Formation (Fortescue Group, Mount Bruce Supergroup)	Australia	MISS	Rasmussen et al. (2009)

(continued)

Table 3.2 (continued)

Age (Ma)	Rock unit	Location	Fossil evidence	References
2723	Tumbiana Formation (Fortescue Group, Mount Bruce Supergroup)	Australia	Microfossils—filaments Lacustrine stromatolites	Schopf and Walter (1983), Buick (1992), Schopf (2006), Lepot et al. (2008), Awramik and Buchheim (2009)
2600	Monte Cristo Formation (Chuniespoort Group, Transvaal Supergroup)	South Africa	Microfossils—filamentous, coccoid, rod shaped	Schopf (2006)
2560	Lime Acres Member, Ghaap Plateau Dolomite (Campbell Group, Transvaal Supergroup)	South Africa	Microfossils—coccoid, ellipsoid, filamentous, tubular sheaths	Altermann and Schopf (1995), Schopf (2006)
2516	Tsineng Member, Gamohaan Formation (Ghaap Group, Transvaal Supergroup)	South Africa	Microfossils—tubular sheaths	Klein et al. (1987), Schopf (2006)

[a]The biogenicity of microfossils in the Dresser Formation is equivocal (see Wacey 2009)
[b]Precise locality unknown (Schopf 2006; Wacey 2009; Brasier and Wacey 2012; see Sect. 3.4.4)
[c]The biogenicity of microfossils in the Apex Chert is debated (see Sect. 3.4.4.1)

been described from several stratigraphic levels and widely spread locations. These are summarized in Table 3.2 which contains a list of reported occurrences; it is worth noting that the biogenicity of some of the microfossils included in the table is debated. A few of these controversial occurrences are outlined later in this section, but in-depth assessments of the biogenicity of most of the reported Archean fossil assemblages have been provided by Wacey (2009).

Microfossils from the Kaapvaal Craton are found in the Swaziland and Transvaal Supergroups. The oldest of these fossils originate from the Swaziland Supergroup, which ranges from 3550 to 3220 Ma in age and can be divided into three stratigraphic intervals: the lowermost Onverwacht Group (~3550–3300 Ma), the Fig Tree Group (~3260–3225 Ma), and the Moodies Group (3220 Ma) (Van Kranendonk et al. 2007). Spheroidal or coccoid microfossils have been reported from each of these groups. Additional diversity of morphological types is known from the Onverwacht Group, where filamentous, "sausage"-shaped, and spindle-shaped microfossils have also been documented (Walsh 1992; Westall et al. 2001, 2006; Glikson et al. 2008; Wacey 2009). Younger microfossils are known from the Transvaal Supergroup and include filamentous, coccoid, rod-shaped, and ellipsoid forms (Altermann and Schopf 1995; Schopf 2006).

The microfossil-bearing units of the Pilbara Craton in Western Australia include the Pilbara Supergroup (3530–3170 Ma) in the East Pilbara Terrane and the overlying De Grey Supergroup (~3020–2930 Ma) (Van Kranendonk et al. 2007; Wacey 2009). The Pilbara Supergroup consists of volcano-sedimentary greenstone belts and is subdivided into four groups: the Warrawoona Group (~3520–3427 Ma), the Kelly Group (~3350–3315 Ma), the Sulphur Springs Group (~3270–3230 Ma), and the Soansville Group (~3230–3170 Ma). Microfossils are also known from the Dixon Island Formation (3200 Ma, Kiyokawa et al. 2006) in the West Pilbara Subterrane and the Tumbiana Formation (2723 Ma, Schopf 2006). Morphological types reported from the Pilbara Craton are similar to those of South Africa and include branched and unbranched filaments, septate filaments, tubular sheaths, coccoid or spheroidal forms, rod-shaped, ellipsoid, and lenticular or spindle-shaped microfossils (reviewed by Schopf 2006; Wacey 2009; and Wacey 2012).

Aside from microfossils, other types of evidence of life have been identified in these Archean cratons. Compelling examples include microbially induced structures that have been reported from the Moodies and Onverwacht Groups in South Africa, as well as the 3.48 Ga Dresser Formation (Warrawoona Group), and the 3.35 Ga Euro Basalt and ~3.42–3.35 Ga Strelley Pool Formation (Kelly Group) in Australia. In rocks of the Moodies Group, MISS are considered promising biosignatures and include wrinkle structures, desiccation cracks, and roll-up structures in sandstone (Noffke et al. 2006; Wacey 2009). Microborings in the rims of pillow lavas, consisting of mineralized tubular structures, from the Onverwacht Group and the Euro Basalt, are thought to have formed by the corrosion of the volcanic glass by endolithic microbes and are similar to those found in modern oceanic crust (Furnes et al. 2004; Banerjee et al. 2007; Wacey 2009; McLoughlin et al. 2012). The diverse MISS reported from the Dresser Formation (~3480 Ma) include sedimentary structures that are interpreted as originating from microbial mats in an ancient coastal sabkha (Noffke et al. 2013a). Microborings have also been described in pyrite grains of the Strelley Pool sandstone (Wacey 2009; Wacey et al. 2011b).

The later part of the Archean may have witnessed the advent of oxygenic photosynthesis with the evolution of cyanobacteria. Atmospheric oxygen concentrations rose to relatively stable, moderate to high levels (the "Great Oxidation Event") in the early Paleoproterozoic, ca. 2.4–2.3 Ga ago (Bekker et al. 2004). It is generally though that this was due to the evolution of cyanobacteria, which is placed somewhere toward the end Archean, ca. 2.8–2.7 Ga ago, based on several lines of evidence (Buick 2012; Konhauser and Riding 2012):

1. Presence of biomarkers associated with cyanobacteria (2α-methylhopanes) in the 2.72–2.6 Ga Fortescue and Hamersley Groups (Australia) (Brocks et al. 1999; Brocks et al. 2003a, b)
2. The stromatolites of the 2.7 Ga Tumbiana Formation (Fortescue Group, Australia) wherein fabrics indicating construction by microbes and the absence of iron- and sulfur-rich sediments indicate oxygenic photosynthesis (Buick 1992, 2012)

3. Assemblages of filamentous microfossils described from the 2.6 to 2.5 Ga peritidal carbonates of the Campbell Group (Transvaal Supergroup, South Africa), which include forms similar to modern oscillatoriacean cyanobacteria, such as *Lyngbya* and *Phormidium* (Altermann and Schopf 1995).

However, it is worth noting that the cyanobacterial affinities of the Campbell Group microfossils are not unequivocal (Knoll 2012) and the oldest unambiguous cyanobacteria to date are mid-Paleoproterozoic and comprise colonial forms described from the ca. 2 Ga Belcher Supergroup (Canada) by Hofmann (1976) and Golubic and Hofmann (1976).

Interestingly, the Tumbiana Formation stromatolites represent the oldest such structures formed in a freshwater lacustrine system, as indicated by the type of evaporite mineral association present, which includes carbonate and halite, with no sulfate present (Buick 1992; Awramik and Buchheim 2009). Another Fortescue Group unit, the 2.75 Ga Hardey Formation, hosts an occurrence of laminated microstructures of probably biogenic origin, which have been interpreted as the products of photosynthetic and methanotrophic prokaryotes forming microbial biofilms in cavities of lake-bottom sediments (Rasmussen et al. 2009).

3.4.4 Reevaluating Archean Fossil Datapoints

Establishing the biogenicity of early microfossils is a complex process requiring multiple lines of evidence. As the criteria used for identifying bona fide Archean microfossils keep being updated in the wake of successive debates over questionable fossils, many older discoveries may require reevaluation. As discussed in depth in a previous section, microfossils that are widely accepted as biogenic meet criteria for biogenicity that consider the geologic context, morphology, patterns of association and taphonomy, as well as the geochemistry of the fossils. Historically, geochemical analyses were not widely used in studying Archean microfossils and many methods of analysis were recently developed or employed in this field—Knoll (2012) provides a good summary of chemistry and microscopy techniques of more recent use in studies of Precambrian microfossils.

A technique that is not so new to the field anymore, laser Raman spectroscopy, has already seen ebbs and tides of usage (Pflug and Jaeschke-Boyer 1979; Ueno et al. 2001a; Schopf et al. 2002; Pasteris and Wopenka 2003; Brasier et al. 2002, 2004; Marshall et al. 2010). Raman spectroscopy has been used primarily for fine-scale mineralogy and is also a reliable method for determining the crystallinity of reduced carbon, a good proxy for the metamorphic and diagenetic history of carbonaceous material (Marshall et al. 2010; Knoll 2012; Ohtomo et al. 2014). Fourier transform infrared (FTIR) spectroscopy has been used in combination with Raman spectroscopy at microscale to determine the molecular composition and structure of Proterozoic eukaryotes and putative eukaryotes (Marshall et al. 2005). Secondary ion mass spectrometry (SIMS) can be used to determine elemental and isotopic compositions and map them on samples at high resolution (House et al. 2000; De Gregorio et al. 2009). The molecular composition of microfossils has been assessed using laser pyrolysis gas chromatography—mass spectrometry (Arouri et al. 2000). Furthermore, several types of electron microscopy (transmission electron microscopy [TEM], scanning-transmission electron microscopy [STEM], and high-resolution transmission electron microscopy [HRTEM]) have been employed to examine the morphology and nanostructure of carbon (Ohtomo et al. 2014) and microfossils (Javaux et al. 2004, 2010). Fossil imaging is also performed using confocal laser microscopy (Schopf et al. 2006; Lepot et al. 2008). Synchrotron-based scanning-transmission X-ray microscopy (STXM) has been used in combination with X-ray absorption near-edge structure spectroscopy (XANES) for imaging, as well as elemental and molecular mapping (Boyce et al. 2002; De Gregorio et al. 2009).

These techniques represent important sources of new information in the reevaluation of older discoveries. This is the case for some of the fossils listed in Table 3.2, including microfossils from the Kromberg Formation and some of the microfossils reported from the Hooggenoeg Formation (Onverwacht Group) (Wacey 2009), as well as those from the Tumbiana Formation (Buick 2001). Moreover, in order to identify the oldest evidence of life, the age of the rock and syngenicity of the microfossils must be firmly established, for which thorough understanding of the geologic context of the locality is key. When microfossils from the ~3470 Ma Mount Ada Basalt were reported in 1983, these were the oldest convincing microfossils known at the time (Awramik et al. 1983). However, the precise collection locality remains unknown and has never been resampled, which casts reasonable doubts on the age of these fossils, as they could originate from the much younger Fortescue Group (Schopf 2006; Wacey 2009; Brasier and Wacey 2012).

3.4.4.1 The Apex Chert Debate

Microfossils from cherts in the Apex Basalt (Warrawoona Group, Australia) (Fig. 3.5a) were first reported by Schopf and Packer (1987) and later described by Schopf (1993). Eleven taxa of filamentous prokaryotes were circumscribed based on various aspects of their morphology. They were interpreted as bona fide Archean microfossils based on their frequent presence in the rocks, the early Archean age of the fossiliferous cherts, the occurrence of the fossils within clasts of the brecciated chert and absence from the surrounding matrix, and their morphological complexity and similarity to younger prokaryotes (Schopf 1993). Data from Laser-Raman analysis was later presented in support of biogenicity (Schopf et al. 2002), showing that the microfossils contained kerogen (geologically transformed organic matter). The Apex fossils were widely accepted as the oldest and best-preserved microbial fossils (Marshall et al. 2011) until Brasier et al. (2002) disputed their biogenicity, sparking a debate that has made them the most controversial Archean microfossils.

Arguments against the biogenicity of the Apex microfossils focus on the microfossils themselves and the carbonaceous material associated with them. Evidence inconsistent with biogenicity and suggesting the microfossils are instead artifacts includes (1) the fact that many of the microfossils branch extensively or are connected to crystals (like in ambient inclusion trails) when examined in multiple focal planes (Brasier et al. 2002), (2) the discovery of similar structures throughout the unit and in younger crosscutting features (Brasier et al. 2002; Marshall et al. 2011), and (3) a study demonstrating that comparable microfossil-like structures are actually quartz and hematite-filled fractures (Marshall et al. 2011). Extensive debate has centered on the nature of the carbonaceous material detected by Schopf et al. (2002). Also using Raman spectroscopy, Brasier and colleagues (2002) contended the carbonaceous material was abiogenic graphite and not kerogen. Subsequently, others have pointed out that Raman spectroscopy cannot be used to determine the origin of carbonaceous material or to distinguish different types of biogenic carbon (Pasteris and Wopenka 2003; Brasier et al. 2004; Marshall et al. 2010). An independent study using a suite of different methods (transmission electron microscopy, synchrotron-based scanning-transmission X-ray microscopy, and secondary ion mass spectrometry) concluded that biogenic origin is more likely, without entirely ruling out abiogenic origin (De Gregorio et al. 2009). However, a recent study using high-resolution transmission electron microscopy found four different populations of carbonaceous material in the Apex Chert. This indicates that the carbonaceous material was deposited at four separate times, most likely by postdepositional hydrothermal fluid flow, and that the carbonaceous material is not syngenetic with the original rock (Olcott Marshall et al. 2014). As the debate on the Apex Chert is currently ongoing without an end in sight (Wacey 2012), the search for traces of the earliest life should turn to other assemblages in the meanwhile.

3.4.4.2 The Isua and Akilia Debates

The oldest supracrustal rocks on Earth are located in rocks of the 3.81–3.7 Ga Isua Supracrustal Belt and 3.83 Ga Akilia Island of West Greenland (Lepland et al. 2002; Pons et al. 2011; Wacey 2009). The age and sedimentary origin of these rocks make them potential sources of the oldest biosignatures. However, intensive deformation and metamorphism of these units have presented substantial challenges in determining the validity of putative biomarkers and even the sedimentary nature of the protoliths (source rocks).

At Akilia, the discovery of ^{13}C-depleted graphite associated with apatite was interpreted as evidence of life earlier than 3.83 Ga (Mojzsis et al. 1996). This interpretation was later challenged, and a recent study has suggested an abiotic origin for the carbon, from fluid deposition during metamorphism (Lepland et al. 2011). Nonetheless, in light of debates over the sedimentary nature of the protolith, any claims of biogenic graphite from Akilia should be treated cautiously (Wacey 2009). The small outcrop containing the graphite in question was initially interpreted as originating from sedimentary iron formations (McGregor and Mason 1977; Mojzsis et al. 1996; Papineau et al. 2010). However, a separate team revisited the outcrop and concluded that the protolith was igneous and not sedimentary, casting significant doubts on the biological origin of the graphite (Fedo and Whitehouse 2002). Yet another study presented evidence in support of a sedimentary origin (Dauphas et al. 2004) and the debate is ongoing, while the nature of the protolith remains unclear (Wacey 2009; Papineau et al. 2010).

The sedimentary origins of rocks in the Isua Supracrustal Belt are, in contrast, unequivocal (Van Zuilen et al. 2003). As the oldest confirmed sedimentary rocks on Earth, studies seeking biological signals within these strata have been ongoing since Moorbath and colleagues discovered the age of these rocks in 1973 (Moorbath et al. 1973; Appel et al. 2003). Putative microfossils, consisting of small, black spherical objects in quartz grains, were later discovered and named *Isuasphaera isua* (Pflug 1978a, b) (Fig. 3.5b, c). Their biogenicity, however, was subsequently contested and the structures reinterpreted as limonite-stained fluid inclusions or cavities (Bridgewater et al. 1981; Roedder 1981). Upon reexamination of the locality, an independent group concluded that the extreme stretching and deformation of the host rocks could not have preserved syndepositional spherical objects (Appel et al. 2003).

The presence of graphite in Isua rocks has been reported by a number of studies and has been proposed as evidence for biological activity (Schidlowski 1988; Mojzsis et al. 1996; Rosing 1999). Such claims are debated and a number of considerations have been raised regarding these graphite inclusions. Some authors have proposed the thermal decomposition of ferrous carbonate (siderite) as an abiotic mechanism for graphite formation at Isua (Van Zuilen et al. 2003). Others point out that many of the carbonate rocks at Isua, which contain some of the graphite in question, are not sedimentary in origin, and therefore graphite originating from such samples should be treated with caution (Lepland et al. 2002). Further, some of the carbon in the Isua rocks was found to originate from recent endolithic organisms infiltrating cracks and fissures in the rock. This indicates that studies using bulk-sampling methods may have detected carbon from these nonindigenous sources (Westall and Folk 2003). Although the evidence discussed above is inconsistent with an ancient biological origin for the Isua graphite, none of these studies have been able to firmly reject biogenicity.

In the latest twist, Ohtomo et al. (2014) analyzed graphite from black-gray schists at Isua and concluded that it represented traces of early life. Using Raman spectroscopy and geochemical analyses, the team determined that the schists formed from clastic marine sediments and that the carbonaceous material was present in the rock prior to prograde metamorphism. They further ruled out thermal degradation of ferrous carbonate as an abiotic formation mechanism, as proposed by others (i.e., Van Zuilen et al. 2003), and electron microscopy revealed structural characteristics consistent with biogenic graphite (Ohtomo et al. 2014). These results provide the most compelling evidence to date for the presence of life ca. 3.7 billion years ago. However, the contentious nature of biomarkers from Isua, the recent publication date of this study, and the brief life span of unchallenged claims in the field of early biosignature research caution against immediate adoption of these findings which await scientific consensus and independent corroboration.

3.4.4.3 The Alan Hills 84001 Martian Meteorite Controversy

Although only indirectly relevant to questions about early life on Earth, the Alan Hills—ALH84001 Martian meteorite controversy is included here because of its relevance to issues of biogenicity and their implications for astrobiology. In 1996, McKay et al. argued for the presence of traces of Martian life in meteorite ALH 84001. The evidence they presented in support of their claim came from several types of data: (1) polycyclic aromatic hydrocarbons interpreted as diagenetic products of microorganisms and considered to be indigenous to the meteorite, (2) chemistry and mineral composition and structures suggestive of (microbial) biogenic products or biologic behavior, (3) magnetite particles similar to magnetofossils (magnetosomes) left by magnetotactic bacteria (Fig. 3.5d), and (4) submicron bacteriomorphs similar to nanobacteria (Folk 1993) and bacterial fossils.

McKay et al.'s (1996) claim that these ALH 84001 features represented evidence of ancient Martian life was met with a great deal of skepticism (Bradley et al. 1997 and others). Multiple research groups have verified the presence of polycyclic aromatic hydrocarbons in ALH84001 (Clemett et al. 1998; Steele et al. 2012). However, mechanisms for abiotic formation of polycyclic aromatic hydrocarbons have been documented (Zolotov and Shock 2000; McCollum 2003; Treiman 2003a). Additionally, Steele et al. (2012) demonstrated that abiotically formed organic compounds are present on multiple Martian meteorites (including ALH84001) and that these compounds were not contaminants from Earth. These findings are inconsistent with the original claims of biogenic polycyclic aromatic hydrocarbons in ALH84001. The claim that the mineral composition—magnetite, iron sulfide, and siderite—is indicative of life and its byproducts has also been scrutinized. This suite of minerals is well known in low-temperature, aqueous systems (Anders 1996; Golden et al. 2000) and has not been shown to be exclusively indicative of biologic behavior (Treiman 2003b).

There is disagreement over how similar the magnetite grains in ALH84001 are to the magnetosomes of Earth bacteria (such as the marine magnetotactic vibrio *Magnetovibrio blakemorei* strain MV-1) (Clemett et al. 2002; Treiman 2003b). While some authors point to close similarity (Thomas-Keprta et al. 2001), others have documented differences in morphology between magnetosomes of MV-1 and other bacterial strains, and the magnetite crystals in ALH84001 (Buseck et al. 2001; Golden et al. 2004). Golden et al. (2004) demonstrated that the morphology of ALH84001 magnetite crystals is replicated abiogenically and that the most common crystal morphology for biogenic magnetite is different from that in both ALH84001 and abiogenic magnetite, concluding that rather than representing a compelling biosignature, the morphology of the ALH84001 magnetite crystals is consistent with an abiogenic origin. Finally, the submicron bacteriomorphs from

ALH84001 are largely discounted as biogenic structures (Treiman 2003b) based on the fact that at < 100 nm they are just below the size of the smallest free-living prokaryotes (Gorbushina and Krumbein 2000) and that a number of abiotic mechanisms have been proposed for the formation of such structures, including mineral precipitation from solution (Bradley et al. 1998; Kirkland et al. 1999; Vecht and Ireland 2000; Grasby 2003).

Overall it is the inability to reject the null hypothesis of abiotic formation (Brasier and Wacey 2012) that does not allow the features described in ALH84001 to be considered biogenic (Treiman 2003b). Since the polycyclic aromatic hydrocarbons have been shown to be abiotic (Steele et al. 2012), the mineral composition of the carbonate globules is not a definitive indicator of biological activity, and both the magnetite grains and nano-bacteriomorphs shaped could have formed abiotically, it is unlikely that ALH84001 contains biogenic material. However, it will be interesting to see the results of the application of Gehring et al.'s (2011) method of distinguishing between bacterial magnetosomes and abiogenic magnetite to the ALH84001 magnetite grains.

3.4.5 Oldest Traces of Life

In the context of the findings summarized above, it is clear that there is a large body of evidence suggesting that life has existed since the Paleoarchean (3600–3200 Ma), despite the questionable or equivocal status of some of these fossils (Knoll 2012). Although not all stromatolite and other microbially induced structure occurrences are included in Table 3.2, it is worth noting that the stratigraphic extent of stromatolites, MISS, and microborings mirrors closely that of microfossils (e.g., Hofmann 2000; Awramik and Grey 2005; Brasier et al. 2006; Fliegel et al. 2010). While the oldest reported putative microfossils originate from the ~3480 Ma Dresser Formation (Australia) (Fig. 3.6a), their biogenicity remains equivocal. The most convincing of these microfossils are carbonaceous filaments reported by Ueno et al. (2001a, b), but their biogenic interpretation based on morphology and stable carbon isotopes awaits further verification (Wacey 2009). Microbial presence in the Dresser Formation is also supported by putative stromatolites (Buick et al. 1981; Awramik and Grey 2005; Van Kranendonk 2006; Wacey 2009) (Fig. 3.6b). Irrespective of the verdict on the microfossils and stromatolites, unequivocal MISS described by Noffke et al. (2013a) in the Dresser Formation demonstrate presence of microbial life very close to 3.5 Ga ago. The Dresser Formation MISS formed in shallow-water, low-energy evaporitic coastal environments (Buick and Dunlop 1990; Noffke et al. 2013a).

Convincing microfossils, only slightly younger than the Dresser Formation fossils, are known from the ~3466 Ma Kitty's Gap Chert (Warrawoona Group) (Westall et al. 2006) (Fig. 3.7a, b); the ~3450 Ma Hooggenoeg Formation (Onverwacht Group), particularly those reported by Glikson et al. (2008) (Fig. 3.7c, e, f); and the ~3426–3350 Strelley Pool Formation (Kelly Group)

(Wacey et al. 2011a; Sugitani et al. 2013) (Fig. 3.8a–e). Dresser Formation microfossils notwithstanding, the Kitty's Gap Chert, and Hooggenoeg Formation host the oldest unequivocal microbial body fossils known to date. Although slightly younger, the Strelley Pool Formation microfossils provide very good evidence for life corroborated by independent studies, with good geochemical evidence supporting their biogenicity, and the presence of other types of fossils within the formation, including microborings and stromatolites (Hofmann et al. 1999; Brasier et al. 2006; Allwood et al. 2007; Wacey et al. 2006, 2011a, b; Sugitani et al. 2013) (Fig. 3.8f–h). Aside from the Strelley Pool Formation microborings, which are hosted in sedimentary rocks and the oldest known to date in the fossil record, the oldest microborings in volcanic glass have been reported from 3.34 Ga in the Hooggenoeg Formation (Furnes et al. 2004; Fliegel et al. 2010; McLoughlin et al. 2012) (Fig. 3.7d, g) and from the 3.42–3.31 Ga Euro Basalt of the Kelly Group (Banerjee et al. 2007) (Fig. 3.8i).

3.5 Microbial Eukaryotes: Recognition and Early Fossil Record

Irrespective of the detailed circumstances of their evolution, as proposed by several competing hypotheses, eukaryotes arose from prokaryotic stock (Knoll and Bambach 2000; Lang and Burger 2012). Whereas in the case of the earliest prokaryotes proof of biogenicity is one of the biggest hurdles, in early eukaryotes their very eukaryotic affinities are challenging to ascertain. Because of their prokaryotic origins, it is to be expected that early eukaryotes were unicellular organisms and that they will be difficult to distinguish from prokaryotes. Indeed, the earliest bona fide eukaryotes and even older putative eukaryotes are unicellular dispersed microfossils (Javaux et al. 2010; Knoll 2014). From the perspective of the paleontologist, most early eukaryotes fall in the category of acritarchs (e.g., Figs. 3.9a–d, 3.10a–d, g, 3.11a, c), an artificial group of organic-walled unicellular microfossils of large size (50 μm or more) and uncertain biological affinities which comprise the most abundant and widely distributed record of Proterozoic protists (Knoll et al. 2006; Buick 2010).

3.5.1 Recognizing Early Eukaryotes

Known from dispersed unicellular microfossils in Proterozoic rocks (and possibly going as far back in time as the Archean; Javaux et al. 2010; Knoll et al. 2006; Knoll 2014), early eukaryotes have to pass the same tests of indigenousness and syngenicity as their prokaryotic counterparts (e.g., Bengtson et al. 2009; Javaux et al. 2010), in addition to satisfying eukaryote-specific criteria. Putative eukaryotes

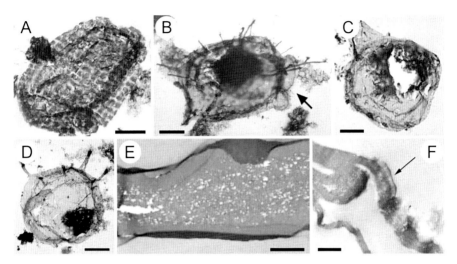

Fig. 3.9 Recognizing early unicellular eukaryotes—acritarchs. (**a**) *Satka favosa*, showing surface ornamentation (cell wall consisting of interlocking panels), Mesoproterozoic Roper Group, Australia. (**b**) *Tappania plana*, showing asymmetrically distributed long processes (some of which are branched) protruding from the cell wall and bulbous protrusions (*arrow*) potentially indicative of vegetative reproduction by "budding," Roper Group. (**c**) *Tappania plana*, showing possible excystment structure at the apex of a necklike extension (at *top left*), Roper group. (**d**) *Tappania plana*, showing asymmetrically distributed long cell surface processes, Roper Group. (**e**) *Leiosphaeridia jacutica*, showing complex cell wall structure: two electron-dense, homogeneous layers that sandwich a thick central layer with electron-dense, porous texture, Roper Group. (**f**) *Leiosphaeridia crassa*, showing complex cell wall organization consisting of as many as four structurally distinct layers preserved at places (*arrow*), Roper Group. Scale bars: (**a**) 40 µm, (**b**) 35 µm, (**c**) 20 µm, (**d**) 35 µm, (**e**) 1 µm, (**f**) 1.6 µm. Credits—images used with permission from (**a**), (**b**), and (**d**) Nature Publishing Group (Javaux et al. 2001). (**c**), (**e**), and (**f**) Blackwell Publishing Ltd. (Javaux et al. 2004)

have generally been recognized based on the fact that they display complexity unknown in prokaryotes (Schopf and Klein 1992; Javaux 2007). This is expressed in the interrelated abilities to synthesize complex polymers and produce complex structures (e.g., Javaux et al. 2004; Javaux 2007). The production of complex polymers is reflected in the recalcitrant nature of cell walls or their ability to withstand acid maceration, a feature often considered important in distinguishing taxonomic affinities at the domain level (e.g., Javaux et al. 2004; Knoll et al. 2006; Javaux 2007). In turn, recalcitrant cell walls preserve complex cell surface structures. Aside from these, large size (> ca. 50 µm) is often treated as indicative of eukaryotic affinities (e.g., Schopf and Klein 1992). However, size is not diagnostic by itself, as large bacteria and cyanobacterial sheaths are known (e.g., Waterbury and Stanier 1978; Schulz et al. 1999), as well as eukaryotes smaller than 1 µm in diameter (e.g., Courties et al. 1994).

The hallmark of complex cell structure, cellular organelles, has yet to be unequivocally substantiated in the early fossil record. Early reports of acritarchs

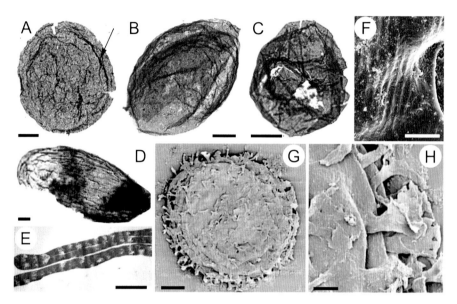

Fig. 3.10 Earliest eukaryotic fossils–acritarchs. (**a**) Microfossil representing a putative eukaryote, showing surface covered with very fine granules and concentric folds, Archean Moodies Group, South Africa. (**b**) Ellipsoidal acritarch with a medial split possibly indicating excystment. Paleoproterozoic Changzhougou Formation, China. (**c**) Acritarch with multilayered wall structure observed as sequential differences in contrast and brightness visible in successive planes of focus, Changzhougou Formation. (**d**) Ellipsoidal acritarch displaying cell wall ornamentation (longitudinal striations), Changzhougou Formation. (**e**) Section through acritarch wall showing ultrastructure that consists of alternating electron-dense and electron-tenuous bands spaced at 0.2–0.3 μm, Paleoproterozoic Changlinggou Formation, China. (**f**) Striated ornamentation of inner surface of acritarch wall, consisting of ridges spaced at 0.2–0.3 μm, Changlinggou Formation. (**g**) *Shuiyousphaeridium macroreticulatum*, acritarch showing flaring furcating cell wall processes; not very conspicuous, a reticulated pattern also characterizes the cell wall of this acritarch, Mesoproterozoic Ruyang Group, China. (**h**) Detail of (**g**). Scale bars: (**a**) 50 μm, (**b**) 20 μm, (**c**) and (**d**) 20 μm, (**e**) 1 μm, (**f**) 1 μm, (**g**) 50 μm, (**h**) 10 μm. Credits—images used with permission from (**a**) Nature Publishing Group (Javaux et al. 2010). (**b**), (**c**), and (**d**) Elsevier Science Publishers (Lamb et al. 2009). (**e**) and (**f**) Elsevier Science Publishers (Peng et al. 2009). (**g**) and (**f**) Blackwell Publishing Ltd. (Javaux et al. 2004)

with internal organelle-like structures have originated from the 800 Ma Bitter Springs Formation (Australia) (Schopf 1968; Oehler 1976, 1977). Initially interpreted as organelles, pyrenoids or pyrenoid-like bodies, their nature was questioned both at the time of publication (e.g., Knoll and Barghoorn 1975) and recently (Pang et al. 2013). Experimental results indicate that some of the "cells" preserved in the Bitter Springs Formation represent prokaryotic mucilaginous sheaths containing collapsed and condensed cell contents (reported as "pyrenoids") (Pang et al. 2013). Nevertheless, based on the same set of experiments, Pang et al. (2013) concluded that the internally preserved structures of some microfossils in an older rock unit, the 1.60–1.25 Ga Ruyang Group (China), represent in vivo

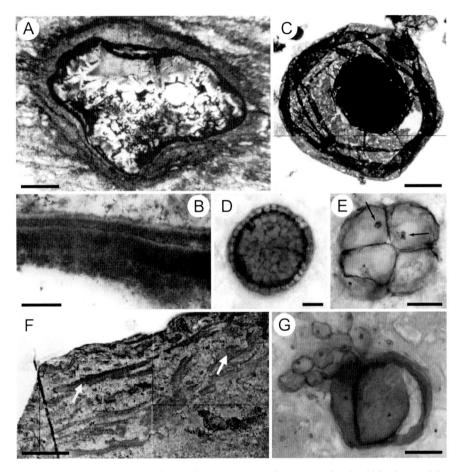

Fig. 3.11 Early eukaryotic fossils. (**a**) *Crassicorium pendjariensis* acritarch within chert nodule, showing internal exfoliation of wall lamellae, Mesoproterozoic Bangemall Group, Australia. (**b**) *Crassicorium pendjariensis*—detail of multilayered wall structure, Bangemall Group. (**c**) Large acritarch with a dense central body (at *center*), basal Neoproterozoic Diabaig Formation, Torridon Group, Scotland. (**d**) Spherical ball of cells enclosed within a complex wall (thin section from phosphatic nodule), Diabaig Formation. (**e**) Cell cluster exhibiting mutually appressed cells with internal "spots" (*arrowed*) (thin section from phosphatic nodule), Diabaig Formation. (**f**) Siltstone containing microbially induced sedimentary structures (roll-up structures, dark-toned, at arrows); this sample also yielded microfossils, composite image, Mesoproterozoic Dripping Spring Quartzite, Apache Group, Arizona. (**g**) Complex microfossil in phosphatic nodule, showing the emergence of an adjoined cluster of at least ten light-walled ellipsoidal coccoids from a single point of a larger, dark-walled coccoid structure, basal Neoproterozoic Cailleach Head Formation, Torridon Group, Scotland. Scale bars: (**a**) 200 μm, (**b**) 20 μm, (**c**) 25 μm, (**d**) 5 μm, (**e**) 10 μm, (**f**) 1 cm, (**g**) 10 μm. Credits—images used with permission from (**a**) and (**b**). The Paleontological Society (Buick and Knoll 1999). (**c**), (**d**), and (**e**) Nature Publishing Group (Strother et al. 2011). (**f**) Society for Sedimentary Geology (Beraldi-Campesi et al. 2014). (**g**) Elsevier Science Publishers (Battison and Brasier 2012)

protoplasm condensation corresponding to an encystment stage common in some algae, thus supporting their eukaryotic affinities.

3.5.1.1 Morphological Criteria

Dispersed unicellular microfossils are most compellingly recognized as eukaryotes based on morphological criteria (Javaux et al. 2003) although geochemical data on biomarkers have also been used, sometimes exclusively (e.g., Brocks et al. 1999), in support of eukaryote presence. In terms of morphology, three categories of features are considered diagnostic of eukaryotic nature, especially when co-occurring, in studies of microfossils (Javaux et al. 2003): complex cell surface features, excystment structures, and complex wall organization.

1. *Complex cell surface features*: these include ornamentation (Figs. 3.9a and 3.10a, d, f) and processes protruding from the cell surface (Figs. 3.9b, d and 3.10 g, h). The use of complex surface features to recognize eukaryotic cells stems from Cavalier-Smith's (2002) reasoning that production of such surface features would require that the cell possesses an endomembrane system and cytoskeleton, which are fundamental components of the eukaryotic cell. Javaux et al. (2004) point out that while prokaryotic organisms can synthesize both cell wall ornament and preservable structures, wall ornamentation rarely occurs on the size scale observed in Precambrian candidate eukaryotes and is seldom found on preservable structures.

   Regular polygonal patterns occurring at the micron scale are seen on the surfaces of Proterozoic eukaryotic microfossils (Javaux et al. 2003, 2004). Although many prokaryotes possess oblique, square, or hexagonal crystalline arrays on their surfaces, these shapes occur at the nanometer scale. Furthermore, their easy removability by chemicals in culture studies suggests that they are unlikely to preserve or to survive the acid maceration processes used to extract dispersed carbonaceous microfossils from the rock (Javaux et al. 2003). Other examples of surface ornamentation include striations and lineations (Fig. 3.10d), chagrinate (covered with very fine granules) surfaces (Fig. 3.10a), cruciform structures, or internal (inner wall surface) striations (Fig. 3.10f) (Javaux et al. 2001, 2004; Butterfield 2009). Sometimes, larger bulbous protrusions suggestive of vegetative reproduction through budding are present, as in *Tappania plana* (Javaux et al. 2001, 2004) (Fig. 3.9b).

   Cell surface processes documented in early eukaryotes have varied morphologies, from spiny or thin and unbranched to bulbous or branched (Javaux et al. 2003). Some can be long (e.g., up to 60 μm in *Tappania plana* of the 1.50–1.45 Ga Roper Group, Australia; Javaux et al. 2001) (Fig. 3.9b, d), while others are septate, branched (Fig. 3.9b and 3.10 g, h), and anastomosing (e.g., the putative basal fungus "*Tappania*" from the 850 to 750 Ma Wynniatt Formation, Canada; Butterfield 2009). Processes with irregular length, exhibiting heteromorphism within one cell or asymmetric distribution on the cell surface (Javaux et al. 2001; Knoll et al. 2006) (Fig. 3.9b, d), can only be accounted for by active growth and remodeling of the cell which requires a dynamic cytoskeletal architecture and regulatory networks that characterize eukaryotes (Javaux et al. 2001, 2004).

2. *Excystment structures*: these are openings in the walls of microfossils that are thought to be produced by liberation of internal vegetative cells from the wall of a resting stage, another eukaryotic feature (Fig. 3.9c and 3.10b). Commonly recognized in Phanerozoic microfossils, excystment structures are much more equivocal in Precambrian fossils (Javaux et al. 2003). For instance, in the Roper Group fossils, at least two taxa show possible excystment structures. Because these structures are perforations or splits that run some length of the cell wall, they are not regularly occurring (there are few occurrences in the population), and the microfossils generally lack evidence of vegetative cells having been present, interpretation as degradational ruptures or breaks in the microfossil cell wall seems equally likely. More complex structures, such as those seen in *Tappania plana* which possesses a slit at the apex of a necklike extension (Javaux et al. 2003) (Fig. 3.9c), are more credible as eukaryotic excystment structures.

3. *Complex cell wall organization*: the structural complexity of eukaryotic cell walls can be preserved in microfossils and distinguished from acetolysis-resistant structures formed by bacteria (Javaux et al. 2003). Thus, ultrastructural features (Figs. 3.9e, f, 3.10e, and 3.11b) can provide evidence for eukaryotic affinities, even in microfossils in which morphology is not diagnostic (Javaux et al. 2004). These authors undertook a careful TEM study of cell wall ultrastructure demonstrating that while some eukaryotes (recognized based on morphological characters observed in light microscopy and SEM) share nondiagnostic wall ultrastructures consisting of single, homogeneous, electron-dense layers of variable thickness, in others cell wall organization is more complex (Fig. 3.9e, f). In the latter microfossils, some of which do not display other features indicative of eukaryotic affinities, cell walls exhibit at least four distinct types of structures characterized by multiple layers of different density and organization (e.g., dense, porous, laminar, fibrous) (Javaux et al. 2004) (Fig. 3.9f).

Discussing the multilayered structure of *Leiosphaeridia crassa* from the Roper Group, Javaux et al. (2004) point out that such structures occur in the acetolysis-resistant walls of many green algae. Several authors have discussed similarities between the cell wall structure of different early eukaryotes and those of extant chlorococcalean green algae (Arouri et al. 1999), prasinophyte green algae (Loeblich 1970), or volvocalean green algae (Moczydlowska and Willman 2009). Javaux et al. (2004) also discuss the possibility that acetolysis-resistant walls structurally similar to those described in the early putative eukaryotes are produced by prokaryotic organisms; this seems not to be the case, as few bacteria make spores with comparable size, surface ornamentation, and preservation potential. For example, myxobacteria (which are mostly terrestrial) produce spore-enclosing structures (sporangioles) up to 50 μm in diameter, but these have smooth walls and unknown chemical composition and are not known to preserve in sediments. Actinobacteria have spores up to 3 μm in diameter that can be ornamented, but their surface ornaments are very small, proteinaceous, and are unlikely to survive fossilization processes. In contrast, the extracellular polysaccharide sheaths that envelop coccoidal cyanobacterial colonies can exceed 100 μm in size and are commonly fossilized (Bartley 1996); however, these envelopes are smooth walled, and their ultrastructure consists of fibrous layers distinct from those of eukaryotes (Javaux et al. 2004).

3.5.1.2 Geochemistry: Hydrocarbon Biomarkers

Steranes are molecular fossils representing the geologically stable derivatives of sterols, which are eukaryote-specific compounds (Summons and Lincoln 2012). As a result, steranes in the rock record have been used as biomarkers for eukaryote presence. The oldest claims for eukaryotes are based exclusively on the presence of steranes, unaccompanied by body fossils, in Archean rocks dated at 2.72–2.60 Ma in Australia (Fortescue and Hamersley Groups; Brocks et al. 1999; Brocks et al. 2003a, b) and at 2.67–2.46 Ga in South Africa (Transvaal Supergroup; Waldbauer et al. 2009). However, when not backed by microfossil evidence, the use of hydrocarbon biomarkers as evidence for eukaryotes is contentious for several reasons (Buick 2010; Knoll 2014). One of the reasons is that sterol biosynthesis requires molecular oxygen, which was not present in the Archean, except maybe for rare, very localized "oxygen oases" produced by photosynthetic cyanobacteria in microbial mats (Knoll 2014). Another reason is that although very rare, some prokaryotic organisms are known to synthesize sterols; this reason is relatively easy to dismiss as known prokaryotic sterols are simple and, thus, could not have generated the complex steranes that have been reported from Archean rocks (Knoll 2014). The most contentious issues are those of indigenousness and syngenicity, because hydrocarbons are known to migrate through rock over significant distances and thicknesses from their layers of origin (geological contaminants) or can be introduced in drill core samples by the drilling process (modern contaminants) (Knoll 2014). Indeed, based on very different carbon isotope ratios measured in different organic fractions of the same rocks, Rasmussen et al. (2008) showed that the biomarkers reported by Brocks et al. (1999) are not indigenous. To date, the oldest unequivocally dated steranes come from >2.2 Ga fluid inclusions trapped by metamorphic processes in closed systems, and therefore contamination-proof, in the Matinenda Formation (Canada; Dutkiewicz et al. 2006; George et al. 2008).

3.5.2 Oldest Evidence for Eukaryotes

As pointed out by Knoll (2014), when it comes to the oldest evidence for eukaryotes, there is no clear-cut age limit—rather, the geological record presents us with evidence that falls along a sliding scale of certainty that ranges from bona fide, confidently interpreted fossils down to around 1.8–2.0 Ga ago to increasingly more equivocal fossils or chemical biosignatures in older rocks. At least ten rock units around the world have yielded Paleoproterozoic and Mesoproterozoic acritarchs of eukaryotic affinities to date (Table 3.3); the morphological diversity of acritarchs in most of these units has been carefully summarized by Knoll et al. (2006). Some of these units also host putative multicellular eukaryotes.

3.5.2.1 Acritarchs

The oldest acritarchs interpreted with reasonable confidence as eukaryotes have been reported from the late Paleoproterozoic Changzhougou Formation and the overlying Changlinggou Formation (Chang Cheng Group, China; Lamb et al. 2009; Peng et al. 2009) dated between 1.80 and 1.65 Ga (Table 3.3) (Fig. 3.10b–f). They include several distinct forms with complex cell wall structure (Fig. 3.10c, e), wall ornamentation (Fig. 3.10d, f), or putative excystment structures (Fig. 3.10b), indicating that eukaryotes had already started diversifying before the end of the Paleoproterozoic (Lamb et al. 2009). The ca. 1.65 Ga Mallapunyah Formation (Australia) has also yielded acritarchs with ornamented cell walls interpreted as eukaryotes (Javaux et al. 2004). In the Mesoproterozoic, several rock units dated between 1.6 and 1.1 Ga host diverse eukaryotic acritarch assemblages—the Ruyang Group in China (Xiao et al. 1997; Yin 1997; Javaux et al. 2004) (Fig. 3.10g, h), the Sarda and Avadh Formations in India (Knoll et al. 2006), and the Bangemall Group in Australia (Buick and Knoll 1999) (Fig. 3.11a, b). Detailed studies of both the morphology and the distribution of eukaryotic acritarchs in the Roper Group of Australia (Fig. 3.9a–f) have revealed high morphological diversity and suggest niche partition among early eukaryotes 1.50–1.45 Ga ago, at the beginning of the Mesoproterozoic (Javaux et al. 2001, 2003, 2004). Another early Mesoproterozoic unit containing diverse acritarchs is the Chamberlain Shale of Montana (1.47–1.42 Ga; Horodyski 1980). Considerable eukaryotic acritarch diversity has been reported from the middle and late Mesoproterozoic Thule Supergroup (1.3–1.2 Ga; Greenland), which includes several fossiliferous subunits (Samuelsson et al. 1999), and the Lakhanda Group (1.1–1.0 Ga; Siberia; Knoll et al. 2006).

The oldest evidence for eukaryotic life on land may be preserved in the ca. 1.2 Ga Dripping Spring Quartzite (Apache Group, Arizona; Beraldi-Campesi et al. 2014) (Fig. 3.11f). Here, diverse and abundant MISS preserved on paleosurfaces that display desiccation features and bear strong morphological resemblance to modern terrestrial biocrusts co-occur with eukaryote- and prokaryote-like microfossils in river floodplain deposits. These have been interpreted as evidence that microbial communities, including eukaryotic components, were already adapted to live in dry habitats and formed biological soil crust-like communities long before the advent of land plants (Beraldi-Campesi et al. 2014). Previously, the ca. 1 Ga Torridon Group in Scotland was considered to provide the oldest evidence for eukaryotic life on continents represented by

Table 3.3 Mesoproterozoic and older eukaryotic fossil record

Age (Ga)	Rock unit	Location	Eukaryotic fossils	References
~2.1	Negaunee Iron Formation	USA (Michigan)	*Grypania* macrofossils	Han and Runnegar (1992)
1.80–1.65	Changzhougou and Chuanlinggou Formation	China	Acritarchs, multicellular filaments	Yan and Liu (1993), Knoll et al. (2006), Lamb et al. (2009), Peng et al. (2009)
1.7–1.6	Lower Vindhyan Supergroup	India	*Grypania*-like macrofossils, multicellular filaments	Han and Runnegar (1992), Bengtson et al. (2009)
~1.65	Mallapunyah Formation	Australia	Acritarchs	Javaux et al. (2004); Knoll et al. (2006)
1.60–1.25	Ruyang Group	China	Acritarchs	Xiao et al. (1997), Yin (1997), Knoll et al. (2006)
1.6–1.0	Sarda and Avadh Formation	India	Acritarchs	Knoll et al. (2006)
1.6–1.0	Bangemall Group	Australia	Acritarchs and *Horodyskia* macrofossils	Grey and Williams (1990), Buick and Knoll (1999), Grey et al. (2010), Knoll et al. (2006)
1.50–1.45	Roper Group	Australia	Acritarchs	Javaux et al. (2001, 2003, 2004)
1.47–1.42	Chamberlain Shale	USA (Montana)	Acritarchs	Horodyski (1980), Knoll et al. (2006)
~1.4	Gaoyuzhuang Formation	China	*Grypania* macrofossils	Walter et al. (1990)
1.4–1.3	Greyson Shale/ Appekunny Argillite	USA (Montana)	*Grypania* and *Horodyskia* macrofossils	Walter et al. (1976, 1990), Horodyski (1982)
1.3–1.2	Thule Supergroup	Greenland	Acritarchs	Samuelsson et al. (1999)
~1.2	Hunting Formation	Canada (Somerset Island)	*Bangiomorpha* and two other putative multicellular eukaryotes	Butterfield (2000, 2001)
~1.2	Dripping Spring Quartzite	USA (Arizona)	Putative eukaryote microfossils	Beraldi-Campesi et al. (2014)
1.1–1.0	Lakhanda Group	Russia (Siberia)	Acritarchs and diverse macrofossils	Knoll et al. (2006)
~1.0	Torridon Group	Scotland	Acritarchs (lacustrine environment)	Strother et al. (2011), Battison and Brasier (2012)

diverse acritarchs described from lacustrine deposits (Strother et al. 2011; Battison and Brasier 2012) (Fig. 3.11c–e, g).

3.5.2.2 Macrofossils

Enigmatic macrofossils reported from the Paleoproterozoic and Mesoproterozoic may also represent eukaryotic organisms. The most common of these are *Grypania* and *Horodyskia*. Their regular morphology and macroscopic size suggest a eukaryotic origin (Knoll 2014). *Grypania* occurs as strap-shaped compressions of originally cylindrical organisms that form coils up to 24 mm across (Walter et al. 1976, 1990) and has been interpreted as a sessile algal eukaryote that was most likely multinucleate (coenocytic or multicellular) (Han and Runnegar 1992). *Horodyskia* fossils consist of 1–4 mm spheroidal (or sometimes conical, ovoid or rectangular) bodies connected by thin threads to form uniseriate structures (Yochelson and Fedonkin 2000). Eukaryotic affinity of *Horodyskia* is considered probable, but not beyond debate (Knoll et al. 2006). The oldest *Grypania* are known from mid-Paleoproterozoic rocks older than the earliest eukaryotic acritarchs (2.1 Ga; Negaunee Iron Formation, Michigan; Han and Runnegar 1992). The genus is also known from several younger rock units which include the ca. 1.4 Ga Gaoyuzhuang Formation in China (Walter et al. 1990), the 1.4–1.3 Ga Greyson Shale (Appekunny Argillite) of Montana (Walter et al. 1976, 1990; Horodyski 1982), and possibly from the 1.7–1.6 Ga lower Vindhyan Supergroup in India (Rohtas Formation, where it has been described as *Katnia*; Tandon and Kumar 1977; Han and Runnegar 1992). The oldest *Horodyskia* fossils have been reported from the Mesoproterozoic Bangemall Group in Australia (Grey and Williams 1990) but the genus is also known from the Greyson Shale of Montana (Horodyski 1982).

Tubular objects 100–180 μm in diameter with walls replaced by apatite are preserved in the lower Vindhyan Supergroup of India (Tirohan Dolomite; Bengtson et al. 2009). These fossils date from around the Paleoproterozoic-Mesoproterozoic boundary (1.7–1.6 Ga) and display regular annulation (shallow transverse grooves) on the outer surface corresponding to internal septa. Bengtson et al. (2009) interpret these fossils as filamentous algae (multicellular eukaryotes). Similar tubular or filamentous fossils are described by Yan and Liu (1993) from the late Paleoproterozoic in the 1.8–1.65 Ga Chang Cheng Group (China) and by Butterfield (2001) from the mid-Mesoproterozoic in the 1.2 Ga Hunting Formation (Somerset Island, arctic Canada). These fossils are associated with two other multicellular types: flat, layered units with internal differentiation and a stratified cellular structure (compared to phylloid algae; Butterfield 2001) and *Bangiomorpha*, the first unequivocal occurrence of complex multicellularity in the fossil record, the oldest reported occurrence of sexual reproduction, and the oldest record for an extant phylum (Rhodophyta, the red algae) and an extant family (Bangiaceae) (Butterfield 2000). Finally, in the late Mesoproterozoic, the Lakhanda Group hosts, along with diverse acritarchs, larger eukaryotic fossils compared to xanthophyte algae, fungi, and metazoans (Knoll et al. 2006; Hermann and Podkovyrov 2006; German and Podkovyrov 2009).

3.5.2.3 Eukaryotes in the Archean?

Some microfossils older than the Chang Cheng Group biota have been discussed as potential eukaryotes. The most prominent case is that of the spheroidal microfossils described as *Eosphaera* by Barghoorn and Tyler (1965) from a diverse biota in the Gunflint Iron Formation of Ontario (ca. 1.88 Ga; Schneider et al. 2002). *Eosphaera* fossils are small (up to 15 µm) and comprised of two concentric spherical envelopes enclosing up to 15 small spheroidal tubercles in the circular space between the two envelopes. Although *Eosphaera* has been compared to volvocalean green algae (Kazmierczak 1979), currently most authors agree that the Gunflint microfossil assemblage is entirely prokaryotic (e.g., Awramik and Barghoorn 1977) and that *Eosphaera* does not preserve sufficient detail in support of eukaryotic affinities (Knoll 1992). However, other lines of evidence indicate that eukaryotic life might have originated before the late Paleoproterozoic and even as early as the Archean (older than 2.5 Ga).

The shales of the Francevillian B Formation of Gabon, dated at 2.1 Ga, have yielded pyritized macrofossils consisting of elongated to isodiametric flattened specimens up to 12 cm in size, with a thicker central area and thinner radially patterned, undulate, and lobed margins (El Albani et al. 2010). The morphology of the fossils has been interpreted as reflecting growth that requires cell-to-cell signaling and coordinated growth responses indicative of well-integrated colonial organization or multicellular life in the early Paleoproterozoic, an interpretation bolstered by the detection of steranes in the same layers (El Albani et al. 2010).

Both the Gabonese macrofossils and *Grypania*—if the latter indeed represents a multicellular (or multinucleate) eukaryote—with its oldest occurrence at 2.1 Ga (Han and Runnegar 1992), predate the oldest unicellular eukaryotes (acritarchs) of the Chang Cheng Group (Lamb et al. 2009; Peng et al. 2009) by at least 300 million years. Assuming that unicellular eukaryotes evolved before multicellular forms, this implies that the prototypical unicellular eukaryote evolved prior to 2.1 Ga and we should expect to find eukaryotic acritarchs older than that age. Javaux et al. (2010) have reported organic-walled microfossils from shallow-marine deposits in the 3.2 Ga Moodies Group of South Africa. These microfossils are older than the oldest claims for sterane biomarkers (2.7–2.5 Ga; Brocks et al. 1999; Waldbauer et al. 2009). They are the oldest and largest Archean organic-walled microfossils reported to date (Javaux et al. 2010) and co-occur with microbially induced sedimentary structures (Noffke et al. 2006). The fossils display features consistent with eukaryotic affinities, such as good organic preservation, which is indicative of cell walls containing recalcitrant polymers and large sizes (ca. 30–300 µm) (Fig. 3.10a). These microfossils could, in principle, be eukaryotic, but their lack of cell surface ornamentation and their simple cell wall structure also make them easily comparable to the extracellular envelopes of some bacteria, thus precluding unequivocal interpretation as eukaryotes (Knoll 2014). It will be interesting to see if geochemical studies of the Moodies Group (not published to date) ascertain the presence of sterane biomarkers.

3.5.3 Salient Patterns in the Early Eukaryotic Fossil Record

The oldest bona fide unicellular eukaryotes are known from 1.80 to 1.65 Ga (mid-Paleoproterozoic) rocks, but the fossil record provides strong evidence that organisms capable of producing macroscopic bodies by growth processes that required coordinated responses, whether in a coenocyte or an aggregation of cells, had evolved by 2.1 Ga. This mode of development is not known in prokaryotes but is consistent with coenocytic or multicellular organization in eukaryotes, in which multicellularity has been achieved independently and to different extents in different clades (Niklas and Newman 2013; Niklas et al. 2013). Since complex intracellular organization evolved in eukaryotes well before the appearance of metazoans (Knoll et al. 2006), the earliest unicellular eukaryotes should be sought after in rocks older than 2.1 Ga (early Paleoproterozoic and Archean), ideally using approaches that combine morphological and ultrastructural characterization with geochemical studies.

Precambrian eukaryotes have been reported predominantly from rocks deposited in marine environments, and only starting with the ca. 1.2 Ga Dripping Spring Quartzite (Beraldi-Campesi et al. 2014) do we see eukaryotes on continents. Considering the living environments of the early marine acritarchs, Knoll et al. (2006) echo the views of Butterfied (2005a, b) that not all acritarchs represent reproductive cysts of planktonic algae—some must be the remains of benthic heterotrophic organisms. Javaux et al. (2001) documented, in their study of the Roper Group acritarchs, an onshore-offshore pattern in fossil distribution between depositional environments ranging from marginal marine to basinal. Overall, their data show seaward decrease in abundance and decline in diversity, as well as changing dominance among different species. These are interpreted as reflecting the effects of natural selection by physical habitat variables on species distributions by ca. 1.5 Ga, contributing to the rise of biological diversity (Javaux et al. 2001).

Late Proterozoic and Mesoproterozoic rocks contain abundant eukaryotic fossils, but morphological diversity (disparity; Wills 2001) of these fossil assemblages maintains low to moderate levels (summarized by Knoll et al. 2006). Among these, microfossils of large size and with complex wall structure and surface ornamentation are frequent in rocks up to ca. 1.6 Ga, but in older rocks, the cell surface ornamentation and ultrastructure are less distinctive, leading some to posit some residual uncertainty concerning taxonomic assignments at the domain level (Knoll 2014). A significant leap in the level of eukaryote disparity becomes apparent around 800 Ma, marking the end of the long interval between the evolution of the first eukaryotes and their taxonomic radiation in the second half of the Neoproterozoic (Knoll et al. 2006).

The morphologies of Proterozoic eukaryotes are varied, from simple unornamented cells, morphologically complex ornamented unicells, to three-dimensionally complex multicellular organisms displaying cellular differentiation, but for many of these fossils, phylogenetic placement is difficult because the dearth and generality of characters preserved do not support unequivocal taxonomic affinities (Knoll et al. 2006). For example, Knoll et al. (2006) point out that filamentous fossils document the early evolution of a molecular capacity for simple multicellularity which was subsequently exploited by multiple clades. Branching of structures, thought to require a more sophisticated intracellular organization (Knoll et al. 2006), cell-to-cell signaling, and polarity of cell development and division, is first seen in the bifurcating processes of unicellular *Tappania* of the Roper Group (Javaux et al. 2001) and in ca. 1 Ga filamentous fossils (Knoll et al. 2006). By 1.2 Ga, multicellularity and polarity of cell divisions had led to the evolution of complex body plans—e.g., *Bangiomorpha* (Butterfield 2000).

3.6 Roles of Microbes in Taphonomic Pathways

Another dimension of the microbial contribution to the fossil record has to do with the role of microbes in processes leading to fossilization (fossil preservation). Taphonomy, a field of paleontological investigation pioneered by Ivan Efremov (1940), is concerned with elucidating decay, disarticulation, transport, and burial pathways that result in fossilization—or not. Taphonomic studies document biostratinomy (physical and chemical changes incurred after death) and diagenesis (physical and chemical changes incurred after burial). Most investigations of taphonomy are biased toward those fossilization pathways leading to Konservat-Lagerstätten, deposits in which fossilized organisms exhibit exceptional preservation (Schiffbauer and Laflamme 2012). Sometimes these even include the preservation of soft tissues which may have resulted from carcass burial in an anoxic environment. Microbes play substantial roles in the formation of most Konservat-Lagerstätten, where they act as agents of degradation, as well as, ironically, preservation. Much of what is known about the roles of microbes in taphonomic pathways derives from studies of Konservat-Lagerstätten.

Another important body of evidence contributing to understanding of microbial roles in taphonomic pathways results from experimental investigations of taphonomy in living organisms and modern environments; these experiments are sometimes referred to as "actualistic" (not to be confused with the philosophic position, although the association is enlightening). By comparing the outcomes of actualistic experiments with observations from the fossil record, paleontologists have gained a strong appreciation for the necessity and ubiquity of microbial contributions to exceptional preservation. First and foremost, microbes facilitate authigenic mineralization, which is a special type of "cast and mold" formation, acting as nucleation sites and by altering local geochemical gradients. Microbes can also entirely replicate soft tissue by consuming original tissue and replacing it with extracellular polymeric substances (EPS). Finally, microbial mats can form "death masks" over specimens, which inhibit decay and disarticulation in addition to promoting mineralization, as well as stabilizing sediment interfaces, which has important implications for studies of fossil trackways. It is important to note that many of the roles that microbes play in taphonomic pathways occur concurrently (e.g., Darroch et al. 2012; Cosmidis et al. 2013; Iniesto et al. 2013). As such, although we have broken the following discussion into several major categories of microbe-mineral-fossil interactions, it should be understood that these are artificial and intergrading.

3.6.1 Cautionary Tales

As microbial contribution to exceptional preservation is most often implicated in preservation of ephemeral structures and soft tissue, it is useful to briefly review what is presently known about the organic structures that can, and (thus far) cannot, be found in the fossil record. This is particularly germane as many of the debates surrounding these structures invoke or include fossil or living bacteria. For instance, a putative dinosaurian heart has been refuted as such, but the cemented sediment does bear evidence of microbiogenic texture (Cleland et al 2011). Structures preserved within the body cavities of three early Cretaceous birds from the Jehol biota have been interpreted as ovarian follicles (Zheng et al. 2013), but this is unlikely, as gonads are among the first visceral organs to suffer autolytic and bacterial decay (Mayr and Manegold 2013). Finally, the sensational report of nonmineralized soft tissue in a tyrannosaurid (Schweitzer et al. 2005) was later reinterpreted as evidence that many fossil specimens are colonized by microbes after fossilization and exhumation (Kaye et al. 2008), in addition to during biostratinomy. Schweitzer et al. (2005) originally reported transparent hollow tubules obtained from acid digestion of dinosaur bone. The tubules were interpreted as blood vessels and contained spherical microstructures that were interpreted as nuclei from endothelial cells (Schweitzer et al. 2005). In their actualistic reinvestigation, Kaye et al. (2008) acid-digested a variety of fossil bones and recovered structures very similar to those observed by Schweitzer et al. (2005). SEM-EDX analyses indicated that the microspheres were likely to have originally been framboidal pyrite, diagenetically altered to iron oxides (Kaye et al. 2008), consistent with an earlier hypothesis put forth by Martill and Unwin (1997). Fourier transform infrared (FTIR) spectroscopy indicated that as opposed to reflecting diagenetically altered collagen, the material was comparable to modern biofilms; their lack of antiquity was confirmed by carbon dating. Kaye et al. (2008) suggested that recent biofilms had lined voids in fossil bones, thus producing the hollow structures morphologically similar to blood vessels and osteocytes.

Unlike the aforementioned putative soft tissues, some reports of original biopolymers in dinosaurian tissue are more compelling, having been assessed with FTIR spectroscopy. These include collagen peptides (San Antonio et al. 2011) and amide groups (Manning et al. 2009) reported from Cretaceous bones and in embryonic Jurassic bone, which presently constitutes the oldest evidence of in situ preservation of complex organics in vertebrate remains (Reis et al. 2013). Other biopolymers known to be highly resistant with good fossilization potential include chitin, cellulose, sporopollenin (Briggs 1999; Wolfe et al. 2012), and the compounds found in melanosomes (Liu and Simon 2003; Wilson et al. 2007; Vinther et al. 2008, 2010), the latter of which have been previously interpreted as fossilized bacteria (Wuttke 1983a). Not only can melanosomes be used to reconstruct color patterning (Li et al. 2010a; Knight et al. 2011), their presence is tightly correlated with the preservation potential of integument (Zhang et al. 2010). Maganins, which are peptides found in dermal secretions of some amphibians, also contribute to preservation of integument (McNamara et al. 2009). Maganins can permeate and lyse microbial cell membranes, and because they persist post mortem, they inhibit microbial decay to the extent that *Rana* skin remains intact more than 50 days, during which time all contents of the body cavity entirely decay (Wuttke 1983b).

Finally, reports of nuclei in the fossil record (e.g., Baxter 1950; Millay and Eggert 1974; Brack-Hanes and Vaughn 1978; Martill 1990; Schweitzer et al. 2007) have been problematic. Actualistic experiments have demonstrated pseudonucleic structures (Westall et al. 1995), but intracellular mineralization of legitimate nuclei remains elusive, with a sole exception to date—the recently published nuclei and chromosomes preserved in the root cells of a Jurassic (ca. 187 Ma) *Osmunda* fern rhizome from Scania (Sweden; Bomfleur et al. 2014). In the Miocene Clarkia flora, where plant fossils are preserved as compressions and ultrastructural TEM investigations have substantiated reports of nuclei, they are preserved at much lower abundance than chloroplasts or mitochondria (Niklas 1983), which typically autosenesce before nuclear disorganization occurs (Woolhouse 1984). This seeming discrepancy may result from partitioning of protoplast into nucleate and enucleate microprotoplasts via osmotic shock (Niklas et al. 1985). By isolating plastids from nuclei, nuclear-controlled autosenescence (Yoshida 1962) could have been circumvented (Niklas 1983). While this model would account for some reports of fossil nuclei in the absence of plastids, it is possible that other reports of fossil nuclei in plants represent condensed cellular contents (Niklas 1983). For instance, osmotic stress can cause protoplast shrinkage away from the cell walls within 10 min (Munns 2002). The shrunken protoplast, which may be as little as 1/3 or less of the cell volume, remains centrally positioned by attachment of plasma membrane strands (Fig. 5, Munns 2002). Plasma shrinkage has also been invoked (Pang et al. 2013) in refuting putative nuclei in Ediacaran Doushantuo Formation (ca. 555 Ma, China) specimens, which have become the focus of taxonomic disputes (e.g., Huldtgren et al. 2011). Because nuclei are composed of water, proteins, and nucleic acids, they are easily degraded in comparison to lipids (Eglinton et al. 1991); in animal fossils at least, all lipid bilayers could be expected to fossilize, and the lack of mitochondria in the Doushantuo specimens suggests the nucleus-like bodies are not what they seem (Pang et al. 2013).

3.6.2 Authigenic Mineralization

Authigenic minerals are defined simply as those minerals which are formed in situ and, contra Schopf et al. (2011), are not implicitly biogenic, although many biologically mediated pathways induce the precipitation of iron, carbonate, phosphate, sulfur, silicate, and clay minerals (e.g., Konhauser 1998; Bazylinski and Frankel 2003; Briggs 2003a, b; Martin et al. 2004; Peckmann and Goedert 2005; Konhauser et al. 2011; Darroch et al. 2012). Carbonaceous polymers alone may be sufficient to precipitate many minerals (McNamara et al. 2009; Schopf et al. 2011; Roberts et al. 2013)—this process has been termed *organomineralization* (Trichet and Defarge 1995) and is considered abiogenic, although the polymers may be derived from living tissue. *Biomineralization*, on the other hand, encompasses both active and passive precipitations of minerals and is defined as occurring in the presence of living cells (Trichet and Defarge 1995).

Active biomineralization occurs when the precipitation of minerals is under direct control of the cell (biologically controlled biomineralization; Konhauser and Riding 2012), as in the case of magnetite crystals formed within magnetotactic, microaerophilic bacteria (Bazylinski 1996). Passively induced precipitation (biologically induced biomineralization; Konhauser and Riding 2012) can occur through metabolic processes that alter local chemistry permitting stoichiometric precipitation (Lovley and Phillips 1986; Lovley et al. 1987; Roh et al. 2003; Straub et al. 2004) and may also occur through mineral nucleation on EPS or microbial surfaces (e.g., Ferris et al. 1987; Thompson and Ferris 1990; Fortin et al. 1997; Léveillé et al. 2000; Van Lith et al. 2003). EPS are chemically complex bacterial exudates that contain mucus, sugars, extracellular DNAs, and proteins (Wingender et al. 2012) and are anionic, resulting in metal ion sorption (Beveridge 1989). Bacterial cells are also characterized by the electric double layer immediately external to the cell walls, which results from proton shuffling to establish a proton motive force, forming a highly localized electrochemical gradient (Poortinga et al. 2002).

3.6.2.1 Organomineralization

Biomineralization has been generally recognized as a key factor in exceptional preservation (Briggs et al. 1993; Briggs and Kear 1993; Briggs and Wilby 1996; Martin et al. 2004; Darroch et al. 2012), but it is important to emphasize that fossilization may readily result from organomineralization as well. For instance, sulfurization, a common diagenetic phenomenon (Sinninghe Damsté and de Leeuw 1990), can preserve soft tissue as sulfur-rich organic residue (McNamara et al. 2009, 2010). Similarly, diagenetic (post-burial) processes often reduce recalcitrant biomolecules like chitin, cellulose, and lignin into geologically stable macromolecules, termed kerogen (Briggs 2003a, b); kerogen is composed of interlinked polycyclic aromatic hydrocarbons, and organomineralization could proceed via hydrogen bonding between metal ions and peripheral hydrogen atoms (Schopf et al. 2011).

The precipitation of dolomite (Ca(Mg)CO$_3$) and disordered high-Mg carbonates can also occur via organomineralization (Roberts et al. 2013), and Klymiuk et al. (2013a) have suggested that some kerogenous plant fossils preserved in calcium carbonate concretions are at least partially mineralized with disordered high-Mg carbonates. The lack of an actualistic explanation for the precipitation, under ambient temperatures, of the geologically significant amounts of dolomite found in the rock record has often been referred to as "the dolomite problem," as it was only known to occur under a very narrow set of conditions, in rarely occurring environment—hypersaline sabkhas, in association with sulfate-reducing bacteria (Wright and Wacey 2004). Roberts et al. (2013) achieved abiotic nucleation of high-Mig carbonates on carboxylated polystyrene, and they suggest that Mg^{2+} ions are complexed and dewatered by surface-bound carboxyl groups. Because the model does not require metabolic activity, any type of organic matter—including microbial cell walls, which are rich in carboxyl groups—could serve as nucleation sites for dolomite precipitation (Roberts et al. 2013).

Fossils of 63 adult frogs, constituents of the late Miocene Libros biota of northeast Spain, provide evidence that tissue composition itself can control precipitation of authigenic phosphates (McNamara et al. 2009). In some of the specimens, the presence of distinctive microfabrics (Wilby and Briggs 1997; Briggs 2003a, b; Chin et al. 2003; Briggs et al. 2005) indicates that phosphate nucleated directly upon soft tissue without bacterial intermediates (McNamara et al. 2009). Bacterial contributions are often implicated in phosphatization, as calcium carbonate (CaCO$_3$) precipitation is thermodynamically favored over calcium phosphate (CaPO$_4$) at pH below 6.38 (Briggs 2003a, b), and bacterial decay leads to localized acidification. Phosphatization of the Libros frogs, however, was probably not significantly enhanced by diffusion of acidic microbial metabolites from internal biofilms, but was instead facilitated primarily by release of SO$_4^{2-}$ during degradation of glycosaminoglycans (GAGs), which are abundant in extant adult *Rana* skin where they are chemically bound to collagen fibers (McNamara et al. 2009). Collagen contains abundant hydroxyl groups, to which Ca^{2+} readily adsorbs; calcium phosphate precipitates are extremely localized to dermal collagen fibers in the fossil frogs and absent elsewhere in the specimens (McNamara et al. 2009). The Libros frogs also serve as an opportunity to make an important distinction: mineral precipitation was *itself* not bacterially mediated, but the decay of collagen, which increases hydroxyproline and therefore the abundance of hydroxyl nucleation sites (Lawson and Czernuszka 1998), was almost certainly facilitated by bacterial collagenases (McNamara et al. 2009). The distinction as to whether mineralization is abiogenic or biogenically mediated is one that explicitly concerns *how the mineral is precipitated* and does not preclude subsidiary microbial interactions. Because decaying organic matter comprises heterogeneous microenvironments (Briggs and Wilby 1996), it is probable that most taphonomic pathways include both organomineralization and biomineralization.

3.6.2.2 Biomineralization

Burgess Shale-Type Preservation

Burgess Shale-type preservation is ubiquitous in Ediacaran and Cambrian deposits (Gabbott et al. 2004; Xiao et al. 2002; Zhu et al. 2008; Anderson et al. 2011; Cai et al. 2012). Factors thought to be crucial in the formation of Burgess Shale-type deposits include dysoxia (low oxygen content) or anoxia (Gaines et al. 2012; Cai et al. 2012), the lack of biogenic reworking of sediment (bioturbation) (Orr et al. 2003), and inhibition of autolysis. Many authors have hypothesized that early diagenetic mineralization is an intrinsic component of Burgess Shale-type taphonomic pathways. Butterfield (1995) suggested that authigenic clay minerals inhibit autolytic decay, whereas Orr et al. (1998) suggested that they adsorb to organic matter and create templates. Gabbott et al. (2004) considered pyritization more relevant for the Chengjiang biota, where it was precipitated on organic remains during bacterial sulfate and iron reduction. Finally, Petrovich (2001) has suggested that Fe^{2+} ions adsorbed to organic tissues, promoting nucleation of iron-rich clays. Still others (Gaines et al. 2008; Page et al. 2008; Butterfield et al. 2007) have maintained that clays are metamorphic in nature and play no role in early diagenesis.

Analyses of 53 fossils from 11 Burgess Shale-type deposits using scanning electron microscopy—energy-dispersive X-ray spectroscopy (SEM-EDX)—indicate that all share a principal taphonomic pathway (Gaines et al. 2008). According to Gaines et al. (2008), conservation of organic tissues (i.e., as kerogen), and not early diagenetic authigenic mineralization, is the primary mechanism responsible for preservation. Fossils from all sites display carbonaceous preservation, ranging from continuous robust through degraded films, but all are enriched with C and depleted of Al, Si, and O with respect to the surrounding matrix (Gaines et al. 2008). This principal pathway may be occasionally augmented by authigenic mineral replication of some tissues (Gaines et al. 2008), including pyrite precipitated in gut tracts of trilobites (Lin 2007), and calcium phosphate replication of some soft tissues (Butterfield 2002, 2007). More recent analyses, however, have examined unmetamorphosed fossils preserved in the Doushantuo Formation, which also exhibit Burgess Shale-type preservation (Anderson et al. 2011). All of the Doushantuo fossils examined are closely associated with clay and pyrite minerals, which appears to confirm the role of these two components in early diagenesis (Anderson et al. 2011).

The presence of authigenic clay minerals can be explained in several ways: reducing conditions at the sediment-water interface could be deflocculating aluminosilicate grains (Gaines et al. 2005); clay minerals could be attracted and adsorbed onto organic surfaces (Gabbott 1998), a process which is facilitated by sulfate reduction and reduced pH (Martin et al. 2004); clay minerals could be authigenically precipitated from pore waters, in which localized zones of reduced pH resulted in dissolution of metals from resident sediments (Gabbott et al. 2001);

or clay minerals could conceivably be bound to organic surfaces by bacterial biofilms (Konhauser et al. 1998; Toporski et al. 2002). By invoking reducing conditions and low pH, these models find at least a subsidiary role for sulfate-reducing bacteria, and actualistic studies of decay of lobster eggs indicate that quartz and clay mineral attachment to decaying tissue is enhanced in the presence of sulfate-reducing bacteria, precipitating in proximity to decaying tissue, while iron sulfides precipitate in surrounding sediments (Martin et al. 2004).

Phosphate Mineralization of Soft Tissue

In order to preserve details of soft tissue, mineralization must necessarily outpace decay (Allison 1988a); thus phosphatization, the replication of soft tissues by calcium phosphate (apatite, Ca_3PO_4), represents a biostratinomic or early diagenetic taphonomic pathway and a discrete temporal window for preservation. Phosphatization of soft tissues is relatively common in vertebrate fossils (Martill 1987, 1988, 1989, 1991; Martill and Unwin 1989; Kellner 1996a, b; Briggs et al. 1997; Frey et al. 2003), and, with only few exceptions (e.g., McNamara et al. 2009), it is generally conceded that bacteria have a vital role in soft tissue mineralization by Ca_3PO_4. Anaerobic decay generates steep geochemical gradients favoring Ca_3PO_4 precipitation over $CaCO_3$ (Briggs and Kear 1993, 1994; Hof and Briggs 1997; Sagemann et al. 1999; Briggs 2003a, b), and carbonate precipitation is inhibited at pH < 6.38 (Briggs 2003a, b) or when PO_4^{3-} concentrations are high (Sagemann et al. 1999). Chemical microenvironments in the vicinity of decaying carcasses geochemically differ from surrounding sediments in terms of ion concentration, pH, and Eh (oxidation potential) (Sagemann et al. 1999), and these differences become more extreme toward the interior of the carcass (Briggs and Wilby 1996). If PO_4^{3-} ions are sourced from the decaying tissue itself, mineral precipitation will be highly localized, whereas if PO_4^{3-} is derived from external sediments, or from microbial consortia engaged in active phosphate precipitation (Wilby et al. 1996; Cosmidis et al. 2013), phosphatization is likely to be extensive (Wilby and Briggs 1997).

Invertebrates tend to be less prone to widespread soft tissue phosphatization. In arthropods, muscle, hepatopancreas, gills, nerve ganglia (Briggs and Kear 1993, 1994; Hof and Briggs 1997; Sagemann et al. 1999), and midgut glands (Butterfield 2002) can be selectively mineralized. Actualistic experiments produce only partial phosphatization (Briggs and Kear 1994; Hof and Briggs 1997; Sagemann et al. 1999). For shrimp carcasses inoculated with sulfate-reducing bacteria and sulfide-oxidizing and fermenting bacteria, decay was most intense and phosphatization most extensive under anaerobic sulfate reduction (Sagemann et al. 1999). While this implies a prominent role for sulfate-reducing bacteria in invertebrate soft tissue phosphatization, Briggs et al. (2005) report organomineralization in a specimen of *Mesolimulus*, a horseshoe crab known from the Jurassic Solnhofen and Nusplingen biotas. Muscle fibers with distinct banding have been preserved by apatite precipitation directly onto fibers, although structures interpreted as cyanobacterial body fossils have also been observed (Briggs et al. 2005). The preserved midgut of a trilobite (Lerosey-Aubril et al. 2012) was probably also organomineralized; the surrounding sediment is low in phosphate, and although microbial mats can concentrate phosphates (Wilby et al. 1996), none are present within the surrounding sediments. Instead, Lerosey-Aubril et al. (2012) suggest that epithelial cells of the midgut probably contained mineral calcium phosphate concretions, a form of mineral storage in preparation for ecdysis (molting).

Microbial activity can contribute phosphates to a microenvironment through active and passive mineralization. Calcium phosphate (apatite) is actively precipitated intra- and extracellularly by a number of bacteria, including *Bacterionema matruchotii*, *Chromohalobacter marismortui*, *Escherichia coli*, *Providencia rettgeri*, *Ramlibacter tataouinensis*, and *Serratia* sp. (reviewed in Cosmidis et al. 2013). Microbial metabolic activities alone may, however, be sufficient to generate amounts of free phosphate high enough to passively precipitate large quantities of laminar calcium phosphate (Arning et al. 2009). Lipid biomarkers associated with sulfate-reducing bacteria and abundance of the giant sulfur bacteria *Thiomargarita namibiensis* are correlated with PO_4^{3-} concentrations in modern phosphate-rich sediments (Schulz and Schulz 2005; Arning et al. 2008). Phosphatized fossils have not been assessed for biomarkers, but coccoid, spiral, and bacillus-like structures interpreted as bacterial body fossils are occasionally observed in zones of soft-tissue fossilization in vertebrates and invertebrates (Briggs et al. 1997; Toporski et al. 2002; Briggs et al. 2005; Skawina 2010; Pinheiro et al. 2012). While morphology cannot be used to circumscribe phylogenetic affinities of these putative fossil bacteria, actualistic experiments in freshwater decay of invertebrate tissue (Skawina 2010) reveal a succession of bacterial morphotypes that are generally correlative with the stage of decay: cocci predominate over bacilli when pH >7 (i.e., at the beginning and end of decay; Briggs and Kear 1994; Hof and Briggs 1997; Skawina 2010), whereas bacilli are more prevalent when pH is reduced <7. These observations thus provide a tentative link between bacillus-type structures and sulfate-reducing agents of anaerobic decay.

Iron Minerals and Pyritization

Although some iron oxides can be precipitated abiotically in steep geochemical gradients, like those resulting from collagen decay (Kremer et al. 2012b), precipitation of many iron oxides, iron carbonates, and iron sulfide (pyrite) is bacterially mediated (Ferris et al. 1987; Beveridge 1989; Ferris 1993; Konhauser 1998; Bazylinski and Frankel 2003; Châtellier et al. 2004; Konhauser et al. 2011). For instance, direct precipitation of iron oxides onto bacterial EPS releases a proton into the extracellular microenvironment; by acidifying their immediate surroundings, aerobic iron-oxidizing bacteria can enhance the proton motive force, thus increasing the energy-generating potential of a cell (Chan et al. 2004). Alternately, authigenic pyrite formation is frequently a result of bacterial metabolisms employing sulfate reduction. Sulfate-reducing bacteria utilize sulfate as a terminal electron acceptor under anaerobic conditions; H_2S results as a metabolic by-product and reacts with dissolved iron, precipitating iron sulfide (Frankel and Bazylinski 2003). In some cases, the cellular membrane itself can serve as an anionic matrix, immobilizing Fe^{2+}, which then reacts with metabolic H_2S, autolithifying the bacterial cell in the process (Ferris et al. 1988). Despite the ubiquity of these processes in modern environments, there are relatively few reports of pyritized bacterial body fossils (Southam et al. 2001; Schieber and Riciputi 2005; Cosmidis et al. 2013; and possibly those reported by Tomescu et al. 2008).

Bacteriogenic pyrite is generally depleted in ^{34}S (Canfield and Thamdrup 1994), and when crystals do not nucleate directly upon cell surfaces, they are framboidal, or lacking distinct crystal faces (Popa et al. 2004). Although the presence of framboidal pyrite is not in itself positively indicative of sulfate-reducing metabolisms (Butler and Rickard 2000; Pósfai and Dunin-Borkowski 2006), there is a strong association between bacterial EPS and the formation of pyrite framboids (Maclean et al. 2008). Framboidal pyrite commonly occurs below the reduction-oxidation transition zone in subaqueous microbial mats (Popa et al. 2004). Focused ion beam SEM-EDX provides a novel view of the interior of low-temperature framboidal pyrite aggregates formed in microbial mats: organic matrix occurs between individual pyrite crystals, suggesting that nanocrystals nucleate directly within the organic matrix of bacterial EPS (MacLean et al. 2008). Extensive precipitation of pyrite, whether on cell surfaces or within EPS, can result in high-fidelity preservation of fossil morphology (Grimes et al. 2001, 2002; Gabbott et al. 2004; Brock et al. 2006; Darroch et al. 2012; Wang et al. 2012). In some cases, pyrite framboids form inside bacterial cells, replacing them, as is the case with the cyanobacteria reported by Tomescu et al. (2006, 2009) from the Early Silurian Massanutten Sandstone. Pyrite often completely replaces organics, as in the Jehol biota, where insects were initially preserved in pyrite that was later weathered to iron oxides (Wang et al. 2012), and cellularly preserved Devonian, Mississippian, and Eocene fossil plants that have been replaced with pyrite (e.g., Allison 1988b; Rothwell et al. 1989; Tomescu et al. 2001).

An actualistic experiment in organomineralization of a celery petiole (Grimes et al. 2001) has yielded important insights into authigenic pyrite formation in plant tissues. Grimes et al. (2001) demonstrated that pyrite readily precipitates on inner plant cell wall surfaces, within cell walls, and in the middle lamella between cells. Pyrite initially nucleates on the inner walls of parenchyma cells, before penetrating inward (Grimes et al. 2001). Inward penetration occurs through successive nucleation upon previously precipitated crystals, rather than by continued crystal growth (Grimes et al. 2001). Pyrite precipitates only between fibrils of cellulose and not on lignified surfaces; minerals do not replace the plant tissue but rather nucleate within fluid-filled spaces between cellulose fibrils (Grimes et al. 2001). Thus, pyritized plant tissues probably represent middle lamella regions, as opposed to replaced cellulose, and if present, lignin will be coalified (Grimes et al. 2001, 2002).

Taphonomic pathways hypothesized for the London Clay (one of the more diverse Eocene floras in Europe; Collinson et al. 2010), which invoke pyritic replacement of plant tissues in the presence of sulfate-reducing bacteria, have also been experimentally investigated (Grimes et al. 2001; Brock et al. 2006). In both studies, the experimental system rapidly became driven by anaerobic respiration, with diffuse precipitation of iron sulfides into sediment surrounding *Platanus* twigs. Local pH also declined, but as decay tapered off—at 12 and 5 weeks (Grimes et al. 2001; Brock et al. 2006 respectively)—pH increased, reflecting a metabolic shift to sulfide oxidation (Brock et al. 2006). Although pyrite precipitated on plant surfaces, few of the twigs exhibited internal sulfide minerals (Grimes et al. 2001; Brock et al. 2006); these results may have been due to inherent heterogeneity of the system (Brock et al. 2006) or hydrophobic moieties in lignin molecules (Jung and Deetz 1993; Grimes et al. 2001).

Doushantuo fossils preserved in small cm-sized chert nodules represent a more specialized pyritization pathway. The chert nodules, which contain microbial mat fragments at their core, are surrounded by silica cortex and inner pyrite rim, with an exterior rim of late diagenetic blocky calcite (Xiao et al. 2010). Because pyrite crystals are immersed in groundmass silica, which exhibits no concentric growth zones, the two minerals were probably syngenetic and swiftly precipitated (Xiao et al. 2010). Formation of pyrite-silica rims was likely facilitated by local pH changes related to bacterial sulfate reduction (Xiao et al. 2010), consistent with diffusion-precipitation models that posit a spherical precipitation front formed at the boundary between diffusing H_2S and a surrounding Fe^{2+} reservoir (Raiswell et al. 1993; Coleman and Raiswell 1995). Pervasive marine anoxia and substantially higher levels of dissolved iron during the Ediacaran are thought to have encouraged proliferation of sulfate-reducing bacteria (Canfield et al. 2008; Li et al. 2010b). Sulfate reduction generates alkalinity (HCO_3^-), thereby promoting $CaCO_3$ precipitation, but because sulfate reduction and pyrite precipitation have a net increase of protons, pH declines (Xiao et al. 2010). Within a narrow window (pH 9.0–10.0), the solubilities of carbonate and silica behave inversely, and silica precipitates at the same time $CaCO_3$ enters solution; therefore, proton generation by sulfate-reducing bacteria led to silicification, while H_2S generated during metabolism led to pyrite precipitation in the Doushantuo nodules (Xiao et al. 2010).

Compression-Impression Leaf Fossilization

The presence of biofilms may be integral to preservation of leaves as compression-impression fossils. Two taphonomic pathways invoking microbially mediated preservation have been proposed for compression-impression plant macrofossils. The exceptional preservation of plant and insect biota at the Eocene Florissant locality is probably a special case and not widely replicated in the paleobotanical record: fossilization is thought to have been facilitated by extensive diatom blooms in response to increased levels of dissolved silica (derived from periodic volcanic ashfall) within a lacustrine system (O'Brien et al. 2002). When exposed along bedding planes, it is apparent that the fossils are encased within diatomaceous laminae; O'Brien et al. (2002) suggest that diatom biofilms were established on floating plant and insect debris, with subsequent sinking and incorporation into the sediment.

The second taphonomic pathway implicating biofilms was first proposed by Spicer (1977) who recognized that iron was often precipitated on submerged surfaces of leaves prior to burial, and leaf compression fossils are often encrusted with iron oxides (Dunn et al. 1997). Leaf surfaces, however, are covered by hydrophobic cuticles that inhibit metal binding. Degradation of cuticle followed by authigenic mineralization of cellulose could account for this seeming oxymoron, but although exceptional instances of cellulose preservation have been demonstrated (Wolfe et al. 2012), cellulose typically degrades rapidly in comparison to cuticle (Spicer 1981). Dissimilatory iron-reducing bacteria have also been invoked to explain this oxymoron (Spicer 1977), but they are restricted to aerobic environments with abundant Fe^{3+} (Dunn et al. 1997). By contrast, leaf surfaces host a variety of microbes capable of forming biofilms (Morris et al. 1997). The anionic nature of bacterial EPS facilitates metal binding (Beveridge 1989), and experiments conducted by Dunn et al. (1997) show that mineral precipitation does not occur in the absence of biofilm formation.

3.6.3 Doushantuo "Embryos"

The Doushantuo biota, preserved in Ediacaran marine deposits, has been recently subject to intense debates, centering on the presence of microscopic multiloculate structures that have been interpreted as having bilateral symmetry and thus representing bilaterian embryos in varying stages of development (Xiao et al. 1998; Huldtgren et al 2011; Yin et al. 2014). A recent analysis using backscattered electron imaging, electron probe microanalysis, and synchrotron X-ray tomography microscopy (SRXTM) compared the so-called embryos to preserved cells of other Doushantuo fossils with uncontested affinities to distinguish between crystal structures of mineralized phases preserving (or replacing) original structure and those attributable to later diagenetic effects, including void filling (Cunningham et al. 2012a). Supposed nuclei exhibit crystal textures characteristic of void filling, and cells purported to represent later developmental stages are instead diagenetic artifacts (Cunningham et al. 2012a). In light of these analyses, the Doushantuo specimens are unlikely to actually represent fossilized embryos (trace fossil evidence, however, suggests that bilaterians may have evolved by this time; see Pecoits et al. 2012). Nevertheless, these enigmatic specimens triggered an intense period of research into taphonomic pathways that could replicate the structures.

Under abiotic conditions, invertebrate egg surfaces could be mineralized in solutions of calcium carbonate and calcium phosphate in as little as 1–2 weeks (Hippler et al. 2012) to 1 month (Martin et al. 2005). Sediments could also bind to the egg surfaces, similarly replicating exterior morphology (Martin et al. 2005). Although eggs may not exhibit substantial decay for up to a year, neither experiment was able to induce internal mineralization (Martin et al. 2005; Hippler et al. 2012).

Unlike abiotic mineralization, bacterial pseudomorphing does replicate internal structure. A bacterial pseudomorph can be formed by establishment of a surface biofilm and then invasion of bacteria into the interior, where they consume cytoplasm, replacing it with EPS and bacterial biomass (Raff et al. 2008, 2013). Local biofilms are initially formed at structural boundaries (Stoodley et al. 2002) and then act as scaffolds, conjoining to form the full pseudomorphs (Raff et al. 2013). Localized surface heterogeneities promote generation of very small biofilms conformed to the local shape, and therefore a fully pseudomorphed structure is composed of numerous local biofilms (Raff et al. 2013). If autolysis of embryos is blocked by anoxia or reducing conditions (Raff et al. 2006), they are reliably pseudomorphed, in aerobic and anaerobic conditions, by natural seawater bacterial populations dominated by gammaproteobacteria (Raff et al. 2008, 2013). Furthermore, Raff et al. (2013) were able to identify the single species that could each replicate one of three taphonomic pathways observed in embryos exposed to natural seawater populations (Raff et al. 2013).

Because pseudomorphs initiate as minute biofilms, microbial flora may be heterogeneous across a specimen. Single-taxon experiments demonstrate that different colonizers will yield different taphonomic outcomes. *Pseudoalteromonas tunicata* produced high-fidelity pseudomorphs that replicated both external and internal structure within 2–3 days, a timeline comparable to pseudomorph generation in natural seawater. *Vibrio harveyi* generated pseudomorphs replicating external surfaces only, while *Pseudoalteromonas luteoviolacea* resulted in complete degradation within a few days (Raff et al. 2013). Finally, although *Pseudoalteromonas atlantica* is known to form surface biofilms, it did not interact with the embryos, suggesting that not all biofilm formers are competent pseudomorphers (Raff et al. 2013). Species identity may not be foremost in determining taphonomic outcomes (Raff et al. 2013), but rather may depend on the suite of genetic capabilities (Burke et al. 2011).

Competition experiments (Raff et al. 2013) illustrated that the products of an initial pseudomorphing strain could be obliterated by tissue-destructive strains, but once formed, bacterial pseudomorphs of *Artemia* embryos and nauplius larvae were stable for up to 19 months. These experiments demonstrated that if *Pseudoalteromonas luteoviolacea* comprised more than 5 % of a mixed population, pseudomorphing was inhibited. This suggests that preservation of soft tissue may depend upon favorable competitive outcomes between closely related species.

Hypotheses regarding the taxonomic affinities of the Doushantuo "embryos" have also been tested using actualistic experiments. Bailey et al. (2007) suggested that the Doushantuo fossils could represent fossilized giant sulfur bacteria similar to *Thiomargarita*, in which a large central vacuole accounts for 98 % of the cell volume. *Thiomargarita* reproduces by reductive cell division (Kalanetra et al. 2005) and thus provides a morphological analogue to "cleaving cells" seen in the putative embryos. During laboratory decay of *Thiomargarita*, however, cell membranes decayed in advance of mucus sheaths, which may remain stable for months or years prior to degradation (Cunningham et al. 2012b). Furthermore, when *Thiomargarita* cells were inoculated with pseudomorphing bacteria of the type used in experiments by Raff et al. (2008), the cells collapsed into their central vacuole (Cunningham et al. 2012b). Despite similar morphologies, it is thus unlikely that the Doushantuo specimens represent giant sulfur bacteria.

3.6.4 Microbial Mats

Microbial mats, which are composed of biofilms with microbial cells spatially organized in EPS (Stoodley et al. 2002), are vertically stratified communities defined by light penetration and vertical redox gradients generated by microbial metabolic activities (Wierzchos et al. 1996). The upper surfaces are dominated by photosynthetic cyanobacteria, with aerobic heterotrophic microbes in the oxidized upper layer (Visscher and Stolz 2005). These overlie a deeper anoxic layer characterized by anoxygenic photosynthetic bacteria, fermenters, and chemolithoautotrophic sulfur bacteria; lowermost layers contain dissimilatory sulfate and sulfur-reducing bacteria and methanogens (Dupraz and Visscher 2005; Visscher and Stolz 2005). Transitions between these zones may occur within millimeters. Oxygenation and carbon dynamics within mats are also subject to temporal shifts: in daylight, photosynthesis results in supersaturated O_2 concentrations and high carbon production. After dark, microbes employing aerobic respiratory pathways rapidly consume the accumulated carbon and render much of the mat anoxic; thereafter, respiration switches to sulfate reduction, and sulfides accumulate, peaking near dawn (Canfield and Des Marais 1991; Visscher et al. 1991; Canfield et al. 2004). Microbial mats induce precipitation of both iron sulfides and carbonates, and because they grow continually, mats also trap sedimentary particles and organic remains (Krumbein 1979; Visscher and Stolz 2005).

3.6.4.1 Death Masks

First proposed by Gehling (1999), the microbial death mask model of fossil preservation invokes anaerobic sulfate-reducing metabolisms, which induce formation of pyritic shrouds that drape the decaying carcass and preserve features as mineralized casts (Gehling 1999; Gehling et al. 2005; Callow and Brasier 2009). Pyrite precipitation through death masks formed beneath active microbial mats constitutes the leading hypothesis for Ediacaran-type preservation (Gehling 1999; Gehling et al. 2005; Darroch et al. 2012). In Ediacaran-type deposits, framboidal pyrite is found in direct association with some three-dimensionally preserved fossils (Laflamme et al. 2011), and beds commonly contain oxidized weathering products of pyrite (goethite and limonite) as well as sedimentary textures characteristic of preserved microbial mats (Gehling 1999; Laflamme et al. 2012). It has also been suggested that some Ediacaran morphologies may even be taphomorphs, preservational variants of structures produced during postmortem microbial decay (Liu et al. 2011).

Microbial mats were probably extensive across Proterozoic seafloors, where they substantially contributed to sediment lithification (Gehling et al. 2005; Droser et al. 2006). Although the rise of metazoan grazers diminished the extent to which microbial mats influenced sedimentary structure on global scale (Orr et al. 2003), they remained important agents in the formation of Konservat-Lagerstätten (Gall et al. 1985; Gall 1990; Seilacher et al. 1985; Briggs 2003a, b; Fregenal-Martínez and Buscalioni 2010). Mats contribute to fossilization through biostratinomic and early diagenetic processes. The former include envelopment of carcasses in which the microbial mat can replicate the body surface on the underside of the mat and protect the body from scavengers and disarticulation. Anoxic conditions within the mat can also inhibit microbial decay, and mineral precipitation can stabilize the specimen (Gall 1990; Wilby et al. 1996; Gehling 1999; Briggs 2003a, b; Briggs et al. 2005; Martill et al. 2008; Iniesto et al. 2013). Although iron sulfide precipitation is usually invoked in death mask preservation, microbial mats also precipitate carbonates and phosphates (Reid et al. 2000; Decho and Kawaguchi 2003; Dupraz et al. 2009; Cosmidis et al. 2013).

In the first experiment to test death mask hypotheses, Darroch et al. (2012) followed the taphonomic pathways of lepidopteran larvae placed on top of microbial mats, marine sand, and sterilized sand, which were allowed to decay over a 6-week period. By the end of the experiment, only specimens from microbial mat arrays consistently exhibited Ediacaran-type epirelief structure that replicated morphology. Iron sulfides precipitated within a day, reaching their maximum extent within 2 weeks and reentering solution by the end of the experiment (Darroch et al. 2012), likely due to a shift to sulfide-oxidizing metabolism. These results suggest that the temporal window in which Ediacaran-type preservation can occur is short. The formation of abundant iron sulfides, despite negative stoichiometric bias due to the use of low-sulfate freshwater, provides substantial support for the importance of sulfate-reducing bacteria in Ediacaran-type preservation. However, there is little evidence for iron sulfides precipitating as finely grained cements capable of replicating morphology (sensu Gehling 1999). Instead, sediments and some clay minerals appear to have been stabilized by microbial EPS (Darroch et al. 2012), which may play a critical role in early cementation of grains and preservation of morphological detail (Briggs and Kear 1994; Wilby et al. 1996; Briggs et al. 2005; Laflamme et al. 2011).

The progression of decay in vertebrates has also been assessed in the context of microbial mats. Neon tetra carcasses were maintained on microbial mats and control sediment over a 27-month period (Iniesto et al. 2013). Control fish exhibited advanced decay by 15 days, and by day 50 the entire specimen was readily disarticulated. By comparison, structural integrity of mat fish was stable between 7 and 30 days; organized scales and tegument persisted at least 3 months in the mat fish, whereas control fish were almost entirely decomposed to a few fragmentary skeletal remains by the end of three months (Iniesto et al. 2013). The proliferation of the microbial mat over decaying fish had significant implications for preservation: by day 30, carcasses had been almost entirely covered by cyanobacterial filaments, which thickened over time such that by day 270, fish carcasses were immersed in the mat to the depth of transition zone between oxic and anoxic layers, with full incorporation into the anoxic layer by day 540 (Iniesto et al. 2013). SEM-EDX analyses also demonstrated that despite Ca- and Mg-enriched water chemistry, the major mineral precipitated in experimental mats was calcium carbonate, with spherules appearing in localized patches by day 7 and a thin film of calcium carbonate covering the whole carcass by day 15. Iniesto et al. (2013) also examined several decayed microbial mat carcasses using magnetic resonance imaging, which revealed that internal skeletal organization and soft organs were readily apparent even at day 270, indicating that immersion in the mat inhibited decomposition. This experiment illustrated that microbial mats directly contribute to preservation of fish in two major ways: formation of a cyanobacterial sarcophagus which prevents disarticulation and inhibits decay and early biostratinomic precipitation of a calcium-rich film (Iniesto et al. 2013).

3.6.4.2 Trackways

Microbial mats have long been understood as integral to preservation of vertebrate trackways and footprints (Thulborn 1990; Conti et al. 2005; Marty et al. 2009). However, there has been only one actualistic examination of track preservation in modern microbial mats (Marty et al. 2009), despite the fact that debate in vertebrate ichnology (study of trace fossils) often hinges on whether a track is a primary imprint, an underprint, or an overprint. Primary imprints, discernable by the presence of skin or claw impressions, are relatively rare in the fossil record. By contrast, underprints, where the act of impression distorts underlying sediments which then lithify, are thought to have been readily incorporated into the fossil record (Lockley 1991). Overtracks, on the other hand, are formed when tracks are stabilized by microbial mats and are then infilled with sediments; as this may happen multiple times, with each new lamina stabilized by a successive mat, a number of smaller-perimeter and less-detailed surfaces will develop (Thulborn 1990). Tracks preserved in this fashion could break along any fissile plane, and the exposed surface might therefore represent a later infill and not the original impression surface. Observations of footprints along modern tidal flats suggest that most of the vertebrate fossil track record comprises modified true tracks and overtracks (Marty et al. 2009). Furthermore, footprints are generally only produced during wet periods; when dry, microbial mats are resistant to pressures, and even large tracemakers will leave no impression. Because microbial mats are densely consolidated once dried, they rarely disintegrate even under heavy rainfall, and surfaces which retain impressions are therefore those which are subject to repeated inundations, rarely drying (Marty et al. 2009). This suggests that tracks represent narrow taphonomic windows—those on the same bedding plane are likely to have been emplaced within hours to a few days of each other. Understanding essential factors leading to fossilization of tracks has important implications for biomechanical modeling and behavioral interpretations (e.g., Currie and Sarjeant 1979; Lockley 1986).

3.6.5 Microbial Interactions with Bone

Microbial degradation of buried bones is not necessarily immediate (Hedges 2002), especially in waterlogged and humid sites (Jans et al. 2004). Because most environments are not geochemically stable with respect to the apatite (calcium phosphate) they contain, bones are not in thermodynamic equilibrium with soil solution, and biological attack generally precedes demineralization (Collins et al. 2002). Under normal environmental parameters, collagen is recalcitrant (Collins et al. 2002). Deterioration of collagen, which is enhanced by a variety of microbial collagenases (McDermid et al. 1988; Harrington 1996), follows demineralization (Collins et al. 2002). Although the low permeability and high mineral content of bone can initially inhibit substantial decomposition prior to demineralization (Child 1995), microbial "bone taint" can begin as early as 6 h postmortem (Roberts and Mead 1985), owing to proliferation of gut and muscle flora, which initially propagate along foramina housing vessels and nerves (Child 1995). Soft tissue decay reduces pH, accelerating demineralization that continues as bones are incorporated into sediment (Child 1995). In addition to bacteria, common soil fungi, including *Mucor*, are also capable of producing microscopic focal destructions (Child 1995). Mycelial proliferation may facilitate bacterial proliferation within bone because fungal hyphae readily cross voids in cancellous bone (Daniel and Chin 2010). Fungal proliferation in bone is often inhibited by accumulating NH_4^+ and nitrogen which result from autolysis, microbial degradation, and acid hydrolysis of collagen (Child 1995).

Microbial decay is the ultimate fate of most bone incorporated into sediment, and several studies have demonstrated that biofilm establishment is imperative for preservation. Carbonates can substitute in bone lattice for PO_4^{3-} (Timlin et al. 2000). Carpenter (2005) demonstrated that carbonate minerals develop on bone when it is exposed to calcium carbonate solutions in the presence of bacteria, whereas $CaCO_3$ did not precipitate in sterile solution. Daniel and Chin (2010) built on this study, defining the temporal window in which precipitation commences. Cubes of bone which were not surface-sterilized prior to suspension in a sand matrix exhibited extensive mineral deposition, particularly within interior cancellous bone. Samples that had been washed, but not sterilized, showed a similar pattern, although interior deposition was lower. By contrast, sterilized samples showed little deposition that was limited to the exterior surfaces (Daniel and Chin 2010). These studies confirm the importance of biofilms that both induce mineral precipitation and trap sediment around bone and also indicate that early diagenetic mineralization is more likely to initiate in cancellous bone than compact bone (Daniel and Chin).

Bones deposited in marine settings are subject to colonization by anaerobic-oxidizing archaeans and sulfate-reducing bacteria (Shapiro and Spangler 2009). High-sulfide concentrations occur in the proximity of skeletons as a result of oxidation of lipids by sulfate-reducing bacteria, and carbonates precipitate readily (Allison et al. 1991; Shapiro and Spangler 2009). Epifluorescence bacterial counts show that highest concentrations of bacteria occur at bone surfaces, with no statistical difference between bone buried in sediment or exposed at the sediment–water interface (Deming et al. 1997). Rich mats of microbial biofilms cover up to 50 % of the bone surface, and destruction of the outer edges of bones is facilitated by bacterial boring (Allison et al. 1991). Features consistent with microboring and bacteriogenic precipitation of carbonates and iron sulfides have been observed in fossil whale falls from the Eocene through Pleistocene and in some Cretaceous plesiosaurs (Amano and Little 2005; Shapiro and Spangler 2009). Aragonite with botryoidal fabrics, associated with methane oxidation (Campbell et al. 2002), also occurs in some specimens (Shapiro and Spangler 2009).

3.7 Microbial Symbioses in the Fossil Record

Symbiotic associations of different trophic consequences (mutualistic, parasitic, etc.) involving microbes continue to be discovered in high numbers in all modern ecosystems, and they are probably even more widely spread than we could imagine. Therefore, it is not surprising that their fossil counterparts are also being discovered at high rates in the rock record. As a result, many symbiotic associations known from the modern biota have been reported from the fossil record. Some of these associations fit into well-circumscribed categories, such as the lichen symbioses and mycorrhizal associations, which are very briefly addressed here because of the imminent publication of a book that will provide an exhaustive treatment of the fungal fossil record (Taylor et al. 2014). Other associations described in the fossil record belong to more loosely circumscribed types, such as the endophytic syndrome or the bacterial-cyanobacterial mat consortia (the oldest record of which has been described in ca. 440 Ma Silurian cyanobacterial colonies; Tomescu et al. 2008).

3.7.1 Lichen Symbioses

Lichens are symbioses between a phylogenetically heterogeneous assemblage of mycobionts—predominantly ascomycete fungi (Gargas et al. 1995; Grube and Winka 2002; Prieto and Wedin 2013)—and photobionts, the majority of which are chlorophyte algae (Honegger 2009). A smaller number of mycobionts, perhaps 10 %, form symbioses with cyanobacteria (Honegger 2009). The lichenized habit has been independently gained and lost multiple times (Lutzoni et al. 2001, 2004;

Nelsen et al. 2009), and few characters used in traditional morphological classification are phylogenetically informative (e.g., Stenroos and DePriest 1998; Grube and Kroken 2000). Thus, although the fossil record of lichens [comprehensively reviewed by Rikkinen and Poinar (2008) and Matsunaga et al. (2013)] is stratigraphically extensive, the phylogenetic affinities of most fossil lichen symbionts cannot be determined with confidence. Several Paleogene specimens preserved in amber have been attributed to living lineages (e.g., Peterson 2000; Poinar et al. 2000; Rikkinen 2003; Rikkinen and Poinar 2002, 2008), but few fossils exhibit the key reproductive features that would allow identification of the mycobiont.

The oldest record of a putative lichen is a ca. 585 Ma phosphatized algal mat from the Doushantuo Formation containing several 0.5–0.9 µm wide filaments that are interpreted as coenocytic hyphae with terminal sporulation, thought to be comparable to those of the glomeromycete *Geosiphon* (Yuan et al. 2005), which forms an endosymbiotic association with *Nostoc* cyanobacteria (Gehrig et al. 1996; Kluge et al. 2003). Yuan et al. (2005) interpreted these Ediacaran hyphae as evidence of a presymbiotic syndrome defined by facultative use of algal products by a marine fungus. The fungal affinities of the filaments are, however, rather suspect due to their small diameter: hyphae of *Geosiphon* are significantly larger, as are those of Mucoromycota (Schüssler and Kluge 2001; Deacon 2006). New techniques, including Raman spectroscopy or time-of-flight mass spectrometry, and confocal laser scanning microscopy may provide more insight into the nature of these earliest lichen-like specimens. No other evidence exists for lichen symbioses prior to the Devonian: suggestions that some Ediacaran fossils conventionally considered metazoans instead represent lichens (Retallack 1994, 2013) have been refuted (Antcliffe and Hancy 2013a, b; Xiao et al. 2013).

Other contenders for the earliest lichens include those described from early Devonian of Scotland and Wales. Honneger et al. (2013a, b) have described two fossil lichens preserved in siltstone from the Welsh borderlands; the specimens consist of septate hyphae of probable ascomycete affinity, which hosted cyanobacteria and unicellular chlorophyte algae. As in some extant lichens (Cardinale et al. 2006; Grube and Berg 2009), the latter also appears to have been colonized by non-photosynthetic bacteria, including actinomycetes (Honegger et al. 2013a). More important to our understanding of the fossil record of lichen symbioses, both Welsh lichens are dorsiventrally organized with internal stratification consistent with modern lichens. This stands in contrast to another Devonian structure, the enigmatic *Winfrenatia reticulata* of Scotland's Rhynie Chert. *Winfrenatia* was described as a morphologically primitive crustose cyanolichen (Taylor et al. 1995a, 1997). It lacks internal stratification, and cyanobacteria are thought to have been housed in depressions pocking the surface of the thallus (Taylor et al. 1995a, 1997).

As described, the architecture of *Winfrenatia* is unlike that of any living lichen (Taylor et al. 1997; Honegger et al. 2013b). Additional specimens of *Winfrenatia*, described by Karatygin et al. (2009), suggest that much of the thallus is composed of sheaths of extracellular polymeric substances of filamentous biofilm-forming cyanobacteria. While some modern lichens contain multiple photobiont species, several authors have suggested that *Winfrenatia* instead represents opportunistic fungal parasitism of cyanobacterial colonies (Poinar et al. 2000; Karatygin et al. 2009). While this interpretation accounts for multiple cyanobacteria, and the lack of internal stratification, the three-dimensional complexity of the specimens remains problematic. We suggest instead that *Winfrenatia* represents a microbial mat with intrinsic fungal biota. Cantrell and Duval-Pérez (2012) have isolated 43 species of fungi from a hypersaline microbial mat, including *Aspergillus*, *Cladosporium*, and *Acremonium* species. Microbial mats are composed of aggregated biofilms, which are themselves highly structured multispecies communities with complex internal architecture: most cells are arranged in sessile microcolonies surrounded by EPS and separated by minute water channels (Costerton et al. 1995; Stoodley et al. 2002). Such channelization would have aided percolation of silica-rich water of the hydrothermal pools in which the Rhynie Chert was deposited (Rice et al. 2002). Furthermore, microbial mats have complex three-dimensional morphology resulting from desiccation, gas production, and water flow (Gerdes et al. 1993), accounting for the pocket-like depressions evident in *Winfrenatia*.

3.7.2 Mycorrhizal Symbioses

Of the microbes that colonize root tissue, mycorrhizal fungi in particular are thought to have been integral to the evolution of land plants and their successful exploitation of terrestrial soils (Pirozynski and Malloch 1975; Humphreys et al. 2010; Wang et al. 2010; Bidartondo et al. 2011). Three highly conserved genes (DMI1, DMI3, and IPD3) found in all major land plant lineages are necessary for mycorrhizal formation, suggesting that this symbiotic relationship evolved with the common ancestor of liverworts and vascular plants (Wang et al. 2010). Arbuscular mycorrhizal fungi are known from at least the Ordovician (Redecker et al. 2000) and were highly diverse by the Devonian (e.g., Remy et al. 1994; Taylor et al. 1995b; Taylor and Taylor 2000; Dotzler et al. 2006, 2009; Garcia Massini 2007; Krings et al. 2012; Strullu-Derrien et al. 2014). Despite the ephemeral nature of absorptive arbuscules, they have been observed in numerous fossil plants, including the Triassic *Antarcticycas* (Phipps and Taylor 1996) and the seed fern *Glossopteris* (Harper et al. 2013), as well as the Eocene conifer *Metasequoia milleri* (Stockey et al. 2001). The fossil record for ectomycorrhizae, on the other hand, is exceedingly sparse: a *Suillus*- or *Rhizopogon*-like fungus is known from the Eocene Princeton Chert, where it formed an ectomycorrhizal association with the extinct pine, *Pinus arnoldii* (LePage et al. 1997; Klymiuk et al. 2011).

3.7.3 Microbial Endophytes

In paleobotanical and paleomycological literature, it has become common practice to use the term "endophyte" to refer to any fossil microbe occurring within plant tissues (Krings et al. 2009). This usage is intended to be purely descriptive (Krings pers. comm.) and does not imply ecology of the microbe in question. Mycologists and microbial ecologists, however, use the term in an explicitly ecological context, defining an endophyte as a microbe that grows asymptomatically within its host plant (for a discussion of endophyte definitions, see Stone et al. 2000); mycorrhizal fungi and nitrogen-fixing bacteria, while endophytic, are not always classified within the "endophyte catch-all." While some endophytes may be engaged in cryptic mutualism with their hosts, as has been hypothesized of some dark septate endophytes (a heterogeneous assemblage of predominantly ascomycetous fungi), others may be latent pathogens or become saprotrophs upon the death of their host (Jumpponen and Trappe 1998; Saikkonen et al. 1998; Jumpponen 2001; Rodriguez et al. 2009; Maciá-Vicente et al. 2009; Newsham 2011). Some fossil fungi described as endophytes (as defined by Krings et al. 2009) elicited host responses characteristic of infection or parasitism (e.g., Krings et al. 2007; Schwendemann et al. 2010; Taylor et al. 2012). Recently, a study of the Early Devonian vascular plant *Horneophyton lignieri*, eliciting comparisons with modern basal land plants (embryophytes), has demonstrated the presence of endophytes belonging to two major fungal lineages, the Glomeromycota and Mucoromycotina, and revealed previously undocumented diversity in the fungal associations of basal embryophytes (Strullu-Derrien et al. 2014).

Endophytes, excluding mycorrhizal symbionts, have yet to be conclusively demonstrated in the fossil record. A cyanobacterial "endophyte" has also been described, from the Rhynie chert: cyanobacterial colonization of *Aglaophyton major*, a nonvascular plant that was extensively colonized by arbuscular mycorrhizae, has been observed in sections cut from two blocks of chert (Krings et al. 2009). The presence of aquatic species, including the charophyte alga *Palaeonitella*, indicates that these blocks represent a part of the system that experienced sustained inundation. Cyanobacteria within the specimens have morphology consistent with living Oscillatoriales and appear to have colonized the tissue by invading via stomata (Krings et al. 2009). While acknowledging that the specimens exhibited no explicit evidence for mutualism, Krings et al. (2009) suggest that they may represent a model for precursory or initial stages of a mutualistic interaction. Although some extant plants like cycads are known to form stable mutualisms with N_2-fixing cyanobacteria, these photobionts are usually *Anabaena* or *Nostoc* (Rai 1990; Costa et al. 1999; Adams and Duggan 2008). By contrast, a number of oscillatorian cyanobacteria are known to produce toxins (Chorus and Bartram 1999), including microcystins, which have inhibitory effects on plant growth, photosynthetic capacity, and seedling development (McElhiney et al. 2001).

In the course of paleomycological investigations of the Eocene Princeton Chert of British Columbia, Canada, Klymiuk et al. (2013b) described vegetative mycelia and microsclerotia characteristic of some dark septate endophytes (e.g., Currah et al. 1988; Ahlich and Sieber 2006; Fernández et al. 2008; Stoyke and Currah 1991). Intracellular microsclerotia were found in the outer cortex of the aquatic angiosperm *Eorhiza arnoldii* (Klymiuk et al. 2013b); extant dark septate endophytes also produce microsclerotia within host cortex, typically in response to stress or host senescence (Fernando and Currah 1995; Jumpponen and Trappe 1998; Barrow 2003). Because the host-fungus interface of living dark septate endophytes involves a network of nonchitinous mucilaginous hyphae intimately associated with host sieve elements (Barrow 2003), Klymiuk et al. (2013b) indicated that it is unlikely that this interface will be observed in the fossil record. Although conidiogenesis can be diagnostic for a number of root endophytes (Fernando and Currah 1995; Addy et al. 2005), Klymiuk et al. (2013b) did not observe conidia in association with the putative endophytes. Furthermore, at the time of preservation, the host tissue was probably moribund, and it is possible that the fungi represent saprotrophs.

3.8 Future Directions

Studies of microbial fossils are poised to reveal the timing of the advent of cellular life and to contribute to understanding of the early evolution of life and its role as a component of Earth systems; additionally, they can illuminate the origin and early evolution of eukaryotes, as well as clarify aspects of the genesis of fossils as records of past life and of the evolution of symbioses involving microbial participants. These contributions are important as the geologic record provides the only direct evidence and, thus, independent tests for hypotheses on the timing and tempo of events and processes that otherwise can only be inferred based on the modern earth systems and biota.

Looking into the future, it is immediately apparent that continued work in the field and in the lab to document in more detail known fossil occurrences and to identify new fossil localities, as well as to recognize more potential fossils and confirm them as bona fide microfossils, will always have their place in the study of the microbial fossil record. In discussing the fossil record of Archean microbial life, Knoll (2012) reiterated the general acceptance of the fact that life existed at least as far back as 3.5 Ga and suggested two major areas of inquiry for future research. One of these involves continued discovery and application of analytical tools to resolve the biogenicity of increasingly older putative fossils and to elucidate the physiology or phylogenetic relationships of the earliest life forms. But studies coming from the opposite direction, that of modern microbes in their host ecosystems, are also needed. Such studies will lead to better understanding of the roles and products of microbial components in the chemical cycles of different ecosystems and in different geologic or petrologic contexts. The findings can lead to the development of new methods to more reliably assess indigenousness, syngenicity, and biogenicity of putative body fossils and to unequivocally identify microbial fossils of all types (stromatolites, alteration textures, etc.) even in the absence of body fossils, which will ultimately improve our ability to trace the trajectory of microbial life through the rock record. Such actualistic studies also benefit from the application of cutting-edge analytical tools and can point the way to applications in fossil contexts. For example, Schmid et al. (2014) used a combination of advanced complementary three-dimensional microscopy tomography techniques (focused ion beam—scanning electron microscopy tomography, transmission electron microscopy tomography, scanning-transmission X-ray microscopy tomography, and confocal laser scanning microscopy) to characterize bacterial cell-(iron) mineral aggregates formed during Fe(II) oxidation by nitrate-reducing bacteria. Their study showed that only in combination did the different techniques provide a comprehensive understanding of structure and composition of the various precipitates and their association with bacterial cells and EPS; such an approach is directly applicable to the discovery and characterization of similar structures and relationships in the fossil record.

Another major area of inquiry identified by Knoll (2012) concerns the rise of cyanobacteria and aerobic photosynthesis in terms of the timing of these events, as they relate to the oxygenation of the atmosphere to stable levels. In this context, Knoll asks whether cyanobacteria (aerobic photosynthesizers) could be counted among the primary producers in Neoarchean (2.8–2.5 Ga) ecosystems. This is relevant to the issue of small positive oscillations recorded in atmospheric oxygen concentrations toward the end of the Neoarchean, before the 2.5–2.3 Ga Great Oxidation Event, in a context in which documented biosignatures (stromatolites, stable isotopes, hydrocarbon biomarkers) don't exclude the presence of cyanobacteria, but they don't require it either (Knoll 2012). Some answers may come from studies such as that recently published by Planavsky et al. (2014) who document chemical biosignatures for oxygenic photosynthesis in 2.95 Ga Sinqeni Formation (Pongola Supergroup, South Africa), at least a half billion years before the Great Oxidation Event. In the Sinqeni Formation, rocks deposited in a nearshore environment yielded molybdenum isotopic signatures consistent with interaction with manganese oxides, which imply presence of oxygen produced through oxygenic photosynthesis.

More precise constraints on the dating of environmental changes are also needed for the Proterozoic, to draw less tentative conclusions on the causes and mechanisms of early eukaryote evolution and diversification. This was pointed out by Javaux (2007), who reviewed the different ideas proposed to explain the observed pattern of diversification of early eukaryote-like fossils, concluding that no event in particular explained it. For example, some unanswered questions are whether early eukaryotes diversifying in the marine realm displaced a preexisting cyanobacterial biota or evolved in an ecologically undersaturated environment and whether eukaryotes remained in minority while diversifying or they quickly formed eukaryote-dominated communities (Knoll and Awramik 1983), or when eukaryotes did invade the continents. In terms of recognition of early (unicellular) eukaryotes, Peat et al. (1978) discussed critically the value of an actualistic approach which can bias interpretations of ancient microfossils under the assumption that prokaryotes in the Precambrian were morphologically similar to modern prokaryotes and suggested that it is not out of the realm of possibility to discover disproportionately large (eukaryote-like) prokaryote microfossils in Precambrian rocks.

Like in the case of prokaryote evolution, answers to questions concerning early eukaryotes will come both from continued studies of the fossil record and from better understanding of their taphonomy and modes of preservation based on actualistic studies. Early on, Golubic and Barghoorn (1977) emphasized the need for ultrastructural studies of diagenetic alteration in the cell walls of extant microorganisms, for application in the microbial fossil record. Conversely yet convergently, Javaux et al. (2003) pointed out that taphonomic studies [like those conducted by Knoll and Barghoorn (1975) or Bartley (1996) on prokaryotes] are needed to elucidate the probability of preservation of intracellular—components such as pyrenoids, starch, and cytoplasm—which, contrary to conventional wisdom, is far from vanishingly small (e.g., Bomfleur et al. 2014). By improving understanding of fossils and their mode of formation, such studies will lead to more detailed and accurate interpretations of the fossils' implications for the taxonomy, ecology, and evolution of ancient microbes.

Microbially induced sedimentary structures have entered the sphere of interest of Precambrian biological evolution studies relatively recently, benefiting fully from well-developed and articulated concepts of sedimentology, an outlook emphasizing actualistic studies, and the breadth of modern environments in which they are formed. Recently, Wilmeth et al. (2014) documented domal sand structures of putative microbial origin in the 1.09 Ga Copper Harbor Conglomerate of Michigan, expanding the fossil record of continental microbialites to siliciclastic fluvial environments.

In contrast to MISS, stromatolites, although recognized and studied for more than a century, lack extensive modern analogues and have ranked among the more contentious Precambrian fossil biosignatures. It is therefore exciting to note a resurgence of actualistic studies addressing stromatolite structures and the microbial communities that build them from several perspectives. Kremer et al.'s (2012a) studies of microbial taphonomy and fossilization potential in modern stromatolites in Tonga are such a study (see "Stromatolites" in sect. 3.2.2.1). Related to this, a study by Knoll et al. (2013) documented in detail both the composition of microfossil assemblages and the sedimentary structures (petrofabrics) in Mesoproterozoic carbonate platform microbialites of the Angmaat/Society Cliffs Formation (Baffin and Bylot Islands). Their study revealed covariation of microfossil assemblages with petrofabrics, supporting hypotheses that link stromatolite microstructure to the composition and diversity of microbial mat communities.

Mirroring Knoll et al.'s (2013) work in modern microbialites, Russell et al. (2014) studied the microbial communities building microbialites in Pavilion Lake (Canada). Using molecular analyses, Russell et al. (2014) documented diverse communities including phototrophs (cyanobacteria) as well as heterotrophs and photoheterotrophs. They also showed that the microbialite-building communities are more diverse than the non-lithifying microbial mats in the lake and that microbial community composition does not correlate with depth-related changes in microbialite morphology, suggesting that microbialite structure may not be under strict control of microbial community composition. Finally, in a study of extracellular polymeric substances and functional gene diversity within biofilm communities of modern oolitic sands (analogous in genesis to stratiform stromatolites) from Great Bahama Bank, Diaz et al. (2014) suggest that carbonate precipitation in marine oolitic biofilms is spatially and temporally controlled by a consortium of microbes with diverse physiologies (photosynthesizers, heterotrophs, denitrifiers, sulfate reducers, and ammonifiers) and point to a role of EPS-mediated microbial calcium carbonate precipitation in the formation of the microlaminated oolitic structures.

Long time considered ill positioned to make independent contributions beyond merely documenting historical confirmation of events and processes, paleontology (including Precambrian paleobiology; Schopf 2009) witnessed in the 1970s–1980s the "paleobiological revolution" that reinstated it at the "high table" of evolutionary biology (Sepkoski and Ruse 2009). Building on this newfound identity and adding to it a developmental anatomy - comparative morphology twist, today paleontology is poised to make meaningful contributions to the field of evolutionary-developmental biology by documenting in fossils and tracing through time the anatomical and morphological fingerprints of genetic pathways that regulate development (e.g., Rothwell et al. 2014). In this context, a recent study by Flood et al. (2014) inspires some exciting ideas. These authors used comparative genomics to study phylogenetically distant bacteria that induce formation of wrinkle structures (a type of MISS) in modern sediments. Their results suggest that horizontally transferred genes may code for phenotypic traits that underlie similar biostabilizing influences of these organisms on sediments. On the one hand, this implies that the ecological utility of some phenotypic traits such as the construction of mats and biofilms, along with the lateral mobility of genes in the microbial world, render inferences of phylogenetic relationships from gross morphological features preserved in the rock record uncertain (Flood et al. 2014). On the other hand, this study expands the range of phenotypic traits that can be used as morphological fingerprints for shared genetic pathways, to the realm of micro- and macroscale sedimentary structures.

Deep in the rock record, the fossil evidence may not look spectacular by most standards—tiny microfossils, wrinkles on a rock face, readings of chemical composition on a computer screen, or a spectrometer curve. Yet the studies of the microbial fossil record can have tremendous outcomes, as they can bring key contributions to addressing two of the most profound and perennial questions that have puzzled humanity and science. On the one hand, tracing the microbial fossil record is our only direct way of catching a glimpse of the beginnings and early evolution of life, and this chapter has attempted to provide an introduction and overview of the paradigms that underpin such studies. On the other hand, the methods and ideas developed as a result of studies of the microbial fossil record for recognizing ancient and inconspicuous traces of life, as well as the knowledge and experience accumulated in the process, have crystallized in an approach that is directly applicable to the search for traces of life on other planets (e.g., Brasier and

Wacey 2012). Although the connections of paleomicrobiological work with astrobiology are not explored in this chapter, it is noteworthy that this relatively new field of research has already produced an impressive body of publications, including several books (e.g., Seckbach and Walsh 2009), and has two dedicated journals (*Astrobiology* and *International Journal of Astrobiology*) which host some of the references cited throughout this chapter. It is only fitting to conclude, then, that from the deepest reaches of time and of Earth's crust, to the landscapes of other worlds, paleomicrobiology can open unprecedented perspectives in the study of life.

Acknowledgments We are indebted to Emma Fryer for momentous help with obtaining permissions from publishers to use copyrighted material.

References

Adams DG, Duggan PS (2008) Cyanobacteria-bryophyte symbioses. J Exp Bot 59:1047–1058
Addy HD, Piercey MM, Currah RS (2005) Microfungal endophytes in roots. Can J Bot 83:1–13
Ahlich K, Sieber TN (2006) The profusion of dark septate endophytic fungi in non-ectomycorrhizal fine roots of forest trees and shrubs. New Phytol 132:259–270
Aitken JD (1967) Classification and environmental significance of cryptalgal limestones and dolomites, with illustrations from the Cambrian and Ordovician of southwestern Alberta. J Sed Petrol 37:1163–1178
Allison PA (1988a) The role of anoxia in the decay and mineralization of proteinaceous macro–fossils. Paleobiology 14:139–154
Allison PA (1988b) Taphonomy of the Eocene London clay biota. Palaeontology 31:1079–1100
Allison PA, Smith CR, Kukert H et al (1991) Deep-water taphonomy of vertebrate carcasses: a whale skeleton in the bathyal Santa Catalina Basin. Paleobiology 17:78–89
Allwood AC, Walter MR, Burch IW et al (2007) 3.4 billion-year-old stromatolite reef from the Pilbara Craton of Western Australia: ecosystem-scale insights to early life on Earth. Precambrian Res 158:198–227
Alt JC, Mata P (2000) On the role of microbes in the alteration of submarine basaltic glass: a TEM study. Earth Planet Sci Lett 181:301–313
Altermann W, Schopf JW (1995) Microfossils from the Neoarchean Campbell Group, Griqualand West Sequence of the Transvaal Supergroup, and their paleoenvironmental and evolutionary implications. Precambrian Res 75:65–90
Altermann W, Kazmierczak J, Oren A et al (2006) Cyanobacterial calcification and its rock-building potential during 3.5 billion years of Earth history. Geobiology 4:147–166
Amano K, Little CT (2005) Miocene whale-fall community from Hokkaido, northern Japan. Palaeogeogr Palaeoclimat Palaeoecol 215:345–356
Anders E (1996) Evaluating the evidence for past life on Mars. Science 274:2119–2121
Anderson EP, Schiffbauer JD, Xiao S (2011) Taphonomic study of Ediacaran organic–walled fossils confirms the importance of clay minerals and pyrite in Burgess Shale-type preservation. Geology 39:643–646
Antcliffe JB, Hancy AD (2013a) Critical questions about early character acquisition—comment on Retallack 2012: Some Ediacaran fossils lived on land. Evol Dev 15:225–227

Antcliffe JB, Hancy AD (2013b) Reply to Retallack (2013): Ediacaran characters. Evol Dev 15:389–392
Antcliffe JB, McLoughlin N (2009) Deciphering fossil evidence for the origin of life and the origin of animals: common challenges in different worlds. In: Seckbach J, Walsh M (eds) From fossils to astrobiology. Springer, Dordrecht, pp 211–229
Appel PWU, Moorbath S, Myers JS (2003) *Isuasphaera isua* (Pflug) revisited. Precambrian Res 126:309–312
Arning ET, Birgel D, Brunner B et al (2009) Bacterial formation of phosphatic laminites off Peru. Geobiology 7:295–307
Arning ET, Birgel D, Schulz-Vogt HN et al (2008) Lipid biomarker patterns of phosphogenic sediments from upwelling regions. Geomicrobiol J 25:69–82
Arouri K, Greenwood PF, Walter MR (1999) A possible chlorophycean affinity of some Neoproterozoic acritarchs. Org Geochem 30:1323–1337
Arouri KR, Greenwood PF, Walter MR (2000) Biological affinities of Neoproterozoic acritarchs from Australia: microscopic and chemical characterisation. Org Geochem 31:75–89
Awramik SM (2006) Respect for stromatolites. Nature 441:700–701
Awramik SM, Buchheim HP (2009) A giant, Late Archean lake system: the Meentheena Member (Tumbiana Formation; Fortescue Group), Western Australia. Precambrian Res 174:215–240
Awramik SM, Barghoorn ES (1977) The Gunflint microbiota. Precambrian Res 5:121–142
Awramik SM, Grey K (2005) Stromatolites: biogenicity, biosignatures, and bioconfusion. In: Gladstone GR, Hoover RB, Levin GV et al. (eds) Astrobiology and Planetary Missions. SPIE Proceedings 5906:1–9
Awramik SM, Margulis L (1974) Stromatolite Newslett 2:5
Awramik SM, Riding R (1988) Role of algal eukaryotes in subtidal columnar stromatolite formation. Proc Nat Acad Sci USA 85:1327–1329
Awramik SM, Schopf JW, Walter MR (1983) Filamentous fossil bacteria from the Archean of Western Australia. Precambrian Res 20:357–374
Bailey JV, Joye SB, Kalanetra KM et al (2007) Palaeontology: undressing and redressing Ediacaran embryos (Reply). Nature 446:E10–E11
Banerjee NR, Simonetti A, Furnes H et al (2007) Direct dating of Archean microbial ichnofossils. Geology 35:487–490
Barghoorn ES, Tyler SA (1965) Microorganisms from the Gunflint chert. Science 147:563–577
Barrow JR (2003) Atypical morphology of dark septate fungal root endophytes of *Bouteloua* in arid southwestern USA rangelands. Mycorrhiza 13:239–247
Bartley JK (1996) Actualistic taphonomy of cyanobacteria: implications for the Precambrian fossil record. Palaios 11:571–586
Battison L, Brasier MD (2012) Remarkably preserved prokaryote and eukaryote microfossils within 1 Ga-old lake phosphates of the Torridon Group, NW Scotland. Precambrian Res 196–197:204–217
Baxter RW (1950) *Peltastrobus reedae*: a new sphenopsid cone from the Pennsylvanian of Indiana. Bot Gaz 112:174–182
Bazylinski DA (1996) Controlled biomineralization of magnetic minerals by magnetotactic bacteria. Chem Geol 132:191–198
Bazylinski DA, Frankel RB (2003) Biologically controlled mineralization in prokaryotes. Rev Miner Geochem 54:217–247
Bekker A, Holland HD, Wang P-L et al (2004) Dating the rise of atmospheric oxygen. Nature 427:117–120
Bengtson S, Belivanova V, Rasmussen B et al (2009) The controversial "Cambrian" fossils of the Vindhyan are real but more than a billion years older. Proc Nat Acad Sci USA 106:7729–7734
Beraldi-Campesi H, Garcia-Pichel F (2011) The biogenicity of modern terrestrial roll-up structures and its significance for ancient life on land. Geobiology 9:10–23
Beraldi-Campesi H, Farmer JD, Garcia-Pichel F (2014) Modern terrestrial sedimentary biostructures and their fossil analogs in Mesoproterozoic subaerial deposits. Palaios 29:45–54

Beveridge TJ (1989) Role of cellular design in bacterial metal accumulation and mineralization. Ann Rev Microbiol 43:147–171

Bidartondo MI, Read DJ, Trappe JM et al (2011) The dawn of symbiosis between plants and fungi. Biol Lett 7:574–577

Boal D, Ng R (2010) Shape analysis of filamentous Precambrian microfossils and modern cyanobacteria. Paleobiology 36:555–572

Bomfleur B, McLoughlin S, Vajda V (2014) Fossilized nuclei and chromosomes reveal 180 million years of genomic stasis in royal ferns. Science 343:1376–1377

Boyce CK, Hazen RM, Knoll AH (2002) Non-destructive, in situ, cellular-scale mapping of elemental abundances including organic carbon in permineralized fossils. Proc Nat Acad Sci USA 98:5970–5974

Brack-Hanes SD, Vaughn JC (1978) Evidence of Paleozoic chromosomes from lycopod microgametophytes. Science 200:1383–1385

Bradley JP, Harvey RP, McSween HY Jr (1997) No 'nannofossils' in martian meteorite. Nature 390:454–455

Bradley JP, McSween HY Jr, Harvey RP (1998) Epitaxial growth of nanophase magnetite in Martian meteorite ALH 84001: implications for biogenic mineralization. Meteorit Planet Sci 33:765–773

Brasier MD, Wacey D (2012) Fossils and astrobiology: new protocols for cell evolution in deep time. Int J Astrobiol 11:217–228

Brasier MD, Green OR, Jephcoat AP et al (2002) Questioning the evidence for Earth's oldest fossils. Nature 417:76–81

Brasier M, Green O, Lindsay J et al (2004) Earths oldest (~3.5 Ga) fossils and the 'early eden hypothesis': questioning the evidence. Origins Life Evol B 34:257–269

Brasier M, McLoughlin N, Green O et al (2006) A fresh look at the fossil evidence for Early Archaean cellular life. Phil Trans R Soc B 361:887–902

Bridgewater D, Allaart JH, Schopf JW et al (1981) Microfossil-like objects from the Archean of Greenland: a cautionary note. Nature 289:51–53

Briggs DE (1999) Molecular taphonomy of animal and plant cuticles: selective preservation and diagenesis. Phil Trans R Soc Lond B354:7–17

Briggs DE (2003a) The role of decay and mineralization in the preservation of soft-bodied fossils. Ann Rev Earth Planet Sci 31:275–301

Briggs DE (2003b) The role of biofilms in the fossilization of non-biomineralized tissues. In: Krumbein WE, Paterson DM, Zavarzin GA (eds) Fossil and recent biofilms. Kluwer, Dordrecht, pp 281–290

Briggs DE, Kear AJ (1993) Fossilization of soft tissue in the laboratory. Science 259:439–1442

Briggs DE, Kear AJ (1994) Decay and mineralization of shrimps. Palaios 9:431–456

Briggs DE, Wilby PR (1996) The role of the calcium carbonate-calcium phosphate switch in the mineralization of soft-bodied fossils. J Geol Soc 153:665–668

Briggs DEG, Kear AJ, Martill DM et al (1993) Phosphatization of soft-tissue in experiments and fossils. J Geol Soc 150:1035–1038

Briggs DE, Moore RA, Shultz JW et al (2005) Mineralization of soft-part anatomy and invading microbes in the horseshoe crab *Mesolimulus* from the Upper Jurassic Lagerstätte of Nusplingen, Germany. Proc R Soc B: Biol Sci 272:627–632

Briggs DE, Wilby PR, Pérez-Moreno BP et al (1997) The mineralization of dinosaur soft tissue in the Lower Cretaceous of Las Hoyas, Spain. J Geol Soc 154:587–588

Brock F, Parkes RJ, Briggs DE (2006) Experimental pyrite formation associated with decay of plant material. Palaios 21:499–506

Brocks JJ, Logan GA, Buick R et al (1999) Archean molecular fossils and the early rise of eukaryotes. Science 285:1033–1036

Brocks JJ, Buick R, Logan GA et al (2003a) Composition and syngeneity of molecular fossils from the 2.78 to 2.45 billion-year-old Mount Bruce Supergroup, Pilbara Craton, Western Australia. Geochim Cosmochim Acta 67:4289–4319

Brocks JJ, Buick R, Logan GA et al (2003b) A reconstruction of Archean biological diversity based on molecular fossils from the 2.78 to 2.45 billion-year-old Mount Bruce Supergroup, Hamersley Basin, Western Australia. Geochim Cosmochim Acta 67:4321–4335

Büdel B, Weber B, Kühl M et al (2004) Reshaping of sandstone surfaces by cryptoendolithic cyanobacteria: bioalkalization causes chemical weathering in arid landscapes. Geobiology 2:261–268

Buick R (1984) Carbonaceous filaments from North Pole, Western Australia: are they fossil bacteria in Archean stromatolites? Precambrian Res 24:157–172

Buick R (1990) Microfossil recognition in Archean rocks: an appraisal of spheroids and filaments from the 3500 m.y. old chert-barite unit at North Pole, Western Australia. Palaios 5:441–459

Buick R (1992) The antiquity of oxygenic photosynthesis: evidence from stromatolites in sulphate-deficient Archaean lakes. Science 255:74–77

Buick R (2001) Life in the Archean. In: Briggs DEG, Crowther PR (eds) Palaeobiology II. Blackwell Science, Oxford, pp 13–21

Buick R (2010) Ancient acritarchs. Nature 463:885–886

Buick R (2012) Geobiology of the Archean Eon. In: Knoll AH, Canfield DE, Konhauser KO (eds) Fundamentals of geobiology. Wiley-Blackwell, Chichester, pp 351–370

Buick R, Dunlop J (1990) Evaporitic sediments of Early Archaean age from the Warrawoona Group, North Pole, Western Australia. Sedimentology 37:247–277

Buick R, Knoll AH (1999) Acritarchs and microfossils from the Mesoproterozoic Bangemall Group, Northwestern Australia. J Paleontol 73:744–764

Buick R, Dunlop JSR, Groves DI (1981) Stromatolite recognition in ancient rocks: an appraisal of irregularly laminated structures in an Early Archaean chert-barite unit from North Pole, Western Australia. Alcheringa 5:161–181

Burke C, Steinberg P, Rusch D et al (2011) Bacterial community assembly based on functional genes rather than species. Proc Nat Acad Sci USA 108:14288–14293

Burne RV, Moore L (1987) Microbialites: organosedimentary deposits of benthic microbial communities. Palaios 2:241–254

Buseck PR, Dunin-Borkowski RE, Devouard B et al (2001) Magnetite morphology and life on Mars. Proc Nat Acad Sci USA 98:13490–13495

Butler IB, Rickard D (2000) Framboidal pyrite formation via the oxidation of iron (II) monosulfide by hydrogen sulphide. Geochim Cosmochim Acta 64:2665–2672

Butterfield NJ (1995) Secular distribution of Burgess-Shale-type preservation. Lethaia 28:1–13

Butterfield NJ (2000) *Bangiomorpha pubescens* n. gen., n. sp.: implications for the evolution of sex, multicellularity and the Mesoproterozoic/Neoproterozoic radiation of eukaryotes. Paleobiology 26:386–404

Butterfield NJ (2001) Paleobiology of the Late Mesoproterozoic (ca. 1200 Ma) hunting formation, Somerset Island, arctic Canada. Precambrian Res 111:235–256

Butterfield NJ (2002) *Leanchoilia* guts and the interpretation of three-dimensional structures in Burgess Shale-type fossils. Paleobiology 28:155–171

Butterfield NJ (2005a) Probable Proterozoic fungi. Paleobiology 31:165–182

Butterfield NJ (2005b) Reconstructing a complex early Neoproterozoic eukaryote, Wynniatt Formation, arctic Canada. Lethaia 38:155–169

Butterfield NJ, Balthasar U, Wilson LA (2007) Fossil diagenesis in the Burgess Shale. Palaeontology 50:537–543

Butterfield NJ (2009) Modes of pre-Ediacaran multicellularity. Precambrian Res 173:201–2011

Cai Y, Schiffbauer JD, Hua H et al (2012) Preservational modes in the Ediacaran Gaojiashan Lagerstätte: Pyritization, aluminosilicification, and carbonaceous compression. Palaeogeogr Palaeoclimat Palaeoecol 326:109–117

Callot G, Maurette M, Pottier L et al (1987) Biogenic etching of microfractures in amorphous and crystalline silicates. Nature 328:147–149

Callow RH, Brasier MD (2009) Remarkable preservation of microbial mats in Neoproterozoic siliciclastic settings: implications for Ediacaran taphonomic models. Earth Sci Rev 96:207–219

Campbell SE (1982) Precambrian endoliths discovered. Nature 299:429–431

Campbell KA, Farmer JD, Des Marais D (2002) Ancient hydrocarbon seeps from the Mesozoic convergent margin of California: carbonate geochemistry, fluids and palaeoenvironments. Geofluids 2:63–94

Canfield DE, Des Marais DJ (1991) Aerobic sulfate reduction in microbial mats. Science 251:1471–1473

Canfield DE, Thamdrup B (1994) The production of ^{34}S-depleted sulfide during bacterial disproportionation of elemental sulfur. Science 266:1973–1975

Canfield DE, Poulton SW, Knoll AH et al (2008) Ferruginous conditions dominated later Neoproterozoic deep-water chemistry. Science 321:949–952

Canfield DE, Sørensen KB, Oren A (2004) Biogeochemistry of a gypsum-encrusted microbial ecosystem. Geobiology 2:133–150

Cantrell SA, Duval-Pérez L (2012) Microbial mats: an ecological niche for fungi. Front Microbiol 3:1–7

Cardinale M, Puglia AM, Grube M (2006) Molecular analysis of lichen-associated bacterial communities. FEMS Microbiol Ecol 57:484–495

Carpenter K (2005) Experimental investigation of the role of bacteria in bone fossilization. N Jb Geol Palaont Mh 11:83–94

Cavalier-Smith T (2002) The neomuran origin of archaebacteria: the negibacteria root of the universal tree and bacteria megaclassification. Int J Syst Microbiol 52:7–76

Chan CS, De Stasio G, Welch SA et al (2004) Microbial polysaccharides template assembly of nanocrystal fibers. Science 303:1656–1658

Châtellier X, West MM, Rose J et al (2004) Characterization of iron-oxides formed by oxidation of ferrous ions in the presence of various bacterial species and inorganic ligands. Geomicrobiol J 21:99–112

Child AM (1995) Towards and understanding of the microbial decomposition of archaeological bone in the burial environment. J Archaeol Sci 22:165–174

Chin K, Eberth DA, Schweitzer MH et al (2003) Remarkable preservation of undigested muscle tissue within a Late Cretaceous tyrannosaurid coprolite from Alberta, Canada. Palaios 18:286–294

Chorus I, Bartram J (1999) Toxic cyanobacteria in water: a guide to their public health consequences, monitoring and management. Routledge, London

Cleland TP, Stoskopf MK, Schweitzer MH (2011) Histological, chemical, and morphological reexamination of the "heart" of a small Late Cretaceous *Thescelosaurus*. Naturwissenschaften 98:203–211

Clemett SJ, Dulay MT, Gilette JS et al (1998) Evidence for the extraterrestrial origin of polycyclic aromatic hydrocarbons (PAHs) in the Martian meteorite ALH 84001. Faraday Discuss 109:417–436

Clemett SJ, Thomas-Keprta KL, Shimmin J et al (2002) Crystal morphology of MV-1 magnetite. Am Mineral 87:1727–1730

Cloud P (1973) Pseudofossils: a plea for caution. Geology 1:123–127

Cloud PE Jr, Hagen H (1965) Electron microscopy of the Gunflint microflora. Proc Nat Acad Sci USA 54:1–8

Cloud P, Morrison K (1979) On microbial contaminants, micropseudofossils, and the oldest records of life. Precambrian Res 9:81–91

Coleman ML, Raiswell R (1995) Source of carbonate and origin of zonation in pyritiferous carbonate concretions; evaluation of a dynamic model. Am J Sci 295:282–308

Collins MJ, Nielsen-Marsh CM, Hiller J (2002) The survival of organic matter in bone: a review. Archaeometry 44:383–394

Collinson ME, Manchester SR, Wilde V et al (2010) Fruit and seed floras from exceptionally preserved biotas in the European Paleogene. Bull Geosci 85:155–162

Conti MA, Morsilli M, Nicosia U et al (2005) Jurassic dinosaur footprints from southern Italy: footprints as indicators of constraints in paleogeographic interpretation. Palaios 20:534–550

Cosmidis J, Benzerara K, Menguy N et al (2013) Microscopy evidence of bacterial microfossils in phosphorite crusts of the Peruvian shelf: Implications for phosphogenesis mechanisms. Chem Geol 359:10–22

Costerton JW, Lewandowski Z, Caldwell DE et al (1995) Microbial biofilms. Ann Rev Microbiol 49:711–745

Costa JL, Paulsrud P, Lindblad P (1999) Cyanobiont diversity within coralloid roots of selected cycad species. FEMS Microb Ecol 28:85–91

Courties C, Vaquer A, Troussellier M et al (1994) Smallest eukaryotic organism. Nature 370:255

Cunningham JA, Thomas CW, Bengtson S et al (2012a) Distinguishing geology from biology in the Ediacaran Doushantuo biota relaxes constraints on the timing of the origin of bilaterians. Proc R Soc B: Biol Sci 279:2369–2376

Cunningham JA, Thomas CW, Bengtson S et al (2012b) Experimental taphonomy of giant sulphur bacteria: implications for the interpretation of the embryo-like Ediacaran Doushantuo fossils. Proc R Soc B: Biol Sci 279:1857–1864

Currah RS, Hambleton S, Smreciu A (1988) Mycorrhizae and mycorrhizal fungi of *Calypso bulbosa*. Am J Bot 75:739–752

Currie PJ, Sarjeant WA (1979) Lower Cretaceous dinosaur footprints from the Peace River Canyon, British Columbia, Canada. Palaeogeogr Palaeoclim Palaeoecol 28:103–115

Daniel JC, Chin K (2010) The role of bacterially mediated precipitation in the permineralization of bone. Palaios 25:507–516

Darroch SA, Laflamme M, Schiffbauer JD et al (2012) Experimental formation of a microbial death mask. Palaios 27:293–303

Dauphas N, Van Zuilen M, Wadhwa M et al (2004) Clues from Fe isotope variations on the origin of early Archean BIFs from Greenland. Science 306:2077–2080

De Gregorio BT, Sharp TG, Flynn GJ et al (2009) Biogenic origin for Earth's oldest putative microfossils. Geology 37:631–634

Deacon JW (2006) Fungal biology. Wiley-Blackwell, Oxford

Decho AW, Kawaguchi T (2003) Extracellular polymers (EPS) and calcification within modern marine stromatolites. In: Krumbein WE, Paterson DM, Zavarzin GA (eds) Fossil and recent biofilms. Springer, Dordrecht, pp 227–240

Deming JW, Reysenbach AL, Macko SA et al (1997) Evidence for the microbial basis of a chemoautotrophic invertebrate community at a whale fall on the deep seafloor: bone-colonizing bacteria and invertebrate endosymbionts. Microsc Res Tech 37:162–170

Diaz MR, van Norstrand JD, Eberli GP et al (2014) Functional gene diversity of oolitic sands from Great Bahama Bank. Geobiology 12:231–49. doi:10.1111/gbi.12079

Dotzler N, Krings M, Taylor TN et al (2006) Germination shields in *Scutellospora* (Glomeromycota: Diversisporales, Gigasporaceae) from the 400 million-year-old Rhynie chert. Mycol Progr 5:178–184

Dotzler N, Walker C, Krings M et al (2009) Acaulosporoid glomeromycotan spores with a germination shield from the 400-million-year-old Rhynie chert. Mycol Progr 8:9–18

Droser ML, Gehling JG, Jensen SR (2006) Assemblage palaeoecology of the Ediacara biota: the unabridged edition? Palaeogeogr Palaeoclim Palaeoecol 232:131–147

Duck LJ, Glikson M, Golding SD et al (2007) Microbial remains and other carbonaceous forms from the 3.24 Ga Sulphur Springs black smoker deposit, Western Australia. Precambrian Res 154:205–220

Dunlop JSR, Milne VA, Groves DI et al (1978) A new microfossils assemblage from the Archean of Western Australia. Nature 274:676–678

Dunn KA, McLean RJC, Upchurch GR et al (1997) Enhancement of leaf fossilization potential by bacterial biofilms. Geology 25:1119–1122

Dupraz C, Visscher PT (2005) Microbial lithification in marine stromatolites and hypersaline mats. Trends Microbiol 13:429–438

Dupraz C, Reid RP, Braissant O et al (2009) Processes of carbonate precipitation in modern microbial mats. Earth Sci Rev 96:141–162

Dutkiewicz A, Volk H, George SC et al (2006) Biomarkers from Huronian oil-bearing fluid inclusions: an uncontaminated record of life before the Great Oxidation Event. Geology 34:437–440

Efremov IA (1940) Taphonomy: a new branch of paleontology. Pan-Am Geol 74:81–93

Eglinton G, Logan GA, Ambler RP et al (1991) Molecular preservation [and discussion]. Phil Trans R Soc Lond B 333:315–328

El Albani A, Bengtson S, Canfield DE et al (2010) Large colonial organisms with coordinated growth in oxygenated environments 2.1 Gyr ago. Nature 466:100–104

Fang HW, Shang QQ, Chen MH et al (2014) Changes in the critical erosion velocity for sediment colonized by biofilm. Sedimentology 61:648–659

Fedo CM, Whitehouse MJ (2002) Metasomatic origin of quartz-pyroxene rock, Akilia, Greenland, and implications for Earth's earliest life. Science 296:1448–1452

Fernando AA, Currah RS (1995) *Leptodontidium orchidicola* (*Mycelium radicis atrovirens* complex): aspects of its conidiogenesis and ecology. Mycotaxon 54:287–294

Fernández N, Messuti MI, Fontenla S (2008) Arbuscular mycorrhizas and dark septate fungi in *Lycopodium paniculatum* (Lycopodiaceae) and *Equisetum bogotense* (Equisetaceae) in a Valdivian temperate forest of Patagonia, Argentina. Am Fern J 98:117–127

Ferris FG (1993) Microbial biomineralization in natural environments. Earth Sci 47:233–250

Ferris FG, Fyfe WS, Beveridge TJ (1987) Bacteria as nucleation sites for authigenic minerals in a metal-contaminated lake sediment. Chem Geol 63:225–232

Ferris FG, Fyfe WS, Beveridge TJ (1988) Metallic ion binding by *Bacillus subtilis*: implications for the fossilization of microorganisms. Geology 16:149–152

Fisk MR, Giovannoni SJ, Thorseth IH (1998) Alteration of oceanic volcanic glass: textural evidence of microbial activity. Science 281:978–979

Fletcher BJ, Beerling DJ, Chaloner WG (2004) Stable carbon isotopes and the metabolism of the terrestrial Devonian organism *Spongiophyton*. Geobiology 2:107–119

Fliegel D, Kosler J, McLoughlin N et al (2010) In-situ dating of the Earth's oldest trace fossil at 3.34 Ga. Earth Planet Sci Lett 299:290–298

Flood BE, Bailey JV, Biddle JF (2014) Horizontal gene transfer and the rock record: comparative genomics of phylogenetically distant bacteria that induce wrinkle structure formation in modern sediments. Geobiology 12:119–132

Folk RL (1993) SEM imaging of bacteria and nannobacteria in carbonate sediments and rocks. J Sed Petrol 63:990–999

Fortin D, Ferris FG, Beveridge TJ (1997) Surface-mediated mineral development by bacteria. In: Banfield JF, Nealson KH (eds) Geomicrobiology: interactions between microbes and minerals: Reviews in Mineralogy 35. Mineralogical Society of America, Washington, pp 161–180

Frankel RB, Bazylinski DA (2003) Biologically induced mineralization by bacteria. Rev Mineral Geochem 54:95–114

Fregenal-Martínez MA, Buscalioni AD (2010) A holistic approach to the palaeoecology of Las Hoyas Konservat-Lagerstätte (La Huérguina Formation, Lower Cretaceous, Iberian Ranges, Spain). J Iberian Geol 36:297–326

Frey E, Martill DM, Buchy MC (2003) A new species of tapejarid pterosaur with soft–tissue head crest. Geol Soc Lond Spec Publ 217:65–72

Furnes H, Muehlenbachs K (2003) Bioalteration recorded in ophiolitic pillow lavas. Geol Soc Lond Spec Publ 218:415–426

Furnes H, Staudigel H, Thorseth IH et al (2001) Bioalteration of basaltic glass in the oceanic crust. Geochem Geophys Geosyst 2:1049. doi:10.1029/2000GC000150

Furnes H, Banerjee NR, Muehlenbachs K et al (2004) Early life recorded in Archean pillow lavas. Science 304:578–581

Furnes H, Banerjee NR, Staudigel H et al (2007) Comparing petrographic signatures of bioalteration in recent to Mesoarchean pillow lavas: tracing subsurface life in oceanic igneous rocks. Precambrian Res 158:156–176

Gabbott SE (1998) Taphonomy of the Ordovician Soom Shale Lagerstätte: an example of soft tissue preservation in clay minerals. Palaeontology 41:631–668

Gabbott SE, Norry MJ, Aldridge RJ et al (2001) Preservation of fossils in clay minerals; a unique example from the Upper Ordovician Soom Shale, South Africa. Proc Yorkshire Geol Polytech Soc 53:237–244

Gabbott SE, Xian-Guang H, Norry MJ et al (2004) Preservation of Early Cambrian animals of the Chengjiang biota. Geology 32:901–904

Gaines RR, Briggs DE, Yuanlong Z (2008) Cambrian Burgess Shale-type deposits share a common mode of fossilization. Geology 36:755–758

Gaines RR, Kennedy MJ, Droser ML (2005) A new hypothesis for organic preservation of Burgess Shale taxa in the middle Cambrian Wheeler Formation, House Range, Utah. Palaeogeogr Palaeoclim Palaeoecol 220:193–205

Gall JC, Bernier P, Gaillard C et al (1985) Influence du développement d'un voile algaire sur la sédimentation et la taphonomie des calcaires lithographiques. Exemple du gisement de Cerin (Kimméridgien supérieur, Jura méridional français). Mém Phys Chim Sci l'univers, Sci de la Terre 301:547–552

Gall JC (1990) Les voiles microbiens. Leur contribution à la fossilisation des organismes au corps mou. Lethaia 23:21–28

Gaines RR, Hammarlund EU, Hou X, Qi C, Gabbott SE, Zhao Y, Peng J, Canfield DE (2012) Mechanism for Burgess Shale-type preservation. Proc Natl Acad Sci USA 109:5180–5184

Garcia Massini JL (2007) A glomalean fungus from the Permian of Antarctica. Int J Plant Sci 168:673–678

Garcia-Ruiz JM, Hyde ST, Carnerup AM et al (2003) Self-assembled silica-carbonate structures and detection of ancient microfossils. Science 302:1194–1197

Gargas A, DePriest PT, Grube M et al (1995) Multiple origins of lichen symbioses in fungi suggested by SSU rDNA phylogeny. Science 268:1492–1495

Gehling JG (1999) Microbial mats in terminal Proterozoic siliciclastics; Ediacaran death masks. Palaios 14:40–57

Gehling JG, Droser ML, Jensen SR et al (2005) Ediacara organisms: relating form to function. In: Evolving form and function: fossils and development, Symposium Proceedings, Peabody Museum of Natural History, Yale University, New Haven, p 43–67

Gehrig H, Schüssler A, Kluge M (1996) *Geosiphon pyriforme*, a fungus forming endocytobiosis with *Nostoc* (Cyanobacteria), is an ancestral member of the glomales: evidence by SSU rRNA Analysis. J Mol Evol 4:371–81

Gehring AU, Kind J, Charilaou M et al (2011) The detection of magnetotactic bacteria and magnetofossils by means of magnetic anisotropy. Earth Planet Sci Lett 309:113–117

George SC, Volk H, Dutkiewicz A et al (2008) Preservation of hydrocarbons and biomarkers in oil trapped inside fluid inclusions for >2 billion years. Geochim Cosmochim Acta 72:844–870

Gerdes G, Claes M, Dunajtschik-Piewak K et al (1993) Contribution of microbial mats to sedimentary surface structures. Facies 29:61–74

German TN, Podkovyrov VN (2009) New insights into the nature of the Late Riphean *Eosolenides*. Precambrian Res 173:154–162

Giovannoni SJ, Fisk MR, Mullins TD et al (1996) Genetic evidence for endolithic microbial life colonizing basaltic glass/seawater interfaces. In: Alt JC, Kinoshita H, Stokking LB et al (eds) Proceedings of the Ocean Drilling Program. Sci Results 148:207–214

Glikson M, Duck LJ, Golding SD et al (2008) Microbial remains in some earliest Earth rocks: comparison with a potential modern analogue. Precambrian Res 164:187–200

Golden DC, Ming DW, Schwandt CS et al (2000) An experimental study on kinetically-driven precipitation of Ca-Mg-Fe carbonates from solution: implications for the low temperature formation of carbonates in Martian meteorite Allan Hills 84001. Meteorit Planet Sci 35:457–465

Golden DC, Ming DW, Morris RV et al (2004) Evidence for exclusively inorganic formation of magnetite in Martian meteorite ALH84001. Am Mineral 89:681–695

Golubic S, Barghoorn ES (1977) Interpretation of microbial fossils with special reference to the Precambrian. In: Flügel E (ed) Fossil algae: recent results and developments. Springer, Berlin, pp 1–14

Golubic S, Hofmann HJ (1976) Comparison of modern and mid-Precambrian Entophysalidaceae (Cyanophyta) in stromatolitic algal mats: cell division and degradation. J Paleontol 50:1074–1092

Golubic S, Knoll AH (1993) Prokaryotes. In: Lipps JH (ed) Fossil prokaryotes and protists. Blackwell, Boston, pp 51–76

Golubic S, Friedmann I, Schneider J (1981) The lithobiontic ecological niche, with special reference to microorganisms. J Sed Petrol 51:475–478

Gorbushina AA, Krumbein WE (2000) Subaerial microbial mats and their effects on soil and rock. In: Riding RE, Awramik SM (eds) Microbial sediments. Springer, Berlin, pp 161–170

Grasby SE (2003) Naturally precipitating vaterite (μ-$CaCO_3$) spheres: unusual carbonates formed in an extreme environment. Geochim Cosmochim Acta 67:1659–1666

Grey K, Williams IR (1990) Problematic bedding-plane markings from the Middle Proterozoic Manganese Supergroup, Bangemall Basin, Western Australia. Precambrian Res 46:307–327

Grey K, Yochelson EL, Fedonkin MA et al (2010) *Horodyskia williamsii* new species, a Mesoproterozoic macrofossil from Western Australia. Precambrian Res 180:1–17

Grimes ST, Brock F, Rickard D et al (2001) Understanding fossilization: experimental pyritization of plants. Geology 29:123–126

Grimes ST, Davies KL, Butler IB et al (2002) Fossil plants from the Eocene London Clay: the use of pyrite textures to determine the mechanism of pyritization. J Geol Soc 159:493–501

Grotzinger JP, Rothman DH (1996) An abiotic model for stomatolite morphogenesis. Nature 383:423–425

Grube M, Berg G (2009) Microbial consortia of bacteria and fungi with focus on the lichen symbiosis. Fungal Biol Rev 23:72–85

Grube M, Kroken S (2000) Molecular approaches and the concept of species and species complexes in lichenized fungi. Mycol Res 104:1284–1294

Grube M, Winka K (2002) Progress in understanding the evolution and classification of lichenized ascomycetes. Mycologist 16:67–76

Han T-M, Runnegar B (1992) Megascopic eukaryotic algae from the 2.1 billion-tear-old Negaunee Iron Formation, Michigan. Science 257:232–235

Harper CJ, Taylor TN, Krings M et al (2013) Mycorrhizal symbiosis in the Paleozoic seed fern *Glossopteris* from Antarctica. Rev Palaeobot Palynol 192:22–31

Harrington DJ (1996) Bacterial collagenases and collagen-degrading enzymes and their potential role in human disease. Infect Immun 64:1885–1891

Hedges RE (2002) Bone diagenesis: an overview of processes. Archaeometry 44:319–328

Helm RF, Huang Z, Edwards D et al (2000) Structural characterization of the released polysaccharide of desiccation-tolerant *Nostoc commune* DRH-1. J Bacteriol 182:974–982

Hermann TN, Podkovyrov VN (2006) Fungal remains from the Late Riphean. Paleontol J 40:207–214

Hickman AH, Van Kranendonk MJ (2012) Early Earth evolution: evidence from the 3.5–1.8 Ga geologic history of the Pilbara region of Western Australia. Episodes 35:283–297

Hippler D, Hu N, Steiner M et al (2012) Experimental mineralization of crustacean eggs: new implications for the fossilization of Precambrian-Cambrian embryos. Biogeosciences 9:1765–1775

Hof CH, Briggs DE (1997) Decay and mineralization of mantis shrimps (Stomatopoda; Crustacea); a key to their fossil record. Palaios 12:420–438

Hofmann HJ (1972) Precambrian remains in Canada: fossils, dubiofossils and pseudofossils. In: Proceedings of the 24th International Geological Congress, Section 1, p 20–30

Hofmann HJ (1976) Precambrian microflora, Belcher Islands, Canada—significance and systematics. J Paleo 50:1040–1073

Hofmann HJ (2000) Archean stromatolites as microbial archives. In: Riding RE, Awramik SM (eds) Microbial sediments. Springer, Berlin, pp 315–327

Hofmann HJ, Grey K, Hickman AH et al (1999) Origin of 3.45 Ga coniform stromatolites in Warrawoona Group, Western Australia. Geol Soc Am Bull 111:1256–1262

Honegger R (2009) Lichen-forming fungi and their photobionts. In: Deising H (ed) The Mycota, vol 5, Plant relationships. Springer, Berlin, pp 305–333

Honegger R, Axe L, Edwards D (2013a) Bacterial epibionts and endolichenic actinobacteria and fungi in the Lower Devonian lichen *Chlorolichenomycites salopensis*. Fungal Biol 117:512–518

Honegger R, Edwards D, Axe L (2013b) The earliest records of internally stratified cyanobacterial and algal lichens from the Lower Devonian of the Welsh Borderland. New Phytol 197:264–275

Horodyski RJ (1980) Middle Proterozoic shale-facies microbiota from the Lower Belt Supergroup, Little Belt Mountains, Montana. J Paleo 54:649–663

Horodyski RJ (1982) Problematic bedding-plane markings from the Middle Proterozoic Appekunny Argillite, Belt Supergroup, Northwestern Montana. J Paleo 56:882–889

Horodyski RJ, Knauth LP (1994) Life on land in the Precambrian. Science 263:494–498

House CH, Schopf JW, McKeegan KD et al (2000) Carbon isotopic composition of individual Precambrian microfossils. Geology 28:707–710

Huldtgren T, Cunningham JA, Yin C et al (2011) Fossilized nuclei and germination structures identify Ediacaran "animal embryos" as encysting protists. Science 334:1696–1699

Humphreys CP, Franks PJ, Rees M et al (2010) Mutualistic mycorrhiza-like symbiosis in the most ancient group of land plants. Nat Commun 1:103

Iniesto M, Lopez-Archilla AI, Fregenal-Martínez M et al (2013) Involvement of microbial mats in delayed decay: an experimental essay on fish preservation. Palaios 28:56–66

Jans MME, Nielsen-Marsh CM, Smith CI et al (2004) Characterisation of microbial attack on archaeological bone. J Archaeol Sci 31:87–95

Javaux EJ (2007) Patterns of diversification in early eukaryotes. Carnets de Geologie/Notebooks on Geology 2007(01):38–42

Javaux EJ, Knoll AH, Walter MR (2001) Morphological and ecological complexity in early eukaryotic ecosystems. Nature 412:66–69

Javaux EJ, Knoll AH, Walter MR (2003) Recognizing and interpreting the fossils of early eukaryotes. Orig Life Evol Biosph 33:75–94

Javaux EJ, Knoll AH, Walter MR (2004) TEM evidence for eukaryotic diversity in mid-Proterozoic oceans. Geobiology 2:121–132

Javaux EJ, Marshall CP, Bekker A (2010) Organic-walled microfossils in 3.2-billion-year-old shallow-marine siliciclastic deposits. Nature 463:934–938

Jumpponen A (2001) Dark septate endophytes–are they mycorrhizal? Mycorrhiza 11:207–211

Jumpponen A, Trappe JM (1998) Dark septate endophytes: a review of facultative biotrophic root-colonizing fungi. New Phytol 140:295–310

Jung HG, Deetz DA (1993) Cell wall lignification and degradability. In: Jung HG, Buxton DR, Hatfield RD et al (eds) Forage cell wall structure and digestibility. American Society of Agronomy, Crop Science Society of America, Soil Science Society of America, p 315–346

Kalanetra KM, Joye SB, Sunser NR et al (2005) Novel vacuolate sulfur bacteria from the Gulf of Mexico reproduce by reductive division in three dimensions. Environ Microbiol 7:1451–1460

Kalkowsky E (1908) Oolith und Stromatolith im norddeutschen Buntsandstein. Z Dtsch Geol Ges 60:68–125

Karatygin IV, Snigirevskaya NS, Vikulin SV (2009) The most ancient terrestrial lichen *Winfrenatia reticulata*: a new find and new interpretation. Paleontol J 43:107–114

Kaye TG, Gaugler G, Sawlowicz Z (2008) Dinosaurian soft tissues interpreted as bacterial biofilms. PLoS One 3, e2808

Kazmierczak J (1979) The eukaryotic nature of *Eosphaera*-like ferriferous structures from the Precambrian Gunflint Iron Formation, Canada: a comparative study. Precambrian Res 9:1–22

Kazmierczak J, Kempe S (2006) Genuine modern analogues of Precambrian stromatolites from caldera lakes of Niuafo'ou Island, Tonga. Naturwissenschaften 93:119–126

Kellner AW (1996a) Reinterpretation of a remarkably well preserved pterosaur soft tissue from the Early Cretaceous of Brazil. J Vert Paleo 16:718–722

Kellner AWA (1996b) Fossilized theropod soft tissue. Nature 379:32

Kennard JM, James NP (1986) Thrombolites and stromatolites; two distinct types of microbial structures. Palaios 1:492–503

Kirkland BL, Lynch FL, Rahnis MA et al (1999) Alternative origins for nannobacteria-like objects in calcite. Geology 27:347–350

Kiyokawa S, Ito T, Ikehara M et al (2006) Middle Archean volcano-hydrothermal sequence: bacterial microfossil-bearing 3.2 Ga Dixon Island Formation, coastal Pilbara terrane, Australia. Geol Soc Am Bull 118:3–22

Klein C, Beukes NJ, Schopf JW (1987) Filamentous microfossils in the early Proterozoic Transvaal Super group: their morphology, significance, and paleoenviron mental setting. Precambrian Res 36:81–94

Kluge M, Mollenhauer D, Wolf E et al (2003) The *Nostoc-Geosiphon* endocytobiosis. In: Rai AN, Bergman B, Rasmussen U (eds) Cyanobacteria in symbiosis. Kluwer, Dordrecht, pp 19–30

Klymiuk AA, Stockey RA, Rothwell GW (2011) The first organismal concept for an extinct species of Pinaceae: *Pinus arnoldii* Miller. Int J Plant Sci 172:294–313

Klymiuk AA, Harper CJ, Moore DM et al (2013a) Reinvestigating Carboniferous "actinomycetes": authigenic formation of biomimetic carbonates provides insight into early diagenesis of permineralized plants. Palaios 28:80–92

Klymiuk AA, Taylor TN, Taylor EL et al (2013b) Paleomycology of the Princeton Chert II. Darkseptate fungi in the aquatic angiosperm Eorhiza arnoldii indicate a diverse assemblage of root-colonizing fungi during the Eocene. Mycologia 105:1100–1109

Knight TK, Bingham PS, Lewis RD, Savrda CE (2011) Feathers of the Ingersoll shale, Eutaw Formation (Upper Cretaceous), eastern Alabama: the largest collection of feathers from North American Mesozoic rocks. Palaios 26:364–376

Knoll AH (1992) The early evolution of eukaryotes: a geological perspective. Science 256:622–627

Knoll AH (2012) The fossil record of microbial life. In: Knoll AH, Canfield DE, Konhauser KO (eds) Fundamentals of geobiology. Wiley-Blackwell, Chichester, pp 297–314

Knoll AH (2014) Paleobiological perspectives on early eukaryotic evolution. Cold Spring Harb Perspect Biol 6:a016121

Knoll AH, Awramik SM (1983) Ancient microbial ecosystems. In: Krumbein WE (ed) Microbial geochemistry. Blackwell, Oxford, pp 287–315

Knoll AH, Bambach RK (2000) Directionality in the history of life: diffusion fom the left wall or repeated scaling on the right? In: Erwin DH, Wing SL (eds) Deep time. Paleobiology's perspective. The Paleontological Society. Supplement to Palaios 26:1–14

Knoll AH, Barghoorn ES (1974) Ambient pyrite in Precambrian chert: new evidence and a theory. Proc Nat Acad Sci USA 71:2329–2331

Knoll AH, Barghoorn ES (1975) Precambrian eukaryotic organisms: a reassessment of the evidence. Science 190:52–54

Knoll AH, Barghoorn ES (1977) Microfossils showing cell division from the Swaziland System of South Africa. Science 198:396–398

Knoll AH, Golubic S (1979) Anatomy and taphonomy of a Precambrian algal stromatolite. Precambrian Res 10:115–151

Knoll AH, Golubic S, Green J et al (1986) Organically preserved microbial endoliths from the late Proterozoic of East Greenland. Nature 321:856

Knoll AH, Javaux EJ, Hewitt D et al (2006) Eukaryotic organisms in Proterozoic oceans. Phil Trans R Soc Lond B 361:1023–1038

Knoll AH, Canfield DE, Konhauser KO (eds) (2012) Fundamentals of geobiology. Wiley-Blackwell, Chichester

Knoll AH, Worndle S, Kah LC (2013) Covariance of microfossil assemblages and microbialite textures across an upper Mesoproterozoic carbonate platform. Palaios 28:453–470

Konhauser KO (1998) Diversity of bacterial iron mineralization. Earth Sci Rev 43:91–121

Konhauser KO, Riding R (2012) Bacterial biomineralization. In: Knoll AH, Canfield DE, Konhauser KO (eds) Fundamentals of geobiology. Wiley-Blackwell, Chichester, pp 105–130

Konhauser KO, Fisher QJ, Fyfe WS et al (1998) Authigenic mineralization and detrital clay binding by freshwater biofilms: the Brahmani River, India. Geomicrobiol J 15:209–222

Konhauser KO, Kappler A, Roden EE (2011) Iron in microbial metabolisms. Elements 7:89–93

Kremer B, Kazmierczak J (2005) Cyanobacterial mats from Silurian black radiolarian cherts: phototrophic life at the edge of darkness? J Sediment Res 75:897–906

Kremer B, Kazmierczak J, Lukomska-Kowalczyk M et al (2012a) Calcification and silicification: fossilization potential of cyanobacteria from stromatolites of Niuafo'ou's caldera lakes (Tonga) and implications for the early fossil record. Astrobiology 12:535–548

Kremer B, Owocki K, Królikowska A et al (2012b) Mineral microbial structures in a bone of the Late Cretaceous dinosaur *Saurolophus angustirostris* from the Gobi Desert, Mongolia—a Raman spectroscopy study. Palaeogeogr Palaeoclim Palaeoecol 358:51–61

Krings M, Taylor TN, Hass H et al (2007) Fungal endophytes in a 400-million-yr-old land plant: infection pathways, spatial distribution, and host responses. New Phytol 174:648–657

Krings M, Hass H, Kerp H et al (2009) Endophytic cyanobacteria in a 400-million-yr-old land plant: a scenario for the origin of a symbiosis? Rev Palaeobot Palynol 153:62–69

Krings M, Taylor TN, Dotzler N (2012) Fungal endophytes as a driving force in land plant evolution: evidence from the fossil record. In: Southworth D (ed) Biocomplexity of plant-fungal interactions. Wiley, New York, pp 5–28

Krumbein WE (1979) Calcification by bacteria and algae. In: Trudinger PA, Swaine DJ (eds) Biogeochemical cycling of mineral-forming element. Elsevier, Amsterdam, pp 47–68

Laflamme M, Schiffbauer JD, Narbonne GM et al (2011) Microbial biofilms and the preservation of the Ediacara biota. Lethaia 44:203–213

Laflamme M, Schiffbauer JD, Narbonne GM (2012) Deep-water microbially induced sedimentary structures (MISS) in deep time: the Ediacaran fossil Ivesheadia. In: Noffke NK, Chafetz H (eds) Microbial mats in siliciclastic depositional systems through time. SEPM Special Publication 101:111–123

Lamb DM, Awramik SM, Chapman DJ et al (2009) Evidence for eukaryotic diversification in the ~1800 million-year-old Changzhougou Formation, North China. Precambrian Res 173:93–104

Lang BF, Burger G (2012) Mitochondrial and eukaryotic origins: a critical review. In: Marechal-Drouard L (ed) Mitochondrial genome evolution. Elsevier—Academic, Amsterdam, pp 1–20

Lawson AC, Czernuszka JT (1998) Collagen-calcium phosphate composites. J Eng Med 212:413–425

LePage B, Currah R, Stockey R et al (1997) Fossil ectomycorrhizae from the Middle Eocene. Am J Bot 84:410–410

Lepland A, Arrhenius G, Cornell D (2002) Apatite in early Archean Isua supracrustal rocks, southern West Greenland: its origin, association with graphite and potential as a biomarker. Precambrian Res 118:221–241

Lepland A, Van Zuilen MA, Philippot P (2011) Fluid-deposited graphite and its geobiological implications in early Archean gneiss from Akilia, Greenland. Geobiology 9:2–9

Lepot K, Benzerara K, Brown GE Jr et al (2008) Microbially influenced formation of 2,724-million-year-old stromatolites. Nat Geosci 1:118–121

Lepot K, Philippot P, Benzerara K et al (2009a) Garnet-filled trails associated with carbonaceous matter mimicking microbial filaments in Archean basalt. Geobiology 7:393–402

Lepot K, Benzerara K, Rividi N et al (2009b) Organic matter heterogeneities in 2.72 Ga stromatolites: alteration versus preservation by sulfur incorporation. Geochim Cosmochim Acta 73:6579–6599

Lerosey-Aubril R, Hegna TA, Kier C et al (2012) Controls on gut phosphatisation: the trilobites from the Weeks Formation Lagerstätte (Cambrian; Utah). PLoS One 7, e32934

Léveillé RJ, Fyfe WS, Longstaffe FJ (2000) Geomicrobiology of carbonate-silicate microbialites from Hawaiian basaltic sea caves. Chem Geol 169:339–355

Li Q, Gao KQ, Vinther J et al (2010a) Plumage color patterns of an extinct dinosaur. Science 327:1369–1372

Li C, Love GD, Lyons TW et al (2010b) A stratified redox model for the Ediacaran Ocean. Science 328:80–83

Liebig K (2001) Bacteria. In: Briggs DEG, Crowther PR (eds) Palaeobiology II. Blackwell, Oxford, pp 253–256

Lin JP (2007) Preservation of the gastrointestinal system in *Olenoides* (Trilobita) from the Kaili Biota (Cambrian) of Guizhou, China. Mem Assoc Australasian Palaeontol 33:179

Liu Y, Simon JD (2003) Isolation and biophysical studies of natural eumelanins: applications of imaging technologies and ultrafast spectroscopy. Pigment Cell Res 16:606–618

Liu AG, Mcilroy D, Antcliffe JB, Brasier MD (2011) Effaced preservation in the Ediacara biota and its implications for the early macrofossil record. Palaeontology 54:607–630

Lockley MG (1986) The paleobiological and paleoenvironmental importance of dinosaur footprints. Palaios 1:37–47

Lockley MG (1991) Tracking dinosaurs: a new look at an ancient world. Cambridge University Press, Cambridge

Loeblich TR (1970) Morphology, ultrastructure and distribution of Palaeozoic acritarchs. In: Proceedings of the North American Palaeontological Convention G, p 705–788

Lovley DR, Stolz JF, Nord GL et al (1987) Anaerobic production of magnetite by a dissimilatory iron-reducing microorganism. Nature 330:252–254

Lovley DR, Phillips EJP (1986) Organic matter mineralization with reduction of ferric iron in anaerobic sediments: a review. Appl Environ Microbiol 51:683–689

Lowe DR (1994) Abiological origin of described stromatolites older than 3.2 Ga. Geology 22:387–390

Lutzoni F, Pagel M, Reeb V (2001) Major fungal lineages are derived from lichen symbiotic ancestors. Nature 411:937–940

Lutzoni F, Kauff F, Cox CJ et al (2004) Assembling the fungal tree of life: progress, classification, and evolution of subcellular traits. Am J Bot 91:1446–1480

Maciá-Vicente JG, Rosso LC, Ciancio A et al (2009) Colonisation of barley roots by endophytic *Fusarium equiseti* and *Pochonia chlamydosporia*: effects on plant growth and disease. Ann Appl Biol 155:391–401

MacLean LC, Tyliszczak T, Gilbert PU, Zhou D, Pray TJ, Onstott TC, Southam G (2008) A high-resolution chemical and structural study of framboidal pyrite formed within a low-temperature bacterial biofilm. Geobiology 6:471–480

Manning PL, Morris PM, McMahon A et al (2009) Mineralized soft–tissue structure and chemistry in a mummified hadrosaur from the Hell Creek Formation, North Dakota (USA). Proc R Soc B: Biol Sci 276:3429–3437

Marshall CP, Javaux EJ, Knoll AH et al (2005) Combined micro-Fourier transform infrared (FTIR) spectroscopy and micro-Raman spectroscopy of Proterozoic acritarchs: a new approach to palaeobiology. Precambrian Res 138:208–224

Marshall CP, Edwards HGM, Jehlicka J (2010) Understanding the application of Raman spectroscopy to the detection of traces of life. Astrobiology 10:229–243

Marshall CP, Emry JR, Olcott Marshall A (2011) Haematite pseudomicrofossils present in the 3.5-billion-year-old Apex Chert. Nat Geosci 4:240–243

Martill DM (1987) Prokaryote mats replacing soft tissues in Mesozoic marine reptiles. Modern Geol 11:265–269

Martill DM (1988) Preservation of fish in the Cretaceous Santana Formation of Brazil. Palaeontology 31:1–18

Martill DM (1989) The Medusa effect: instantaneous fossilization. Geol Today 5:201–205

Martill DM (1991) Organically preserved dinosaur skin: taphonomic and biological implications. Modern Geol 16:61–68

Martill DM (1990) Macromolecular resolution of fossilized muscle tissues from an elopomorph fish. Nature 346:171–172

Martill DM, Unwin DM (1989) Exceptionally well preserved pterosaur wing membrane from the Cretaceous of Brazil. Nature 340:138–140

Martill DM, Unwin DM (1997) Small spheres in fossil bones: blood corpuscles or diagenetic products? Palaeontology 40:619–624

Martill DM, Brito PM, Washington-Evans J (2008) Mass mortality of fishes in the Santana Formation (Lower Cretaceous, ?Albian) of northeast Brazil. Cretceous Res 29:649–658

Martin D, Briggs DE, Parkes RJ (2004) Experimental attachment of sediment particles to invertebrate eggs and the preservation of soft-bodied fossils. J Geol Soc 161:735–738

Martin D, Briggs DE, Parkes RJ (2005) Decay and mineralization of invertebrate eggs. Palaios 20:562–572

Marty D, Strasser A, Meyer CA (2009) Formation and taphonomy of human footprints in microbial mats of present-day tidal-flat environments: implications for the study of fossil footprints. Ichnos 16:127–142

Matsunaga KK, Stockey RA, Tomescu AM (2013) *Honeggeriella complexa* gen. et sp. nov., a heteromerous lichen from the Lower Cretaceous of Vancouver Island (British Columbia, Canada). Am J Bot 100:450–459

Mayr G, Manegold A (2013) Can ovarian follicles fossilize? Nature 499, E1

McCollum TM (2003) Formation of meteorite hydrocarbons from thermal decomposition of siderite ($FeCO_3$). Geochim Cosmochim Acta 67:311–317

McDermid AS, McKee AS, Marsh PD (1988) Effect of environmental pH on enzyme activity and growth of *Bacteroides gingivalis* W50. Infect Immun 56:1096–1100

McElhiney J, Lawton LA, Leifert C (2001) Investigations into the inhibitory effects of microcystins on plant growth, and the toxicity of plant tissues following exposure. Toxicon 39:1411–1420

McGregor VR, Mason B (1977) Petrogenesis and geochemistry of metabasaltic and metasedimentary enclaves in the Amitsoq gneisses, West Greenland. Am Mineral 62:887–904

McKay DS, Gibson EK Jr, Thomas-Keprta KL et al (1996) Search for past life on Mars: possible relic biogenic activity in Martian meteorite ALH 84001. Science 273:924–930

McLoughlin N, Wacey D, Brasier MD et al (2007) On biogenicity criteria for endolithic microborings on early Earth and beyond. Astrobiology 7:10–26

McLoughlin N, Wilson LA, Brasier MD (2008) Growth of synthetic stromatolites and wrinkle structures in the absence of microbes—implications for the early fossil record. Geobiology 6:95–105

McLoughlin N, Furnes H, Banerjee NR et al (2009) Ichnotaxonomy of microbial trace fossils in volcanic glass. J Geol Soc 166:159–169

McLoughlin N, Grosch EG, Kilburn MR et al (2012) Sulfur isotope evidence for Paleoarchean subseafloor biosphere, Barberton, South Africa. Geology 40:1031–1034

McNamara M, Orr PJ, Kearns SL et al (2010) Organic preservation of fossil musculature with ultracellular detail. Proc R Soc B: Biol Sci 277:423–427

McNamara ME, Orr PJ, Kearns SL et al (2009) Soft-tissue preservation in Miocene frogs from Libros, Spain: insights into the genesis of decay microenvironments. Palaios 24:104–117

Millay MA, Eggert DA (1974) Microgametophyte development in the Paleozoic seed fern family Callistophytaceae. Am J Bot 60:1067–1075

Moczydlowska M, Willman S (2009) Ultrastructure of cell walls in ancient microfossils as a proxy to their biological affinities. Precambrian Res 173:27–38

Mojzsis SJ, Arrhenius G, McKeegan KD, Harrison TM, Nutman AP, Friend CRL (1996) Evidence for life on Earth before 3,800 million years ago. Nature 384:55–59

Moorbath S, O'Nions RK, Pankhurst RJ (1973) Early Archaean age for the Isua iron formation, West Greenland. Nature 240:138–139

Morris CE, Monier J, Jacques M (1997) Methods for observing microbial biofilms directly on leaf surfaces and recovering them for isolation of culturable microorganisms. Appl Environ Microbiol 63:1570–1576

Morris PA, Wentworth SJ, Allen CC et al (1999) Methods for determining biogenicity in Archean and other ancient rocks. In: Lunar and Planetary Science Conference XXX, http://www.lpi.usra.edu/meetings/LPSC99/pdf/1952.pdf

Munns R (2002) Comparative physiology of salt and water stress. Plant Cell Environ 25:239–250

Munnecke A, Servais T, Vachard D (2001) *Halysis* Hoeg, 1932—a problematic Cyanophyceae: new evidence from the Silurian of Gotland (Sweden). N Jb Geol Palaont Mh 7:21–42

Nelsen MP, Lücking R, Grube M et al (2009) Unravelling the phylogenetic relationships of lichenised fungi in Dothideomyceta. Stud Mycol 64:135–144

Newsham KK (2011) A meta-analysis of plant responses to dark septate root endophytes. New Phytol 190:783–793

Niklas KJ (1983) Organelle preservation and protoplast partitioning in fossil angiosperm leaf tissues. Am J Bot 70:543–548

Niklas KJ, Newman SA (2013) The origins of multicellular organisms. Evol Dev 15:41–52

Niklas KJ, Brown RM Jr, Santos R (1985) Ultrastructural states of preservation in Clarkia angiosperm leaf tissues: implications on modes of fossilization. In: Smiley CJ (ed) Late Cenozoic History of the Pacific Northwest: interdisciplinary studies of the Clarkia beds of Northern Idaho. Pacific Division of the AAAS, California Academy of Sciences, San Francisco, p 143–159

Niklas KJ, Cobb ED, Crawford DR (2013) The evo-devo of multinucleate cells, tissues, and organisms, and an alternative route to multicellularity. Evol Dev 15:466–474

Nisbet EG (1980) Archean stromatolites and the search for the earliest life. Nature 284:395–396

Noffke N (2009) The criteria for the biogeneicity of microbially induced sedimentary structures (MISS) in Archean and younger, sandy deposits. Earth-Sci Rev 96:173–180

Noffke N (2010) Microbial mats in sandy deposits from the Archean Era to today. Springer, New York

Noffke N, Awramik SM (2013) Stromatolites and MISS—differences between relatives. GSA Today 23:4–9

Noffke N, Chafetz H (eds) (2012) Microbial mats in siliciclastic depositional systems through time. SEPM Special Publicartion no. 101. Society for Sedimentary Geology, Tulsa

Noffke N, Gerdes G, Klenke T et al (1996) Microbially induced sedimentary structures—examples from modern sediments of siliciclastic tidal flats. Zb Geol Palaont 1:307–316

Noffke N, Gerdes G, Klenke T et al (2001) Microbially induced sedimentary structures—a new category within the classification of primary sedimentary structures. J Sediment Res 71:649–656

Noffke N, Hazen R, Nhleko N (2003) Earth's earliest microbial mats in a siliciclastic marine environment (2.9 Ga Mozaan Group, South Africa). Geology 31:673–676

Noffke N, Eriksson KA, Hazen RM et al (2006) A new window into Early Archean life: microbial mats in Earths oldest siliciclastic tidal deposits (3.2 Ga Moodies Group, South Africa). Geology 34:253–256

Noffke N, Christian D, Wacey D et al (2013a) Microbially induced sedimentary structures recording an ancient ecosystem in the ca. 3.48 billion-year-old Dresser Formation, Pilbara, Western Australia. Astrobiology 13:1103–1124

Noffke N, Decho AW, Stoodley P (2013b) Slime through time: the fossil record of prokaryote evolution. Palaios 28:1–5

O'Brien NR, Meyer HW, Reilly K et al (2002) Microbial taphonomic processes in the fossilization of insects and plants in the late Eocene Florissant Formation, Colorado. Rocky Mtn Geol 37:1–11

Oehler DZ (1976) Transmission electron microscopy of organic microfossils from the late Precambrian Bitter Springs Formation, Australia: techniques and survey of preserved ultrastructure. J Paleo 50:90–106

Oehler DZ (1977) Pyrenoid-like structures in late Precambrian algae from the Bitter Springs Formation of Australia. J Paleontol 51:885–901

Ohtomo Y, Kakegawa T, Ishida A et al (2014) Evidence for biogenic graphite in early Archean Isua metasedimentary rocks. Nat Geosci 7:25–28

Olcott Marshall A, Jehlicka J, Rouzaud JN et al (2014) Multiple generations of carbonaceous material deposited in Apex chert by basin-scale pervasive hydrothermal fluid flow. Gondwana Res 25:284–289

Orr PJ, Benton MJ, Briggs DE (2003) Post-Cambrian closure of the deep-water slope-basin taphonomic window. Geology 31:769–772

Orr PJ, Briggs DE, Kearns SL (1998) Cambrian Burgess Shale animals replicated in clay minerals. Science 281:1173–1175

Page A, Gabbott SE, Wilby PR et al (2008) Ubiquitous Burgess Shale-style "clay templates" in low-grade metamorphic mudrocks. Geology 36:855–858

Pang K, Tang Q, Schiffbauer JD et al (2013) The nature and origin of nucleus-like intracellular inclusions in Paleoproterozoic eukaryote microfossils. Geobiology 11:499–510

Papineau D, De Gregorio BT, Cody GD et al (2010) Ancient graphite in the Eoarchean quartz-pyroxene rocks from Akilia in southern West Greenland I: petrographic and spectroscopic characterization. Geochim Cosmochim Ac 74:5862–5883

Pasteris JD, Wopenka B (2003) Necessary, but not sufficient: Raman identification of disordered carbon as a signature of ancient life. Astrobiology 3:727–738

Peach BN, Horne J, Gunn W et al (1907) The Geological Structure of the Northwest Highlands of Scotland. Memoirs of the Geological Survey of Great Britain 1907:1–668

Peat CJ, Muir MD, Plumb KA et al (1978) Proterozoic microfossils from the Roper Group, Northern Territory, Australia. BMR J Aus Geol Geophys 3:1–17

Peckmann J, Goedert JL (2005) Geobiology of ancient and modern methane-seeps. Palaeogeogr Palaeoclim Palaeoecol 227:1–5

Pecoits E, Konhauser KO, Aubet NR, Heaman LM, Veroslavsky G, Stern RA, Gingras MK (2012) Bilaterian burrows and grazing behavior at >585 million years ago. Science 336:1693–1696

Peng Y, Bao H, Yuan X (2009) New morphological observations for Paleoproterozoic acritarchs from the Chuanlinggou Formation, North China. Precambrian Res 168:223–232

Peterson EB (2000) An overlooked fossil lichen (Lobariaceae). Lichenologist 32:298–300

Petrovich R (2001) Mechanisms of fossilization of the soft-bodied and lightly armored faunas of the Burgess Shale and of some other classical localities. Am J Sci 301:683–726

Pflug HD (1978a) Frueheste bisher bekannte Lebewesen *Isuasphaera isua* n. gen. n. spec. aus der Isua-Serie von Groenland (ca. 3800 Mio. J.). Oberhess naturwiss Z 44:131–145

Pflug HD (1978b) Yeast-like microfossils detected in oldest sediments of the Earth. Naturwissenschaften 65:611–615

Pflug HD, Jaeschke-Boyer H (1979) Combined structural and chemical analysis of 3,800-Myr-old microfossils. Nature 280:483–486

Pinheiro FL, Horn BL, Schultz CL et al (2012) Fossilized bacteria in a Cretaceous pterosaur headcrest. Lethaia 45:495–499

Pinti D, Mineau R, Clement V (2013) Comment on "Biogenicity of Earth's oldest fossils: a resolution of the controversy" by J. William Schopf and and Anatoliy B. Kudryavtsev, Gondwana Research 22 (2012), 761–771. Gondwana Res 23:1652–1653

Pirozynski KA, Malloch DW (1975) The origin of land plants: a matter of mycotrophism. BioSystems 6:153–164

Planavsky NJ, Asael D, Hofmann A et al (2014) Evidence for oxygenic photosynthesis half a billion years before the Great Oxidation Event. Nat Geosci 7:283–286

Phipps CJ, Taylor TN (1996) Mixed arbuscular mycorrhizae from the Triassic of Antarctica. Mycologia 88:707–714

Poinar GO, Peterson EB, Platt JL (2000) Fossil *Parmelia* in new world amber. Lichenologist 32:263–269

Pons ML, Quitte G, Fujii T et al (2011) Early Archean serpentine mud volcanoes at Isua, Greenland, as a niche for early life. Proc Nat Acad Sci USA 108:17639–17643

Poortinga AT, Bos R, Norde W et al (2002) Electric double layer interactions in bacterial adhesion to surfaces. Surf Sci Rep 47:1–32

Popa R, Kinkle BK, Badescu A (2004) Pyrite framboids as biomarkers for iron-sulfur systems. Geomicrobiol J 21:193–206

Pósfai M, Dunin-Borkowski RE (2006) Sulfides in biosystems. Rev Miner Geochem 61:679–714

Prieto M, Wedin M (2013) Dating the diversification of the major lineages of Ascomycota (Fungi). PLoS One 8, e65576

Raff EC, Villinski JT, Turner FR, Donoghue PC, Raff RA (2006) Experimental taphonomy shows the feasibility of fossil embryos. Proc Natl Acad Sci USA 103:5846–5851

Raff EC, Andrews ME, Turner FR et al (2013) Contingent interactions among biofilm-forming bacteria determine preservation or decay in the first steps toward fossilization of marine embryos. Evol Dev 15:243–256

Raff EC, Schollaert KL, Nelson DE et al (2008) Embryo fossilization is a biological process mediated by microbial biofilms. Proc Nat Acad Sci USA 105:19360–19365

Rai AN (1990) Handbook of symbiotic cyanobacteria. CRC, Boca Raton, FL

Raiswell R, Whaler K, Dean S et al (1993) A simple three-dimensional model of diffusion-with-precipitation applied to localised pyrite formation in framboids, fossils and detrital iron minerals. Mar Geol 113:89–100

Rasmussen B (2000) Filamentous microfossils in a 3,235-million-year-old volcanogenic massive sulfide deposit. Nature 405:676–679

Rasmussen B, Fletcher IR, Brocks JJ et al (2008) Reassessing the first appearance of eukaryotes and cyanobacteria. Nature 455:1101–1104

Rasmussen B, Blake TS, Fletcher IR et al (2009) Evidence for microbial life in synsedimentary cavities from 2.75 Ga terrestrial environments. Geology 37:423–426

Redecker D, Kodner R, Graham LE (2000) Glomalean fungi from the Ordovician. Science 289:1920–1921

Reid RP, Visscher PT, Decho AW et al (2000) The role of microbes in accretion, lamination and early lithification of modern marine stromatolites. Nature 406:989–992

Reis RR, Huang TD, Roberts EM et al (2013) Embryology of Early Jurassic dinosaur from China with evidence of preserved organic remains. Nature 496:210–214

Reitner J, Quéric N-V, Arp G (eds) (2011) Advances in stromatolite geobiology. Springer, Berlin

Remy W, Taylor TN, Hass H et al (1994) Four hundred-million-year-old vesicular arbuscular mycorrhizae. Proc Nat Acad Sci USA 91:11841–11843

Retallack GJ (2013) Ediacaran life on land. Nature 493:89–92

Retallack GJ (1994) Were the Ediacaran fossils lichens? Paleobiology 20:523–544

Rice CM, Trewin NH, Anderson LI (2002) Geological setting of the Early Devonian Rhynie cherts, Aberdeenshire, Scotland: an early terrestrial hot spring system. J Geol Soc 159:203–214

Riding R (2011) The nature of stromatolites: 3,500 million years of history and one century of research. In: Reitner J, Quéric N-V, Arp G (eds) Advances in stromatolite geobiology. Springer, Berlin, pp 29–74

Rikkinen J (2003) Calicioid lichens from European Tertiary amber. Mycologia 95:1032–1036

Rikkinen J, Poinar GO (2008) A new species of *Phyllopsora* (Lecanorales, lichen-forming Ascomycota) from Dominican amber, with remarks on the fossil history of lichens. J Exp Bot 59:1007–1011

Rikkinen J, Poinar GO Jr (2002) Fossilised *Anzia* (Lecanorales) lichen-forming Ascomycota from European Tertiary amber. Mycol Res 106:984–990

Roberts JA, Kenward PA, Fowle DA et al (2013) Surface chemistry allows for abiotic precipitation of dolomite at low temperature. Proc Nat Acad Sci USA 110:14540–14545

Roberts TA, Mead GC (1985) Involvement of intestinal anaerobes in the spoilage of red meats, poultry and fish. Soc Appl Bacteriol 13:333–349

Rodriguez RJ, White JF Jr, Arnold AE et al (2009) Fungal endophytes: diversity and functional roles. New Phytol 182:314–330

Roedder E (1981) Are the 3,800-Myr-old Isua objects microfossils, limonite-stained fluid inclusions, or neither? Nature 293:459–462

Roh Y, Zhang C-L, Vali H et al (2003) Biogeochemical and environmental factors in Fe biomineralization: magnetite and siderite formation. Clays Clay Min 51:83–95

Rosing MT (1999) 13C-depleted carbon microparticles in >3700-Ma sea-floor sedimentary rocks from West Greenland. Science 283:674–676

Rothwell GW, Scheckler SE, Gillespie WH (1989) *Elkinsia* gen. nov., a late Devonian gymnosperm with cupulate ovules. Bot Gaz 158:170–189

Rothwell GW, Wyatt SE, Tomescu AMF (2014) Plant evolution at the interface of paleontology and developmental biology: an organism-centered paradigm. Am J Bot 101:899–913

Russell JA, Brady AL, Cardman Z et al (2014) Prokaryote populations of extant microbialites along a depth gradient in Pavilion Lake, British Columbia, Canada. Geobiology 12:250. doi:10.1111/gbi.12082

Sagemann J, Bale SJ, Briggs DE et al (1999) Controls on the formation of authigenic minerals in association with decaying organic matter: an experimental approach. Geochim Cosmochim Acta 63:1083–1095

Saikkonen K, Faeth SH, Helander M et al (1998) Fungal endophytes: a continuum of interactions with host plants. Annu Rev Ecol Syst 29:319–343

Samuelsson J, Dawes PR, Vidal G (1999) Organic-walled microfossils from the Proterozoic Thule Supergroup, Northwest Greenland. Precambrian Res 96:1–23

San Antonio JD, Schweitzer MH, Jensen ST et al (2011) Dinosaur peptides suggest mechanisms of protein survival. PLoS One 6, e20381

Schidlowski M (1988) A 3,800-million-year isotopic record of life from carbon in sedimentary rocks. Nature 333:313–318

Schidlowski M (2000) Carbon isotopes and microbial sediments. In: Riding RE, Awramik SM (eds) Microbial sediments. Springer, Berlin, pp 84–95

Schieber J, Riciputi L (2005) Pyrite and marcasite coated grains in the Ordovician Winnipeg Formation, Canada: an intertwined record of surface conditions, stratigraphic condensation, geochemical "reworking", and microbial activity. J Sediment Res 75:907–920

Schieber J, Bose PK, Eriksson PG et al (2007) Atlas of microbial mat features preserved within the siliciclastic rock record. Atlases in Geosciences. Elsevier, Amsterdam

Schiffbauer JD, Laflamme M (2012) Lagerstätten through time: a collection of exceptional preservational pathways from the terminal Neoproterozoic through today. Palaios 27:275–278

Schmid G, Zeitvogel F, Hao L et al (2014) 3-D analysis of bacterial cell-(iron)mineral aggregates formed during Fe(II) oxidation by the nitrate-reducing *Acidovorax* sp. strain BoFeN1 using complementary microsopy tomography approaches. Geobiology 12:340–361

Schneider DA, Bickford ME, Cannon WF et al (2002) Age of volcanic rocks and syndepositional iron formations, Marquette Range Supergroup: implications for the tectonic setting of Paleoproterozoic iron formations of the Lake Superior region. Can J Earth Sci 39:999–1012

Schopf JM (1975) Modes of fossil preservation. Rev Palaeobot Palynol 20:27–53

Schopf JW (1968) Microflora of the Bitter Springs Formation, Late Precambrian, Central Australia. J Paleo 42:651–6881999

Schopf JW (1993) Microfossils of the Early Archean apex chert: new evidence of the antiquity of life. Science 260:640–646

Schopf JW (1999) Fossils and pseudofossils: lessons from the hunt for early life on Earth. In: Size limits of very small microorganisms Conference. National Academy Press, Washington, DC, pp. 88–93

Schopf JW (2006) Fossil evidence of Archean life. Phil Trans R Soc B 361:869–885

Schopf JW (2009) Emergence of Precambrian paleobiology: a new field of science. In: Sepkoski D, Ruse M (eds) The paleobiological revolution. Chicago University Press, Chicago, pp 89–110

Schopf JW, Blacic JM (1971) New microorganisms from the Bitter Springs Formation (Late Precambrian) of the north-central Amadeus Basin, Australia. J Paleo 45:925–961

Schopf JW, Klein C (1992) Times of origin and earliest evidence of major biologic groups. In: Schopf JW, Klein C (eds) The Proterozoic biosphere: a multidisciplinary study. Cambridge University Press, New York, pp 587–593

Schopf JW, Kudryavtsev AB (2012) Biogenicity of Earth's earliest fossils: a resolution of the controversy. Gondwana Res 22:761–771

Schopf JW, Packer BM (1987) Early Archean (3.3- billion to 3.5-billion-year-old) microfossils from Warrawoona Group, Australia. Science 237:70–73

Schopf JW, Sovietov YK (1976) Microfossils in *Conophyton* from the Soviet Union and their bearing on Precambrian biostratigraphy. Science 193:143–146

Schopf JW, Walter MR (1983) Archean microfossils: new evidence of ancient microbes. In: Schopf JW (ed) Earth's earliest biosphere. Princeton University Press, New Jersey, pp 214–239

Schopf JW, Kudryavtsev AB, Agresti DG et al (2002) Laser-Raman imagery of Earth's earliest fossils. Nature 416:73–76

Schopf JW, Tripathi AB, Kudryavtsev AB (2006) Three-dimensional confocal optical imagery of Precambrian microscopic organisms. Astrobiology 6:1–16

Schopf JW, Kudryavtsev AB, Czaja AD et al (2007) Evidence of Archean life: stromatolites and microfossils. Precambrian Res 158:141–155

Schopf JW, Kudryavtsev AB, Sugitani K et al (2010) Precambrian microbe-like pseudofossils: a promising solution to the problem. Precambrian Res 179:191–205

Schopf JW, Kudryavtsev AB, Tripathi AB et al (2011) Three-dimensional morphological (CLSM) and chemical (Raman) imagery of cellularly mineralized fossils. In: Allison PA, Bottjer D (eds) Taphonomy. Springer, Netherlands, pp 457–486

Schulz HN, Schulz HD (2005) Large sulfur bacteria and the formation of phosphorite. Science 307:416–418

Schulz HN, Brinkhoff T, Ferdelman TG et al (1999) Dense populations of a giant sulfur bacterium in Namibian shelf sediments. Science 284:493–495

Schüssler A, Kluge M (2001) *Geosiphon pyriforme*, an endocytosymbiosis between fungus and cyanobacteria, and its meaning as a model system for arbuscular mycorrhizal research. In: Esser K, Lemke PA, Hock B (eds) Fungal associations. Springer, Berlin, pp 151–161

Schweitzer MH, Wittmeyer JL, Horner JR et al (2005) Soft-tissue vessels and cellular preservation in *Tyrannosaurus rex*. Science 307:1952–1955

Schweitzer MH, Wittmeyer JL, Horner JR (2007) Soft tissue and cellular preservation in vertebrate skeletal elements from the Cretaceous to the present. Proc R Soc B: Biol Sci 274:183–197

Schwendemann AB, Taylor TN, Taylor EL, Krings M (2010) Organization, anatomy, and fungal endophytes of a Triassic conifer embryo. Am J Bot 97:1873–1883

Seckbach J, Walsh M (eds) (2009) From fossils to astrobiology. Springer, Dordrecht

Seilacher A, Reif WE, Westphal F et al (1985) Sedimentological, ecological and temporal patterns of fossil Lagerstatten [and Discussion]. Phil Trans R Soc Lond B 311:5–24

Semikhatov SM, Gebelein CD, Cloud P et al (1979) Stromatolite morphogenesis—progress and problems. Can J Earth Sci 16:992–1015

Seong-Joo L, Golubic S (1998) Multi-trichomous cyanobacterial microfossils from the Mesoproterozoic Gaoyuzhuang Formation, China: paleoecological and taxonomic implications. Lethaia 31:169–184

Sepkoski D, Ruse M (2009) Paleontology at the high table. In: Sepkoski D, Ruse M (eds) The paleobiological revolution. Chicago University Press, Chicago, pp 1–11
Sergeev VN (2009) The distribution of microfossil assemblages in Proterozoic rocks. Precambrian Res 173:212–222
Shapiro RS (2000) A comment on the systematic confusion of thrombolites. Palaios 15:166–169
Shapiro RS, Spangler E (2009) Bacterial fossil record in whale-falls: petrographic evidence of microbial sulfate reduction. Palaeogeogr Palaeoclim Palaeocol 274:196–203
Sinninghe Damsté JS, de Leeuw JW (1990) Analysis, structure and geochemical significance of organically-bound sulphur in the geosphere: state of the art and future research. Org Geochem 16:1077–1101
Skawina A (2010) Experimental decay of gills in freshwater bivalves as a key to understanding their preservation in Upper Triassic lacustrine deposits. Palaios 25:215–220
Southam G, Donald R (1999) A structural comparison of bacterial microfossils vs. 'nanobacteria' and nanofossils. Earth-Sci Rev 48:251–264
Southam G, Donald R, Röstad A et al (2001) Pyrite discs in coal: evidence for fossilized bacterial colonies. Geology 29:47–50
Spicer RA (1977) The pre-depositional formation of some leaf impressions. Palaeontology 20:907–912
Spicer RA (1981) The sorting and deposition of allochthonous plant material in a modern environment at Silwood Lake, Silwood Park, Berkshire, England. US Geol Surv Prof Paper 1143:1–77
Staudigel H, Yayanos A, Chastain R et al (1998) Biologically mediated dissolution of volcanic glass in seawater. Earth Planet Sci Lett 164:233–244
Staudigel H, Furnes H, McLoughlin N et al (2008) 3.5 billion years of glass bioalteration: volcanic rocks as basis for microbial life? Earth-Sci Rev 89:156–176
Steele A, McCubbin FM, Fries M et al (2012) A reduced organic Carbon component in Martian basalts. Science 337:212–215
Stenroos SK, DePriest PT (1998) SSU rDNA phylogeny of cladoniiform lichens. Am J Bot 85:1548–1559
Stewart WN, Rothwell GW (1993) Paleobotany and the evolution of plants. Cambridge University Press, New York
Stockey RA, Rothwell GW, Addy HD et al (2001) Mycorrhizal association of the extinct conifer *Metasequoia milleri*. Mycol Res 105:202–205
Stone JK, Bacon CW, White JF Jr (2000) An overview of endophytic microbes: endophytism defined. In: Bacon CW, White JF Jr (eds) Microbial endophytes. Marcel Dekker, New York, pp 3–29
Stoodley P, Sauer K, Davies DG et al (2002) Biofilms as complex differentiated communities. Annu Rev Microbiol 56:187–209
Stoyke G, Currah RS (1991) Endophytic fungi from the mycorrhizae of alpine ericoid plants. Can J Bot 69:347–352
Straub KL, Schünhuber WA, Bucholz-Cleven BEE (2004) Diversity of ferrous iron-oxidizing, nitrate-reducing bacteria and their involvement in oxygen-independent iron cycling. Geomicrobiol J 21:371–378
Strother PK, Battison L, Brasier MD et al (2011) Earth's earliest non-marine eukaryotes. Nature 473:505–509
Strullu-Derrien C, Kenrick P, Pressel S et al (2014) Fungal associations in *Horneophyton ligneri* from the Rhynie Chert (c. 407 million year old) closely resemble those in extant lower land plants: novel insights into ancestral plant-fungus symbioses. New Phytol 203:964–979
Sugitani K, Grey K, Allwood A (2007) Diverse microstructures from Archaean chert from the Mount Goldsworthy–Mount Grant area, Pilbara Craton, Western Australia: microfossils, dubiofossils, or pseudofossils? Precambrian Res 158:228–262

Sugitani K, Lepot K, Nagaoka T et al (2010) Biogenicity of morphologically diverse carbonaceous microstructures from the ca. 3400 Ma Strelley Pool Formation, in the Pilbara Craton, Western Australia. Astrobiology 10:899–920

Sugitani K, Mimura K, Nagaoka T et al (2013) Microfossil assemblage from the 3400 Ma Strelley Pool Formation in the Pilbara Craton, Western Australia: results form a new locality. Precambrian Res 226:59–74

Summons RE, Lincoln SA (2012) Biomarkers: informative molecules for studies in geobiology. In: Knoll AH, Canfield DE, Konhauser KO (eds) Fundamentals of geobiology. Wiley-Blackwell, Chichester, pp 269–296

Tandon KK, Kumar S (1977) Discovery of annelid and arthropod remains from Lower Vindhyan rocks (Precambrian) of central India. Geophytology 7:126–130

Taylor TN, Taylor EL (2000) The Rhynie chert ecosystem: a model for understanding fungal interactions. In: Bacon CW, White JF Jr (eds) Microbial endophytes. Marcel Dekker, New York, pp 31–47

Taylor T, Hass H, Kerp H (1997) A cyanolichen from the Lower Devonian Rhynie chert. Am J Bot 84:992–992

Taylor TN, Hass H, Remy W et al (1995a) The oldest fossil lichen. Nature 378:244

Taylor TN, Remy W, Hass H et al (1995b) Fossil arbuscular mycorrhizae from the Early Devonian. Mycologia 87:560–573

Taylor TN, Krings M, Dotzler N (2012) Fungal endophytes in *Astromyelon*-type (Sphenophyta, Equisetales, Calamitaceae) roots from the Upper Pennsylvanian of France. Rev Palaeobot Palynol 171:9–18

Taylor TN, Krings M, Taylor EL (2015) Fossil fungi. Academic, San Diego

Thomas-Keprta KL, Clemett SJ, Bazylinski DA et al (2001) Truncated hexa-octahedral magnetite crystals in ALH84001: presumptive biosignatures. Proc Nat Acad Sci USA 98:2164–2169

Thompson JB, Ferris FG (1990) Cyanobacterial precipitation of gypsum, calcite, and magnesite from natural alkaline lake water. Geology 18:995–998

Thulborn T (1990) Dinosaur tracks. Chapman and Hall, London

Tice MM, Lowe DR (2004) Photosynthetic microbial mats in the 3,416-Myr-old ocean. Nature 431:549–552

Timlin JA, Carden A, Morris MD et al (2000) Raman spectroscopic imaging markers for fatigue-related microdamage in bovine bone. Anal Chem 72:2229–2236

Tomescu AMF, Rothwell GW, Mapes G (2001) *Lyginopteris royalii* sp. nov. from the Upper Mississippian of North America. Rev Paleobot Palynol 116:159–173

Tomescu AMF, Rothwell GW, Honegger R (2006) Cyanobacterial macrophytes in an Early Silurian (Llandovery) continental biota: Passage Creek, lower Massanutten Sandstone, Virginia, USA. Lethaia 39:329–338

Tomescu AMF, Honegger R, Rothwell GW (2008) Earliest fossil record of bacterial-cyanobacterial mat consortia: the early Silurian Passage Creek biota (440 Ma, Virginia, USA). Geobiology 6:120–124

Tomescu AMF, Rothwell GW, Honegger R (2009) A new genus and species of filamentous microfossil of cyanobacterial affinity from Early Silurian fluvial environments (lower Massanutten Sandstone, Virginia, USA). Bot J Linn Soc 160:284–289

Toporski JKW, Steele A, Westall F et al (2002) Morphologic and spectral investigation of exceptionally well-preserved bacterial biofilms from the Oligocene Enspel formation, Germany. Geochim Cosmochim Acta 66:1773–1791

Thorseth IH, Furnes H, Tumyr O (1991) A textural and chemical study of Icelandic palagonite of varied composition and its bearing on the mechanisms of the glass-palagonite transformation. Geochim Cosmochim Acta 55:731–749

Thorseth IH, Furnes H, Tumyr O (1995) Textural and chemical effects of bacterial activity on basaltic glass: an experimental approach. Chem Geol 119:139–160

Treiman AH (2003a) Submicron magnetite grains and carbon compounds in Martian meteorite ALH84001: inorganic, abiotic formation by shock and thermal metamorphism. Astrobiology 3:369–392

Treiman AH (2003b) Traces of ancient Martian life in meteorite ALH84001: an outline of status in late 2003. Lunar and Planetary Institute, Houston, http://planetaryprotection.nasa.gov/sum mary/alh84001

Trichet J, Defarge C (1995) Non-biologically supported organomineralization. Bull Inst Oceanograph Monaco 14:203–236

Tyler SA, Barghoorn ES (1954) Occurrence of structurally preserved plants in pre-Cambrian rocks of the Canadian Shield. Science 119:606–608

Tyler SA, Barghoorn ES (1963) Ambient pyrite grains in Precambrian cherts. Am J Sci 261:424–432

Ueno Y, Isozaki Y, Yurimoto H et al (2001a) Carbon isotopic signatures of individual Archean microfossils(?) from Western Australia. Int Geol Rev 43:196–212

Ueno Y, Maruyama S, Isozaki Y et al (2001b) Early Archean (ca. 3.5 Ga) microfossils and ^{13}C-depleted carbonaceous matter in the North Pole area, Western Australia: field occurrence and geochemistry. In: Nakashima S, Maruyama S, Brack A et al (eds) Geochemistry and the origin of life. Universal Academy Press, Tokyo, pp 203–236

Van Lith Y, Warthmann R, Vasconcelos C et al (2003) Microbial fossilization in carbonate sediments: a result of the bacterial surface involvement in dolomite precipitation. Sedimentology 50:237–245

Van Kranendonk MJ (2006) Volcanic degassing, hydrothermal circulation and the flourishing of early life on Earth: a review of the evidence from c. 3490–3240Ma rocks of the Pilbara Supergroup, Pilbara Craton, Western Australia. Earth-Sci Rev 74:197–240

Van Kranendonk MJ, Smithies RH, Hickman AH et al (2007) Review: secular tectonic evolution of Archean continental crust: interplay between horizontal and vertical processes in the formation of the Pilbara Craton, Australia. Terra Nova 19:1–38

van Zuilen MA, Lepland A, Teranes J et al (2003) Graphite and carbonates in the 3.8 Ga old Isua Supracrustal Belt, southern West Greenland. Precambrian Res 126:331–348

Vecht A, Ireland TG (2000) The role of vaterite and aragonite in the formation of pseudo-biogenic carbonate structures: implications for Martian exobiology. Geochim Cosmochim Acta 64:2719–2725

Vinther J, Briggs DE, Prum RO et al (2008) The colour of fossil feathers. Biol Lett 4:522–525

Vinther J, Briggs DE, Clarke J et al (2010) Structural coloration in a fossil feather. Biol Lett 6:128–131

Visscher PT, Beukema J, van Gemerden H (1991) In situ characterization of sediments: measurements of oxygen and sulfide profiles with a novel combined needle electrode. Limnol Oceanogr 36:1476–1480

Visscher PT, Stolz JF (2005) Microbial mats as bioreactors: populations, processes, and products. Palaeogeogr Palaeoclim Palaeoecol 219:87–100

Wacey D (2009) Early life on Earth: a practical guide. Springer, New York

Wacey D (2012) Earliest evidence for life on Earth: and Australia perspective. Aust J Earth Sci 59:153–166

Wacey D, McLoughlin N, Green OR et al (2006) The ~3.4 billion-year-old Strelley Pool Sandstone: a new window into early life on Earlt. Int J Astrobiol 5:333–342

Wacey D, Kilburn MR, McLoughlin N et al (2008) Use of NanoSIMS in the search for early life on Earth: ambient inclusion trails in a c. 3400 Ma sandstone. J Geol Soc 165:43–53

Wacey D, Kilburn MR, Saunders M et al (2011a) Microfossils of sulphur-metabolizing cells in 3.4-billion-year-old rocks of Western Australia. Nat Geosci 4:698–702

Wacey D, Saunders M, Brasier MD et al (2011b) Earliest microbially mediated pyrite oxidation in ~3.4 billion-year-old sediments. Earth Planet Sci Let 301:393–402

Waldbauer JR, Sherman LS, Sumner DY et al (2009) Late Archean molecular fossils from the Transvaal Supergroup record the antiquity of microbial diversity and aerobiosis. Precambrian Res 169:28–47

Walsh MM (1992) Microfossils and possible microfossils from the Early Archean Onverwacht Group, Barberton Mountain Land, South Africa. Precambrian Res 54:271–293

Walsh MM, Westall F (2003) Archean biofilms preserved in the Swaziland Supergroup, South Africa. In: Krumbein WE, Paterson DM, Zavarzin GA (eds) Fossil and recent biofilms. Kluwer, Dordrecht, pp 307–316

Walter MR (1983) Archean stromatolites: evidence of the Earth's earliest benthos. In: Schopf JW (ed) Earth's earliest biosphere: its origin and evolution. Princeton University Press, Princeton, pp 187–213

Walter MR, Oehler JH, Oehler DZ (1976) Megascopic algae 1300 million years old from the Belt Supergroup, Montana: a reinterpretation of Walcott's *Helminthoidichnites*. J Paleo 50:872–881

Walter MR, Du RL, Horodyski RJ (1990) Coiled carbonaceous megafossils from the Middle Proterozoic of Jixian (Tianjin) and Montana. Am J Sci 290-A:133–148

Wang B, Yeun LH, Xue JY et al (2010) Presence of three mycorrhizal genes in the common ancestor of land plants suggests a key role of mycorrhizas in the colonization of land by plants. New Phytol 186:514–525

Wang B, Zhao F, Zhang H et al (2012) Widespread pyritization of insects in the Early Cretaceous Jehol Biota. Palaios 27:707–711

Waterbury JB, Stanier RY (1978) Patterns of growth and development in pleurocapsalean cyanobacteria. Microbiol Rev 42:2–44

Westall F (1999) The nature of fossil bacteria: a guide to the search for extraterrestrial life. J Geophys Res—Planets 104(E7):16437–16451

Westall F, Folk RL (2003) Exogenous carbonaceous microstructures in Early Archean cherts and BIFs of the Isua Greenstone Belt: implications for the search for life in ancient rocks. Precambrian Res 126:313–330

Westall F, Boni L, Guerzoni E (1995) The experimental silicification of microorganisms. Palaeontology 38:495–528

Westall F, de Wit MJ, Dann J et al (2001) Early Archean fossil bacteria and biofilms in hydrothermally-influenced sediments from the Barberton greenstone belt, South Africa. Precambrian Res 106:93–116

Westall F, de Vries ST, Nijman W et al (2006) The 3.446 Ga "Kitty's Gap Chert", an early Archean microbial ecosystem. Geol S Am Special Paper 405:105–131

Wierzchos J, Berlanga M, Ascaso C et al (1996) Micromorphological characterization and lithification of microbial mats from the Ebro Delta (Spain). Int Microbiol 9:289–295

Wilby PR, Briggs DE (1997) Taxonomic trends in the resolution of detail preserved in fossil phosphatized soft tissues. Geobios 30:493–502

Wilby PR, Briggs DE, Bernier P et al (1996) Role of microbial mats in the fossilization of soft tissues. Geology 24:787–790

Wills MA (2001) Disparity vs. diversity. In: Briggs DEG, Crowther PR (eds) Palaeobiology II. Blackwell, Oxford, pp 495–500

Wilmeth DT, Dornbos SQ, Isbell JL (2014) Putative domal microbial structures in fluvial siliciclastic facies of the Mesoproterozoic (1.09 Ga) Copper Harbor Conglomerate, Upper Peninsula of Michigan, USA. Geobiology 12:99–108

Wilson AS, Dodson HI, Janaway RC, Pollard AM, Tobin DJ (2007) Selective biodegradation in hair shafts derived from archaeological, forensic and experimental contexts. Br J Dermatol 157:450–457

Wolfe AP, Csank AZ, Reyes AV et al (2012) Pristine Early Eocene wood buried deeply in kimberlite from northern Canada. PLoS One 7, e45537

Woolhouse HW (1984) The biochemistry and regulation of senescence in chloroplasts. Can J Bot 62:2934–2942

Wingender J, Neu TR, Flemming HC (2012) Microbial extracellular polymeric substances: characterization, structure and function. Springer Science & Business Media, Heidelberg, 258p

Wright DT, Wacey D (2004) Sedimentary dolomite: a reality check. Geol Soc Lond Special Pub 235:65–74

Wuttke M (1983a) Weichteilerhaltung durch lithifizierte Mikoorganismen bei mittel-eozänen Vertebraten aus dem Ölschiefer der 'Grube Messel' bei Darmstadt. Senckenbergiana Lethaea 64:503–527

Wuttke M (1983b) Aktuopaläontologische Studien über den Zerfall von Wirbeltieren: Teil 1, Anura. Senckenbergiana Lethaea 64:529–560

Xiao S, Knoll AH, Kaufman AJ et al (1997) Neoproterozoic fossils in Mesoproterozoic rocks? Chamostratigraphic resolution of a biostratigraphic conundrum from the North China Platform. Precambrian Res 84:197–220

Xiao S, Zhang Y, Knoll AH (1998) Three-dimensional preservation of algae and animal embryos in a Neoproterozoic phosphorite. Nature 391:553–558

Xiao S, Yuan X, Steiner M et al (2002) Macroscopic carbonaceous compressions in a terminal Proterozoic shale: a systematic reassessment of the Miaohe biota, South China. J Paleo 76:347–376

Xiao S, Schiffbauer JD, McFadden KA et al (2010) Petrographic and SIMS pyrite sulfur isotope analyses of Ediacaran chert nodules: implications for microbial processes in pyrite rim formation, silicification, and exceptional fossil preservation. Earth Planet Sci Lett 297:481–495

Xiao S, Droser M, Gehling JG et al (2013) Affirming life aquatic for the Ediacara biota in China and Australia. Geology 41:1095–1098

Yan Y, Liu Z (1993) Significance of eukaryotic organisms in the microfossil flora of Changcheng system. Acta Micropalaeontol Sin 10:167–180

Yin L-M (1997) Acanthomorphic acritarchs from Meso-Neoproterozoic shales of the Ruyang Group, Shanxi, China. Rev Palaeobot Palynol 98:15–25

Yin Z, Liu P, Li G et al (2014) Biological and taphonomic implications of Ediacaran fossil embryos undergoing cytokinesis. Gondwana Res 25:1019–1026

Yochelson EL, Fedonkin MA (2000) A new tissue-grade organism 1.5 billion years old from Montana. Proc Biol Soc Washington 113:843–847

Yoshida Y (1962) Nuclear control of chloroplast activity in *Elodea* leaf cells. Protoplasma 54:476–492

Yuan X, Xiao S, Taylor TN (2005) Lichen-like symbiosis 600 million years ago. Science 308:1017–1020

Zegers TE, de Wit MJ, Dann J, White SH (1998) Vaalbara, Earth's oldest assembled continent? A combined structural, geochronological, and palaeomagnetic test. Terra Nova 10:250–259

Zhang F, Kearns SL, Orr PJ et al (2010) Fossilized melanosomes and the colour of Cretaceous dinosaurs and birds. Nature 463:1075–1078

Zheng X, O'Connor J, Huchzermeyer F et al (2013) Preservation of ovarian follicles reveals early evolution of avian reproductive behaviour. Nature 495:507–511

Zhu M, Gehling JG, Xiao S et al (2008) Eight-armed Ediacara fossil preserved in contrasting taphonomic windows from China and Australia. Geology 36:867–870

Zolotov MY, Shock EL (2000) An abiotic origin for hydrocarbons in the Allan Hills 84001 Martian meteorite through cooling of magmatic and impact-generated gases. Meteorit Planet Sci 35:629–638

Chapter 4
Endolithic Microorganisms and Their Habitats

Christopher R. Omelon

Abstract Endolithic microorganisms are widespread in desert biomes, where hostile environmental conditions limit the majority of life to rock habitats. In these habitats, microorganisms receive light for photosynthesis, moderated and warmer temperatures, protection from UV radiation, and prolonged exposure to liquid water. In general, these microbial communities are composed of phototrophic microorganisms as well as fungi and heterotrophic bacteria. Microbial composition is distinct from soil communities, suggesting these habitats select for microorganisms best suited to this environment. The habitat is not nutrient limited, which explains why these microbial communities colonize a wide range of lithic substrates with different mineralogies; however, greater environmental pressures select for those able to tolerate increasingly harsh conditions. Growth rates vary primarily as a function of moisture availability, resulting in long-lived communities in the driest deserts. While most microorganisms require liquid water for growth, some lichens with an algal phycobiont can photosynthesize with water vapor alone, a significant advantage in these water-limited biomes. Additional strategies against stress include synthesis of pigments, EPS, and osmoprotectants, which significantly offsets the growth of biomass. Microbial activity leads to physical and geochemical weathering, but can also result in stabilization of the lithic habitat. Identification of endolithic biosignatures and microbial fossils has resulted in their study from an astrobiological perspective in the search for life on other planets.

4.1 Introduction

The abundance and diversity of microorganisms in the biosphere reflect their capacity to harvest energy from diverse organic and inorganic substrates and ability to grow under a wide range of natural conditions. Often, however, the complexity of

---

C.R. Omelon (✉)
Department of Geological Sciences, The University of Texas at Austin, 2275 Speedway, Mail Stop C9000, Austin, TX 78712, USA
e-mail: omelon@jsg.utexas.edu

a microbial community depends upon the suitability of a given habitat for colonization and the availability of energy and carbon. In many cases, habitats exhibiting extremes in climatic or environmental conditions effectively limit the number of species that can exist within a given ecological niche. An example of this natural selection is found in *endolithic* ("endo-" = within; "-lithic" = rock) habitats (Goublic et al. 1981), where microbial colonization is restricted to those organisms able to acquire the necessary resources for growth within the physical confines beneath rock surfaces. The two most common forms of endolithic microbial communities are found as interstitial colonizers of cracks and fissures (chasmoendolithic) or in pore spaces between mineral grains (cryptoendolithic). Other lithobiontic habitats include the ventral surfaces of translucent or opaque rocks (hypolithic), within the underside of translucent rocks (hypoendolithic), or within porosity in rocks created through active boring by microorganisms into the substrate (euendolithic).

This endolithic habitat is most commonly colonized by microorganisms in hot and cold deserts throughout the world, where extremes in temperature, moisture, and radiation prevail at the Earth's surface. Colonization and development of diverse microbial assemblages results from preferential microenvironmental conditions within these endolithic habitats where extremes are moderated (Friedmann 1980; Warren-Rhodes et al. 2006). However, this life must also utilize specific strategies to ameliorate stresses such as desiccation and rapid temperature fluctuations. The metabolic activity of endolithic microorganisms can subsequently alter the local geochemical environment through microbe-mineral interactions, which can be constructive or destructive depending upon the nature of a particular mechanism or substrate. While destructive mechanisms such as leaching and mineral solubilization can lead to accelerated weathering of the endolithic habitat, constructive development of crusts can provide further protection from stresses. In addition, the formation and preservation of microbial fossils or biosignatures is of current interest in the context of studies in astrobiology and the search for evidence of life on other planets.

4.2 Global Distribution of Endolithic Microorganisms

Endolithic microorganisms have been reported from temperate regions where climatic extremes limit epilithic ("epi-" = upon) colonization (Bell et al. 1986; Casamatta et al. 2002; Ferris and Lowson 1997; Gerrath et al. 2000; Tang et al. 2012); however, endolithic and other lithobiontic microorganisms are more common in the vast arid desert climates of the world that make up the largest terrestrial biome on the planet (Pointing and Belnap 2012). These regions are identified by their low ratios of precipitation to potential evaporation (normally < 1), highlighting the fact that they are moisture limited. Lithobiontic microorganisms have been studied in hot deserts including the Mojave and Sonora (USA), Atacama (Chile), Gobi (China, Mongolia), Negev (Israel), Namib

(Namibia, Angola), the Al-Jafr Basin (Jordan), and Turpan Depression (China) and as well as polar deserts in the Arctic and Antarctic Dry Valleys (Bell 1993; Bungartz et al. 2004; Cockell and Stokes 2004; Cockell et al. 2003; Cowan et al. 2010; Dong et al. 2007; Friedmann 1980; Friedmann et al. 1987; Hughes and Lawley 2003; Lacap et al. 2011; McKay et al. 2003; Omelon et al. 2006a; Schlesinger et al. 2003; Smith et al. 2000; Stomeo et al. 2013; Warren-Rhodes et al. 2007a). They are also documented from high altitudes in mountainous regions of the world, including Europe and Asia (Hoppert et al. 2004; Sigler et al. 2003; Walker and Pace 2007b; Wong et al. 2010b). Given the scarcity of plant or animal life as well as low levels of soil nutrients, it is thought that these microbial communities are the dominant form of biomass in desert environments (Cary et al. 2010; Cockell and Stokes 2004; de la Torre et al. 2003; Pointing et al. 2009; Walker and Pace 2007a; Warren-Rhodes et al. 2006). Due to the abundance of desert pavements in arid climates that are colonized by hypolithic microorganisms, these have been intensively studied to determine what controls the limits to life on Earth as well as microbial diversity in these extreme habitats (Makhalanyane et al. 2013a, b). While outside the scope of this work, it will be interesting to observe how these studies are translated to future evaluations of the diversity of microorganisms within endolithic habitats and if these same findings hold true.

4.3 Microbial Diversity of Endolithic Habitats

In contrast to microorganisms inhabiting the deep subsurface where activity is driven by chemolithotrophic metabolisms such as sulfate reduction, iron reduction, or methanogenesis (Lovely and Chapelle 1995), endolithic communities at the Earth's surface consist of photoautotrophic primary producers such as algae, cyanobacteria, and lichens, as well as consumers and decomposers including fungi and heterotrophic bacteria (Cockell and Stokes 2004; Cowan et al. 2010; de la Torre et al. 2003; Friedmann et al. 1980, 1981; Friedmann 1982; Friedmann and Ocampo-Friedmann 1984a; Hirsch et al. 1988, 2004b; Omelon et al. 2007; Selbmann et al. 2005). Archaea have also been documented in endolithic habitats around the world (de los Ríos et al. 2010; Horath and Bachofen 2009; Khan et al. 2011; Walker and Pace 2007b; Wong et al. 2010b).

Being one of the first discoveries of widespread life in an extreme polar desert, microorganisms from cryptoendolithic habitats in the Dry Valleys of Antarctica have received much attention. Two primary assemblages are defined: (1) eukaryotic communities that include lichenized fungi and algae and (2) prokaryotic communities dominated by cyanobacteria, both containing heterotrophic bacteria and fungi (de los Ríos et al. 2014). The cyanobacterium *Chroococcidiopsis*, found in the Dry Valleys, has been widely reported from lithic habitats in desert biomes around the world (Schlesinger et al. 2003; Smith et al. 2000; Warren-Rhodes et al. 2006). The likely reason for their apparent cosmopolitan presence in these habitats is due to their ability to withstand long periods of desiccation (Potts 1999) and radiation (Billi et al. 2000) as well as their ability to produce unique survival cells under nitrogen-limited conditions (Fewer et al. 2002). In addition, they are capable of quickly reactivating photosynthesis after these long periods of desiccation when liquid water is available (Hawes et al. 1992).

Early work to characterize the microbial diversity of lithobiontic habitats focused on culturing techniques, whereby microorganisms including phototrophic cyanobacteria and algae as well as fungi and heterotrophic bacteria were identified by morphology. More recent culture-independent molecular techniques have led to a more complete understanding of the diversity of microorganisms in these habitats, which often include in addition to the *Cyanobacteria* a wide range of other microorganisms such as the *Acidobacteria*, *Bacteroidetes*, *Chloroflexi*, *Proteobacteria*, *Actinobacteria*, *Verrucomicrobia*, *Firmicutes*, and others. Numerous studies have shown that typically they are distinct from soil communities, despite the most notable fact that hypolithic microbial assemblages are in direct in contact with soils (Pointing et al. 2007, 2009; Schlesinger et al. 2003; Warren-Rhodes et al. 2006; Wong et al. 2010a). The argument has been made that these habitats worldwide select for microorganisms best suited for this environment (Sigler et al. 2003); however, a growing understanding of what "extreme" means from a microbial ecology standpoint appears to pinpoint the hyperarid core of the Atacama Desert as having conditions beyond which life cannot survive (Warren-Rhodes et al. 2006).

4.4 Geology of Endolithic Habitats

While endolithic microbial communities are generally restricted to terrestrial surface habitats, they colonize a diverse range of geologic substrates. In most cases, the lithobiontic environment provides a protective environment while still permitting enough light penetration to support primary productivity by photosynthesis (Warren-Rhodes et al. 2007b). Physical and chemical properties of the rock substrate such as mineralogy, porosity and permeability, capacity for moisture uptake and retention, pH, and access to nutrients and protection from climatic extremes are all important, especially in endolithic habitats (Cockell et al. 2009a, b; Herrera et al. 2009; Kelly et al. 2011; Omelon et al. 2007). In contrast to hypolithic habitats, endolithic microorganisms are provided addition protection from physical weathering such as wind abrasion due to the rigid framework of the rock matrix, which contrasts to ecosystems living at the soil-rock interface.

The most common occurrences are found in porous sedimentary rocks such as quartz-rich sandstone; however, weathered limestone and dolomites are also colonized (Ferris and Lowson 1997; Friedmann 1980; Norris and Castenholz 2006; Omelon et al. 2006a; Saiz-Jimenez et al. 1990; Sigler et al. 2003; Tang et al. 2012; Wong et al. 2010b). In the case of hypolithic habitats, both translucent and opaque rocks can host microbial life (Cockell and Stokes 2004). It has recently been shown that contemporary calcite precipitating from groundwater spring discharge is colonized by endolithic microorganisms as the lithic habitat become drier and more cemented (Starke et al. 2013).

Lithic microorganisms are found in other rock types including evaporites such as halite and gypsum (Boison et al. 2004; Cockell et al. 2010; Dong et al. 2007; Hughes and Lawley 2003; Stromberg et al. 2014; Wierzchos et al. 2006; Ziolkowski et al. 2013), weathered granite (Ascaso and Wierzchos 2003; de los Ríos et al. 2002, 2005, 2007; Wierzchos et al. 2003), and marble (Büdel et al. 2009; Sterflinger et al. 1997). Porosity in volcanic rocks including rhyolitic ignimbrites has been shown to support endolithic microbial communities (Wierzchos et al. 2013) as have deposits altered by meteorite impacts on Earth (Parnell et al. 2004; Pontefract et al. 2014) and silica associated with hot spring sinter (Phoenix et al. 2006; Walker et al. 2005). Although normally found in intertidal zones, euendolithic microorganisms have also been observed in terrestrial micrite (Hoppert et al. 2004).

4.5 Light Regime Within Endolithic Habitats

Phototrophic microorganisms dominate as primary producers and are the largest community in these near-surface environments. Given that they require sunlight for growth, they are limited to the maximum depth of light penetration into the lithic habitat, be it through a translucent pebble as in hypolithic habitats or into a rock matrix such as sandstones. Light diminishes rapidly with depth on the order of 70–90 % for each millimeter beneath the surface; however, measurements of the light regime show that these photosynthetic microorganisms can grow under very low light levels, down to 0.08 % and 0.005 % of the incident light flux in hypolithic and cryptoendolithic habitats, respectively (Nienow et al. 1988b; Schlesinger et al. 2003).

Degree of light penetration varies as a function of grain size, the presence of opaque minerals, as well as microbial biofilms. Temporal changes on various scales including the development of a surface crust or varnish (long term) and the presence or absence of water can significantly change light conditions within the endolithic habitat. Accumulation of allochthonous dust composed of iron oxides and clays on the rock surface will dramatically reduce light penetration; in one study, reduction by up to 90 % at a depth less than 0.5 mm beneath the rock surface was noted, further diminished to only 0.005 % of the incident light flux at the deepest point of colonization (Nienow et al. 1988b). In contrast, rocks having little to no surface crust are colonized to greater depths as light penetrates much deeper into the rock (Omelon et al. 2007). These differences, however, may be constrained to certain rock types or environments as limestone rocks inhabited by endolithic microorganisms showed no correlation between degree of light penetration and colonization depth (Matthes et al. 2001). In contrast, water in pore spaces enhances light penetration due to its high refractive index compared to air and combined with the reflective properties of certain substrates such as quartz grains can greatly increase the amount of light in the cryptoendolithic habitat (Nienow et al. 1988b).

The attenuating properties of the overlying rock mean that these habitats receive less incident UV and photosynthetically active radiation. While potentially beneficial for life by providing some protection from photoinhibition, analysis of the full spectrum of solar radiation penetrating into the endolithic habitat of colonized rock from the Antarctic Dry Valleys shows that the presence of water diminishes protection against UV radiation under saturated conditions (McKay 2012), leading to short-term higher doses of radiation until this water has evaporated. Despite the rarity of liquid water in these habitats, intense UV exposure to these slow-growing microorganisms for even short periods of time may limit microbial diversity to those that can tolerate higher UV levels.

4.6 Nutrients

Endolithic habitats are composed of rocks that vary in solubility but are generally not considered a source of nutrients for microorganisms. Studies of nutrient conditions within endolithic habitats from the Dry Valleys show colonized regions having adequate supplies of inorganic nitrogen as nitrates and ammonium (Friedmann and Kibler 1980; Greenfield 1988). In addition, microbial growth is not stimulated when these or other nutrients including phosphate or manganese are added to the system (Johnston and Vestal 1986, 1991; Vestal 1988a), suggesting that they are not lacking. Although in some cases it is possible that nutrients required for microbial growth are obtained directly by in situ weathering of the host rock (Siebert et al. 1996), there does not appear to be a general selection by endolithic microorganisms for a specific type of mineralogy (e.g., quartz, calcite, granite). This suggests that nutrients required for energy and growth are more likely derived from allochthonous sediments such as dust (Johnston and Vestal 1989; Omelon et al. 2007; Pontefract et al. 2014; Tang et al. 2012; Walker and Pace 2007a). Another reason that these microbial communities may not suffer from a lack of nutrients is due to the fact that they are active for only very short periods of time, either due to freezing temperatures (Friedmann and Kibler 1980) or desiccation (Wierzchos et al. 2013).

4.7 Microclimatic Conditions Within Endolithic Habitats

As primary productivity in lithobiontic habitats is based upon photosynthetic life, microorganisms inhabiting these environments are in close proximity to surface desert conditions, which range from cold to hot and semiarid to hyperarid. While their temperature and moisture regimes are dictated generally by the local climate, speculation that microenvironmental conditions in lithobiontic habitats differed from those outside the lithic environment was first raised in the 1960s in studies of algae in the Negev Desert (Friedmann et al. 1967). Since then, many studies have shown how these habitats provide necessary advantages that permit microbial life to flourish in an otherwise hostile environment. Characterization of the microclimate—most notably temperature and water availability—has provided important information regarding the activity of these microbial communities. Comparison of these microenvironmental conditions to the local climate provides insight not only into how these habitats provide refuge for lithobiontic microorganisms but also sources of water that are required to support life in severely harsh conditions.

Although not strictly endolithic, studies of phototrophic microorganisms in hot desert hypolithic habitats have shown that they can remain active over a wide range of temperatures (~0–50 °C) (Tracy et al. 2010), with tolerance up to >90 °C (Schlesinger et al. 2003). In contrast, seasonal darkness in polar environments results in long periods of subzero temperatures that effectively arrest microbial activity. Warming of cryptoendolithic habitats occurs through solar radiation reaching the rock surface; due to the heat capacity and transparency of rock, elevated temperatures are generated in the subsurface compared to the overlying air (Friedmann et al. 1987, 1993; Omelon et al. 2006a). Temperature differences between air and the endolithic habitat can be significant, with measured differences up to 20 °C (Cockell et al. 2003; Friedmann 1977; Kappen et al. 1981; McKay and Friedmann 1985; Omelon et al. 2006a). This enhanced heating of the endolithic habitat increases the time during which microorganisms can be metabolically active (Friedmann et al. 1987, 1993; Omelon et al. 2006a), which is especially important in polar deserts such as the Dry Valleys where temperatures are >0 °C for only 50–500 h year^{-1} (Friedmann et al. 1987, 1993; Kappen et al. 1981). The heat capacity of rocks can also moderate high-frequency rapid temperature fluctuations, which when oscillating around 0 °C generate freeze-thaw cycles that can lead to uncontrolled intracellular ice formation within microorganisms. While such conditions are thought to be at least partially responsible for controlling epilithic colonization of rock surfaces, it has been suggested that temperature fluctuations are moderated within endolithic habitats (McKay and Friedmann 1985; Nienow et al. 1988a).

Precipitation in the Antarctic Dry Valleys is rare and only in the form of snow, much of which sublimates back into the atmosphere and is therefore mostly unavailable for life (Friedmann et al. 1987). Water that does occasionally melt on rock surfaces by solar heating enters the endolithic habitat via percolation into pore spaces (Friedmann 1978; Friedmann and McKay 1985). Vital for sustaining life in these endolithic habitats is the fact that this infiltrated moisture persists in the subsurface long after precipitation events have passed (Friedmann et al. 1987, 1988; Kappen et al. 1981; Omelon et al. 2006a). This water is effectively trapped within the endolithic habitat and permits reactivation of desiccated biomass, with subsequent loss back into the atmosphere varying as a function of porosity, permeability, and the degree of surface crust formation that can severely retard evaporation rates (Friedmann and Ocampo-Friedmann 1984a).

This combination of cold temperatures and scarce water results in short periods of time for microbial activity, during which growth rates are slow; measurements of carbon turnover times of these microbial communities suggest that they are 10^3–10^4 years old (Johnston and Vestal 1991; Sun and Friedmann 1999). The remarkable age of this community points also to extreme stability of the endolithic habitat, as would be necessary to support and protect endolithic microorganisms over long time periods, and is shown by the rock surface having an estimated atmospheric exposure period of several thousand years (Friedmann and Weed 1987). In contrast, similar habitats at the same latitude in the Arctic experience warmer temperatures and much longer active growth periods of ~2400 h year^{-1} (Omelon et al. 2006a). Combined with higher numbers of precipitation events including summer rainfall, carbon turnover times of 10^1 years show that these endolithic microorganisms are much younger (Ziolkowski et al. 2013). It is possible that this relative abundance of liquid water combined with longer periods of microbial activity in the Arctic leads to higher rates of erosion due to enhanced chemical weathering of the endolithic habitat.

Studies in search of life in the most extreme deserts on Earth have revealed that water is the most important determinant for survival in all lithobiontic habitats. In hot deserts, water can occur as sporadic rainfall but more commonly as fog or dew that condenses at night on rock surfaces, only to evaporate shortly after sunrise (Kappen et al. 1980; Kidron 2000; Warren-Rhodes et al. 2006). In some instances both rain and fog can contribute moisture, with fog in one case being more important to support life in these habitats (Warren-Rhodes et al. 2013). Semiarid desert conditions in Australia can support hypolithic photosynthetic activity for ~942 h year^{-1} (Tracy et al. 2010), similar to time periods in other deserts such as China (200–922 h year^{-1}) (Warren-Rhodes et al. 2007a, b) and the Negev Desert (1400 h year^{-1}). These estimates are far greater than those calculated for the hyperarid core of the Atacama Desert, where photosynthesis is restricted to <75 h year^{-1} (Warren-Rhodes et al. 2006). Such limited water availability and therefore short growth periods under extremely arid conditions result in long-lived communities, comparable in age to those found in cryptoendolithic habitats in the Antarctic Dry Valleys (Warren-Rhodes et al. 2006).

In addition to hypolithic habitats, endolithic microorganisms inhabit soil gypsum and anhydrite crusts that form by water migration and evaporation in semiarid and arid regions including the Atacama, Mojave, and Al-Jafr Deserts (Dong et al. 2007; Wierzchos et al. 2006). Halite in the hyperarid core of the Atacama also hosts endolithic microorganisms (de los Ríos et al. 2010; Gramain et al. 2011; Parro et al. 2011), which is thought to select for those microorganisms that can tolerate or manage osmotic stresses associated with hypersalinity. The Atacama has received much attention in recent years due to the extreme aridity of this hyperarid region, which is considered to be the driest place on Earth. Studies of hypolithic microbial communities in this area were thought to reveal the absolute limits to photosynthetic life (Navarro-González et al. 2003; Warren-Rhodes et al. 2006); however, that has since been challenged by confirmation of cryptoendolithic microorganisms colonizing volcanic rhyolitic ignimbrites (Wierzchos et al. 2013). As in all cases, these endolithic communities include photosynthetic microorganisms and heterotrophic bacteria and are afforded the same protective characteristics as in other endolithic habitats. Water in this part of the hyperarid core is observed as rainfall that occurs only sporadically, with estimates of photosynthetic activity at <100 h year^{-1} (Wierzchos et al. 2013) being comparable to that in hypolithic habitats reported previously from the Atacama (Warren-Rhodes et al. 2006).

Studies of calcium sulfate crusts in the Atacama showed that endolithic microorganisms were present in areas that experienced a cumulative relative humidity of $RH > 60\ \%$ but were absent in areas that were below this threshold (Wierzchos et al. 2011), suggesting a possible moisture limit to life in these types of endolithic habitats. Endolithic microorganisms inhabiting halite crusts benefit from condensation of water vapor in pore spaces at an $RH > 75\ \%$ that corresponds to halite deliquescence (Davila et al. 2008). Moisture measurements showed that this occurred within the endolithic habitat far more frequently than in the overlying air to create conditions amenable to photosynthesis for ~215 h year^{-1}, in contrast to 6 h year^{-1} in the latter. Perhaps even more remarkable was the discovery of endolithic microorganisms surviving in nanoporous halite crusts where water forms by capillary condensation at humidities lower than halite deliquescence, with the capacity to retain moisture far exceeding that in other endolithic habitats (Wierzchos et al. 2012).

4.8 Microbial Activity in the Absence of Liquid Water

While studies have attempted to define the minimum limits of moisture required for supporting growth of cyanobacteria such as *Chroococcidiopsis* in lithic habitats, much of that work has focused on lichens due to their poikilohydric nature, meaning that these microorganisms can tolerate low cell water content caused by long periods of dryness that allows them to live in such extreme arid climates without suffering damage. As has been observed in microclimatic studies of endolithic habitats in hot deserts, activity is often dictated by the presence of water at night that permits hydration and dark respiration followed by CO_2 fixation associated with net photosynthesis in the early part of the day. This activity subsequently ceases as temperatures rise and humidity levels drop, leading to desiccation due to water loss through evaporation (Lange et al. 1990, 2006).

Under more extreme arid conditions, it has been shown that lichens are metabolically active in the absence of liquid water, down to 70 % relative humidity (Lange 1969; Lange et al. 1970, 1994; Nash et al. 1990; Palmer and Friedmann 1990; Lange and Redon 1983; Redon and Lange 1983). Interestingly, lichens with algal phycobionts appear to function at these lower relative humidities, whereas those with cyanobacterial phycobionts do not, having a higher threshold cutoff near 90 % (Hess 1962; Palmer and Friedmann 1990). While all can revert back to active metabolism through contact with liquid water (Potts and Friedmann 1981), it has been shown that uptake of water vapor alone can reactivate photosynthesis in lichens with an algal phycobiont (Butin 1954; Lange and Bertsch 1965; Lange and Kilian 1985; Nash et al. 1990; Schroeter 1994). In contrast, lichens with cyanobacteria as phycobionts do not exhibit the same universal capacity and appear to require liquid water to activate photosynthesis (Lange and Kilian 1985; Lange et al. 1986, 1990, 1993, 2001; Lange and Ziegler 1986; Schroeter 1994). Microscopic examination of both types of lichens showed this to be due to the inability of cyanobacteria to attain turgidity when hydrated with water vapor alone (Büdel and Lange 1991). However, a cyanobacterial phycobiont isolated in the laboratory has been shown to achieve turgor and photosynthesize under conditions of high humidity (Lange et al. 1994). Such work brings validity to earlier studies showing that cyanobacteria can photosynthesize under arid conditions, including endolithic habitats (Brock 1975; Palmer and Friedmann 1990; Potts and Friedmann 1981).

While photosynthetic activity in the absence of liquid water has been documented in arid climates of temperate regions where local humidity can be high (Lange and Redon 1983; Redon and Lange 1983), metabolic activity of these microorganisms can also occur at subzero temperatures where water exists in a solid phase as snow or ice, often under snow cover (Kappen et al. 1986, 1990; Kappen 1989, 1993a; Kappen and Breuer 1991; Pannewitz et al. 2003; Schroeter and Scheidegger 1995). Melting of snow and ice can lead to moistening (Lange 2003), but water vapor by itself can support metabolic activity under cold temperatures (Kappen et al. 1995). This is believed to occur whereby a vapor gradient forms between ice and the dry lichen thallus (Kappen and Schroeter 1997).

The ability to attain net photosynthesis using water vapor alone as well as survive long periods of desiccation is an important survival strategy for microorganisms in desert habitats. Lichens with algal phycobionts appear to attain positive net photosynthesis under lower relative humidity conditions than those with a cyanobacterial phycobiont and experience much higher rates of photosynthesis when exposed to higher humidity levels. This suggests that they are the best opportunists to survive under the most arid conditions on the planet and only benefit from living in endolithic habitats that can provide advantageous moisture conditions.

4.9 Adaptation to Stress

Climatic stresses in desert environments are greatest on exposed surfaces, which is why endolithic habitats are normally the last refuges for life as they provide protection to some degree. However, the pressures of desiccation and excessive radiation are only partially offset by periodic moisture availability and radiative shielding. Various photoprotective-screening and photoprotective-quenching pigments are synthesized by cyanobacteria, algae, and fungi in cryptoendolithic habitats, which often results in a clear vertical zonation of these different microorganisms in response to light and UV levels. Examples include scytonemin, mycosporine-like amino acids, carotenoids, and melanin (Jorge Villar et al. 2005a, b; Selbmann et al. 2005; Wynn-Williams et al. 1999).

In addition to pigments, cyanobacteria and fungi in endolithic habitats produce abundant volumes of extracellular polysaccharides (EPS) (de los Ríos et al. 2004; Omelon et al. 2006b; Pointing and Belnap 2012; Selbmann et al. 2005), which help to regulate intracellular water loss (Adhikary 1998; Potts 1999; Tamaru et al. 2005; Wingender et al. 1999) and mediate cell wall damage resulting from shrinking and swelling associated with wide variations in temperatures and moisture (Grilli Caiola et al. 1993; Selbmann et al. 2002). In addition to reducing physiological stress imposed by desiccation, EPS likely helps control rates of cooling that, when occurring rapidly around freeze-thaw temperatures, would help regulate osmotic stress and control intracellular ice formation (Vincent 2007). Alternatively, cyanobacteria have been shown to resist desiccation by accumulating water-stress proteins or producing osmoprotectants such as trehalose and sucrose (Hershkovitz et al. 1991; Scherer and Potts 1989) with abundant free amino acids detected within the endolithic habitat (Greenfield 1988; Siebert et al. 1991). The abundance of these leached materials suggests that only a fraction (0.025 %) of gross productivity supports biomass growth (Friedmann et al. 1993).

4.10 Colonization Extent

Endolithic biomass is highly variable; however, measurements in colonized limestone ranged from 3 to 17 % dry weight that corresponded to rock porosity (Ferris and Lowson 1997). Given the nature of microorganisms inhabiting small spaces within rock and that rock porosity and grain size will vary, it seems reasonable to assume that biomass estimates will vary as well (Büdel et al. 2008; Kappen and Friedmann 1983; Kuhlman et al. 2008; Tuovila and LaRock 1987; Vestal 1988b). Examinations of endolithic microorganisms focus their efforts on studying and highlighting the presence of pigments (Fig. 4.1) associated with either photosynthesis or protection from excessive radiation. While it is true that heterotrophic microorganisms are only a small component of the biomass in these photic zones (Greenfield 1988; Hoppert et al. 2004), nonpigmented fungal hyphae and heterotrophic bacteria are often found penetrating deeper into the lithic habitat (Friedmann 1982; Hoppert et al. 2004; Ruisi et al. 2007). These more deeply penetrating populations are not always accounted for in studies of endolithic communities, and their presence should be determined when estimating biomass.

An example of this oversight stems from studies of microbial activity and carbon dynamics. Measurements of CO_2 concentrations in air in equilibrium with the endolithic habitat have shown that while cryptoendolithic CO_2 concentrations remain similar to atmospheric levels when dry, a net CO_2 flux from the subsurface into the atmosphere occurs upon wetting (Omelon et al. 2013). Analysis of 16S rRNA present both within the photic zone and deeper inside the rock showed a progression from cyanobacteria dominating 0.5 cm beneath the surface to a

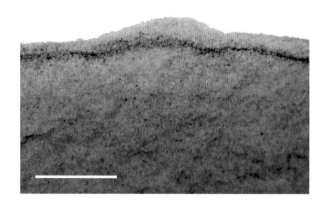

Fig. 4.1 Cryptoendolithic phototrophic microorganisms in sandstone rock from Ellesmere Island, Nunavut, Canada. Scale bar = 1 cm

community composed of *Actinobacteria*, *Alphaproteobacteria*, and *Acidobacteria* at a depth of 3.5 cm, which was beyond the point of light penetration. While phototrophic microorganisms are the primary producers, it would appear that they support a much larger community dominated by heterotrophic microorganisms at depth. These bacteria are an important part of endolithic microbial communities as decomposers (Ferris and Lowson 1997; Hirsch et al. 1988, 2004a, b; Siebert and Hirsch 1988; Siebert et al. 1996), which accelerate nutrient cycling of bioessential elements such as phosphorus (Banerjee et al. 2000). Microbial communities in hypolithic settings are also observed extending beyond the rock-soil interface, with the production of EPS binding soil particles to create a stabilized subsurface microenvironment (Chan et al. 2012) that could enhance biological sequestration of elements as well as provide moisture to support a larger microbial community.

4.11 Controls on Endolithic Microbial Community Structure

Endolithic microbial communities are thought to be some of the simplest ecosystems on Earth (Walker and Pace 2007a, b), but more recent work has shown that microbial diversity is higher in lithic habitats than in surrounding soils in more extreme environments such as the Antarctic Dry Valleys and the hyperarid Namib Desert (Pointing et al. 2009; Stomeo et al. 2013). Early delineation of lichen-dominated and cyanobacteria-dominated endolithic microbial communities in the Dry Valleys (de la Torre et al. 2003; Friedmann 1982) raised the question as to what controls diversity in these habitats and has become a major area of investigational focus given the growing number of studies assessing diversity in a wide range of desert environments. It was proposed earlier that variations in microbial community composition in endolithic habitats of the Antarctic Dry Valleys result either from differences in substrate pH as a function of moisture abundance (Friedmann et al. 1988; Johnston and Vestal 1989) or that moisture itself directly affects species biodiversity by selecting for microorganisms better adapted to aridity (Cockell et al. 2003). While not endolithic per se, a transect study of hypolithic microbial communities in China showed shifts in community structure as a function of liquid water availability that is influenced by the ability of individual microorganisms to tolerate dry conditions (Pointing et al. 2007). In contrast, another study showed the extent of UV transmittance through limestone to be the main driver between lichen-dominated endolithic communities and cyanobacteria-dominated communities, with the former able to survive under higher UV conditions (Wong et al. 2010b). It would seem that in more clement locations, endolithic habitats select for microorganisms that can tolerate these environments to create specific niches, with increasing pressures selecting for specific microorganisms that can tolerate increasingly harsh conditions.

4.12 Microbial Activity, Geochemistry, and Microbe-Mineral Interactions in Endolithic Habitats

Endolithic microorganisms are involved in physical weathering resulting from mechanical forces imposed on the host rock, such as the penetration of fungal hyphae into cracks, the expansion of colonized spaces through biofilm swelling and shrinking associated with hydration-dehydration cycles, or the filling of pore spaces by growth of the biofilm. In contrast, mineral weathering is a geochemical process that is tightly coupled with microbial activity through many different mechanisms. Examples of indirect processes include the production of organic and inorganic acids, EPS, and metal-complexing ligands, as well as respiration and photosynthesis that can alter the pH of the microenvironment, which not only lead to dissolution of minerals but also allow for microbial sequestration of nutrients and trace elements. Bacteria and fungi can also react directly with metals as a result of their reactive cell surfaces or by changing the redox state of elements such as iron or sulfur for use as energy sources to facilitate transport of those elements into the cell for metabolic requirements (Ehrlich 1998; Sterflinger 2000).

While only the basic geology of a rock containing endolithic microorganisms is commonly reported (e.g., sandstone, limestone), metals and minerals are often a component of the rock matrix. Beacon Formation sandstones hosting cryptoendolithic microorganisms in the Dry Valleys include feldspars, clays, and iron oxides within the rock matrix (Blackhurst et al. 2004, 2005; Edwards et al. 2004; Friedmann and Weed 1987; Weed and Ackert 1986), which is probably a common occurrence in sedimentary rocks. These habitats are also geochemically diverse, containing a wide spectrum of macronutrients and metals (Blackhurst et al. 2005; Ferris and Lowson 1997; Friedmann 1982; Johnston and Vestal 1989; Omelon et al. 2007). Analysis of concentrations of metals in Dry Valley uncolonized sandstones showed little change with increasing depth (Blackhurst et al. 2005), but the region colonized by microorganisms was depleted relative to the overlying surface crust and beneath, where metals accumulated (Johnston and Vestal 1989). Variability in metal concentrations between colonized and uncolonized regions of cryptoendolithic habitats has been observed in other endolithic habitats (Ferris and Lowson 1997; Omelon et al. 2007), suggesting this to be a widespread phenomenon. The most commonly reported form of chemical weathering of rocks colonized by endolithic microorganisms is the production of acidity by fungi, such as oxalic acid (Johnston and Vestal 1993; Sterflinger 2000). However, cyanobacteria are known to create high-pH environments that can effectively weather silicates and lead to erosion of these habitats (Büdel et al. 2004). Presence of bioessential metals or nutrients necessary for enzymatic function as a resource within the rock can be of key importance for endolithic microorganisms (Blum et al. 2002; Ferris and Lowson 1997), but this is not always a requirement as shown by cases where the endolithic habitat provides only a structural framework for microorganisms, with requirements for growth coming from external inputs such as allochthonous dust (Cockell et al. 2010; Omelon et al. 2006b; Pontefract et al. 2014).

Many studies have shown evidence for rock weathering by the activity of endolithic microorganisms and their by-products (Ascaso et al. 1998; Burford et al. 2003; Caneva et al. 2014; Danin et al. 1983; Garvie et al. 2008; Gaylarde and Gaylarde 2004; Hirsch et al. 1995; Lian et al. 2010; Omelon et al. 2008; Ortega-Calvo et al. 1995; Palmer and Hirsch 1991; Sand and Bock 1991; Sterflinger et al. 1997; Tang et al. 2012; Weed and Ackert 1986; Wessels and Schoeman 1988). While destructive in nature, rock weathering can contribute to soil development (Tang et al. 2012), and the resulting exposure of the underlying community may result in the dispersal of endolithic microorganisms into soils or at least provide a source of organic matter for other carbon-limited environments (Burkins et al. 2000; Friedmann et al. 1993; Hopkins et al. 2008). Concern for preservation of monuments and buildings against endolithic microorganisms has led to steps to ameliorate colonization through the assessment of different treatments; in one case, 70 % ethanol applied to the surface was the most effective in reducing activity and preventing growth (Rabe et al. 2013).

Weathering, however, does not always lead to erosion. A study of micritic carbonates in glacier forelands characterized rocks sequentially colonized by fungi, algae, and finally lichens (Hoppert et al. 2004). Fungi penetrated the subsurface by chemical dissolution to create cavities in the rock, but this did not lead to destabilization of the endolithic habitat. The mature lichen community reaches steady-state conditions whereby fungal hyphae cease to penetrate deeper into the substrate, creating a network of pores colonized by filaments that strengthens the rock matrix (Hoppert et al. 2004). The authors argue that this situation differs from porous rocks such as sandstones that are susceptible to mechanical and chemical weathering along fissures (Friedmann and Ocampo 1976; Friedmann 1980; Weed and Norton 1991; Wessels and Büdel 1995); however, microbial filaments and EPS production can contribute to surface crust formation that in turn can increase resistance to weathering in the endolithic region (Kurtz and Netoff 2001).

The activity of the lichen *Verrucaria rubrocincta* has been shown to be involved in both destructive and constructive processes in endolithic habitats in caliche (Garvie et al. 2008). As in most reported cases, fungal hyphae weather the rock by deeper penetration into the subsurface that increases porosity and potential for water storage during extended dry periods. However, algal photosynthesis leads to the precipitation of micritic carbonate at the rock-air interface (Garvie et al. 2008). This carbonate not only serves as a barrier to moisture evaporation from the subsurface but also provides a highly reflective coating that shields endolithic microorganisms from intense solar radiation.

Surface crusts are an important constructive component of endolithic microbial systems as they mediate moisture loss, attenuate radiation, and provide protection from wind abrasion. In the Antarctic Dry Valleys, surface crusts of Beacon Formation sandstones are composed of allochthonous dust that builds up on the surface to form a coating rich in clays, feldspars, and iron oxides (Blackhurst et al. 2004; Friedmann and Weed 1987; Weed and Ackert 1986; Weed and Norton 1991). Despite this protection, a combination of biomass accumulation, microbial production of oxalic acids and EPS, and freeze-thaw activity triggers the dissolution of silica cements between quartz grains and subsequent exfoliation of the overlying surface crust and loss of biomass (Johnston and Vestal 1993; Kappen 1993b; Sun et al. 2010; Sun and Friedmann 1999).

This process leads to the formation of distinct exfoliation patterns on rock surfaces (Friedmann 1982; Friedmann and Weed 1987; Nienow and Friedmann 1993; Sun and Friedmann 1999). Similar weathering patterns are observed in other deserts where endolithic microorganisms colonize sandstone outcrops such as the Arctic and South Africa; however, cyanobacteria dominating within these cryptoendolithic habitats drive silica dissolution by generating high-pH conditions during periods of photosynthesis (Büdel et al. 2004; Omelon et al. 2008). Exfoliation of the surface crust exposes the underlying microbial community, of which much is lost to the surrounding landscape due to wind erosion; the remaining biota must then reestablish itself within the cryptoendolithic habitat as a surface crust begins forming on the new rock surface (Fig. 4.2) (Sun and Friedmann 1999).

4.13 Endolithic Biosignatures

The exfoliation of surface crusts is an example of an indirect biosignature or biomarker as it leaves traces of past biological activity (Friedmann and Weed 1987; Kappen 1993b). In contrast to these indirect signatures, detailed examination of the colonized pore spaces of cryptoendolithic habitats in the Dry Valleys identified metals on the surfaces of both living and dead microorganisms (Ascaso and Wierzchos 2003; de los Ríos et al. 2003; Wierzchos et al. 2003). There are also reported occurrences of mineral biomobilization and biotransformation of inorganic deposits, including iron oxyhydroxide nanocrystals and biogenic clays to produce diagenetic biomarkers (Friedmann and Weed 1987; Wierzchos et al. 2003).

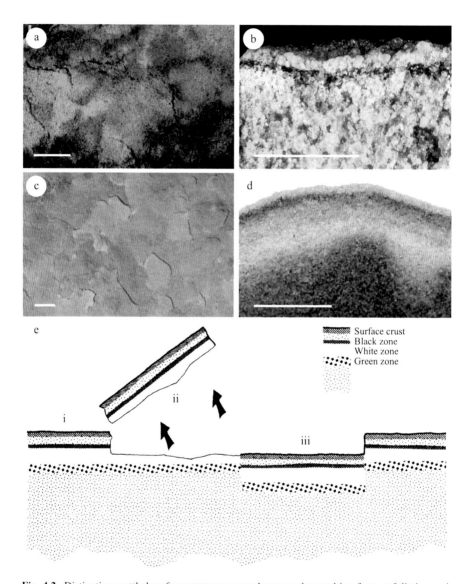

Fig. 4.2 Distinctive mottled surface patterns on sandstone rocks resulting from exfoliative rock weathering by cryptoendolithic microbial communities in the Antarctic Dry Valleys (**a**, **b**) and the Canadian high Arctic (**c**, **d**). Scale bars = 1 cm. Stages for weathering as described by Friedmann (1982), with zones in (**e**) corresponding to colors in (**b**). Stages include (i) initial establishment of endolithic lichen, (ii) exfoliation of crust by biological activity, (iii) reestablishment of lichen in area previously exfoliated. (**a**) From Friedmann and Weed (1987). (**b**, **e**) From Friedmann (1982)

Further progression of mineral encrustation leads to bacterial infilling to produce biosignatures and microfossils, containing either ultrastructural cellular elements or evidence for the transition from living microorganisms to deposits with no diagnostic cellular characteristics (Wierzchos et al. 2005). It is thought that these biosignatures are created through complex interactions between microorganisms and inorganic salts and minerals that preserve microfossils and therefore a record of their presence (Chap. 2 by Tomescu and coauthors). In contrast, similar studies of cryptoendolithic habitats in the Arctic show only the accumulation of clays embedded within EPS generated by microorganisms, with no evidence for in situ biomineralization as metal accumulations on cell surfaces or mineral precipitates, nor the presence of microfossils (Omelon et al. 2006b).

While bacteria can effectively bind metals to their surfaces leading to mineral precipitation, metal accumulation and mineral precipitation associated with microbial activity in natural environments require a source of metals brought near to or directly in contact with capsular material or cell walls. This is most commonly observed in aqueous environments, with water providing the medium to generate supersaturated conditions that drive most mineral precipitation reactions. Although this does not preclude the important activity of microorganisms in metal cycling and microbe-mineral interactions in soils where water is present (Gadd and Sayer 2000; Souza-Egipsy et al. 2002), the scarcity of water in deserts can provide only limited activity as both the solvent and transport agent for the introduction of metals and salts required for biomineralization processes.

The presence of microbial fossils in the Dry Valleys of Antarctica but not in the Arctic, despite both being cold deserts at similar latitudes, suggests that the relative abundance of water plays a crucial role in the formation and preservation of these biosignatures The Arctic receives substantially more precipitation in the form of liquid water than do the Dry Valleys (Omelon et al. 2006a), which enhances transport of dissolved metals and allochthonous debris such as clays into the Arctic cryptoendolithic habitat (Omelon et al. 2006b). This would suggest an increased likelihood for metal accumulation and mineral precipitation around Arctic microbial communities; however, these wetter conditions along with warmer summer temperatures also promote faster rates of biomass growth and turnover (Ziolkowski et al. 2013). In addition, higher Arctic growth rates and associated metabolic activity lead to higher rates of silica dissolution, evidenced by identification of oriented triangular etch pitting on quartz surfaces (Omelon et al. submitted) that can result from the presence of chelating acids or alkaline conditions that increase both quartz and feldspar solubility and dissolution kinetics (Bennett and Siegel 1987; Bennett 1991; Brady and Walther 1989; Brantley et al. 1986; Gratz et al. 1991; Gratz and Bird 1993; Hiebert and Bennett 1992). Evaluation of the upper pH limit at which cyanobacteria isolated from this habitat can fix dissolved inorganic carbon during photosynthesis shows that they can generate high-pH conditions (Omelon et al. 2008). In the presence of water, these high-pH conditions in the endolithic habitat could rapidly dissolve silica cements and quartz surfaces leading to exfoliation of the host rock, exposing the microbial community to harsh aerial conditions and their removal by winds (Omelon et al. 2006a).

In contrast, the colder and drier conditions of the Dry Valleys retard microbial growth rates and may also minimize detrimental stresses such as freeze fracturing associated with ice formation, thereby preserving older viable cryptoendolithic microbial populations (Bonani et al. 1988; Johnston and Vestal 1991). The longer residence time of these communities increases their exposure to aerial deposition of Fe oxides, quartz, and clays that form surface crusts (Friedmann and Weed 1987; Sun and Friedmann 1999; Weed and Ackert 1986; Weed and Norton 1991), which are believed to be the source for biomineralization and fossilization of these microorganisms (Wierzchos et al. 2005).

Previous work showed the varying degrees of viability of microbial communities resulting from the harsh environmental conditions of this region, which can lead to cell damage and mummification (de los Ríos et al. 2004; Friedmann and Koriem 1989; Wierzchos et al. 2004). Biotransformation of minerals was documented as physicochemical weathering of biotite, iron-rich minerals, biogenic clays, and silica as well as calcium oxalate around cells collected from granite rocks along the Ross Sea coastline at Granite Harbour (Ascaso and Wierzchos 2003; Wierzchos et al. 2003). The processes forming these deposits were explained by the inherent capacities of cell surfaces to immobilize cations and produce fine-grained minerals (Fortin et al. 1997, 1998; Fortin and Ferris 1998; Warren and Ferris 1998). While endolithic in nature, the formation of these minerals may be a function of the higher humidity levels experienced along the coast that increase water availability, the production of microbial EPS, and biotransformation rates of elements depositing around microorganisms.

In contrast, studies examining cryptoendolithic pores from the Mount Fleming region of the Dry Valleys observed rocks filled with living microorganisms and no mineral deposits or precipitates, as well as decaying microorganisms filled with clay-like minerals (Wierzchos et al. 2005) derived from airborne dust. Based on these observations, it was determined that inorganic processes occurring after the death of endolithic microorganisms were necessary for microbial fossil formation. These included the infilling of empty microbial molds by clay minerals, as well as mineralization of cell walls and interiors due to varying nucleation rates of silica and coexisting cations within an organic template. It is thought that differences in the elemental composition of cellular components result from multiple episodes of mineralization with intervening organic degradation rather than the primary replacement of organic material at the time of infiltration and that the mineralization variously occurs while cells are biologically active and after their decay to produce biosignatures (Wierzchos and Ascaso 2002). This is made possible through a concentration gradient by which metals or minerals diffuse into decayed microorganisms, resulting in fossilization substitution of organic substances with inorganic material (Wierzchos et al. 2005). The presence of biosignatures so close to the harsh polar desert environment only highlights the remarkable stability of these endolithic habitats, permitting fossilization to occur over timescales yet to be determined.

4.14 Astrobiological Significance

Interest in understanding the limits to life on Earth and potentially on other planets has led to the development of terrestrial analogue studies, which aim to evaluate whether or not life may exist elsewhere in the solar system. These analogues are targeted due to specific physical, chemical, or mineralogical characteristics that are comparable to those identified at extraterrestrial locations on planets including Venus, Mars, Europa, Enceladus, and Titan (Preston and Dartnell 2014). In addition to understanding habitability and the limits to life, analogue studies are also concerned with the identification, characterization, and preservation potential of biosignatures.

Endolithic habitats have been long considered potential locations to look for life on other planets, with those found in the Antarctic Dry Valleys receiving the most attention as terrestrial analogues for Mars ever since the early reporting of these microbial communities (Friedmann and Ocampo-Friedmann 1984b; Friedmann 1986; Friedmann et al. 1986; Friedmann and Koriem 1989; Wharton et al. 1989). Endolithic habitats in evaporite minerals precipitating from groundwater discharge have been considered potentially representative of life-supporting habitats on other planets (Grasby et al. 2003; Grasby and Londry 2007; Rothschild 1990), as are shocked rocks where porosity by impact craters creates new endolithic habitats that could subsequently be colonized (Pontefract et al. 2014).

It has been shown that the common endolithic cyanobacterium *Chroococcidiopsis* can survive damage induced by extreme conditions such as desiccation and radiation, which would be experienced in both space and on the surface of Mars (Billi et al. 2000, 2011). Given such understanding, studies have looked for Earth analogues for conditions found on Mars such as in the Atacama Desert, where discoveries of colonized endolithic habitats in hyperarid, salt-rich environments have provided new analogues for last refuges for life on that planet (Davila et al. 2008; Wierzchos et al. 2011, 2012, 2013).

More recent work has focused on characterizing the preservation of photosynthetic (i.e., fluorescent) pigments such as chlorophyll within these habitats, either by microscopy (Roldán et al. 2014) or spectroscopy (Stromberg et al. 2014). While the various states of preservation could be discerned in contemporary settings by confocal laser scanning microscopy (Roldán et al. 2014), placing endolithic microbial communities under simulated Mars conditions led to varying degrees of success as a function of the type of host mineral (Stromberg et al. 2014). Understanding how molecules degrade or are preserved in various minerals relevant to Mars will help in understanding the likelihood of finding biosignatures on other planets.

4.15 Conclusions

Endolithic microorganisms are afforded necessary respite from climatic stresses in desert habitats due to unique characteristics of the lithic substrate, which provide protection from temperature and radiation extremes as well as prolonged exposure to the rare occurrences of water that is transient at best outside the endolithic habitat. The successful survival of endolithic microorganisms, however, is not solely a function of the surrounding rock architecture, as shown by specific adaptive traits such as pigmentation to protect again UV, synthesis of EPS to fight desiccation, desiccation tolerance when necessary, and in some cases of lichens the ability to photosynthesize using only water vapor.

This activity leads to interactions with the surrounding habitat such as nutrient and element cycling, which can extend below the zone colonized by phototrophic microorganisms. Microbe-mineral interactions often lead to mineral precipitation or dissolution, but most requirements for metabolism and enzymatic function are sourced from outside the endolithic habitat. Some unique and protective aspects of these habitats result from abiotic processes (such as the formation of surface coatings), but microbial activity and microbe-mineral interactions are an important determinant of the success of these communities. In many cases microbial activity leads to weathering and destruction of the very habitat they require for survival, but this does not lead to the extinction of the microbial community. In others, the preservation of endolithic microorganisms or associated biosignatures provides not only a unique opportunity to understand how this terrestrial microbiota survives over long time periods but a fruitful prospect to investigate how biomineralization and fossilization proceed under extreme desert conditions.

References

Adhikary SP (1998) Polysaccharides from mucilaginous envelope layers of cyanobacteria and their ecological significance. J Sci Ind Res 57:454–466

Ascaso C, Wierzchos J (2003) The search for biomarkers and microbial fossils in Antarctic rock microhabitats. Geomicrobiol J 20:439–450

Ascaso C, Wierzchos J, Castello R (1998) Study of the biogenic weathering of calcareous litharenite stones caused by lichen and endolithic microorganisms. Int Biodeterior Biodegrad 42:29–38

Banerjee M, Whitton BA, Wynn-Williams DD (2000) Phosphatase activities of endolithic communities in rocks of the Antarctic Dry Valleys. Microb Ecol 39:80–91

Bell RA (1993) Cryptoendolithic algae of hot semiarid lands and deserts. J Phycol 29:133–139

Bell RA, Athey PV, Sommerfeld MR (1986) Cryptoendolithic algal communities of the Colorado plateau. J Phycol 22:429–435

Bennett PC (1991) Quartz dissolution in organic-rich aqueous systems. Geochim Cosmochim Acta 55:1781–1797

Bennett PC, Siegel DI (1987) Increased solubility of quartz in water due to complexation by dissolved organic compounds. Nature 326:684–687

Billi D, Friedmann EI, Hofer KG, Caiola MG, Ocampo-Friedman R (2000) Ionizing-radiation resistance in the desiccation-tolerant cyanobacterium *Chroococcidiopsis*. Appl Environ Microbiol 66:1489–1492

Billi D, Viaggiu E, Cockell CS, Rabbow E, Horneck G, Onofri S (2011) Damage escape and repair in dried *Chroococcidiopsis* spp. from hot and cold deserts exposed to simulated space and Martian conditions. Astrobiology 11:65–73

Blackhurst RL, Jarvis K, Grady MM (2004) Biologically–induced elemental variations in Antarctic sandstones: a potential test for Martian micro-organisms. Int J Astrobiol 3:97–106

Blackhurst RL, Genge MJ, Kearsley AT, Grady MM (2005) Cryptoendolithic alteration of Antarctic sandstones: pioneers or opportunists? J Geophys Res Planet 110:E12S24

Blum JD, Klaue A, Nezat CA, Driscoll CT, Johnson CE, Siccama TG, Eagar C, Fahey TJ, Likens GE (2002) Mycorrhizal weathering of apatite as an important calcium source in base-poor forest ecosystems. Nature 417:729–731

Boison G, Mergel A, Jolkver H, Bothe H (2004) Bacterial life and dinitrogen fixation at a gypsum rock. Appl Environ Microbiol 70:7070–7077

Bonani G, Friedmann EI, Ocampo-Friedmann R, McKay CP, Wolfli W (1988) Preliminary report on radiocarbon dating of cryptoendolithic microorganisms. Polarforschung 58:199–200

Brady PV, Walther JV (1989) Controls on silicate dissolution rates in neutral and basic pH solutions at 25°C. Geochim Cosmochim Acta 53:2823–2830

Brantley SL, Crane SR, Crerar DA, Hellman R, Stallard R (1986) Dissolution at dislocation etch pits in quartz. Geochim Cosmochim Acta 50:2349–2361

Brock TD (1975) Effect of water potential on a *Microcoleus* (Cyanophyceae) from a desert crust. J Phycol 11:316–320

Büdel B, Lange OL (1991) Water status of green and blue-green phycobionts in lichen thalli after hydration by water vapor uptake: do they become turgid? Bot Acta 104:361–366

Büdel B, Weber B, Kuhl M, Pfanz H, Sultemeyer D, Wessels D (2004) Reshaping of sandstone surfaces by cryptoendolithic cyanobacteria: bioalkalization causes chemical weathering in arid landscapes. Geobiology 2:261–268

Büdel B, Bendix J, Bicker FR, Green TGA (2008) Dewfall as a water source frequently activates the endolithic cyanobacterial communities in the granites of Taylor Valley, Antarctica. J Phycol 44:1415–1424

Büdel B, Schulz B, Reichenberger H, Bicker F, Green TGA (2009) Cryptoendolithic cyanobacteria from calcite marble rock ridges, Taylor Valley, Antarctica. Algol Stud 129:61–69

Bungartz F, Garvie LAJ, Nash TH III (2004) Anatomy of the endolithic Sonoran Desert lichen *Verrucaria rubrocincta* Breuss: implications for biodeterioration and biomineralization. Lichenology 36:55–73

Burford EP, Fomina M, Gadd GM (2003) Fungal involvement in bioweathering and biotransformation of rocks and minerals. Mineral Mag 67:1127–1155

Burkins MB, Virginia RA, Chamberlain CP, Wall DH (2000) Origin and distribution of soil organic matter in Taylor Valley, Antarctica. Ecology 81:2377–2391

Butin H (1954) Physiologisch-ökologische Untersuchungen über den Wasserhaushalt und die Photosynthese bei Flechten. Biol Zbl 73:459–502

Caneva G, Lombardozzi V, Ceschin S, Municchia AC, Salvadori O (2014) Unusual differential erosion related to the presence of endolithic microorganisms (Martvili, Georgia). J Cult Herit 15:538–545

Cary SC, McDonald IR, Barrett JE, Cowan DA (2010) On the rocks: the microbiology of Antarctic Dry Valley soils. Nat Rev Microbiol 8:129–138

Casamatta DA, Verb RG, Beaver JR, Vis ML (2002) An investigation of the cryptobiotic community from sandstone cliffs in southeast Ohio. Int J Plant Sci 163:837–845

Chan Y, Lacap DC, Lau MCY, Ha KY, Warren-Rhodes KA, Cockell CS, Cowan DA, McKay CP, Pointing SB (2012) Hypolithic microbial communities: between a rock and a hard place. Environ Microbiol 14:2272–2282

Cockell C, Stokes MD (2004) Widespread colonization by polar hypoliths. Nature 431:414
Cockell CS, McKay CP, Omelon C (2003) Polar endoliths – an anti-correlation of climate extremes and microbial diversity. Int J Astrobiol 1:305–310
Cockell CS, Olsson K, Knowles F, Kelly L, Herrera A, Thorsteinsson T, Marteinsson V (2009a) Bacteria in weathered basaltic glass, Iceland. Geomicrobiol J 26:491–507
Cockell CS, Olsson-Francis K, Herrera A, Meunier A (2009b) Alteration textures in terrestrial volcanic glass and the associated bacterial community. Geobiology 7:50–65
Cockell CS, Osinski GR, Banerjee NR, Howard KT, Gilmour I, Watson JS (2010) The microbe-mineral environment and gypsum neogenesis in a weathered polar evaporite. Geobiology 8:293–2308
Cowan DA, Khan N, Pointing SB, Cary C (2010) Diverse hypolithic refuge communities in the McMurdo Dry Valleys. Antarct Sci 22:714–720
Danin A, Gerson R, Garty J (1983) Weathering patterns on hard limestone and dolomite by endolithic lichens and cyanobacteria: supporting evidence for eolian contribution to Terra Rosa soil. Soil Sci 136:213–217
Davila AF, Gómez-Silva B, de los Ríos A, Ascaso C, Olivares H, McKay CP, Wierzchos J (2008) Facilitation of endolithic microbial survival in the hyperarid core of the Atacama Desert by mineral deliquescence. J Geophys Res Biogeo 113, G01028
de la Torre JR, Goebel BM, Friedmann EI, Pace NR (2003) Microbial diversity of cryptoendolithic communities from the McMurdo Dry Valleys, Antarctica. Appl Environ Microbiol 69:3858–3867
de los Ríos A, Wierzchos J, Ascaso C (2002) Microhabitats and chemical microenvironments under saxicolous lichens growing on granite. Microb Ecol 43:181–188
de los Ríos A, Wierzchos J, Sancho LG, Ascaso C (2003) Acid microenvironments in microbial biofilms of Antarctic endolithic microecosystems. Environ Microbiol 5:231–237
de los Ríos A, Wierzchos J, Sancho LG, Ascaso C (2004) Exploring the physiological state of continental Antarctic endolithic microorganisms by microscopy. FEMS Microbiol Ecol 50:143–152
de los Ríos A, Sancho LG, Grube M, Wierzchos J, Ascaso C (2005) Endolithic growth of two *Lecidea* lichens in granite from continental Antarctica detected by molecular and microscopy techniques. New Phytol 165:181–190
de los Ríos A, Grube M, Sancho L, Ascaso C (2007) Ultrastructural and genetic characteristics of endolithic cyanobacterial biofilms colonizing Antarctic granite rocks. FEMS Microbiol Ecol 59:386–395
de los Ríos A, Grube M, Sancho LG, Davila AF, Kastovsky J, McKay CP, Gómez-Silva B, Wierzchos J (2010) Comparative analysis of the microbial communities inhabiting halite evaporites of the Atacama Desert. Int Microbiol 13:79–89
de los Ríos A, Wierzchos J, Ascaso C (2014) The lithic microbial ecosystems of Antarctica's McMurdo Dry Valleys. Antarct Sci 26:459–477
Dong H, Rech JA, Jiang H, Sun H, Buck BJ (2007) Endolithic cyanobacteria in soil gypsum: occurrences in Atacama (Chile), Mojave (United States), and Al-Jafr (Jordan) Deserts. J Geophys Res Biogeo 112, G02030
Edwards HGM, Wynn-Williams DD, Jorge-Villar SE (2004) Biological modification of haematite in Antarctic cryptoendolithic communities. J Raman Spectrosc 35:470–474
Ehrlich HL (1998) Geomicrobiology: its significance for geology. Earth-Sci Rev 45:45–60
Ferris FG, Lowson EA (1997) Ultrastructure and geochemistry of endolithic microorganisms in limestone of the Niagara Escarpment. Can J Microbiol 43:211–219
Fewer DJ, Friedl T, Büdel B (2002) *Chroococcidiopsis* and heterocyst-differentiating cyanobacteria are each other's closest living relatives. Mol Phylogenet Evol 23:82–90
Fortin D, Ferris FG (1998) Precipitation of dissolved silica, sulfate and iron on bacterial surfaces. Geomicrobiol J 15:309–324

Fortin D, Ferris FG, Beveridge TJ (1997) Surface-mediated mineral development by bacteria. In: Banfield JF, Nealson KH (eds) Reviews in mineralogy and geochemistry, vol 35. Mineralogical Society of America, Chantilly, VA, pp 161–180

Fortin D, Ferris FG, Scott SD (1998) Formation of Fe-silicates and Fe-oxides on bacterial surfaces in samples collected near hydrothermal vents on the Southern Explorer Ridge in the northeast Pacific Ocean. Am Mineral 83:1399–1408

Friedmann EI (1977) Microorganisms in antarctic desert rocks from dry valleys and Dufek Massif. Antarct J US 12:26–30

Friedmann EI (1978) Melting snow in the dry valleys is a source of water for endolithic microorganisms. Antarct J US 13:162–163

Friedmann EI (1980) Endolithic microbial life in hot and cold deserts. Origins Life Evol B 10:223–235

Friedmann EI (1982) Endolithic microorganisms in the Antarctic cold desert. Science 215:1045–1053

Friedmann EI (1986) The Antarctic cold desert and the search for traces of life on Mars. Adv Space Res 6:167–172

Friedmann EI, Kibler AP (1980) Nitrogen economy of endolithic microbial communities in hot and cold deserts. Microb Ecol 6:95–108

Friedmann EI, Koriem A (1989) Life on Mars: how it disappeared (if it ever was there). Adv Space Res 9:167–172

Friedmann EI, McKay CP (1985) A method for continuous monitoring of snow: application to the cryptoendolithic microbial community of Antarctica. Antarct J US 20:179–181

Friedmann EI, Ocampo R (1976) Endolithic blue-green algae in the dry valleys: primary producers in the Antarctic desert ecosystem. Science 193:1274–1279

Friedmann EI, Ocampo-Friedmann R (1984a) Endolithic microorganisms in extreme dry environments: analysis of a lithobiontic habitat. In: Klug MJ, Reddy CA (eds) Current perspectives in microbiology. American Society of Microbiology, Washington, DC, pp 177–185

Friedmann EI, Ocampo-Friedmann R (1984b) The Antarctic cryptoendolithic ecosystem: relevance to exobiology. Orig Life 14:771–776

Friedmann EI, Weed R (1987) Microbial trace-fossil formation, biogenous, and abiotic weathering in the Antarctic cold desert. Science 236:703–705

Friedmann I, Lipkin Y, Ocampo-Paus R (1967) Desert algae of the Negev (Israel). Phycologia 6:185–200

Friedmann EI, Kappen L, Garty J (1980) Fertile stages of cryptoendolithic lichens in the dry valleys of southern Victoria Land. Antarct J US 15:166–167

Friedmann EI, Friedmann RO, McKay CP (1981) Adaptations of cryptoendolithic lichens in the Antarctic desert. In: Jouventin P, Masse L, Trehen P (eds) Colloque sur les Ecosystemes Subantarctiques. Comite National Francais des Recherches Antarctiques, Paris, pp 65–70

Friedmann EI, Friedmann RO, Weed R (1986) Trace fossils of endolithic microorganisms in Antarctica - a model For Mars. Origins Life Evol B 16:350

Friedmann EI, McKay CP, Nienow JA (1987) The cryptoendolithic microbial environment in the Ross Desert of Antarctica: satellite-transmitted continuous nanoclimate data, 1984 to 1986. Polar Biol 7:273–287

Friedmann EI, Hua M, Ocampo-Friedman R (1988) Cryptoendolithic lichen and cyanobacterial communities of the Ross Desert, Antarctica. Polarforschung 58:251–259

Friedmann EI, Kappen L, Meyer MA, Nienow JA (1993) Long-term productivity in the cryptoendolithic microbial community of the Ross Desert, Antarctica. Microb Ecol 25:51–69

Gadd GM, Sayer JA (2000) Influence of fungi on the environmental mobility of metals and metalloids. In: Lovely DR (ed) Environmental microbe-metal interactions. ASM Press, Washington, DC, pp 237–256

Garvie LAJ, Knauth LP, Bungartz F, Slonowski S, Nash TH III (2008) Life in extreme environments: survival strategy of the endolithic desert lichen *Verrucaria rubrocincta*. Naturwissenschaften 95:705–712

Gaylarde P, Gaylarde C (2004) Deterioration of siliceous stone monuments in Latin America: microorganisms and mechanisms. Corros Rev 22:395–415

Gerrath JF, Gerrath JA, Matthes U, Larson DW (2000) Endolithic algae and cyanobacteria from cliffs of the Niagara Escarpment, Ontario, Canada. Can J Bot Rev Can Bot 78:807–815

Goublic S, Friedmann I, Schneider J (1981) The lithobiontic ecological niche, with special reference to microorganisms. J Sediment Petrol 51:475–478

Gramain A, Diaz GC, Demergasso C, Lowenstein TK, McGenity TJ (2011) Archaeal diversity along a subterranian salt core from the Salar Grande (Chile). Environ Microbiol 13:2105–2121

Grasby SE, Londry KL (2007) Biogeochemistry of hypersaline springs supporting a mid-continent marine ecosystem: an analogue for Martian springs? Astrobiology 7:662–683

Grasby SE, Allen CC, Longazo TG, Lisle JT, Griffin DW, Beauchamp B (2003) Supraglacial sulfur springs and associated biological activity in the Canadian high Arctic – signs of life beneath the ice. Astrobiology 3:583–596

Gratz AJ, Bird P (1993) Quartz dissolution: negative crystal experiments and a rate law. Geochim Cosmochim Acta 57:965–976

Gratz AJ, Manne S, Hansma PK (1991) Atomic force microscopy of atomic-scale ledges and etch pits formed during dissolution of quartz. Science 251:1343–1346

Greenfield LG (1988) Forms of nitrogen in Beacon sandstone rocks containing endolithic microbial communities in Southern Victoria Land, Antarctica. Polarforschung 58:211–218

Grilli Caiola M, Ocampo-Friedmann R, Friedmann EI (1993) Cytology of long-term desiccation in the desert cyanobacterium *Chroococcidiopsis* (Chroococcales). Phycologia 32:315–322

Hawes I, Howard-Williams C, Vincent WF (1992) Desiccation and recovery of Antarctic cyanobacterial mats. Polar Biol 12:587–594

Herrera A, Cockell CS, Self S, Blaxter M, Reitner J, Thorsteinsson T, Arp G, Dröse W, Tindle AG (2009) A cryptoendolithic community in volcanic glass. Astrobiology 9:369–381

Hershkovitz N, Oren A, Cohen Y (1991) Accumulation of trehalose and sucrose in cyanobacteria exposed to matric water stress. Appl Environ Microbiol 57:645–648

Hess U (1962) Uber die hydraturabhangige Entwicklung und die Austrocknungsresistenz von Cyanophyceen. Arch Mikrobiol 44:189–218

Hiebert FK, Bennett PC (1992) Microbial control of silicate weathering in organic-rich ground water. Science 258:278–281

Hirsch P, Hoffmann B, Gallikowski CC, Mevs U, Siebert J, Sittig M (1988) Diversity and identification of heterotrophs from Antarctic rocks of the McMurdo Dry Valleys (Ross Desert). Polarforschung 58:261–269

Hirsch P, Eckhardt FEW, Palmer RJ (1995) Fungi active in weathering of rock and stone monuments. Can J Bot Rev Can Bot 73:1384–1390

Hirsch P, Gallikowski CA, Siebert J, Peissl K, Kroppenstedt RM, Schumann P, Stackebrandt E, Anderson R (2004a) *Deinococcus frigens* sp. nov., *Deinococcus saxicola* sp. nov., and *Deinococcus marmoris* sp. nov., low temperature and draught-tolerating, UV-resistant bacteria from continental Antarctica. Syst Appl Microbiol 27:636–645

Hirsch P, Mevs U, Kroppenstedt RM, Schumann P, Stackebrandt E (2004b) Cryptoendolithic actinomycetes from Antarctic sandstone rock samples: *Micromonospora endolithica* sp. nov. and two isolates related to *Micromonospora coerulea* Jensen 1932. Syst Appl Microbiol 27:166–174

Hopkins DW, Sparrow AD, Gregorich EG, Elberling B, Novis P, Fraser F, Scrimgeour C, Dennis PG, Meier-Augenstein W, Greenfield LG (2008) Isotopic evidence for the provenance and turnover of organic carbon by soil microorganisms in the Antarctic Dry Valleys. Environ Microbiol 11:597–608

Hoppert M, Flies C, Pohl W, Günzel B, Schneider J (2004) Colonization strategies of lithobiontic microorganisms on carbonate rocks. Environ Geol 46:421–428

Horath T, Bachofen R (2009) Molecular characterization of an endolithic microbial community in dolomite rock in the central Alps (Switzerland). Microb Ecol 58:290–306

Hughes KA, Lawley B (2003) A novel Antarctic microbial endolithic community within gypsum crusts. Environ Microbiol 5:555–565
Johnston CG, Vestal JR (1986) Does iron inhibit cryptoendolithic communities? Antarct J US 21:225–226
Johnston CG, Vestal JR (1989) Distribution of inorganic species in two Antarctic cryptoendolithic microbial communities. Geomicrobiol J 7:137–153
Johnston CG, Vestal JR (1991) Photosynthetic carbon incorporation and turnover in Antarctic cryptoendolithic microbial communities: are they the slowest growing communities on earth? Appl Environ Microbiol 57:2308–2311
Johnston CG, Vestal JR (1993) Biogeochemistry of oxalate in the Antarctic cryptoendolithic lichen-dominated community. Microb Ecol 25:305–319
Jorge Villar SE, Edwards HGM, Cockell CS (2005a) Raman spectroscopy of endoliths from Antarctic cold desert environments. Analyst 130:156–162
Jorge Villar SE, Edwards HGM, Worland MR (2005b) Comparative evaluation of Raman spectroscopy at different wavelengths for extremophile exemplars. Origins Life Evol B 35:489–506
Kappen L (1989) Field measurements of carbon dioxide exchange of the Antarctic lichen *Usnea sphacelata* in the frozen state. Antarct Sci 1:31–34
Kappen L (1993a) Plant activity under snow and ice, with particular reference to lichens. Arctic 46:297–302
Kappen L (1993b) Lichens in the Antarctic region. In: Friedman EI (ed) Antarctic microbiology. Wiley, New York, pp 433–490
Kappen L, Breuer M (1991) Ecological and physiological investigations in continental Antarctic cryptogams. II: Moisture relations and photosynthesis of lichens near Casey Station, Wilkes Land. Antarct Sci 3:273–278
Kappen L, Friedmann EI (1983) Ecophysiology of lichens in the dry valleys of Southern Victoria Island, Antarctica. II. CO_2 gas exchange in cryptoendolithic lichens. Polar Biol 1:227–232
Kappen L, Schroeter B (1997) Activity of lichens under the influence of snow and ice. Proc NIPR Symp Polar Biol 10:169–178
Kappen L, Lange OL, Schulze E-D, Buschbom U, Evenari M (1980) Ecophysiological investigations on lichens of the Negev Desert. VII. The influence on the habitat exposure on dew imbibition and photosynthetic productivity. Flora 169:216–229
Kappen L, Friedmann EI, Garty J (1981) Ecophysiology of lichens in the dry valleys of Southern Victoria Land, Antarctica. I. Microclimate of the cryptoendolithic lichen habitat. Flora 171:216–235
Kappen L, Bolter M, Kuhn A (1986) Field measurements of net photosynthesis of lichens in the Antarctic. Polar Biol 5:255
Kappen L, Schroeter B, Sancho LG (1990) Carbon dioxide exchange of Antarctic crustose lichens in situ measured with a CO_2/H_2O porometer. Oecologia 82:311–316
Kappen L, Sommerkorn M, Schroeter B (1995) Carbon acquisition and water relations of lichens in polar regions – potentials and limitations. Lichenologist 27:531–545
Kelly LC, Cockell CS, Herrera-Belaroussi A, Piceno T, Andersen GL, DeSantis T, Brodie E, Thorsteinsson T, Marteinsson V, Poly F, LeRoux X (2011) Bacterial diversity of terrestrial crystalline volcanic rocks, Iceland. Microb Ecol 62:69–79
Khan N, Tuffin M, Stafford W, Cary C, Lacap DC, Pointing SB (2011) Hypolithic microbial communities of quartz rocks from Miers Valley, McMurdo Dry Valleys, Antarctica. Polar Biol 34:1657–1668
Kidron GJ (2000) Dew moisture regime of endolithic and epilithic lichens inhabiting limestone cobbles and rock outcrops, Negev Highlands, Israel. Flora 195:146–153
Kuhlman KR, Venkat P, La Duc MT, Kuhlman GM, McKay CP (2008) Evidence of a microbial community associated with rock varnish at Yungay, Atacama Desert, Chile. J Geophys Res Biogeo 113:G04022

Kurtz HD, Netoff DI (2001) Stabilization of friable sandstone surfaces in a desiccating, windabraded environment of south-central Utah by rock surface microorganisms. J Arid Environ 48:89–100

Lacap DC, Warren-Rhodes KA, McKay CP, Pointing SB (2011) Cyanobacteria and chloroflexidominated hypolithic colonization of quartz at the hyper-arid core of the Atacama Desert, Chile. Extremophiles 15:31–38

Lange OL (1969) Experimentellökologische Untersuchungen an Flechten der Negev Wüste. I. CO2-Gaswechsel von *Ramalina maciformis* (Del.) Bory unter kontrollierten Bedingungen im Laboratorium. Flora 158:324–359

Lange OL (2003) Photosynthetic productivity of the epilithic lichen *Lecanora muralis*: long-term field monitoring of CO_2 exchange and its physiological interpretation. II. Diel and seasonal patterns of net photosynthesis and respiration. Flora 198:55–70

Lange OL, Bertsch A (1965) Photosynthese der Wüstenflechte *Ramalina maciformis* nach Wasserdampfaufnahme aus dem Luftraum. Naturwissenschaften 52:215–216

Lange OL, Kilian E (1985) Reaktivierung der Photosynthese trockener Flechten durch Wasserdampfaufnahme aus dem Luftraum: Artspezifisch unterschiedliches Verhalten. Flora 176:7–23

Lange OL, Redon J (1983) Epiphytische Flechten im Bereich einer chilenischen "Nebeloase" (Fray Jorge). II. Ökophysiologische Charakterisierung von CO_2-Gaswechsel und Wasserhaushalt. Flora 174:245–284

Lange OL, Ziegler H (1986) Different limiting processes of photosynthesis in lichens. In: Marcelle R, Clijsters H, Van Poucke M (eds) Biological control of photosynthesis. Martinus Nijhoff Publishers, Dordrecht, pp 147–161

Lange OL, Kilian E, Ziegler H (1986) Water vapor uptake and photosynthesis of lichens: performance differences in species with green and blue-green algae as phycobionts. Oecologia 71:104–110

Lange OL, Schulze ED, Koch W (1970) Experimentellökologische Untersuchungen an Flechten der Negev Wüste. II. CO2-Gaswechsel und Wasserhaushalt von *Ramalina maciformis* (Del.) Bory am natürlichen Standort während der sommerlichen Trockenperiode. Flora 159:38–62

Lange OL, Meyer A, Zellner H, Ullman I, Wessels DCJ (1990) Eight days in the life of a desert lichen: water relations and photosynthesis of *Teloschistes capensis* in the coastal fog zone of the Namib Desert. Modoqua 17:17–30

Lange OL, Büdel B, Meyer A, Kilian E (1993) Further evidence that activation of net photosynthesis by dry cyanobacterial lichens requires liquid water. Lichenology 25:175–189

Lange OL, Meyer A, Büdel B (1994) Net photosynthesis activation of a desiccated cyanobacterium without liquid water in high air humidity alone. Experiments with *Microcoleus sociatus* isolated from a desert soil crust. Funct Ecol 8:52–57

Lange OL, Green TGA, Heber U (2001) Hydration-dependent photosynthetic production of lichens: what do laboratory studies tell us about field performance? J Exp Bot 52:2033–2042

Lange OL, Green TGA, Melzer B, Meyer A, Zellner H (2006) Water relations and CO_2 exchange of the terrestrial lichen *Teloschistes capensis* in the Namib fog desert: measurements during two seasons in the field and under controlled conditions. Flora 201:268–280

Lian B, Chen Y, Tang Y (2010) Microbes on carbonate rocks and pedogenesis in karst regions. J Earth Sci 21:293–296

Lovely DR, Chapelle FH (1995) Deep subsurface microbial processes. Rev Geophys 33:365–381

Makhalanyane TP, Valverde A, Birkeland NK, Cary SC, Tuffin IM, Cowan DA (2013a) Evidence for successional development in Antarctic hypolithic bacterial communities. ISME J 7:2080–2090

Makhalanyane TP, Valverde A, Lacap DC, Pointing SB, Tuffin MI, Cowan DA (2013b) Evidence of species recruitment and development of hot desert hypolithic communities. Environ Microbiol Rep 5:219–224

Matthes U, Turner SJ, Larson DW (2001) Light attenuation by limestone rock and its constraint on the depth distribution of endolithic algae and cyanobacteria. Int J Plant Sci 162:263–270

McKay CP (2012) Full solar spectrum measurements of absorption of light in a sample of the Beacon Sandstone containing the Antarctic cryptoendolithic microbial community. Antarct Sci 24:243–248

McKay CP, Friedmann EI (1985) The cryptoendolithic microbial environment in the Antarctic cold desert: temperature variations in nature. Polar Biol 4:19–25

McKay CP, Friedmann EI, Gomez-Silva B, Caceres-Villanueva L, Andersen DT, Landheim R (2003) Temperature and moisture conditions for life in the extreme arid region of the Atacama Desert: four years of observations including the El Nino of 1997-1998. Astrobiology 3:393–406

Nash TH III, Reiner A, Demmig-Adams B, Kilian E, Kaiser WM, Lange OL (1990) The effect of atmospheric desiccation and osmotic water stress on photosynthesis and dark respiration of lichens. New Phytol 116:269–276

Navarro-González R, Rainey FA, Molina P, Bagaley DR, Hollen BJ, de la Rosa J, Small AM, Quinn RC, Grunthaner FJ, Cáceres L, Gomez-Silva B, McKay CP (2003) Mars-like soils in the Atacama Desert, Chile, and the dry limit of microbial life. Science 302:1018–1021

Nienow JA, Friedmann EI (1993) Terrestrial lithophytic (rock) communities. In: Friedmann EI, Thistle AB (eds) Antarctic microbiology. Wiley-Liss, New York, NY, pp 343–412

Nienow JAC, McKay CP, Friedmann EI (1988a) The cryptoendolithic microbial environment in the Ross desert of Antarctica: mathematical models of the thermal regime. Microb Ecol 16:253–270

Nienow JAC, McKay CP, Friedmann EI (1988b) The cryptoendolithic microbial environment in the Ross desert of Antarctica: light in the photosynthetically active region. Microb Ecol 16:271–289

Norris TB, Castenholz RW (2006) Endolithic photosynthetic communities within ancient and recent travertine deposits in Yellowstone National Park. FEMS Microbiol Ecol 57:470–483

Omelon CR, Pollard WH, Ferris FG (2006a) Environmental controls on microbial colonization of high Arctic cryptoendolithic habitats. Polar Biol 30:19–29

Omelon CR, Pollard WH, Ferris FG (2006b) Chemical and ultrastructural characterization of high Arctic cryptoendolithic habitats. Geomicrobiol J 23:189–200

Omelon CR, Pollard WH, Ferris FG (2007) Inorganic species distribution and microbial diversity within high Arctic cryptoendolithic habitats. Microb Ecol 54:740–752

Omelon CR, Pollard WH, Ferris FG, Bennett PC (2008) Cyanobacteria within cryptoendolithichabitats: the role of high pH in biogenic rock weathering in the Canadian high Arctic. Paper presented at the 9th international conference on Permafrost, University of Alaska Fairbanks,Fairbanks, 29 June–3 July

Omelon CR, Warden JG, Mykytczuk NCS, Breecker DO, Bennett PC (2013) Microbial respiration in high Arctic cryptoendolithic habitats. In: Paper presented at the 5th international conference on polar and alpine microbiology, Big Sky, Montana, 8–12 Sept 2013

Ortega-Calvo JJ, Arino X, Hernandez-Marine M, Saiz-Jimenez C (1995) Factors affecting the weathering and colonization of monuments by phototrophic microorganisms. Sci Total Environ 167:329–341

Palmer RJ, Friedmann EI (1990) Water relations and photosynthesis in the cryptoendolithic microbial habitat of hot and cold deserts. Microb Ecol 19:111–118

Palmer RJ, Hirsch P (1991) Photosynthesis-based microbial communities on 2 churches in northern Germany – weathering of granite and glazed brick. Geomicrobiol J 9:103–118

Pannewitz S, Schlensog M, Green TGA, Sancho LG, Schroeter B (2003) Are lichens active under snow in continental Antarctica? Oecologia 135:30–38

Parnell J, Lee P, Cockell CS, Osinski GR (2004) Microbial colonization in impact-generated hydrothermal sulphate deposits, Haughton impact structure, and implications for sulphates on Mars. Int J Astrobiol 3:247–256

Parro V, de Diego-Castilla G, Moreno-Paz M, Blanco Y, Cruz-Gil P, Rodríguez-Manfredi JA, Fernández-Remolar D, Gómez F, Gómez MJ, Rivas LA, Demergasso C, Echeverría A, Urtuvia VN, Ruiz-Bermejo M, Garcia-Villadangos M, Postigo M, Sánchez-Román M, Chong-Diaz G,

Gómez-Elvira J (2011) A microbial oasis in the hypersaline Atacama subsurface discovered by a life detector chip: implications for the search for life on Mars. Astrobiology 11:969–996

Phoenix VR, Bennett PC, Engel AS, Tyler SW, Ferris FG (2006) Chilean high-altitude hot-spring sinters: a model system for UV screening mechanisms by early Precambrian cyanobacteria. Geobiology 4:15–28

Pointing SB, Belnap J (2012) Microbial colonization and controls in dryland systems. Nat Rev Microbiol 10:551–562

Pointing SB, Warren-Rhodes KA, Lacap DC, Rhodes KL, McKay CP (2007) Hypolithic community shifts occur as a result of liquid water availability along environmental gradients in China's hot and cold hyperarid deserts. Environ Microbiol 9:414–424

Pointing SB, Chan YK, Lacap DC, Lau MCY, Jurgens JA, Farrell R (2009) Highly specialized microbial diversity in hyper-arid polar desert. Proc Natl Acad Sci USA 106:19964–19969

Pontefract A, Osinski GR, Cockell CS, Moore CA, Moores JE, Southam G (2014) Impact-generated endolithic habitat within crystalline rocks of the Haughton impact structure, Devon Island, Canada. Astrobiology 14:522–533

Potts M (1999) Mechanisms of desiccation tolerance in cyanobacteria. Eur J Phycol 34:319–328

Potts M, Friedmann EI (1981) Effects of water stress on cryptoendolithic cyanobacteria from hot desert rocks. Arch Microbiol 130:267–271

Preston LJ, Dartnell LR (2014) Planetary habitability: lessons learned from terrestrial analogues. Int J Astrobiol 13:81–98

Rabe L, Ernfridsson E, Edlund J, Pedersen K (2013) Evaluation of nine different treatments for reduction of biological biomass on cultural heritage limestone. Paper presented at the 21st international symposium on environmental biogeochemistry, Wuhan, China, October 13–18

Redon J, Lange OL (1983) Epiphytische Flechten im Bereich einer chilenischen "Nebeloase" (Fray Jorge). I. Vegetationskundliche Gliederung und Standortsbedingungen. Flora 174:213–243

Roldán M, Ascaso C, Wierzchos J (2014) Fluorescent fingerprints of endolithic phototrophic cyanobacteria living within halite rocks in the Atacama Desert. Appl Environ Microbiol 80:2998–3006

Rothschild LJ (1990) Earth analogs for Martian life. Microbes in evaporites, a new model system for life on Mars. Icarus 88:246–260

Ruisi S, Barreca D, Selbmann L, Zucconi L, Onofri S (2007) Fungi in Antarctica. Rev Environ Sci Biotechnol 6:127–141

Saiz-Jimenez C, Garcia-Rowe J, Garcia del Cura MA, Ortega-Calvo JJ, Roekens E, Van Grieken R (1990) Endolithic cyanobacteria in Maastricht limestone. Sci Total Environ 94:209–220

Sand W, Bock E (1991) Biodeterioration of mineral materials by microorganisms – biogenic sulfuric and nitric acid corrosion of concrete and natural stone. Geomicrobiol J 9:129–138

Scherer S, Potts M (1989) Novel water stress protein from a desiccation-tolerant cyanobacterium. J Biol Chem 264:12546–12553

Schlesinger WH, Pippen JS, Wallenstein MD, Hofmockel KS, Klepeis DM, Mahall BE (2003) Community composition and photosynthesis by photoautotrophs under quartz pebbles, southern Mojave desert. Ecology 84:3222–3231

Schroeter B (1994) In situ photosynthetic differentiation of the green algal and the cyanobacterial photobiont in the crustose lichen *Placopsis contortuplicata*. Oecologia 98:212–220

Schroeter B, Scheidegger C (1995) Water relations in lichens at subzero temperatures: structural changes and carbon dioxide exchange in the lichen *Umbilicaria aprina* from continental Antarctica. New Phytol 131:273–285

Selbmann L, Onofri S, Fenice M, Federici F, Petruccioli M (2002) Production and structural characterization of the exopolysaccharide of the Antarctic fungus *Phoma herbarum* CCFEE 5080. Res Microbiol 153:585–592

Selbmann L, de Hoog GS, Mazzagalia A, Friedman EI, Onofri S (2005) Fungi at the edge of life: cryptoendolithic black fungi from Antarctic desert. Stud Mycol 51:1–32

Siebert J, Hirsch P (1988) Characterization of 15 selected coccal bacteria isolated from Antarctic rock and soil samples from the McMurdo-Dry Valleys (South Victoria Land). Polar Biol 9:37–44

Siebert J, Palmer RJ, Hirsch P (1991) Analysis of free amino acids in microbially colonized sandstone by precolumn phenyl isothiocyanate derivatization and high-performance liquid chromatography. Appl Environ Microbiol 57:879–881

Siebert J, Hirsch P, Hoffmann B, Gliesche CG, Peissl K, Jendrach M (1996) Cryptoendolithic microorganisms from Antarctic sandstone of Linnaeus Terrace (Asgard range): diversity, properties and interactions. Biodivers Conserv 5:1337–1363

Sigler WV, Bachofen R, Zeyer J (2003) Molecular characterization of endolithic cyanobacteria inhabiting exposed dolomite in central Switzerland. Environ Microbiol 5:618–627

Smith MC, Bowman JP, Scott FJ, Line MA (2000) Sublithic bacteria associated with Antarctic quartz stones. Antarct Sci 12:177–184

Souza-Egipsy V, Ascaso C, Wierzchos J (2002) Ultrastructure and biogeochemical features of microbiotic soil crusts and their implications in a semi-arid habitat. Geomicrobiol J 19:567–580

Starke V, Kirshtein JD, Fogel ML, Steele A (2013) Microbial community composition and endolith colonization at an Arctic thermal spring are driven by calcite precipitation. Environ Microbiol Rep 5:648–659

Sterflinger K (2000) Fungi as geologic agents. Geomicrobiol J 17:97–124

Sterflinger K, De Baere R, de Hoog GS, De Wachter R, Krumbein WE, Haase G (1997) *Coniosporium perforans* and *C. apollinis*, two new rock-inhabiting fungi isolated from marble in the Sanctuary of Delos (Cyclades, Greece). Anton van Lee 72:349–363

Stomeo F, Valverde A, Pointing SB, McKay CP, Warren-Rhodes KA, Tuffin MI, Seely M, Cowan DA (2013) Hypolithic and soil microbial community assembly along an aridity gradient in the Namib desert. Extremophiles 17:329–337

Stromberg JM, Applin DM, Cloutis EA, Rice M, Berard G, Mann P (2014) The persistence of a chlorophyll spectral biosignature from Martian evaporite and spring analogues under Mars-like conditions. Int J Astrobiol 13:203–223

Sun HJ, Friedmann EI (1999) Growth on geological time scales in the Antarctic cryptoendolithic microbial community. Geomicrobiol J 16:193–202

Sun H, Nienow JA, McKay CP (2010) The antarctic cryptoendolithic microbial ecosystem. In: Doran PT, Lyons WB, McKnight DM (eds) Life in Antarctic Deserts and other cold dry environments. Cambridge University Press, Cambridge, pp 110–138

Tamaru Y, Takani Y, Yoshida T, Sakamoto T (2005) Crucial role of extracellular polysaccharides in desiccation and freezing tolerance in the terrestrial cyanobacterium *Nostoc commune*. Appl Environ Microbiol 71:7327–7333

Tang Y, Lian B, Dong H, Liu D, Hou W (2012) Endolithic bacterial communities in dolomite and limestone rocks from the Nanjiang Canyon in Guizhou karst area (China). Geomicrobiol J 29:213–225

Tracy CR, Streten-Joyce C, Dalton R, Nussear KE, Gibb KS, Christian KA (2010) Microclimate and limits to photosynthesis in a diverse community of hypolithic cyanobacteria in northern Australia. Environ Microbiol 12:592–607

Tuovila BJ, LaRock PA (1987) Occurrence and preservation of ATP in Antarctic rocks and its implications in biomass determinations. Geomicrobiol J 5:105–118

Vestal JR (1988a) Carbon metabolism of the cryptoendolithic microbiota from the Antarctic desert. Appl Environ Microbiol 54:960–965

Vestal JR (1988b) Biomass of the cryptoendolithic microbiota from the Antarctic desert. Appl Environ Microbiol 54:957–959

Vincent WF (2007) Cold tolerance in cyanobacteria and life in the cryosphere (from Part 4: Phototrophs in cold environments). In: Algae and Cyanobacteria in extreme environments. Springer, Netherlands, pp 289–301

Walker JJ, Pace NR (2007a) Endolithic microbial ecosystems. Annu Rev Microbiol 61:331–347

Walker JJ, Pace NR (2007b) Phylogenetic composition of rocky mountain endolithic microbial ecosystems. Appl Environ Microbiol 73:3497–3504

Walker JJ, Spear JR, Pace NR (2005) Geobiology of a microbial endolithic community in the Yellowstone geothermal environment. Nature 434:1011–1014

Warren LA, Ferris FG (1998) Continuum between sorption and precipitation of Fe(III) on microbial surfaces. Environ Sci Technol 32:2331–2337

Warren-Rhodes KA, Rhodes KL, Pointing SB, Ewing SA, Lacap DC, Gomez-Silva B, Amundson R, Friedman EI, McKay CP (2006) Hypolithic cyanobacteria, dry limit of photosynthesis, and microbial ecology in the hyperarid Atacama Desert. Microb Ecol 52:389–398

Warren-Rhodes KA, Rhodes KL, Boyle LN, Pointing SB, Chen Y, Liu S, Zhuo P, McKay CP (2007a) Cyanobacterial ecology across environmental gradients and spatial scales in China's hot and cold deserts. FEMS Microbiol Ecol 61:470–482

Warren-Rhodes KA, Rhodes KL, Liu S, Zhou P, McKay CP (2007b) Nanoclimate environment of cyanobacterial communities in China's hot and cold hyperarid deserts. J Geophys Res Biogeo 112:G01016

Warren-Rhodes KA, McKay CP, Boyle LN, Wing MR, Kiekebusch EM, Cowan DA, Stomeo F, Pointing SB, Kaseke KF, Eckardt F, Henschel JR, Anisfeld A, Seely M, Rhodes KL (2013) Physical ecology of hypolithic communities in the central Namib Desert: the role of fog, rain, rock habitat, and light. J Geophys Res Biogeo 118:1451–1460

Weed R, Ackert RPJ (1986) Chemical weathering of Beacon supergroup sandstones and implications for Antarctic glacial chronology. S Afr J Sci 82:513–516

Weed R, Norton SA (1991) Siliceous crusts, quartz rinds and biotic weathering of sandstones in the cold desert of Antarctica. In: Berthelin J (ed) Diversity of Environmental Biogeochemistry (Developments in Geochemistry, Vol. 6). Elsevier, Amsterdam, p 327-339

Wessels DCJ, Büdel B (1995) Epilithic and cryptoendolithic cyanobacteria of Clarens sandstone cliffs in the Golden Gate Highlands National Park, South Africa. Bot Acta 108:220–226

Wessels DCJ, Schoeman P (1988) Mechanism and rate of weathering of Clarens sandstone by an endolithic lichen. S Afr J Sci 84:274–277

Wharton RA Jr, McKay CP, Mancinelli RL, Simmons GM Jr (1989) Early Martian environments: the Antarctic and other terrestrial analogues. Adv Space Res 9:147–153

Wierzchos J, Ascaso C (2002) Microbial fossil record of rocks from the Ross Desert, Antarctica: implications in the search for past life on Mars. Int J Astrobiol 1:51–59

Wierzchos J, Ascaso C, Sancho LG, Green A (2003) Iron-rich diagenetic minerals are biomarkers of microbial activity in Antarctic rocks. Geomicrobiol J 20:15–24

Wierzchos J, de los Ríos A, Sancho LG, Ascaso C (2004) Viability of endolithic micro-organisms in rocks from the McMurdo Dry Valleys of Antarctica established by confocal and fluorescence microscopy. J Microsc 216:57–61

Wierzchos J, Sancho LG, Ascaso C (2005) Biomineralization of endolithic microbes in rocks from the McMurdo Dry Valleys of Antarctica: implications for microbial fossil formation and their detection. Environ Microbiol 7:566–575

Wierzchos J, Ascaso C, McKay CP (2006) Endolithic cyanobacteria in halite rocks from the hyperarid core of the Atacama Desert. Astrobiology 6:415–422

Wierzchos J, Cámara B, de los Ríos A, Davila AF, Sánchez Almazo IM, Artieda O, Wierzchos K, Gómez-Silva B, McKay C, Ascaso C (2011) Microbial colonization of Ca-sulfate crusts in the hyperarid core of the Atacama Desert: implications for the search for life on Mars. Geobiology 9:44–60

Wierzchos J, Davila AF, Sánchez-Almazo IM, Hajnos M, Swieboda R, Ascaso C (2012) Novel water source for endolithic life in the hyperarid core of the Atacama Desert. Biogeosciences 9:2275–2286

Wierzchos J, Davila AF, Artieda O, Cámara-Gallego B, de los Ríos A, Nealson KH, Valea S, García-González MT, Ascaso C (2013) Ignimbrite as a substrate for endolithic life in the hyper-arid Atacama Desert: Implications for the search for life on Mars. Icarus 224:334–346

Wingender J, New TR, Flemming H-C (1999) What are bacterial extracellular polymeric substances? In: Wingender J, New TR, Flemming H-C (eds) Microbial extracellular polymeric substances: characterization, structure and function. Springer, Berlin, pp 1–19

Wong FK, Lacap DC, Lau MC, Aitchison JC, Cowan DA, Pointing SB (2010a) Hypolithic microbial community of quartz pavement in the high-altitude tundra of central Tibet. Microb Ecol 60:730–739

Wong FK, Lau MC, Lacap DC, Aitchison JC, Cowan DA, Pointing SB (2010b) Endolithic microbial colonization of limestone in a high-altitude arid environment. Microb Ecol 59:689–699

Wynn-Williams DD, Edwards HGM, Garcia-Pichel F (1999) Functional biomolecules of Antarctic stromatolitic and endolithic cyanobacterial communities. Eur J Phycol 34:381–391

Ziolkowski LA, Mykytczuk NCS, Omelon CR, Johnson H, Whyte LG, Slater GF (2013) Arctic gypsum endoliths: a biogeochemical characterization of a viable and active microbial community. Biogeosciences 10:7661–7675

Chapter 5
The Snotty and the Stringy: Energy for Subsurface Life in Caves

Daniel S. Jones and Jennifer L. Macalady

Abstract Caves are subterranean environments that support life largely in the absence of light. Because caves are completely or almost completely removed from the photosynthetic productivity of the sunlit realm, most cave ecosystems are supported either by inputs of organic matter from the surface or by in situ sources of inorganic chemical energy. The majority of caves have very low energy and nutrient availability and thus, generally low biological activity and productivity. However, those caves that have abundant inorganic chemical energy or high organic carbon influx represent subterranean oases that support robust microbial communities and diverse animal life. In this chapter, we review the energy resources available to cave microbiota and describe several examples that illustrate the vast diversity of subsurface habits contained within caves.

5.1 Introduction

Microbial life is ubiquitous on Earth. Over the past several decades, we have learned that life is prominent not only at the Earth's surface, but that biological activity extends deep into the subsurface. Active microbial populations abound under kilometers of ocean sediment (Teske 2005; Orsi et al. 2013), in deep terrestrial aquifers (Lin et al. 2006), and in terrestrial sediments (Pedersen 2000) and other crustal rocks (Amend and Teske 2005). By one recent estimate, perhaps half the biomass on the planet resides in the subsurface (Whitman et al. 1998). These deep microorganisms promote crucial carbon and mineral transformations that impact geological processes such as porosity development (Engel and Randall 2011), carbon and nutrient remineralization (Jørgensen 1982; Whitman et al. 1998), methane production and hydrocarbon alteration (Whiticar et al. 1986; Head

D.S. Jones (✉)
Department of Earth Sciences, University of Minnesota, Minneapolis, MN 55455, USA
e-mail: dsjones@umn.edu

J.L. Macalady
Department of Geosciences, Penn State University, University Park, PA 16802, USA

et al. 2003), and the formation of economic mineral deposits (Hill 1995). Furthermore, the subsurface represents a possible haven for life on other planets with inhospitable surface environments (Boston et al. 2001). Yet even on our own planet, the diversity and activity of subsurface life remain largely unexplored.

Caves represent windows into the subsurface through which we can directly access and explore deep microbial diversity. Caves are defined as subterranean voids large enough to be navigable by humans, and the deepest cave has been explored to over 2 km underground (Voronya Cave, Abkhazia; Palmer 2007). Caves represent a diverse suite of microbial habitats, from sparely populated oligotrophic environments with scant energy sources (Barton and Jurado 2007) to energy-rich "sulfidic" caves with flourishing chemosynthetic ecosystems and symbiotic animal life (Sarbu et al. 1996; Dattagupta et al. 2009). Cave microbial communities occur in many unusual and distinctive forms not encountered at the surface, including extremely acidic "snottite" biofilms, anastomosing wall sediments known as vermiculations, sparkling actinomycete "cave jewels," and enigmatic underwater ropes and "microbial mantles" (all discussed below). Furthermore, caves attract the attention of ecologists and biogeochemists, not because of the extraordinary microbial formations per se, but because of their utility as natural laboratories. Caves represent controlled experiments in microbial isolation that are suitable to probe fundamental questions of microbial biogeography and evolution and often provide a constrained setting with reduced environmental variables for investigating microbial diversity and nutrient dynamics (e.g., Engel and Northup 2008; Macalady et al. 2008b, 2013; Engel et al. 2010; Rossmassler et al. 2012; Shabarova et al. 2013). Moreover, caves are an exemplary setting in which to study the impacts of microbe-mineral interactions at scales that range from the microscopic to that of entire aquifers (e.g., Cunningham et al. 1995; Engel and Randall 2011). Through their metabolic activity, microbes themselves may actively contribute to rock dissolution and thus cave enlargement (Hose et al. 2000; Engel et al. 2004) and to the formation of new mineral deposits (Northup and Lavoie 2001; Jones 2001).

5.2 Cavern Development and the Cave Environment

The majority of caves are associated with karst terrains. Karst is a landscape in which the dominant geomorphic process is the dissolution of soluble bedrock such as limestone, dolomite, or gypsum and is characterized by large solutional features including sinkholes, sinking streams, enclosed depressions, and caves (Fig. 5.1). Karst covers roughly 20 % of Earth's surface, the majority of which is developed on carbonate rocks (Ford and Williams 2007). The predominant cave-forming process in these areas is carbonic acid-induced dissolution of carbonate bedrock. Carbonic acid (H_2CO_3) is a weak acid that forms by the reaction of CO_2 and water [Eq. (5.1)], which dissolves limestone ($CaCO_3$) [Eq. (5.2)] to release bicarbonate (HCO_3^-) and calcium (Ca^{2+}) ions.

Fig. 5.1 (**a**) A striking example of a karst landscape, from the Stone Forest in the Yunnan Province, Southern China. Photo by D. Jones. (**b**) A karst spring near the Frasassi cave system, Marche Region, Italy. The emergence is roughly 1 m high. Photo by J. Macalady

$$CO_2 + H_2O \leftrightarrow H_2CO_3 \quad (5.1)$$

$$H_2CO_3 + CaCO_3 \leftrightarrow Ca^{2+} + 2HCO_3^- \quad (5.2)$$

Rainwater in equilibrium with atmospheric CO_2 is mildly acidic (pH ~5.7) due to carbonic acid formation, and soil porewaters typically have orders of magnitude more dissolved CO_2 than rainwater. Waters from the surface thus represent a source of acid for karst development, and cave formation and enlargement occur as long as groundwaters remain undersaturated with respect to the carbonate bedrock. Saturated waters can regain their solutional aggressiveness by groundwater mixing (Wigley and Plummer 1976) or by additional CO_2 input (White 1988).

In some cases, strong acids such as sulfuric acid (H_2SO_4) are important for karstification (Lowe and Gunn 1995; Hose et al. 2000; Engel et al. 2004). Sulfuric acid is most commonly produced in caves and carbonate aquifers where anoxic, hydrogen sulfide (H_2S)-rich groundwaters interact with oxygen [Eq. (5.3)].

$$H_2S + 2O_2 \rightarrow H_2SO_4 \quad (5.3)$$

H_2S-charged waters in such settings are usually sourced from petroleum deposits and other organic-rich sediments or, in some cases, volcanic fluids. Sulfuric acid production results in rapid carbonate dissolution, and H_2S itself is an acid that can drive speleogenesis (Palmer 1991). For a complete treatment of cave formation processes, readers are referred to the excellent discussion by Palmer (1991) or to recent texts by Palmer (2007) and Ford and Williams (2007).

Examples of caves that form by non-karst processes include lava tubes, ice caves, and "pseudokarst" structures in less soluble rocks such as coastal wave erosion features, talus piles, or piping structures (Halliday 2007). With the exception of a brief discussion of lava tube caves, this review is limited to karst caves and also does not address man-made cavities such as mines or tunnels (e.g., Spear et al. 2007). For additional information on microbial communities in non-karst caves, readers are referred to reviews by Northup and Lavoie (2001) and Engel (2010).

5.3 Energy and Carbon in Caves

In the absence of light, cave organisms either require chemical energy to fuel primary production or are dependent on inputs of photosynthetically derived organic matter from the surface. We can thus broadly classify cave ecosystems into two types: (1) those that are largely dependent on allochthonous organic carbon and (2) those with adequate inorganic chemical energy to allow for chemosynthetic primary production in situ. Caves of the first type are usually (but not always) oligotrophic, characterized by sparse microbial populations and low biological activity. Biological activity is generally limited by the influx rate of organic carbon (Simon and Benfield 2002; Simon et al. 2007). Animal life is well known from these oligotrophic caves and is adapted to life under starvation conditions (Culver and Pipan 2009). Caves of the second type are exemplified by inputs of chemical energy in the form of H_2S. Note that, unlike Eqs. (5.1) and (5.2), Eq. (5.3) is an oxidation-reduction ("redox") reaction that represents biologically available energy. Sulfide oxidation [Eq. (5.3)] fuels CO_2 fixation by chemoautotrophic microorganisms (Sarbu et al. 1996), and these "sulfidic" caves are often (but not always) eutrophic and contain abundant and conspicuous microbial life. Other examples of chemical energy for primary production in caves include ammonia, iron, and methane oxidation (Holmes et al. 2001; Hutchens et al. 2004). Overall, chemosynthetic caves represent a relatively small subset of known caves but are important hotspots for subsurface biological activity and subterranean biodiversity (Culver and Sket 2000).

The above classification is useful for the purposes of this review. However, the two types of cave ecosystems described above should be considered end-members of a continuum from surface organic-dominated to chemosynthesis-dominated caves. As with most attempts to dichotomize the natural world, large gray areas exist in this classification, and many cave ecosystems rely on both allochthonous organics and in situ chemosynthesis to some degree.

The Frasassi cave system, Italy, is an exemplary setting with which to illustrate the two types of cave ecosystems described above. The Frasassi Caves are a large cave system with more than 25 km of passages in a massive platform limestone, and the caves are developed along several distinct horizons arranged vertically with older passages above younger ones (Galdenzi 1990; Galdenzi and Maruoka 2003; Mariani et al. 2007). In the lowermost levels, rising anoxic groundwaters supply hydrogen sulfide for chemosynthetic primary production, and life abounds at the sulfidic water table (Fig. 5.2). White biomats of sulfide-oxidizing microbes fill microoxic streams, acidic biofilms (pH 0–2) known as "snottites" hang from cave walls and ceilings, and numerous invertebrates, some with symbiotic sulfide-oxidizing bacteria, thrive in close proximity to sulfidic streams (Fig. 5.3) (Lyon

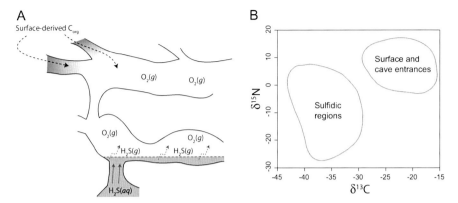

Fig. 5.2 Schematic showing primary energy sources in the Frasassi cave system, Italy. (**a**) Inputs of surface-derived organic carbon support biological activity near cave entrances, whereas microbial sulfide oxidation near the cave water table (*dashed line*) provides energy for primary production in the lower levels. (g = gas, aq = aqueous solution.) (**b**) Stable isotope ratios of organic C and N measured throughout the cave indicate that organic material from the sulfidic cave regions is distinct from surface sources. The *outlined regions* in (**b**) indicate the range of values from Galdenzi and Sarbu (2000), expanded with the inclusion of unpublished data (Jones and Macalady)

et al. 2005; Macalady et al. 2006, 2007; Dattagupta et al. 2009). In contrast, the upper cave levels sit tens of meters above the current water table and are removed from the source of H_2S (Fig. 5.2). Microbial life in the upper levels is generally sparse and much less conspicuous than in the sulfidic regions. However, despite the limited energy in these areas, animal life such as cave salamanders can be found near the cave entrances, and microbial growth can be observed (Figs. 5.2 and 5.3). Stable isotope analyses show that organic matter near the cave entrances is isotopically distinct from the organics produced by chemosynthesis in the sulfidic regions (Fig. 5.2) (Galdenzi and Sarbu 2000).

5.3.1 Cave Ecosystems Based on Surface-Derived Organic Carbon

In the majority of caves, life depends on small inputs of organic carbon from the surface. Organic carbon is transported into caves in dissolved or particulate forms with percolating or flowing waters or to dry regions by falling debris, wind, or animal vectors (Simon et al. 2003, 2007; Culver and Pipan 2009). Because the influx of organic carbon by these mechanisms is generally slow and sporadic, most caves are oligotrophic. However, despite energy limitations, caves contain surprisingly diverse and dense microbial communities inhabiting rock surfaces, sediments, and cave waters (Barton et al. 2007). For example, microorganisms are ubiquitous on wall and speleothem (cave formation) surfaces (Ortiz et al. 2014), in dripwaters

Fig. 5.3 Life in the Frasassi cave system, Italy. Near cave entrances, microbial formations include (**a**) gold and silver "cave jewels," and some animal life such as triton salamanders (**b**). At the sulfidic water table, (**c**) white microbial mats fill sulfidic cave streams, which provide energy for invertebrates such as (**d**) oligochaete worms and (**e**) amphipods (*Niphargus* spp.). Above streams, (**f**) extremely acidic snottites hang from cave walls (*thin arrows*; note the presence of spider webs in the image, *thick arrow*), and (**g**) "biovermiculations" cover exposed limestone walls. Photos by J. Macalady and D. Jones

(Laiz et al. 1999), and in cave streams and lakes (Simon et al. 2003). Microbial communities on the walls of organic-fed caves can be visually striking, such as the yellow and silver "cave gold" and "cave jewels" (Fig. 5.2a) (Pašić et al. 2010; Porca et al. 2012; Hathaway et al. 2014), and microorganisms have made their mark by inducing or impacting the genesis of certain speleothems and cave mineral deposits (see below).

However, not all such caves are oligotrophic. Caves with large bat populations have extensive guano deposits, which constitute an organic-rich resource that harbors thriving heterotrophic microbial communities (e.g., Chroňáková et al. 2009) and supports its own specialized animal fauna (Culver and Pipan 2009). Indeed, invertebrates colonized guano deposits that developed in an artificial bat cave only a few years after the bat population became established (Lavoie and Northup 2009). In addition, many shallow caves are well connected to the surface and thus have a high flux of organic carbon. For example, plant roots are a common feature in lava tubes (see below) and in other shallow caves and highlight a close connection to the surface (Hathaway et al. 2014).

5.3.2 Cave Ecosystems Based on Inorganic Chemical Energy from Sulfide Oxidation

Possibly as many as 5 % of known caves were formed by the oxidation of hydrogen sulfide (Palmer 2007) in a process known as sulfuric acid speleogenesis. Hydrogen sulfide oxidizes to sulfuric acid [Eq. (5.3)] in areas where anoxic sulfidic groundwaters are exposed to oxygenated surface waters and cave air. Sulfuric acid generation promotes rapid carbonate dissolution and aggressive speleogenesis, and the resultant "sulfidic" caves often contain large underground chambers, atypical passage morphologies, and unusual sulfur mineral formations.

Sulfidic caves are hotspots for biological activity in the subsurface. Unlike carbonic acid formation [Eq. (5.1)], sulfuric acid production [Eq. (5.3)] represents a rich source of energy for primary production by microbial chemosynthesis. Sulfidic caves harbor isolated chemotrophic ecosystems that include invertebrate and even vertebrate life (Sarbu et al. 1996; Hose and Pisarowicz 1999; Dattagupta et al. 2009), all ultimately supported by microbial sulfide oxidation. Furthermore, by promoting reaction (5.3), sulfide oxidizers increase rates of acid production and limestone dissolution and may therefore increase rates of cave formation (Hose et al. 2000; Engel et al. 2004).

Sulfidic caves are heterogeneous environments. The water-filled environment below the water table includes anoxic zones, microoxic regions, and oxygen-saturated areas. In contrast, the air-filled areas above the water table are generally fully oxygenated but exhibit a dramatic range in acidities. Exposed limestone surfaces are well buffered and generally have circumneutral pH, but extremely acidic conditions occur where gypsum corrosion residues isolate microbial sulfuric acid generation from carbonate buffering by limestone walls. This diverse geochemical regime and strong pH and redox gradients support an eclectic array of microbial life (Fig. 5.3).

In microoxic sulfidic streams, conspicuous white biofilms with diverse morphologies coat nearly every available sediment surface (Fig. 5.3c, d). These biomats have long received the attention of cave microbiologists (Brigmon et al. 1994; Sarbu et al. 1996; Angert et al. 1998; Engel et al. 2001, 2003, 2010; Hutchens et al. 2004; Macalady et al. 2006; Meisinger et al. 2007; Jones et al. 2010). Large filamentous sulfur-oxidizing bacteria such as *Thiothrix* spp. and *Beggiatoa* spp. often construct the biofilm matrix (e.g., Brigmon et al. 1994), and Engel et al. (2003) identified filamentous *Epsilonproteobacteria* as the dominant biofilm-forming taxa in Lower Kane Cave springs. Macalady et al. (2008b) proposed a niche model in which the dominant mat-forming microbial taxa (*Thiothrix* spp., *Beggiatoa* spp., and filamentous *Epsilonproteobacteria*) can be predicted based on dissolved sulfide and oxygen concentrations and water flow velocity.

Above the water table, viscous biofilms known as "snottites" cling to overhanging gypsum surfaces (Fig. 5.4). Snottites are among the most acidic microbial formations known and commonly reach values as low as pH 0. Snottites occur in close proximity to turbulent sulfidic streams where rapid H_2S degassing provides energy for microbial sulfide oxidation (Jones et al. 2012) and likely achieve such extremely acidic conditions because they hang free from buffering by cave limestone and gypsum. In early work, Vlasceanu et al. (2000), Hose et al. (2000) (see also Hose and Pisarowicz 1999 and Boston et al. 2001), and Macalady et al. (2007) found that snottites from sulfidic caves in Italy and Mexico were dominated by acidophilic sulfide-oxidizing bacteria, with smaller populations of other bacteria, archaea, and eukaryotes. Later metagenomic analyses (Jones et al. 2012, 2014) revealed that the acidophilic sulfide oxidizer *Acidithiobacillus thiooxidans* is the dominant primary producer and likely the "architect" of snottites (Fig. 5.4d).

Much less is known about other sulfidic cave communities. Organic-rich sediments known as biovermiculations form intricate patterns over exposed limestone walls (see below). Other unusual features in sulfidic caves include acidic "ragu" deposits and green rock coatings (e.g., Hose et al. 2000). Little is known about the microbial communities inhabiting gypsum wall crusts, black stream sediments, or the spectacular elemental sulfur precipitations that occur in certain caves (Hose and Pisarowicz 1999; Jones et al. 2010). In anoxic sulfidic lakes, white floating mats sometimes occur at the air-water interface (e.g., Hutchens et al. 2004), while novel ropelike biofilms and other unusual microbial formations have been found in deep anoxic layers (Macalady et al. 2008a; McCauley et al. 2010).

5.3.3 Cave Ecosystems Based on Nitrogen Oxidation: The Nullarbor Caves, Australia

The Nullarbor Plain in Australia represents the world's largest continuous karst landscape. The region includes numerous caves (the Nullarbor Caves) that intersect a saline aquifer. The submerged region of the caves have been explored by cave

5 The Snotty and the Stringy: Energy for Subsurface Life in Caves

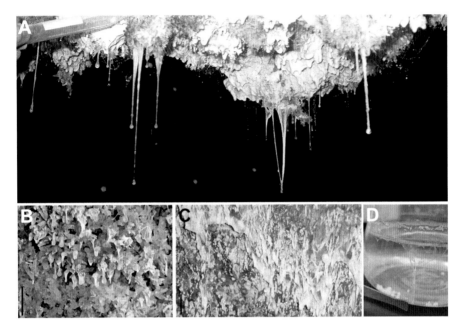

Fig. 5.4 Extremely acidic subaerial formations in sulfidic caves include (**a, b**) snottites and (**c**) "snot curtains." On occasion, bubbles ("snot bubbles") can be observed forming in the biofilm matrix at the location of image (**c**) (see Galdenzi et al. 2010). These biofilms occur in close proximity to the sulfidic water table, where H_2S degasses into the fully oxygenated cave atmosphere and is oxidized by Eq. (5.3) on cave walls and ceilings. Sulfidic cave snottites are almost always associated with gypsum wall crusts (white material in (**a**), crystalline material in (**b**)), which form as a corrosion residue where sulfuric acid interacts with limestone ($H_2SO_4 + CaCO_3 + H_2O \rightarrow CaSO_4 \cdot 2H_2O + CO_2$). Photos by D. Jones and J. Macalady. (**d**) Strains of *Acidithiobacillus thiooxidans* cultured from snottites produce biofilm in vitro. Photo by A. Diefendorf. Photo (**a**) is from Cueva de Villa Luz, Mexico, and (**b**) and (**c**) are from the Acquasanta cave system, Italy. *Black bars* are 1 cm

divers, who noted the occurrence of unusual organic formations commonly known as "microbial mantles" or "slime curtains," which occur above "snowfields" of calcium carbonate deposits (Contos et al. 2001; Holmes et al. 2001). The microbial communities of slime curtains from Weebubbie Cave in the Nullarbor Plain have been characterized using microscopic and molecular techniques.

Microscopic analyses reveal that the Weebubbie slime curtains are composed of a matrix of densely packed microbial filaments. Although the identity of the filaments themselves is currently unknown, molecular analyses have revealed that the primary energy source for lithoautotrophic growth appears to be reduced nitrogen compounds.

$$NH_3 + 1.5O_2 \rightarrow NO_2^- + H_2O + H^+ \tag{5.4}$$

$$NO_2^- + 0.5O_2 \rightarrow NO_3^- \tag{5.5}$$

Unlike the sulfidic caves described above, the groundwaters in the Nullarbor Caves do not contain dissolved sulfide and instead have high concentrations of nitrate, nitrite, and presumably ammonium. Initial work found that nitrite-oxidizing autotrophic bacteria are abundant in the community (Holmes et al. 2001), and subsequent analyses using more detailed metagenomic techniques found that nearly half the community is composed of an ammonia-oxidizing archaea in the Thaumarchaeota clade (Tetu et al. 2013). The Weebubbie archaeon contains genes for carbon fixation and ammonia oxidation and is a close relative of the cultivated ammonia oxidizer *Nitrosopumilus maritimus*. Although much remains to be explored regarding the subsurface microbial processes in the Nullarbor Plain, Weebubbie Cave appears to represent an example of a subsurface ecosystem based on lithoautotrophic ammonia and nitrite oxidation [Eqs. (5.4) and (5.5)]. Similar slime curtains dominated by N and Fe cycling microbial taxa have been identified in flooded caves in the Dominican Republic (Cardman et al. 2015).

5.3.4 Other Inorganic Chemical Energy Sources in Caves

Chemical energy in caves also comes in the form of methane, iron, and manganese. For example, Hutchens et al. (2004) identified active methanotrophs in sulfidic Movile Cave in Romania, indicating that methane oxidation is a major source of energy in this system (in addition to sulfide, ammonia, and nitrite oxidation, Chen et al. 2009). Iron- and manganese-oxidizing microorganisms have been implicated in the formation of ferromanganese crusts in Lechuguilla Cave (Northup et al. 2000; Spilde et al. 2005). Rossi et al. (2010) reported on manganese stromatolites from El Soplao Cave, Spain, that are replete with microbial growth textures and well-preserved microfossils that suggest a microbial role in their formation. Although no longer actively forming, the El Soplao stromatolites are interpreted to have formed by the microbial oxidation of aqueous Mn and Fe in flowing cave waters, likely at redox interfaces near a former water table.

5.3.5 Chemical Energy from the Host Rock: Lava Tube Caves

Lava tube caves represent an entirely different type of subsurface environment from the carbonate-hosted karst caves that are the primary focus of this review. Lava tubes commonly form as the surface of a basaltic lava flow solidifies, while molten lava continues to move underneath (Palmer 2007). Like the majority of karst caves, microbial communities associated with lava tubes are generally fueled by surface sources of organic carbon. In fact, the shallow depths at which lava tubes form facilitate organic inputs, and lava tubes often have dangling plant roots and soil deposits. However, the basalt host rock also offers a source of reduced iron and other metals that provide both reducing power for energy generation and nutrient sources. Popa et al. (2012) found that bacterial isolates from a lava tube in Oregon could grow by aerobic oxidation of reduced iron in olivine, and others have observed iron oxide and other metal deposits associated with lava cave microbial communities (Northup et al. 2011). For recent investigations of microbial communities in lava tube caves, readers are referred to Northup et al. (2011) and Hathaway et al. (2014).

Lava tubes are an attractive analogue for potential subsurface life on Mars and other extraterrestrial bodies. The extreme temperature fluctuations and high ultraviolet radiation of the Martian surface render it inhospitable to life. However, the shallow subsurface may represent a refuge from the harsh surface environment, and basaltic rocks underlie large regions of the Martian surface. Lava tubes or other basaltic cave features are one potential setting for a subsurface biosphere that may harbor life or preserve evidence of biological activity (Boston et al. 2001). Possible cave-like features have been imaged on Mars (Léveillé and Datta 2010), and such subsurface settings may represent a promising target in the search for signatures of past or present Martian life.

5.3.6 Communities Based on Photosynthesis: Artificial Lighting in Show Caves

The cave environment is typically characterized by the absence of light, except in the "twilight zone" where entrances or natural skylights supply limited light (e.g., Secord and Muller-Parker 2005; Azúa-Bustos et al. 2009). However, the introduction of artificial lighting to show caves is sufficient to sustain organic carbon production in situ by photoautotrophy, and microbial phototrophy now occurs in a variety of cave environments where such energy was not previously available (Giordano et al. 2000; Smith and Olson 2007). Certain cyanobacteria that appear to be especially resilient to long periods of darkness often dominate artificially lit regions of show caves (Giordano et al. 2000; Montechiaro and Giordano 2006). The growth of photosynthetic "lamp flora" not only pose an aesthetic concern in show caves, but they may irreversibly impact carbonate speleothems or other cave formations (see below). Different artificial light sources, lowered light intensity, and frequent cleaning are often used to reduce or remove photosynthetic flora in show caves (Faimon et al. 2003).

5.4 Microbial Role in the Formation of Speleothems and Other Cave Deposits

The impact of microbial activity in caves can be recorded via the generation of certain formations and mineral deposits. However, microbial involvement in the genesis of cave formations is difficult to demonstrate convincingly. For example, microbial activity has been implicated in the generation of several types of carbonate speleothems (Cacchio et al. 2004; Blyth and Frisia 2008; Jones 2010). Microorganisms are known to induce calcite precipitation by several mechanisms, including alkalinity production during anaerobic respiration, CO_2 consumption during autotrophic growth, or by liberating trapped calcium ions during the degradation of exopolymeric substances (Dupraz and Visscher 2005). In addition, microbial biomass may provide a passive template upon which carbonate precipitation can occur. Many cave isolates are capable of carbonate dissolution or precipitation (Danielli and Edington 1983; Engel et al. 2001), and microorganisms are commonly found fossilized in carbonates (Melim et al. 2001, 2008). Even given those relationships, however, it remains challenging to directly link microbial activity to speleothem formation. Evidence is often circumstantial and based on the presence of microorganisms, and establishing a direct causal relationship is a formidable challenge due to slow metabolic rates (Jones 2001, 2010; Northup and Lavoie 2001). Here, we review a few examples of cave formations for which microbial activity may be important.

5.4.1 Moonmilk

Many researchers have proposed a microbial role in the formation of a carbonate speleothem known as moonmilk (Cañaveras et al. 2006). Moonmilk is a microcrystalline calcite that generally has a relatively high organic content (e.g., over 0.5 % organic carbon by weight, Engel et al. 2013) and is often colonized by metabolically active microorganisms (Portillo and Gonzalez 2011). Moonmilk has been proposed to form by difference processes including dissolution and reprecipitation of the parent limestone, slow calcite precipitation from seepage water, and other mechanisms (Hill and Forti 1997; Borsato et al. 2000; Jones 2010 and sources therein). Moonmilk is frequently associated with microbial filaments. The calcite crystals in moonmilk often have an elongate habit that resembles filamentous microbial structures, and calcified microbial filaments have been observed (e.g., Cañaveras et al. 1999, 2006; Jones 2010). Based on observations of moonmilk crystal morphology and microbial structures, Cañaveras et al. (2006) proposed a model for moonmilk development in Altamira Cave, Spain, in which the development and evolution of the moonmilk microbial community induces different calcite crystal fabrics. However, abiotic models have also been proposed, and if and how microorganisms impact moonmilk formation remains a subject for debate.

5.4.2 Pool Fingers

Pool fingers are elongate carbonate speleothems that are associated with ponded cave waters. Pool fingers are interpreted to form underwater, perhaps near the air-water interface, and are thought to grow downward and outward with continued carbonate precipitation. A microbial origin was initially proposed for pool fingers in Lechuguilla Cave, in part because they superficially resemble encrusted microbial streamers (Davis 2000). In studies of pool fingers from Hidden Cave and Cottonwood Cave, New Mexico, Melim et al. (2001, 2009) found that pool fingers have internal fabrics that are consistent with other biologically mediated calcite formations, such as stromatolitic textures. Pool fingers contain both sparry calcite cements and micritic lamina, the latter formed by recrystallization of lime mud. The micritic layers contain possible biogenic fabrics, mineralized microbes and are isotopically depleted in carbon-13. Based on petrographic and isotopic evidence, an active microbial role in their formation has been proposed, perhaps associated with a hanging microbial biofilm or other stringy filamentous structure in a subaqueous setting (Melim et al. 2001, 2009; Melim 2011).

5.4.3 Vermiculations

Vermiculations, often known as clay vermiculations, are anastomosing sediments that occur on cave walls and ceilings (Fig. 5.5). Vermiculations are most commonly composed of clay- and silt-sized particles and exhibit a range of morphologies from wall-covering mats to elongated stripes and disconnected spots (Hedges 1993; Hill and Forti 1997). The generally accepted model for vermiculation formation is that of Bini et al. (1978), who proposed that vermiculations form as wall sediments shrink and flocculate while drying.

Vermiculations are especially robust in certain sulfidic caves (Figs. 5.3g and 5.5a), where they are known as "biovermiculations" due to a suspected microbial role in their formation (Hose et al. 2000). In Cueva de Villa Luz, Mexico, and the Frasassi cave system, Italy, biovermiculations have a high organic content (3–45 % organic carbon by weight) and have been observed to form rapidly in close proximity to sulfidic streams (Hose et al. 2000; Jones et al. 2008; unpublished observations). In a study on Frasassi biovermiculations, Jones et al. (2008) identified diverse microbial communities that include some candidate sulfur-oxidizing taxa and used stable isotope ratios of C and N to show the organic material in biovermiculations derives from the sulfidic cave ecosystem rather than from surface sources of organic matter (Fig. 5.2). Imaging of biovermiculations by scanning

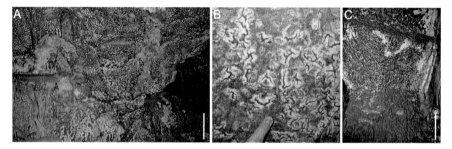

Fig. 5.5 (a) Robust "biovermiculations" cover extensive limestone surfaces in the sulfidic Frasassi cave system, Italy (see also Fig. 5.3g). This image was taken roughly 10 m from a sulfidic spring within the cave. *Scale bar* is approximately 0.5 m. (b) Vermiculations from the Grotta del Mezzogiorno-Grotta di Frasassi complex, located in the Frasassi Gorge but over 300 m above the current sulfidic water table. Vermiculations from this region are isotopically similar to surface-derived organic matter rather that to organics sourced from the sulfidic water table (Fig. 5.2) (Jones et al. 2008). (c) Sediments from the wall of Niagara Cave in Southeastern Minnesota that are starting to form vermiculation-like patterns. *Scale bar* is roughly 20 cm. Photo by J. Steenberg

electron microscopy has shown that the sediment matrix contains extracellular polymeric substances derived from microbial cells (Hose and Northup 2004; Jones et al. 2008).

Much remains unknown regarding vermiculations. Although it seems clear that microorganisms play a role in producing the organic material in sulfidic cave biovermiculations, the role of microorganisms, if any, in influencing their striking, large-scale wall patterns remains to be determined (Boston et al. 2009). Sulfidic cave biovermiculations and other vermiculations are visually similar, and some authors have suggested that microorganisms may play a role in the formation of all vermiculations (Anelli and Graniti 1967; Camassa and Febbroriello 2003). However, additional research will be required to determine if microbial activity is linked to vermiculation formation, and ongoing studies promise to shed more light on the microbiology as well as the patterning mechanisms of these enigmatic and widespread formations (Boston et al. 2009; Strader et al. 2011).

5.4.4 Iron and Manganese Deposits

Precipitates of iron (Fe) and manganese (Mn) oxides, which frequently co-occur as ferromanganese deposits, are a common feature in caves. Fe and Mn oxides can occur as nodules and pebble coatings (White et al. 2009; Carmichael et al. 2013), wall crusts (Northup et al. 2000; Spilde et al. 2005), flowstone (Gradziński et al. 1995), stromatolites (Rossi et al. 2010), and other features including unusual "rusticle" speleothems (Davis 2000). Numerous microorganisms are known to oxidize iron and manganese, and many authors have suggested a microbial role in the formation of ferromanganese deposits in caves (e.g., Peck 1986; Jones 1992;

Gradziński et al. 1995; Northup and Lavoie 2001 and references therein). For example, Rossi et al. (2010) found stromatolitic fabrics and exceptional microfossil preservation in ferromanganese deposits in El Soplao Cave, Spain. In a comprehensive study in the Upper Tennessee River Basin, USA, Carmichael et al. (2013) isolated Mn-oxidizing bacteria from cave ferromanganese deposits and identified relatives of Mn-oxidizing bacteria using culture-independent techniques. The authors also found that mineralized sheaths of the known Fe- and Mn-oxidizing bacteria *Leptothrix* spp. were frequently associated with the deposits.

In Lechuguilla Cave, New Mexico, especially well-developed ferromanganese crusts occur over a large region of the cave (Davis 2000). These crusts are up to 2 cm thick and coat a corroded layer of friable, altered carbonate known as punk rock that grades into less altered material below (see Figs. 1 and 2 in Spilde et al. 2005). The crusts are highly enriched in Fe and Mn relative to the host rock and have been shown to harbor diverse microbial communities, including relatives of known manganese-oxidizing microbes (Northup et al. 2003). Spilde et al. (2005) interpreted these deposits as a corrosion residue left behind following gradual dissolution of the carbonate bedrock and suggested that microorganisms are "mining" the host rock for energy from reduced Fe- and Mn-bearing minerals. Microbial activity in these crusts appears to extend into the altered limestone punk rock below the ferromanganese crusts and Mn oxides formed in enrichment cultures inoculated from the crust (Northup et al. 2003; Spilde et al. 2005). However, the deposits are also associated with humid air currents that likely induce condensation corrosion, and trace gasses in those air currents could also provide a primary energy source to the ferromanganese microbial communities (Cunningham et al. 1995; Davis 2000).

5.5 Microbial Contributions to Cave Formation

In addition to their proposed role in the precipitation of cave carbonates, microorganisms are also implicated in the destruction of cave-forming rocks. Indeed, a role for microbes in cave formation, sometimes termed biokarst or biospeleogenesis, is widely discussed in the literature.

The microbial role in cave formation is most apparent in sulfidic caves. By catalyzing reaction (5.3), microbes directly enhance rates of sulfide oxidation and acid production. Engel et al. (2004) showed that sulfide oxidation in microbe-packed streams in Lower Kane Cave, WY, USA, occurs much faster than would be expected in the absence of biological activity. Engel et al. also provided evidence that sulfide oxidizers in the streams directly contribute to limestone dissolution by localizing acid production at bedrock surfaces (see also Engel and Randall 2011; Steinhauer et al. 2010). In recent work, Jones et al. (2015) used microsensor measurements to show that anaerobic processes are important in acid generation in sulfidic cave streams. Microbial acid generation is clearly important above sulfidic cave water tables, where limestone dissolution is driven by the oxidation of vapor phase $H_2S(g)$ that volatilizes from cave streams. Waters dripping from snottites (Fig. 5.4) have extremely acidic pH values (Macalady et al. 2007), and Hose et al. (2000) described extensive rillenkarren (dissolution channels) and other corrosion features on the walls of snottite-filled Cueva de Villa Luz (Fig. 5.4a).

In carbonate caves, organoheterotrophic organisms may contribute to cave formation by creating locally corrosive microenvironments via CO_2 release from respiration or from the production of organic acids. These microbially generated acids could theoretically impact a wide variety of cave environments (e.g., Bullen et al. 2008). For example, the ferromanganese crusts in Lechuguilla Cave, described above, could result from microbial destruction of the bedrock (Spilde et al. 2005). Anelli and Graniti (1967) proposed that organic acids secreted from fungi and other microorganisms in vermiculations could enhance dissolution of the underlying bedrock. Etched limestone surfaces are often associated with microbial communities on cave walls (Cañaveras et al. 2001), and microbial alteration of calcite and aragonite speleothems is of concern for show caves where artificial lighting and human vectors have introduced photosynthetic and other microbial life (Giordano et al. 2000). Endolithic microbes have also been implicated in cave enlargement. For example, microboring features resulting from cyanobacterial activity are common in the low-light regions near cave entrances in the Cayman Islands (Jones 2010). However, it remains to be seen whether microbial acid generation is more than a locally important phenomenon, and the significance of microbially induced corrosion in caves and karst remains unknown. Further work will be required to determine if microbial-induced corrosion is largely restricted to zones of H_2S oxidation and limited hotspots of belowground microbial activity or if it is a widespread phenomenon associated with slowly growing yet extensive microbial communities that occur throughout oligotrophic cave environments.

5.6 Conclusion

Recent estimates suggest that only 10 % of all caves on Earth have been accessed by humans, and even in Europe and North America, as many as 50 % of caves remain unexplored (Lee et al. 2012 and references therein). Furthermore, some of the most energy-rich caves may be the hardest to find. Caves fed by inorganic chemical energy like H_2S are most commonly associated with rising deep-seated groundwaters (hypogenic caves), rather than descending surface acidity like the majority of carbonic acid-formed caves (epigenic caves). As such, many sulfidic caves lack clear connections to the surface and obvious superficial features such as sinkholes or sinking streams (Palmer 1991) and are frequently discovered by accident while drilling or quarrying (e.g., Sarbu et al. 1994; Por 2007). Cave microbiology explorations have already led to many important discoveries, such as novel chemosynthetically based ecosystems (Sarbu et al. 1996) and the first terrestrial chemoautotrophic animal-microbial symbiosis (Dattagupta et al. 2009). Our increasing knowledge of subterranean environments on Earth is dramatically expanding our knowledge of early Earth-like environments and extraterrestrial analogues for astrobiology research (Boston et al. 2001; Northup et al. 2011). Many exciting discoveries await as we continue to explore the depths of our planet.

Acknowledgments We would like to extend gratitude to everyone who has assisted and supported our cave microbiology research, especially to Alessandro Montanari for continued logistical support and the use of facilities at the Osservatorio Geologico di Coldigioco and to S. Carnevali, S. Cerioni, S. Galdenzi, M. Mainiero, S. Mariani, and the Gruppo Speleologico C.A.I. di Fabriano for technical support and scientific discussions in Italy. We also thank L. Hose, L. Rosales-Lagarde, I. Schaperdoth, S. Dattagupta, E. Lyon, T. Jones, K. Dawson, and R. McCauley for assistance in the field and lab. Our work has been supported by generous funding from the National Science Foundation, the NASA Astrobiology Institute, the Cave Conservancy Foundation, and the Marche Regional Government and the Marche Speleologic Federation. Special thanks to C. Hurst for organizing and editing this volume.

References

Amend JP, Teske A (2005) Expanding frontiers in deep subsurface microbiology. Palaeogeogr Palaeocl 219(1):131–155

Anelli F, Graniti A (1967) Aspetti microbiologici nella genesi delle vermicolazioni argillose delle Grotte di Castellana (Murge di Bari). Le Grotte d, Italia Ser 4:131–138

Angert ER, Northup DE, Reysenbach A-L, Peek AS, Goebel BM, Pace NR (1998) Molecular phylogenetic analysis of a bacterial community in Sulphur River, Parker Cave, Kentucky. Am Mineral 83:1583–1592

Azúa-Bustos A, González-Silva C, Mancilla R, Salas L, Palma R, Wynne J, McKay C, Vicuña R (2009) Ancient photosynthetic eukaryote biofilms in an Atacama Desert coastal cave. Microb Ecol 58(3):485–496

Barton HA, Jurado V (2007) What's up down there? Microbial diversity in caves. Microbe 3:132–138

Barton HA, Taylor NM, Kreate MP, Springer AC, Oehrle SA, Bertog JL (2007) The impact of host rock geochemistry on bacterial community structure in oligotrophic cave environments. Int J Speleol 36(2):93–104

Bini A, Gori MC, Gori S (1978) A critical review of hypotheses on the origin of vermiculations. Int J Speleol 10(1):11–34

Blyth AJ, Frisia S (2008) Molecular evidence for bacterial mediation of calcite formation in cold high-altitude caves. Geomicrobiol J 25(2):101–111

Borsato A, Frisia S, Jones B, Van Der Borg K (2000) Calcite moonmilk: crystal morphology and environment of formation in caves in the Italian Alps. J Sediment Res 70(5):1179–1190

Boston P, Spilde M, Northup D, Melim L, Soroka D, Kleina L, Lavoie K, Hose L, Mallory L, Dahm C, Crossey L, Schelble R (2001) Cave biosignature suites: microbes, minerals, and Mars. Astrobiology 1(1):25–55

Boston P, Curnutt J, Gomez E, Schubert K, Strader B (2009) Patterned growth in extreme environments. In: Proceedings of the third IEEE international conference on space mission challenges for information technology, Citeseer, pp 221–226

Brigmon R, Martin H, Morris T, Bitton G, Zam S (1994) Biogeochemical ecology of *Thiothrix* spp. in underwater limestone caves. Geomicrobiol J 12(3):141–159

Bullen HA, Oehrle SA, Bennett AF, Taylor NM, Barton HA (2008) Use of attenuated total reflectance Fourier transform infrared spectroscopy to identify microbial metabolic products on carbonate mineral surfaces. Appl Environ Microbiol 74(14):4553–4559

Cacchio P, Contento R, Ercole C, Cappuccio G, Martinez MP, Lepidi A (2004) Involvement of microorganisms in the formation of carbonate speleothems in the Cervo Cave (L'Aquila-Italy). Geomicrobiol J 21(8):497–509

Camassa MM, Febbroriello P (2003) Le foval della grotta zinzulusa in Puglia (SE-Italia). Thalassia Salent 26:207–218

Cañaveras J, Hoyos M, Sanchez-Moral S, Sanz-Rubio E, Bedoya J, Soler V, Groth I, Schumann P, Laiz L, Gonzalez I (1999) Microbial communities associated with hydromagnesite and needle-fiber aragonite deposits in a karstic cave (Altamira, Northern Spain). Geomicrobiol J 16(1):9–25

Cañaveras JV, Sloer C, Saiz-Jimenez J (2001) Microorganisms and microbially induced fabrics in cave walls. Geomicrobiol J 18(3):223–240

Cañaveras J, Cuezva S, Sanchez-Moral S, Lario J, Laiz L, Gonzalez JM, Saiz-Jimenez C (2006) On the origin of fiber calcite crystals in moonmilk deposits. Naturwissenschaften 93(1):27–32

Cardman Z, Macalady JL, Schaperdoth I, Broad K, Kakuk B (2015) Fast-growing slime curtains reveal a dynamic nitrogen (and iron?) world in the shallow subsurface. Geological Society of America Abstracts with Programs, Vol 47, No. 7, p 56

Carmichael MJ, Carmichael SK, Santelli CM, Strom A, Bräuer SL (2013) Mn (II)-oxidizing bacteria are abundant and environmentally relevant members of ferromanganese deposits in caves of the upper Tennessee River Basin. Geomicrobiol J 30(9):779–800

Chen Y, Wu L, Boden R, Hillebrand A, Kumaresan D, Moussard H, Baciu M, Lu Y, Murrell JC (2009) Life without light: microbial diversity and evidence of sulfur-and ammonium-based chemolithotrophy in Movile Cave. ISME J 3(9):1093–1104

Chroňáková A, Horák A, Elhottová D, Krištůfek V (2009) Diverse archaeal community of a bat guano pile in Domica Cave (Slovak Karst, Slovakia). Folia Microbiol 54(5):436–446

Contos A, James J, Pitt BHK, Rogers P (2001) Morphoanalysis of bacterially precipitated subaqueous calcium carbonate from Weebubbie Cave, Australia. Geomicrobiol J 18(3):331–343

Culver DC, Pipan T (2009) The biology of caves and other subterranean habitats. Oxford University Press, Oxford

Culver DC, Sket B (2000) Hotspots of subterranean biodiversity in caves and wells. J Cave Karst Stud 62(1):11–17

Cunningham K, Northup D, Pollastro R, Wright W, LaRock E (1995) Bacteria, fungi and biokarst in Lechuguilla Cave, Carlsbad Caverns National Park, New Mexico. Environ Geol 25(1):2–8

Danielli H, Edington M (1983) Bacterial calcification in limestone caves. Geomicrobiol J 3(1):1–16

Dattagupta S, Schaperdoth I, Montanari A, Mariani S, Kita N, Valley JW, Macalady JL (2009) A novel symbiosis between chemoautotrophic bacteria and a freshwater cave amphipod. ISME J 3(8):935–943

Davis DG (2000) Extraordinary features of Lechuguilla Cave, Guadalupe Mountains, New Mexico. J Cave Karst Stud 62(2):147–157

Dupraz C, Visscher PT (2005) Microbial lithification in marine stromatolites and hypersaline mats. Trends Microbiol 13(9):429–438

Engel AS (2010) Microbial diversity of cave ecosystems. In: Barton LL, Mandl M, Loy A (eds) Geomicrobiology. Molecular and Environmental Perspective. Springer, Netherlands, pp 219–238

Engel AS, Northup DE (2008) Caves and karst as model systems for advancing the microbial sciences. In: Martin JB, White WB (eds) Frontiers of Karst Research: Proceedings and recommendations of the workshop held May 3 through 5, 2007, in San Antonio, TX. Karst Waters Institute, Ashland, OH

Engel AS, Randall KW (2011) Experimental evidence for microbially mediated carbonate dissolution from the saline water zone of the Edwards Aquifer, central Texas. Geomicrobiol J 28(4):313–327

Engel AS, Porter ML, Kinkle BK, Kane TC (2001) Ecological assessment and geological significance of microbial communities from Cesspool Cave, Virginia. Geomicrobiol J 18(3):259–274

Engel AS, Lee N, Porter ML, Stern LA, Bennett PC, Wagner M (2003) Filamentous "Epsilonproteobacteria" dominate microbial mats from sulfidic cave springs. Appl Environ Microbiol 69 (9):5503–5511

Engel AS, Stern LA, Bennett PC (2004) Microbial contributions to cave formation: new insights into sulfuric acid speleogenesis. Geology 32(5):369–372

Engel AS, Meisinger DB, Porter ML, Payn RA, Schmid M, Stern LA, Schleifer K, Lee NM (2010) Linking phylogenetic and functional diversity to nutrient spiraling in microbial mats from Lower Kane Cave (USA). ISME J 4(1):98–110

Engel AS, Paoletti MG, Beggio M, Dorigo L, Pamio A, Gomiero T, Furlan C, Brilli M, Dreon AL, Bertoni R (2013) Comparative microbial community composition from secondary 1 carbonate (moonmilk) deposits: implications for the *Cansiliella servadeii* cave hygropetric food web. Int J Speleol 42(3):181–192

Faimon J, Štelcl J, Kubešová S, Zimák J (2003) Environmentally acceptable effect of hydrogen peroxide on cave "lamp-flora", calcite speleothems and limestones. Environ Pollut 122 (3):417–422

Ford DC, Williams PW (2007) Karst hydrogeology and geomorphology. Wiley, West Sussex, England

Galdenzi S (1990) Un modello genetico per la Grotta Grande del Vento. In: Galdenzi S, Menichetti M (eds) Il carsismo della Gola di Frasassi: Memorie Istituto Italiano di Speologia, vol 4. vol 2, pp 123–142

Galdenzi S, Maruoka T (2003) Gypsum deposits in the Frasassi Caves, central Italy. J Cave Karst Stud 65(2):111–125

Galdenzi S, Sarbu S (2000) Chemiosintesi e speleogenesi in un ecosistema ipogeo: I Rami Sulfurei delle Grotte di Frasassi (Italia centrale). Le Grotte d'Italia 1:3–18

Galdenzi S, Cocchioni F, Filipponi G, Morichetti L, Scuri S, Selvaggio R, Cocchioni M (2010) The sulfidic thermal caves of Acquasanta Terme (central Italy). J Cave Karst Stud 72(1):43–58

Giordano M, Mobili F, Pezzoni V, Hein MK, Davis JS (2000) Photosynthesis in the caves of Frasassi (Italy). Phycologia 39(5):384–389

Gradziński M, Banaś M, Uchman A (1995) Biogenic origin of manganese flowstones from Jaskinia Czarna Cave, Tatra Mts., Western Carpathians. Acta Soc Geol Pol 65:19–27

Halliday WR (2007) Pseudokarst in the 21st century. J Cave Karst Stud 69(1):103–113

Hathaway JJM, Garcia MG, Balasch MM, Spilde MN, Stone FD, Dapkevicius MDLN, Amorim IR, Gabriel R, Borges PA, Northup DE (2014) Comparison of bacterial diversity in Azorean and Hawai'ian lava cave microbial mats. Geomicrobiol J 31(3):205–220

Head IM, Jones DM, Larter SR (2003) Biological activity in the deep subsurface and the origin of heavy oil. Nature 426(6964):344–352

Hedges J (1993) A review on vermiculations. Bol Soc Venezolana Espeleol 27:2–6

Hill C (1995) Sulfur redox reactions: hydrocarbons, native sulfur, Mississippi Valley-type deposits, and sulfuric acid karst in the Delaware Basin, New Mexico and Texas. Environ Geol 25(1):16–23

Hill CA, Forti P (1997) Cave minerals of the world, vol 238. National Speleological Society, Huntsville, AL

Holmes AJ, Tujula NA, Holley M, Contos A, James JM, Rogers P, Gillings MR (2001) Phylogenetic structure of unusual aquatic microbial formations in Nullarbor caves, Australia. Environ Microbiol 3(4):256–264

Hose L, Northup D (2004) Biovermiculations: living vermiculation-like deposits in Cueva de Villa Luz, Mexico: Proceedings of the Society: Selected Abstracts, National Speleological Society Convention, Marquette, MI. J Cave Karst Stud 66:112

Hose LD, Pisarowicz JA (1999) Cueva de Villa Luz, Tabasco, Mexico: Reconnaissance study of an active sulfur spring cave and ecosystem. J Cave Karst Stud 61(1):13–21

Hose LD, Palmer AN, Palmer MV, Northup DE, Boston PJ, DuChene HR (2000) Microbiology and geochemistry in a hydrogen-sulfide-rich karst environment. Chem Geol 169:399–423

Hutchens E, Radajewski S, Dumont MG, McDonald IR, Murrell JC (2004) Analysis of methanotrophic bacteria in Movile Cave by stable isotope probing. Environ Microbiol 6 (2):111–120

Jones B (1992) Manganese precipitates in the karst terrain of Grand Cayman, British West Indies. Can J Earth Sci 29(6):1125–1139

Jones B (2001) Microbial activity in caves—a geological perspective. Geomicrobiol J 18 (3):345–357

Jones B (2010) Microbes in caves: agents of calcite corrosion and precipitation. Geol Soc Lond, Spec Publ 336(1):7–30

Jones DS, Polerecky L, Dempsey BA, Galdenzi S, Macalady JL (2015) Fate of sulfide in the Frasassi cave system and implications for sulfuric acid speleogenesis. Chem Geol 410:21

Jones DS, Lyon EH, Macalady JL (2008) Geomicrobiology of biovermiculations from the Frasassi cave system, Italy. J Cave Karst Stud 70(2):78–93

Jones D, Tobler D, Schaperdoth I, Mainiero M, Macalady J (2010) Community structure of subsurface biofilms in the thermal sulfidic caves of Acquasanta Terme, Italy. Appl Environ Microbiol 76(17):5902–5910

Jones D, Albrecht H, Dawson K, Schaperdoth I, Freeman K, Pi Y, Pearson A, Macalady J (2012) Community genomic analysis of an extremely acidophilic sulfur-oxidizing biofilm. ISME J 6 (1):158–170

Jones DS, Schaperdoth I, Macalady JL (2014) Metagenomic evidence for sulfide oxidation in extremely acidic cave biofilms. Geomicrobiol J 31:194–204

Jørgensen BB (1982) Mineralization of organic matter in the sea bed—the role of sulphate reduction. Nature 296:643–645

Laiz L, Groth I, Gonzalez I, Sáiz-Jiménez C (1999) Microbiological study of the dripping waters in Altamira cave (Santillana del Mar, Spain). J Microbiol Methods 36(1):129–138

Lavoie K, Northup D (2009) Invertebrate colonization and deposition rates of guano in a man-made bat cave, the Chiroptorium, Texas USA. Int Cong Speleol Proc 2:1297–1301

Lee NM, Meisinger DB, Aubrecht IR, Kovačik L, Saiz-Jimenez C, Baskar S, Baskar R, Liebl W, Porter ML, Engel AS (2012) Caves and Karst environments. In: Bell EM (ed) Life at extremes: environments, organisms, and strategies for survival, vol 1. CAB International, Oxfordshire, UK, pp 320–344

Léveillé RJ, Datta S (2010) Lava tubes and basaltic caves as astrobiological targets on Earth and Mars: a review. Plan Space Sci 58(4):592–598

Lin L-H, Wang P-L, Rumble D, Lippmann-Pipke J, Boice E, Pratt LM, Lollar BS, Brodie EL, Hazen TC, Andersen GL (2006) Long-term sustainability of a high-energy, low-diversity crustal biome. Science 314(5798):479–482

Lowe D, Gunn J (1995) The role of strong acid in speleo-inception and subsequent cave development. Acta Geographica 34:33–60

Lyon E, Koffman B, Meyer K, Cleaveland L, Mariani S, Galdenzi S, Macalady J (2005) Geomicrobiology of the Frasassi Caves. In: Galdenzi S (ed) Frasassi 1989-2004: Gli sviluppi nella ricerca, pp 152–157

Macalady JL, Lyon EH, Koffman B, Albertson LK, Meyer K, Galdenzi S, Mariani S (2006) Dominant microbial populations in limestone-corroding stream biofilms, Frasassi cave system, Italy. Appl Environ Microbiol 72(8):5596–5609

Macalady JL, Jones DS, Lyon EH (2007) Extremely acidic, pendulous microbial biofilms from the Frasassi cave system, Italy. Environ Microbiol 9(6):1402–1414

Macalady J, Jones D, Schaperdoth I, Bloom D, McCauley R (2008a) Meter-long microbial ropes from euxinic cave lakes. In: AGU Fall Meeting Abstracts, p 0514

Macalady JL, Dattagupta S, Schaperdoth I, Jones DS, Druschel GK, Eastman D (2008b) Niche differentiation among sulfur-oxidizing bacterial populations in cave waters. ISME J 2 (6):590–601

Macalady JL, Hamilton TL, Grettenberger CL, Jones DS, Tsao LE, Burgos WD (2013) Energy, ecology and the distribution of microbial life. Phil Trans R Soc B 368(1622):20120383

Mariani S, Mainiero M, Barchi M, Van Der Borg K, Vonhof H, Montanari A (2007) Use of speleologic data to evaluate Holocene uplifting and tilting: an example from the Frasassi anticline (northeastern Apennines, Italy). Earth Planet Sci Lett 257(1–2):313–328

McCauley R, Jones D, Schaperdoth I, Steinberg L, Macalady J (2010) Metabolic strategies in energy-limited microbial communities in the anoxic subsurface (Frasassi Cave System, Italy). In: AGU Fall Meeting Abstracts, p 0317

Meisinger DB, Zimmermann J, Ludwig W, Schleifer KH, Wanner G, Schmid M, Bennett PC, Engel AS, Lee NM (2007) In situ detection of novel Acidobacteria in microbial mats from a chemolithoautotrophically based cave ecosystem (Lower Kane Cave, WY, USA). Environ Microbiol 9(6):1523–1534

Melim LA (2011) Stable isotopic evidence for microbial precipitation of calcite in cave pool fingers. Geological Society of America Abstracts with Programs 43(5):329

Melim LA, Shinglman KM, Boston PJ, Northup DE, Spilde MN, Queen JM (2001) Evidence for microbial involvement in pool finger precipitation, Hidden Cave, New Mexico. Geomicrobiol J 18(3):311–329

Melim LA, Northup DE, Spilde MN, Jones B, Boston PJ, Bixby RJ (2008) Reticulated filaments in cave pool speleothems: microbe or mineral? J Cave Karst Stud 70(3):135–141

Melim L, Liescheidt R, Northup D, Spilde M, Boston P, Queen J (2009) A biosignature suite from cave pool precipitates, Cottonwood Cave, New Mexico. Astrobiology 9(9):907–917

Montechiaro F, Giordano M (2006) Effect of prolonged dark incubation on pigments and photosynthesis of the cave-dwelling cyanobacterium *Phormidium autumnale* (Oscillatoriales, Cyanobacteria). Phycologia 45(6):704–710

Northup DE, Lavoie KH (2001) Geomicrobiology of caves: a review. Geomicrobiol J 18(3):199–222

Northup DE, Dahm CN, Melim LA, Spilde MN, Crossey LJ, Lavoie KH, Mallory LM, Boston PJ, Cunningham KI, Barns SM (2000) Evidence for geomicrobiological interactions in Guadalupe caves. J Cave Karst Stud 62(2):80–90

Northup DE, Barns SM, Yu LE, Spilde MN, Schelble RT, Dano KE, Crossey LJ, Connolly CA, Boston PJ, Natvig DO (2003) Diverse microbial communities inhabiting ferromanganese deposits in Lechuguilla and Spider Caves. Environ Microbiol 5(11):1071–1086

Northup D, Melim L, Spilde M, Hathaway J, Garcia M, Moya M, Stone F, Boston P, Dapkevicius M, Riquelme C (2011) Lava cave microbial communities within mats and secondary mineral deposits: implications for life detection on other planets. Astrobiology 11(7):601–618

Orsi WD, Edgcomb VP, Christman GD, Biddle JF (2013) Gene expression in the deep biosphere. Nature 499(7457):205–208

Ortiz M, Legatzki A, Neilson JW, Fryslie B, Nelson WM, Wing RA, Soderlund CA, Pryor BM, Maier RM (2014) Making a living while starving in the dark: metagenomic insights into the energy dynamics of a carbonate cave. ISME J 8(2):478–491

Palmer AN (1991) Origin and morphology of limestone caves. Geol Soc Am Bull 103(1):1–21

Palmer AN (2007) Cave geology. Cave Books, Dayton, OH

Pašić L, Kovče B, Sket B, Herzog-Velikonja B (2010) Diversity of microbial communities colonizing the walls of a Karstic cave in Slovenia. FEMS Microbiol Ecol 71(1):50–60

Peck S (1986) Bacterial deposition of iron and manganese oxides in North American caves. NSS Bull 48(1):26–30

Pedersen K (2000) Exploration of deep intraterrestrial microbial life: current perspectives. FEMS Microbiol Lett 185(1):9–16

Popa R, Smith AR, Popa R, Boone J, Fisk M (2012) Olivine-respiring bacteria isolated from the rock-ice interface in a lava-tube cave, a Mars analog environment. Astrobiology 12(1):9–18

Por FD (2007) Ophel: a groundwater biome based on chemoautotrophic resources. The global significance of the Ayyalon cave finds. Israel Hydrobiol 592(1):1–10

Porca E, Jurado V, Žgur-Bertok D, Saiz-Jimenez C, Pašić L (2012) Comparative analysis of yellow microbial communities growing on the walls of geographically distinct caves indicates a common core of microorganisms involved in their formation. FEMS Microbiol Ecol 81(1):255–266

Portillo MC, Gonzalez JM (2011) Moonmilk deposits originate from specific bacterial communities in Altamira Cave (Spain). Microb Ecol 61(1):182–189

Rossi C, Lozano RP, Isanta N, Hellstrom J (2010) Manganese stromatolites in caves: El Soplao (Cantabria, Spain). Geology 38(12):1119–1122

Rossmassler K, Engel AS, Twing KI, Hanson TE, Campbell BJ (2012) Drivers of epsilonproteobacterial community composition in sulfidic caves and springs. FEMS Microbiol Ecol 79(2):421–432

Sarbu S, Kinkle B, Vlasceanu L, Kane T, Popa R (1994) Microbiological characterization of a sulfide-rich groundwater ecosystem. Geomicrobiol J 12(3):175–182

Sarbu SM, Kane TC, Kinkle BK (1996) A chemoautotrophically based cave ecosystem. Science 272(5270):1953–1955

Secord D, Muller-Parker G (2005) Symbiont distribution along a light gradient within an intertidal cave. Limnol Oceanogr 50(1):272–278

Shabarova T, Widmer F, Pernthaler J (2013) Mass effects meet species sorting: transformations of microbial assemblages in epiphreatic subsurface karst water pools. Environ Microbiol 15(9):2476–2488

Simon KS, Benfield EF (2002) Ammonium retention and whole-stream metabolism in cave streams. Hydrobiologia 482(1–3):31–39

Simon K, Benfield E, Macko S (2003) Food web structure and the role of epilithic biofilms in cave streams. Ecology 84(9):2395–2406

Simon KS, Pipan T, Culver DC (2007) A conceptual model of the flow and distribution of organic carbon in caves. J Cave Karst Stud 69(2):279–284

Smith T, Olson R (2007) A taxonomic survey of lamp flora (Algae and Cyanobacteria) in electrically lit passages within Mammoth Cave National Park, Kentucky. Int J Speleol 36(2):105–114

Spear JR, Barton HA, Robertson CE, Francis CA, Pace NR (2007) Microbial community biofabrics in a geothermal mine adit. Appl Environ Microbiol 73(19):6172–6180

Spilde MN, Northup DE, Boston PJ, Schelble RT, Dano KE, Crossey LJ, Dahm CN (2005) Geomicrobiology of cave ferromanganese deposits: a field and laboratory investigation. Geomicrobiol J 22(3-4):99–116

Steinhauer ES, Omelon CR, Bennett PC (2010) Limestone corrosion by neutrophilic sulfur-oxidizing bacteria: a coupled microbe-mineral system. Geomicrobiol J 27(8):723–738

Strader B, Schubert K, Quintana M, Gomez E, Curnutt J, Boston P (2011) Estimation, modeling, and simulation of patterned growth in extreme environments. In: Software tools and algorithms for biological systems. Springer, pp 157–170

Teske AP (2005) The deep subsurface biosphere is alive and well. Trends Microbiol 13(9):402–404

Tetu SG, Breakwell K, Elbourne LD, Holmes AJ, Gillings MR, Paulsen IT (2013) Life in the dark: metagenomic evidence that a microbial slime community is driven by inorganic nitrogen metabolism. ISME J 7:1227–1236

Vlasceanu L, Sarbu SM, Engel AS, Kinkle BK (2000) Acidic cave-wall biofilms located in the Frasassi Gorge, Italy. Geomicrobiol J 17(2):125–139

White WB (1988) Geomorphology and hydrology of karst terrains. Oxford University Press, New York

White WB, Vito C, Scheetz BE (2009) The mineralogy and trace element chemistry of black manganese oxide deposits from caves. J Cave Karst Stud 71(2):136–143

Whiticar MJ, Faber E, Schoell M (1986) Biogenic methane formation in marine and freshwater environments: CO_2 reduction *vs* acetate fermentation—Isotope evidence. Geochim Cosmochim Acta 50(5):693–709

Whitman WB, Coleman DC, Wiebe WJ (1998) Prokaryotes: the unseen majority. Proc Natl Acad Sci USA 95(12):6578–6583

Wigley T, Plummer L (1976) Mixing of carbonate waters. Geochim Cosmochim Acta 40(9):989–995

Chapter 6
Microbiology of the Deep Continental Biosphere

Thomas L. Kieft

Abstract Subsurface microbial communities in sediment and fractured rock environments beneath continental surface environments and beneath the ocean floor comprise a significant but largely unexplored portion of the Earth's biosphere. The continental subsurface is highly geologically varied, and so the abundance, diversity, and metabolic functions of its inhabitant microbes are even more widely ranging than those of marine systems. Microbial ecosystems in relatively shallow groundwater systems are largely fueled by organic carbon derived from photosynthesis, whereas deeper groundwater ecosystems are fueled by molecular hydrogen, methane, and short-chain hydrocarbons ("geogas"), produced by abiotic water–rock interactions, e.g., serpentinization and radiolysis of water. The abundances of microbes generally decline with depth, with deep fracture waters containing $\sim 10^3$–10^4 cells ml^{-1}; many of these microbes are metabolically active, albeit at very slow rates The depth limit of the biosphere may be controlled by a combination of temperature and other factors such as energy availability and pressure. Diverse bacteria and archaea appear to be adapted for life under the extremes posed by subterranean conditions. Further research is needed to explore a wider range of subsurface continental geologic settings, to constrain the rates of microbial metabolism, and to understand mechanisms of evolution in the subsurface.

6.1 Introduction

This chapter focuses on our current understanding of and major outstanding research questions regarding indigenous microorganisms in deep groundwater environments underlying the Earth's continents. Together with marine sediments and crust, these habitats are termed the "subsurface." In the deep relatively inaccessible regions that harbor life, they can be termed the "deep biosphere" or even "dark life." The study of deep subsurface habitats and their inhabitant microbes is

T.L. Kieft (✉)
New Mexico Institute of Mining and Technology, Socorro, NM 87801, USA
e-mail: tkieft@nmt.edu

relatively young, but has matured such that we no longer present it in terms of novel glimpses of a shadowy world; the mass of accumulated data has made it easier to defend the findings of microbes as not merely contaminant artifacts but as truly indigenous microbes functioning in active subsurface ecosystems (Fredrickson and Balkwill 2006). The existence of the deep biosphere is now well entrenched in the textbooks (e.g., Madigan et al. 2013). Indeed, the portion of the biosphere comprising subsurface continental as well as marine environments is now recognized as containing a significant proportion of the Earth's microbial cells, possibly even the majority (Whitman et al. 1998; Kallmeyer et al. 2012; Røy et al. 2012). Early studies, primarily in the vicinities of petroleum reservoirs beginning in the 1920s (Bastin et al. 1926), set the stage for concerted programs, e.g., by the U.S. Environmental Protection Agency and the U.S. Department of Energy in the 1980s, which probed subsurface aquifers, at first to only a few meters (Wilson et al. 1983), but then followed by increasingly deep drilling to depths approaching 3 km (Onstott et al. 1998). In addition to drilling from the surface as a means of accessing the deep biosphere, microbiologists began descending into deep mines to sample groundwater via boreholes drilled from within the mines and extending outward into pristine, often ancient groundwater (Kieft et al. 1999; Onstott et al. 2003). Both of these approaches are now widely used and are unearthing new discoveries on a regular basis. There is now a history of reviews on the topic (e.g., Fredrickson and Onstott 1996; Amy and Haldeman 1997; Chapelle 2000; Pedersen 2000; Amend and Teske 2005; Onstott et al. 2009; Colwell and D'Hondt 2013), which serve as a background for this chapter. While there are many practical motivations for studying subsurface microbiology (e.g., understanding contaminant fate and transport, developing hazardous waste repositories, etc.), this review focuses on recent findings and prospects in the basic science of mostly uncontaminated continental subsurface environments.

Much of our current understanding of the microbiology of groundwater was succinctly summarized by T.C. Onstott in a diagram created for the National Science Foundation's EarthLab report (NSF 2003) (Fig. 6.1). Energy sources in shallow subsurface environs are seen to be primarily photosynthetically generated organic carbon that can occur as detritus buried in sediments and as organic matter transported via diffusion and advection in groundwater to depth. The quantity and quality of this organic carbon available for microbial metabolism decline with depth, and so the abundance of the dominantly heterotrophic bacteria in these aquifers also declines with depth. The general patterns of abundance and biogeochemical activities of microbes in these relatively shallow aquifers were reasonably well characterized 20 years ago; regional groundwater flow results in a predictable sequence of terminal electron-accepting processes along a flow path, from aerobic metabolism near the recharge zone to anaerobic processes such as methanogenesis in the most distal regions (Murphy et al. 1992; Lovley et al. 1994) (Fig. 6.2). Also, microbial metabolism of buried organic matter in fine-textured aquitards results in diffusion of fermentation products to adjacent sandy aquifers, where they serve as electron donors stimulating microbial activities at boundaries between layers. Rates of activity generally decline as one proceeds from surface environments to more

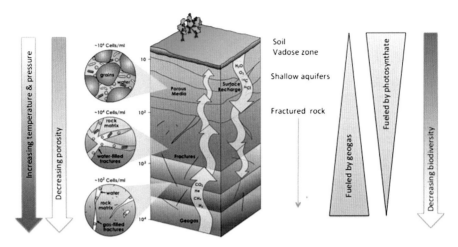

Fig. 6.1 Overview of subsurface continental environments. Microbial abundance in pore water generally declines with depth, from shallow aquifers to underlying fractured rock environments. The quantity and quality of photosynthetically derived organic C decline with depth, and the importance of "geogas" energy sources (e.g., H_2, CH_4) generally increases with depth. Temperature increases with depth according to the local geothermal gradient; pressure increases with depth (~10 MPa/km). (modified from EarthLab, National Science Foundation 2003)

static subsurface systems (Phelps et al. 1994; Kieft and Phelps 1997) (Fig. 6.3). Proceeding further into the subsurface (Fig. 6.1), one encounters the bedrock underlying surface sediments and the porosity generally diminishes with depth, as well. As the availability of photosynthate declines with depth, microbial abundance declines, and H_2 generated abiotically through water–rock interactions, termed "geogas" by Pedersen (1997), is the dominant energy source in the deepest regions of the biosphere (Fig. 6.4). Some of the most exciting current work in the subsurface, both continental and marine, involves determining the rates and extent of these H_2-fueled ecosystems and characterizing the communities therein.

Although the field has matured, there are still many outstanding research questions. Broad research questions include the following, as organized by an U.S. National Science Foundation-sponsored group (Fredrickson et al. 2006):

- How deeply does life extend into the Earth?
- What fuels the deep biosphere?
- How does the interplay between biology and geology shape the subsurface?
- What are subsurface genomes telling us?
- Did life on the earth's surface originate underground?
- Is there life in the subsurface as we don't know it?

More specific questions that are being or should be addressed include:

- What is the physiological state of subsurface microbes, i.e., are they slowly metabolizing, "healthy" cells that are well adapted to the rigors of subsurface life, or are they ill-adapted, moribund cells in slow decline?

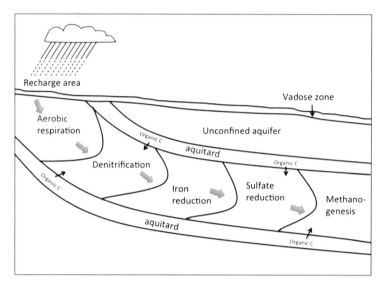

Fig. 6.2 Idealized sequence of terminal electron-accepting processes occurring along a flow path over tens of km in a confined aquifer. Buried organic carbon in confining aquitards is slowly degraded and fermentation end products diffuse into the aquifer, creating zones of high microbial activity at the aquitard–aquifer interfaces. (After Smith and Harris 2007)

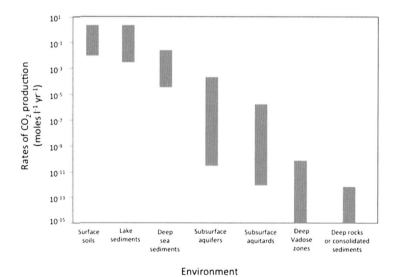

Fig. 6.3 Ranges of rates of in situ CO_2 production for various environments. Subsurface rates were estimated from groundwater chemical analyses and geochemical modeling. (After Kieft and Phelps 1997)

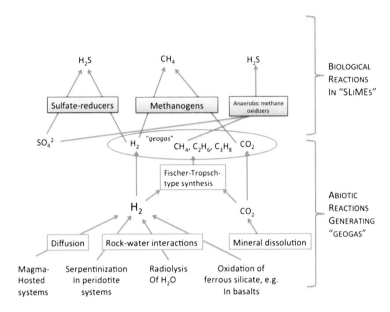

Fig. 6.4 Abiotic subsurface processes that generate H_2, CH_4, and CO_2 ("geogas") and example subsurface lithoautotrophic ecosystems ("SLiMEs") that are fueled by geogas

- By what means and at what rates are microbes transported to the subsurface and within subsurface environments?
- What's the role of lateral gene transfer among subsurface populations?
- How do the planktonic cells sampled from the bulk water phase differ from sessile, biofilm communities?
- What are the interactions within subsurface microbial communities?

6.2 Comparison of Deep Continental and Marine Biospheres

While continental and marine subsurface environments have much in common (dark anaerobic ecosystems, slow metabolic rates, heterotrophic metabolism of buried and transported organic C, and evidence for chemolithoautotrophic ecosystems), the study of the two has remained largely separate (Colwell and Smith 2004). We now know a great deal about the marine deep biosphere—both deep hot (vent related) subsurface marine systems (Huber et al. 2007; Kelley et al. 2005; Wang et al. 2009) and not so hot subseafloor sediments (D'Hondt et al. 2004; Kallmeyer et al. 2012). However, the vent and vent-related systems are essentially localized, hot, chemoautotrophic systems and the sediments are vast, cold heterotrophic systems functioning on old, buried photosynthate. The deep continental biosphere in contrast contains what could be termed "cool" chemoautotrophic ecosystems, cool in the geochemical context meaning anything <100 °C. The rock–water interactions that fuel these subsurface ecosystems are not dependent upon localized magmatic input, e.g., spreading centers, volcanic activity, or hot spots, but instead are influenced by regional, topographically driven fluid flow, e.g., in the cases of the elevated plateau containing the Witwatersrand Basin in South Africa (Gihring et al. 2006; Onstott et al. 2006), the "cooler," low elevation subsurface in granitic aquifers of the Fennoscandian Precambrian shield (Pedersen 1997, 2000; Itavaara et al. 2011), and Columbia River basaltic aquifers (Stevens and McKinley 1995, 2000).

6.3 Methods for Sampling the Subsurface

The deep biosphere is challenging to access, generally requiring drilling, either from the surface or from a preexisting subsurface site, e.g., in deep mines. Drilling is inherently messy, usually involving drilling fluids with potential for chemical and microbiological contamination. Techniques for minimizing contamination and for tracing and quantifying contaminant fluids and particulates were devised during the U.S. Department of Energy's Subsurface Science Program (Phelps et al. 1989; Colwell et al. 1992; Russell et al. 1992), and these have been adapted for use elsewhere, including the Integrated Ocean Drilling Program (Smith et al. 2000). Subsurface sampling approaches have been reviewed by Moser et al. (2001), Kieft et al. (2007), and Kieft (2010).

Drilling and coring to depths greater than ~300 m require rotary drilling using a drilling fluid, either liquid or gas, to lubricate and cool the drill bit and to remove the cuttings. These fluids can be problematic, especially when drilling muds with organic additives are used, because they favor the growth of contaminating microbes. Air or an inert gas, e.g., Ar, can be used and these can be filtered, although this requires a massive filter (Colwell et al. 1992). Water can be used, but denser fluids are generally required for very deep drilling. If possible, organic additives and petroleum-based lubricants should be avoided. Online gas analyses of the drilling fluid are commonly used in the oil and gas industry and can be employed in scientific drilling to identify biologically active zones (Erzinger et al. 2006). When cores are to be used for microbiological analyses, the usual approach is to deploy solute and particulate tracers. Solute tracers added to the drilling fluid include fluorescent dyes, LiBr, and perfluorinated hydrocarbons. The latter can be quantified over a broad concentration range by gas chromatography. Particulate tracers include microbe-sized (0.5 or 1.0 µm diameter) fluorescent microbeads that are carboxylated to mimic the negative surface charge on most bacteria. These are added in a plastic bag at the bottom of the core barrel such that the bag is broken on contact with the formation and the beads are mixed with fluid that contacts the core. A subcore is removed from the interior of the core and tracers are quantified in the parings from the core perimeter and in the subcore. Ideally, the subcore should have $\geq$10,000-fold lower concentration of tracers than the parings.

Microbial communities in the drilling fluid and in the subcore can also be compared as a further test for drilling-induced contamination (Lehman et al. 1995; Dong et al. 2014). Sidewall coring is another option for collecting solids (Colwell et al. 1997; Dong et al. 2014), but the volume is very limited. Samples are known to change their microbiological composition very shortly after collection, so they should be processed as soon as possible (Brockman et al. 1998). Samples intended for analysis of nucleic acids, proteins, lipids, etc., should be frozen immediately, if possible. Samples intended for cultivation of microbes should be handled in a glove bag containing an inert atmosphere to preserve oxygen-sensitive anaerobes.

Following drilling, boreholes can be sampled for groundwater as long as the borehole is flushed sufficiently to remove contaminating solutes and microbes in the open borehole. Discrete depth intervals can be targeted using packers (Dong et al. 2014). Multilevel samplers can sample from more than one depth interval. A U-tube system (Freifeld 2009) can also be used for collecting from depth. Microbes suspended in groundwater can be collected and concentrated onto filters (Moser et al. 2005; Gihring et al. 2006). Wireline formation testers can monitor borehole water chemistry and also sample water from central and outer portions of the rock formation before coring (Dong et al. 2014). Devices can also be inserted into the borehole for long-term monitoring and collection of samples and to enrich for subsurface microbes (Orcutt et al. 2010, 2011: Silver et al. 2010).

Drilling from existing deep underground sites, e.g., from within mines, confers the advantages of lower costs and also facilitates drilling to great depth without having to begin with a wide diameter collar. This approach has enabled collection of some of the deepest fluids so far collected from the subsurface (Moser et al. 2003, 2005; Kieft et al. 2005; Borgonie et al. 2011; Lippmann-Pipke et al. 2011b). Boreholes, drilled into surrounding pristine rock, generally as part of mine operations, can intersect ancient fracture water that then flows from the borehole, flushing drilling contaminants and carrying inhabitant microbes that can be collected by filtration. Sterile packers connected to a manifold system can be used to transfer water and gas samples for filters, sample vials, etc., without exposure to mine air (Fig. 6.5). Unfortunately, mining interests rarely overlap with the interests of geomicrobiologists, so opportunities for sampling in mines tend to be few and short lived. For this reason, sampling and long-term monitoring and experimentation in dedicated underground laboratories are an attractive option (Pedersen 1997, 2000; Edwards et al. 2006; Onstott et al. 2009; Fukuda et al. 2010).

6.4 Microbial Abundance

One of the simplest questions to ask is "How many are there?" Shallow sedimentary aquifers may harbor nearly 10^7 cells ml^{-1} of groundwater (Sinclair and Ghiorse 1989), nearly all of these being prokaryotic (bacteria and archaea). Microbial abundance generally declines with depth (Onstott et al. 1999; Itavaara et al. 2011), and the abundance of cells in fracture water collected from deep

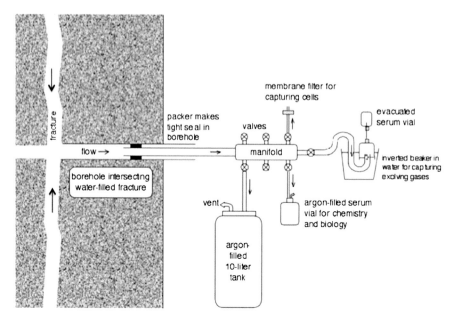

Fig. 6.5 Packer-manifold system for collecting deep fracture waters from boreholes drilled in mines. The borehole intersects a fracture filled with pressurized fluid. The flowing fracture water flushes the borehole of contaminants. The sterilized packer and manifold allow collection of water and gases without contamination from mine air. (Modified from Kieft et al. 2007, with permission from ASM Press)

crystalline rock is markedly lower than in sedimentary aquifers. Abundance of microbes in continental subsurface environments appears to decline more slowly than in the pattern described by Parkes for marine sediments (Onstott et al. 1999; Parkes et al. 1994). Flow cytometric counts of cells in deep fracture water collected from boreholes at depths of ~2–3.5 km in mines from across the Witwatersrand Basin in South Africa ranged from 2.5×10^2 to 5.9×10^4 cells ml^{-1} (Onstott et al. 2006). A more recent paper reports 9.0×10^1 to 2.2×10^3 cells ml^{-1} in fracture water at 1.8 km depth (Davidson et al. 2011). Pedersen (1997) reported 1.2×10^4 to 9.2×10^4 bacteria ml^{-1} in granitic fracture water at 450 m depth in the Äspö Hard Rock Lab in Sweden. Itavaara et al. (2011) published cell numbers declining from ~4.6×10^5 at 100 m depth to ~6×10^4 at 1500 m in crystalline rock aquifers in Finland. Fukuda et al. (2010) counted 1.1×10^4 to 5.2×10^4 cells ml^{-1} in ~1.1-km depth fracture water accessed in the Mizunami Underground Research Laboratory in Japan. These low numbers of subsurface microbes reflect the sluggish energy fluxes and possibly other limiting factors, and they make further microbial characterization especially challenging.

6.5 What Controls the Depth Limit of the Biosphere?

The deepest limit of the continental biosphere has yet to be clearly delineated at any site and so the factors that control that lower depth limit are not well understood. Temperature is the least forgiving of environmental parameters and therefore a first approximation of the deepest extent of the biosphere can be set at the ~121 °C isotherm. Reports have pushed the upper temperature for microbial proliferation to 121 °C (Kashefi and Lovley 2003) and even 122 °C when combined with elevated pressure (Takai et al. 2008), and the record may even be broken again, but most investigators would likely agree that it won't be by much. Combining this upper temperature limit for life with geothermal gradients that range from ~8 to 30 °C/km, one can estimate the deep limit of life to range from 2 to 12 km below land surface (kmbls) for mean surface temperatures from 0 to 25 °C. The Witwatersrand Basin of South Africa is a stable cratonic region where the geothermal gradient is low at 8–10 °C km^{-1} (Omar et al. 2003), so the biosphere could theoretically exceed 10 km, more than twice as deep and hot as has been probed thus far. A hole drilled in the Songliao Basin in China and cored for microbiological analysis may actually have probed beyond the lower limit of the biosphere (Dong 2009). A sharp drop in microbial biosignatures at a depth corresponding to ~120 °C supports temperature as the ultimate arbiter of microbial distribution. However, evidence from other sites suggests that factors besides temperature may also be at work; these include pressure, availability of pore space, energy flux, and inorganic nutrient availability.

Patterns of biodegradation of petroleum hydrocarbons with depth and temperature within reservoirs have been invoked to suggest an upper limit for microbial activities of ~80 °C (Wilhelms et al. 2001; Head et al. 2003). Petroleum reservoirs commonly harbor indigenous microbes, and in fact, the first cultivation of subsurface microbes, in this case, sulfate reducers, was from a petroleum reservoir (Bastin et al. 1926). Evidence of petroleum biodegradation includes absence of low-molecular-weight constituents and a preponderance of heavy oil, concentration of metals such as Va and Ni, accumulation of biogenic methane, and accumulation of isotopically heavy CO_2; these signatures of biological degradation are common in reservoirs where the temperature is below 80 °C, but appear not to occur at warmer temperatures. Head et al. (2003) attribute the lack of biodegradation at elevated temperatures to such factors as low fluxes of electron acceptors and inorganic nutrients and concomitant rates of metabolism that are so slow as to be unable to keep pace with the thermal breakdown of cellular constituents. Indeed, the currently described bacteria and archaea functioning at temperatures exceeding 80 °C are all from thermal springs and deep sea hydrothermal vents, where high concentrations of readily metabolized substrates (e.g., H_2, H_2S) interface with favorable electron acceptors, e.g., O_2. Slow energy fluxes are likely the rule in deep subsurface habitats other than petroleum reservoirs, as well, especially those in which electron donor generation is exclusively via rock–water interactions. The maintenance demands of life at temperatures exceeding 80 °C may simply be too great for the "slow-lane" lifestyle of subsurface microbes.

Onstott et al. (2014) reported that the combination of energy limitation and high temperature can limit microbes in subsurface environments due to the racemization of amino acids, specifically aspartate. Amino acids spontaneously racemize at higher rates with increasing temperature and aspartate racemizes more rapidly than other amino acids. This racemization requires replacement of proteins, which may demand more energy than is available in most subsurface environments with elevated temperature. Protein turnover times, estimated from amino acid D/L ratios, were found to be shorter in deep groundwater than expected: ~27 years at 1 km depth and 27 °C and 1–2 years at 3 km and 54 °C in the Witwatersrand Basin, South Africa. Amino acid racemization may be a factor explaining the ~80–90 °C limit to petroleum biodegradation in deep, hot petroleum reservoirs (Wilhelms et al. 2001; Head et al. 2003).

Although the pressures at depth in the continental subsurface are less than those in the deepest regions of the marine environment, they may nonetheless influence microbial activities. For example, the combined effect of hydrostatic and geostatic pressures at 3 kmbls where studies have been focused in South Africa can be estimated at ~30 Mpa, which is nowhere near the extreme pressures found in the ocean's deepest trenches, ~100 MPA. Nonetheless, these deep continental groundwater pressures may be sufficient to negatively impact growth rates, especially when combined with elevated temperature. Alternatively, in situ pressures may select for piezophilic, continental fracture water microbes, as has been shown for deep marine habitats (Bartlett. 2009). If obligate barophiles exist, then our current efforts at cultivation are missing them, as collection and incubation under pressure have so far not been conducted. There's a definite need for such studies in deep continental environs.

The effects of high pressure have been well studied in deep marine environments (Yayanos 1995, 2001; DeLong et al. 1997; Lauro and Bartlett 2008; Nagata et al. 2010); however, relatively little is known of the responses and adaptations of deep terrestrial microorganisms to high temperature. To date, there have been no published studies of the responses of deep subsurface terrestrial microorganisms to high pressure. While many of the responses are likely similar to those of marine microbes, the other in situ parameters associated with high pressure in the deep terrestrial biosphere can be very different. Temperatures in the deep ocean are cold (away from spreading centers), whereas the deep continental realm is warm (Yayanos 1995; Nagata et al. 2010). Moreover, high pressure in the deep Earth can be accompanied by very high partial pressures of various dissolved gases, e.g., H_2, CH_4. These can serve as energy sources, but may also affect metabolic processes in other ways. Piezophilic marine microbes are now being better characterized by high-throughput metagenomic and metatranscriptomic sequencing (Eloe et al. 2011; Wu et al. 2013). Similar approaches that are ongoing with deep continental groundwater samples may reveal similar or contrasting responses to elevated pressure.

Although porosity generally declines with depth, water-filled fractures do exist even at extreme depths (Stober and Bucher 2004), although the tortuosity of the fluid phase may be so great as to preclude transport of microbes. Some fracture fluids may even be totally sequestered from the biosphere, with no opportunity for colonization by microbes. In some cases, deep fractures may contain ancient water that has been geohydrologically sequestered from other groundwater for millions of years (Lippmann-Pipke et al. 2003), although these are bulk ages and mixing with younger, paleometeoric water appears to be common. Microbes that persist in such ancient, sequestered waters must either survive on endogenous energy reserves or metabolize exogenous substrates that become biologically available in these environments. A flux of exogenous, energy-rich substrates, even if sluggish or sporadic, is required for truly long-term persistence.

6.6 Geogas and SLiMES

While the majority of subsurface ecosystems studied to date are powered by organic carbon derived from photosynthesis at the surface, exciting studies have revealed chemosynthetic subterranean ecosystems (termed SLiMEs, subsurface lithoautotrophic microbial ecosystems) that gain energy from geochemically generated inorganic energy sources, e.g., H_2 (Pedersen 1993, 1997; Stevens and McKinley 1995; Chapelle et al. 2002; Nealson et al. 2005). Thomas Gold (1992) first speculated on a vast subsurface, chemolithoautotrophically driven "deep, hot biosphere," and subsequent field studies have clearly demonstrated these ecosystems in isolated locations such as the groundwater feeding Liddy hot springs in Idaho (Chapelle et al. 2002), the Lost City vents in the Atlantic Ocean (Kelley et al. 2005), and deep fracture waters in the Witwatersrand, South Africa (Lin et al. 2006; Chivian et al. 2008). As these are generally anaerobic systems, they're totally independent of the products of photosynthesis, both O_2 and organic carbon, unlike deep sea hydrothermal vent ecosystems. Abiotic rock–water interactions that generate H_2 include serpentinization of ultramafic rocks (Schrenk et al. 2013), oxidation of ferrous silicate minerals in basaltic aquifers (Stevens and McKinley 2000), and radiolysis of water in environments with significant radiation flux (Lin et al. 2006). The H_2 in turn generates CH_4 and short-chain hydrocarbons via Fischer–Tropsch type synthesis reactions (Sherwood Lollar et al. 2002, 2007). Carbon monoxide is also commonly found in deep subsurface waters and is thought to be geochemical in origin (Kieft et al. 2005; Gihring et al. 2006; Onstott et al. 2006). Together, the H_2, CH_4, short-chain hydrocarbons, and CO (geogas) make for relatively energetic anaerobic ecosystems that appear to be dominated by a few species of prokaryotes, including methanogens and sulfate reducers. Considering the widespread occurrences of basaltic crust, serpentinizing low-silicate ultramafic rocks, and sources of gamma irradiation in the subsurface, SLiMEs may underlie large areas of the Earth, possibly approaching the ubiquity (but not the extreme depth) posited by Gold (1992).

High concentrations of H_2 have been measured in Precambrian Shield groundwaters in Canada, South Africa, and Finland, with values as high as 7.4 mM in the Witwatersrand Basin (Sherwood Lollar et al. 2007). The question then arises as to how such high concentrations of an easily metabolized energy source can accumulate without being oxidized by microorganisms. These high concentrations are found in the deepest, most saline fracture waters, with bulk water ages of millions to tens of millions of years (Lippmann-Pipke et al. 2003). While limitation by inorganic nutrients may curb microbial activity in some cases (Kieft et al. 2005), Sherwood Lollar et al. (2007) suggested that ancient H_2-rich saline waters are hydrogeologically isolated and are released to mix with other waters only sporadically, possibly due to widening of pore-throat diameters following tectonic shifts. Lippmann-Pipke et al. (2011a) simultaneously monitored mining-induced seismicity (blasting) and geogas concentrations, including H_2 and CH_4, at 3.54 km depth in TauTona mine in South Africa, and found that spikes in geogas concentrations coincided with the daily blasting schedule in the mine. This finding has important implications for the metabolism of geogas by SLiMEs in fractured rock.

6.7 Subsurface Biodiversity

Microbial communities in deep subsurface continental habitats can be diverse and they vary with geochemical conditions (Gihring et al. 2006). Many of the microbes detected in culture-independent surveys represent novel lineages (Takai et al. 2001a; Gihring et al. 2006; Chivian et al. 2008; Sahl et al. 2008), many of which appear to be unique to subsurface environs. Some of these are cosmopolitan in their distribution, being detected as very similar small subunit rRNA gene sequences in water from boreholes that are widely separated geographically, even on different continents. The Firmicute *Candidatus Desulforudis audaxviator* is a case in point, having been detected in borehole waters from across the Witwatersrand Basin in South Africa (Moser et al. 2005; Lin et al. 2006; Chivian et al. 2008), as well as in Finland (Itavaara et al. 2011), and beneath Death Valley in California (Moser 2012). Other sequences have been detected in only a single borehole, suggesting that these represent microbes that are uniquely adapted to conditions at that site or that they're part of the so-called "rare biosphere" (Sogin et al. 2006). Surveys of microbial diversity in deep subsurface environs frequently reveal novel taxa, including previously unknown microbial phyla and phyla with no known cultivated representatives (Gihring et al. 2006; Chivian et al. 2008; Sahl et al. 2008; Dong et al. 2014). In one case, a previously unknown prokaryotic cell morphology, rod-shaped cells that are five-pointed and six-pointed stars in cross section, was discovered in a deep platinum mine in South Africa (Wanger et al. 2008). Clearly the deep continental biosphere is expanding our understanding of biodiversity.

A general pattern of decreasing diversity with depth has been observed (Gihring et al. 2006; Lin et al. 2012a). Explanations for the decreasing biodiversity include the extreme physical-chemical conditions (alkaline pH, elevated temperature, hydrostatic pressure), the limited number of available substrates, and the limited opportunities for colonization by immigrant microbes (Gihring et al. 2006). The ultimate in low diversity may be found in the simple, one-species ecosystem reported by Chivian for planktonic microbes collected from fracture water at 2.8 km depth in Mponeng gold mine in South Africa (Lin et al. 2006; Chivian et al. 2008). The dominance (95 % of the community) of hydrogenotrophic methanogens in the groundwater feeding Lidy Hot Springs in Idaho (Chapelle et al. 2002) may be another example. In both of these cases, chemolithotrophs are metabolizing H_2 generated by abiotic water–rock interactions, without the need for fermentation or syntrophs.

Archaea are commonly detected in deep continental fracture waters (Takai et al. 2001a; Moser et al. 2005; Gihring et al. 2006; Davidson et al. 2011; Nyssönen et al. 2012). Takai et al. (2001a) detected novel groups of Euryarchaeota (SAGMEG 1 and 2) and Crenarchaeota SAGMCG 1 and 2). The Crenarchaeota may be primarily drilling fluid contaminants, whereas the methanogenic Euryarchaeota may be indigenous groundwater archaea (Gihring et al. 2006; Davidson et al. 2011). So far, we have hints of anaerobic methane-oxidizing archaea (ANME) in the deep continental surface (Gihring et al. 2006) in the form of ANME-related 16S rDNA sequences, but they appear not to be dominant, and the energetics of anaerobic methane oxidation coupled to sulfate reduction appears to be overshadowed by other more favorable reactions (Moser et al. 2005; Kieft et al. 2005).

Sequences representing diverse bacterial phyla have been detected in the subsurface, with Proteobacteria and Firmicutes often predominating (Gihring et al. 2006; Moser et al. 2003; Itavaara et al. 2011). Alpha-, beta-, and gammaproteobacteria are commonly found. In some cases, these appear to comprise a greater proportion of the community when water is first sampled from a borehole, but then later diminish as the borehole community becomes dominated by indigenous fracture water microbes (Sahl et al. 2008; Moser et al. 2003; Davidson et al. 2011). In other cases, the Proteobacteria appear to truly dominate (Dong et al. 2014). Sulfate-reducing bacteria of the deltaproteobacteria are also reported (Onstott et al. 2003; Itavaara et al. 2011; Davidson et al. 2011; Nyssönen et al. 2012). Proteobacteria may be more likely to dominate in shallower aquifers (Itavaara et al. 2011; Lin et al. 2012a). Firmicute sequences frequently encountered include sulfate-reducing, spore-forming genera, e.g., *Cand. D. audaxviator* and *Desulfotomaculum* spp., as well as *Thermoanaerobacter* spp. (Gihring et al. 2006; Davidson et al. 2011; Nyssönen et al. 2012; Aüllo et al. 2013).

The sulfate-reducing Firmicute *Cand. D. audaxviator* deserves special attention. Culture-independent surveys of deep fracture water communities in the Witwatersrand Basin, South Africa, have repeatedly found closely related ($\geq$99 % homology) 16S rRNA sequences of a novel Firmicute with the most closely related cultured organisms belonging to the genus *Desulfotomacum* (Baker et al. 2003; Moser et al. 2003, 2005; Gihring et al. 2006). At 2.8 km depth in fracture water accessed via a borehole in Mponeng mine, it was found to comprise >99 % of the community of suspended cells. All evidence indicates that this organism can function as a chemoautotrophic sulfate reducer in a simple ecosystem using radiolytically generated H_2 and sulfate chemically oxidized from pyrite by radiolytically generated oxygen species. Genomic sequencing revealed hygrogenase, a complete dissimilatory sulfate reduction pathway, and capability for fixing CO_2 via the Wood–Ljungdahl pathway, thereby confirming its role in this simple ecosystem (Fig. 6.6) (Chivian et al. 2008). Other features revealed in the genome include genes for endospores, flagella, nitrogen fixation, and evidence for horizontal gene transfer from other bacteria as well as archaea, and it was described as the new candidate species *D. audaxviator*. Despite numerous attempts and knowledge of its full genome, no one has successfully isolated a member of this species (however, a related organism has now been grown in enrichment culture (Duane Moser, personal communication). Clearly, *D. audaxviator* is a highly successful organism, adapted to a wide range of subsurface physical and geochemical conditions. As described above, it's been detected in groundwater from three continents, but so far not detected in surface environments. A related sequence was found in deep marine crustal fluid (Cowen et al. 2003). Its ubiquity in widely spaced, hydrogeologically isolated, ancient fracture waters presents a biogeographical conundrum. Presumably, the genomes of the widely flung versions of this organism vary to a greater degree than is evident in their 16S rRNA genes. This hypothesis is being tested by Ramunas Stepauskas using the single cell genomics approach (personal communication). As such, the various versions of this species may be analogous to Darwin's finches, being among the few organisms to survive transport to remote deep environs and then adapting to the specific and varied conditions there, an idea put forth by Tullis Onstott and Ramunas Stepanauskas (personal communication).

The finding of large numbers of sequences from spore-forming Firmicutes raises the question of whether or not they're present as actively metabolizing vegetative cells or as inactive endospores. Arguments in favor of active cells include the observation by SEM of long, rod-shaped cells rather than spores in filtrate from the 2.8 km deep South African Mponeng mine water containing the simple, one-species *Cand. D. audaxviator* ecosystem (Chivian et al. 2008) and sulfur isotope fractionation in deep South African water samples that also have high proportions of sulfate-reducing Firmicute sequences (Lau et al. 2013). Evidence suggesting that the Firmicutes are present mainly as endospores lies in a comparison of DNA- and RNA-based surveys of microbial communities in deep South African fracture water communities. Several samples showed markedly higher proportions of Firmicute sequences in the DNA than in the RNA. This is a question that needs to be addressed further, e.g., using fluorescent in situ hybridization, analyses of mRNA as well as rRNA, stable isotope probing, and quantification of the endospore constituent calcium dipicolinate.

Eukarya are found only seldom in the deep continental biosphere and then generally not in the deepest, most anoxic groundwater. Sinclair and Ghiorse (1989) reported protozoa in relatively shallow (50–250 m) eastern coast plain aquifers of the United States. Lin et al. (2012b) reported a variety of protists in a shallow, unconfined aquifer at the Hanford Site, in Eastern Washington State, with a seasonal influence from the nearby Columbia River. Fungi were also reported by Sinclair and Ghiorse (1989) at 1–50 propagules per g dry weight of sediment; however, in deeper continental samples, fungi are generally not present and may

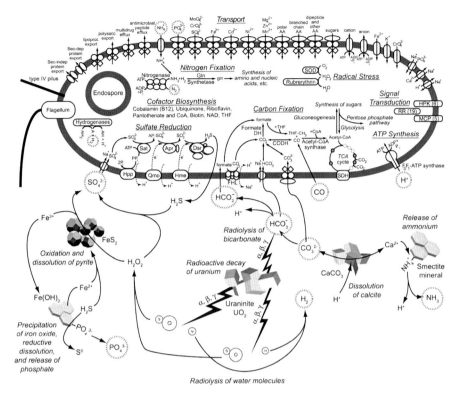

Fig. 6.6 Diagram of the genetically coded attributes of Candidatus *Desulforudis audaxviator* and its interactions with chemical components of its environment. (From Chivian et al. 2008, with permission)

even be considered to be serendipitous tracers of contamination (Onstott et al. 2003). Fungi have been reported in marine sediments (Orsi et al. 2013), so this may be another difference between continental and marine biospheres. Surprisingly, metazoan predators in the form of nematodes, including a novel genus, have been discovered feeding on bacteria in groundwater at depths of 0.9–3.6 km in the Witwatersrand Basin (Borgonie et al. 2011). These roundworms have been reported only from relatively young fracture waters (3000–12,000 years) containing low concentrations of dissolved O_2 (13–72 µM). Presumably, groundwater that's deeper, older, and anoxic does not contain metazoans.

Viruses occur in deep fracture water environments even though their host populations are relatively sparse. Kyle et al. (2008) reported 10^5–10^7 viral-like particles per ml of fracture water at the Äspö Hard Rock Laboratory (HRL) in Sweden at depths of 69–450 m, tenfold higher than the abundance of prokaryotes (10^4–10^6 cells ml^{-1}). Morphology indicated at least four different bacteriophage groups. Eydal et al. (2009) went on to isolate bacteriophages from these waters that specifically infected *Desulfovibrio aespoeensis*, which was previously isolated from these waters. Other evidence of subsurface bacteriophages is less direct. The genome of *Cand. D. audaxviator* contains CRISPR-cas system genes (clustered regularly spaced short palindromic repeats, genomic sequences, which in combination with Cas proteins, serve as a prokaryotic acquired immune system important in resistance to foreign genetic elements such as plasmids and phages), possibly indicating past exposure to bacteriophages. Labonté et al. (2015) reported abundant temperate phage sequences in the genomes of subsurface bacteria. Viruses may be major controllers of subsurface bacterial and archaeal populations and they may also be important vectors for horizontal gene transfer among subsurface microbes (Anderson et al. 2013). Viruses in the subsurface deserve greater attention.

Diverse subsurface bacteria and archaea have been isolated in culture, although as in most environments, these are often not the important players in culture-independent, sequence-based surveys. The U.S. Department of Energy's Subsurface Science Program isolated thousands of strains of bacteria, mostly aerobic heterotrophs from subsurface samples (Balkwill and Boone 1997; Balkwill et al. 1997). Deeper sampling has produced primarily anaerobes (Table 6.1). Spore-forming Firmicutes, metal reducers, and iron reducers are well represented. Few methanogens have been isolated, likely due to sampling and cultivation challenges associated with their extreme O_2 sensitivity. Bonin and Boone (2004) cultivated a methanogen from a South African mine sample. These isolates can serve as useful model organisms that share metabolic characteristics with the more numerically dominant microbes. For example, a *Desulfotomaculum putei* from the Taylorsville Basin in Virginia (Liu et al. 1997) was used to quantify sulfur isotope fractionation ($\Delta^{34}S$) in a biomass-recycling turbidostat that modeled energy-limiting conditions in the subsurface; the magnitude of $\Delta^{34}S$ increased with energy limitation (Davidson et al. 2009). Many novel subsurface isolates have potential for biotechnological applications. Metal-reducing *Thermoanaerobacter* strains from deep sedimentary basins have been put to work synthesizing specialty minerals and immobilizing metal and radionuclide contaminants (Moon et al. 2007; Yeary et al. 2011; Madden et al. 2012). *Thermus scotoductus* strain SA-01 also has potential applications as a metal reducer (Opperman et al. 2010; Cason et al. 2012). Other cultures have been and will be isolated, with the potential for uses as model organisms and with potential for biotechnological applications.

6.8 Outlook for the Deep Continental Biosphere

The deep biosphere, both continental and marine, offers huge potential for discovery, and important revelations are made with each new opportunity to probe the subsurface. Nonetheless, the sobriquet "dark life" remains very appropriate. The volume of the subsurface that has been sampled is minuscule compared to the total volume of the deep continental biosphere, and more importantly, there remains a tremendous diversity of geological settings and habitat types that have yet to be examined. At present, the factor that is most limiting to progress in the continental subsurface is the ability to gain access to deep environments for sample collection,

6 Microbiology of the Deep Continental Biosphere

Table 6.1 Selected deep subsurface microbial isolates. Selection was based upon the criteria of anaerobic growth or natural presence at depths >1 km

Isolate	Environment	Depth (km)	Characteristics	References
Bacillus infernus	Taylorsville Basin, Virginia sediment	2.7	Thermophilic (61 °C), strictly anaerobic, nitrate- and metal-reducing, endospore-forming Firmicute	Boone et al. (1995)
Desulfotomaculum putei	Taylorsville Basin, Virginia sediment	2.7	Thermophilic (50 °C) sulfate-reducing, endo-spore-forming Firmicute	Liu et al. (1997)
Desulfovibrio aespoeensis	Äspö Hard Rock Laboratory, Sweden	0.6	Sulfate-reducing deltaproteobacterium (25–30 °C)	Motamedi and Pedersen (1998)
Thermus scotoductus strain SA-01	Deep fracture water collected via Mponeng mine shaft 5, South Africa	3.2	Thermophile (65 °C), aerobic, and also nitrate and metal reducer	Kieft et al. (1999), Balkwill et al. (2004)
Alkaliphilus transvaalensis	Driefontein mine 5 shaft dam water	3.2	Thermophilic (20–50 °C,), alkaliphilic (pH 8–12.5) Firmicute; endospore former, strict anaerobe; heterotrophic nitrate, thiosulfate, and fumarate reducer	Takai et al. (2001b)
Thermoanaerobacter ethanolicus	Piceance Basin, Colorado	2.1	Thermophilic (60 °C) metal-reducing bacteria, produce magnetite	Roh et al. (2002)
Geobacillus thermoleovorans strain GE-7	Deep fracture water collected via Driefontein mine, South Africa	3.2	Thermophile (65 °C), aerobic, and also nitrate reducer	DeFlaun et al. (2007)
Halanaerocella petrolearia	Hypersaline oil reservoir, Gabon	1.15	Halophilic (growth at 6–126 % NaCl) fermentative Firmicute	Gales et al. (2011)
Desulfocurvus vexinensis	Deep aquifer, Paris Basin, France	0.83	Sulfate-reducing deltaproteobacterium (37 °C)	Klouche et al. (2009)

monitoring, and experimentation. While the necessary sampling equipment isn't quite as specialized and expensive as the drill ships used by the International Ocean Discovery Program, drilling on and in land is expensive, especially when the targets are >1 km deep. Drilling to significant depth generally requires millions of dollars and thus requires either major funding or the opportunity to piggyback a scientific investigation onto drilling carried out for other purposes, e.g., by the extractive industries and by entities seeking to dispose of hazardous waste (e.g., radionuclides,

CO_2). The International Scientific Drilling Program (ICDP) partially funds scientific drilling projects and has identified the deep biosphere as an important driver for future drilling expeditions (Harms et al. 2007), so it can be hoped that opportunities will expand. Drilling from underground platforms needs to be advanced via dedicated underground facilities, e.g., ones used for high-energy particle physics (Fredrickson et al. 2006; Onstott et al. 2009). Permanent facilities can be especially useful for long-term monitoring and experimentation.

While sampling opportunities have been limiting, the technologies for analyzing microbial communities and their metabolic activities have grown enormously. The entire "-omics" revolution (genomics, transcriptomics, proteomics, metabolomics) is being applied to the subsurface as well as nearly every other conceivable environment, and so one can expect the next decade to bring major breakthroughs. As we gain more and more data on the genetics and metabolism of subsurface microorganisms, the challenge will be to use these data to answer major questions such as how are the microbes interacting in situ, what are their rates of activity, and how did they get there and adapt to subsurface conditions in the first place? Better insight into Earth's dark life may also give insights to the consideration of life beneath the surfaces of other planets or even into the origin and early evolution of life on our own planet. There's no shortage of questions, and these days there's not even much limitation by available technology; we need only to expand the scientific will to drill.

References

Amend JP, Teske A (2005) Expanding frontiers in deep subsurface microbiology. Palaeogeogr Palaeoclimatol Palaeoecol 219(1–2):131–155

Amy PS, Haldeman DL (eds) (1997) The microbiology of the terrestrial subsurface. CRC, Boca Raton, FL

Anderson RE, Brazelton WJ, Baross JA (2013) The deep viriosphere: assessing the viral impact on microbial community dynamics in the deep subsurface. In: Hazen RM, Jones AP, Baross JA (eds) Carbon in Earth. Rev Miner Geochem 75:649–675

Aüllo T, Ranchu-Peyruse A, Olivier B, Mogot M (2013) *Desulfotomaculum* spp. and related gram-positive sulfate reducing bacteria in deep subsurface environments. Front Microbiol 4:1–12

Baker BJ, Moser DB, MacGregor BJ, Fishbain S, Wagner M, Fry NK, Jackson B, Speolstra N, Loos S, Takai K, Sherwood Lollar B, Fredrickson J, Balkwill D, Onstott TC, Wimpee CF, Stahl DA (2003) Related assemblages of sulphate-reducing bacteria associated with ultradeep gold mines of South Africa and deep basalt aquifers of Washington State. Environ Microbiol 5:1168–1191

Balkwill DL, Boone DR (1997) Identity and diversity of microorganisms cultured from subsurface environments. In: Amy PS, Haldeman DL (eds) The microbiology of the terrestrial deep subsurface. CRC Lewis, Boca Raton, pp 105–117

Balkwill DL, Reeves RH, Drake GR, Reeves JG, Crocker FH, King MB, Boone DR (1997) Phylogenetic characterization of bacteria in the subsurface microbial culture collection. FEMS Microbiol Rev 20:201–216

Balkwill DL, Kieft TL, Tsukuda T, Kostandarithes HM, Onstott TC, Macnaughton S, Bownas J, Fredrickson JK (2004) Identification of iron-reducing *Thermus* strains as *Thermus scotoductus*. Extremophiles 8:37–44

Bartlett DH (2009) Microbial life in the trenches. Mar Technol Soc J 43:128–131

Bastin ES, Greer FE, Merritt CA, Moulton G (1926) The presence of sulphate reducing bacteria in oil field waters. Science 63:21–24

Bonin AS, Boone DR (2004) Microbial isolations and characterizations from the deep terrestrial subsurface of the South African gold mines. Proceedings of the 104th General Meeting of the American Society for Microbiology. New Orleans, LA. May 2004

Boone DR, Liu YT, Zhao ZJ, Balkwill DL, Drake GR, Stevens TO, Aldrich HC (1995) *Bacillus infernus* sp. nov., an Fe(III)- and Mn(IV)-reducing anaerobe from the deep terrestrial subsurface. Int J Syst Bacteriol 45:441–448

Borgonie G, García-Moyano A, Litthauer D, Bert W, Bester A, van Heerden E, Onstott TC (2011) Nematoda from the terrestrial deep subsurface of South Africa. Nature 474:79–82

Brockman FJ, Li SW, Fredrickson JK, Ringelberg DB, Kieft TL, Spadoni CM, White DC, McKinley JP (1998) Post-sampling changes in microbial community composition and activity in a subsurface paleosol. Microb Ecol 36:152–164

Cason ED, Piater LA, van Heerden E (2012) Reduction of U(VI) by the deep subsurface bacterium, Thermus scotoductus SA-01, and the involvement of the ABC transporter protein. Chemosphere 86:572–577

Chapelle FH (2000) Ground-water microbiology and geochemistry, 2nd edn. Wiley, New York

Chapelle FH, O'Neill K, Bradley PM, Methe BA, Ciufo SA, Knobel LL, Lovley DR (2002) A hydrogen-based subsurface microbial community dominated by methanogens. Nature 415:312–315

Chivian D, Alm E, Brodie E, Culley D, Dehal P, DeSantis T, Gihring T, Lapidus A, Lin L-H, Lowry S, Moser D, Richardson P, Southam G, Wanger G, Pratt L, Andersen G, Hazen T, Brockman F, Arkin A, Onstott T (2008) Environmental genomics reveals a single species ecosystem deep within the Earth. Science 322:275–278

Colwell FS, D'Hondt S (2013) Nature and extent of the deep biosphere. In: Hazen RM, Jones AP, Baross JA (eds) Carbon in Earth. Rev Miner Geochem 75:547–574

Colwell FS, Onstott TC, Delwiche ME, Chandler D, Fredrickson JK, Yao QJ, McKinley JP, Boone DR, Griffiths R, Phelps TJ, Ringelberg D, White DC, LaFreniere L, Balkwill D, Lehman RM, Konisky J, Long PE (1997) Microorganisms from deep, high temperature sandstones: constraints on microbial colonization. FEMS Microbiol Rev 20:425–435

Colwell FS, Stormberg GJ, Phelps TJ, Birnbaum SA, McKinley J, Rawson SA, Veverka C, Goodwin S, Long PE, Russell BF, Garland T, Thompson D, Skinner P, Grover S (1992) Innovative techniques for collection of saturated and unsaturated subsurface basalts and sediments for microbiological characterization. J Microbiol Methods 15:279–292

Colwell FS, Smith RP (2004) Unifying principles of the deep terrestrial and deep marine biospheres. biospheres. In: Wilcock WSD, Delong EF, Kelley DS, Baross JA, Cary SC (eds) Subseafloor biosphere at mid-ocean ridges, vol 104, Geophysical Monograph Series. American Geophysical Union, Washington, DC, pp 355–367

Cowen JP, Giovannoni SJ, Kenig F, Johnson HP, Butterfield D, Rappé MS, Hutnak M, Lam P (2003) Fluids from aging ocean crust that support microbial life. Science 299:120–123

Davidson MM, Bisher ME, Pratt LM, Fong J, Southam G, Pfiffner SM, Reches Z, Onstott TC (2009) Sulfur isotope enrichment during maintenance metabolism in the thermophilic sulfate-reducing bacterium *Desulfotomaculum putei*. Appl Environ Microbiol 75:5621–5630

Davidson MM, Silver BJ, Onstott TC, Moser DP, Gihring TM, Pratt LM, Boice EA, Sherwood Lollar B, Lippmann-Pipke J, Pfiffner SM, Kieft TL, Symore W, Ralston C (2011) Capture of planktonic microbial diversity in fractures by long-term monitoring of flowing boreholes, Evander Basin, South Africa. Geomicrobiol J 28:275–300

D'Hondt S, Jorgensen BB, Miller DJ, Batzke A, Blake R, Cragg BA, Cypionka H, Dickens GR, Ferdelman T, Hinrichs KU, Holm NG, Mitterer R, Spivack A, Wang GZ, Bekins B, Engelen B,

Ford K, Gettemy G, Rutherford SD, Sass H, Skilbeck CG, Aiello IW, Guerin G, House CH, Inagaki F, Meister P, Naehr T, Niitsuma S, Parkes RJ, Schippers A, Smith DC, Teske A, Wiegel J, Padilla CN, Acosta JLS (2004) Distributions of microbial activities in deep subseafloor sediments. Science 306:2216–2221

DeFlaun MF, Fredrickson JK, Dong H, Pfiffner SM, Onstott TC, Balkwill DL, Streger SH, Stackebrandt E, Knoessen S, van Heerden E (2007) Isolation and characterization of a *Geobacillus thermoleovorans* strain from an ultra-deep South African gold mine. Syst Appl Microbiol 30:152–164

DeLong EF, Franks DG, Yayanos AA (1997) Evolutionary relationships of cultivated psychrophilic and barophilic deep-sea bacteria. Appl Environ Microbiol 63:2105–2108

Dong H, Zhang G, Huang L, Dai X, Wang Y, Lu G, Dong Z, Dong X (2009) The deep subsurface microbiology research in China: results from Chinese Continental Scientific Drilling Project, AGU Fall meeting, San Francisco, CA, December 2009

Dong Y, Kumar CG, Chia N, Kim P-J, Miller PA, Price ND, Can IKO, Flynn TM, Sanford RA, Krapac IG, Locke RA, Hong P-Y, Tamaki H, Liu W-T, Mackie RI, Hernandez AG, Wright CL, Mikel MA, Walker JL, Sivaguru M, Fried G, Yannarell AC, Fouke BW (2014) *Halomonas sulfidaeris*-dominated microbial community inhabits a 1.8 km-deep subsurface Cambrian Sandstone reservoir. Environ Microbiol 16:1695–708. doi:10.1111/1462-2920.12325

Edwards RA, Rodrigues-Brito B, Wegley L, Haynes M, Breitbart M, Peterson D, Saar M, Alexander S, Alexander EC, Rohwer F (2006) Using pyrosequencing to shed light on deep mine microbial ecology. BMC Genomics 7:57–70

Eloe EA, Fadrosh DW, Novotny M, Allen LZ, Kim M, Lombardo MJ, Yee-Greenbaum J, Yooseph S, Alen EE, Lasken R, Williamson SJ, Bartlett DH (2011) Going deeper: metagenome of a hadopelagic microbial community. PLoS One 6, e20388

Erzinger J, Wiersberg T, Zimmer M (2006) Real-time mud gas logging and sampling during drilling. Geofluids 6:225–233

Eydal HSC, Jagevall S, Hermansson M, Pedersen K (2009) Bacteriophage lytic to *Desulfovibrio aespoeensis* isolated from deep groundwater. ISME J 3:1139–1147

Fredrickson JK, Balkwill DL (2006) Geomicrobiological processes and diversity in the deep terrestrial subsurface. Geomicrobiol J 23:345–356

Fredrickson JK, Kieft TL, Moran N, Moser DP, Onstott TC, Phelps TJ, Teidje JM (2006) DUSEL: Window to the Subsurface Biosphere. National Science Foundation Report. http://www.deepscience.org/TechnicalDocuments/Final/deepbiology_final.pdf

Fredrickson JK, Onstott TC (1996) Microbes deep inside the Earth. Sci Am 275(4):68–73

Freifeld B (2009) The U-tube: a new paradigm for borehole fluid sampling. Scientific Drill 8:41–45

Fukuda A, Haigiwara H, Ishimura T, Kouduka M, Ioka S, Amano Y, Tsunogai U, Suzuki Y, Mizuno T (2010) Microb Ecol 60:214–225

Gales G, Cehider N, Joulian C, Battaglia-Brunet F, Cayol J-L, Postec A, Borgomano J, Neria-Gonzales I, Lomans BP, Ollivier B, Alazard D (2011) Characterization of *Haloanaerocella petroleara* gen. nov., sp. nov., a new anaerobic moderately halophilic fermentative bacterium isolated from a deep subsurface hypersaline oil reservoir. Extremophiles 15:565–571

Gihring TM, Moser DP, Lin L-H, Davidson M, Onstott TC, Morgan L, Millesson M, Kieft TL, Trimarco E, Balkwill DL, Dollhopf ME (2006) The distribution of microbial taxa in the subsurface water of the Kalahari Shield, South Africa. Geomicrobiol J 23:415–430

Gold T (1992) The deep, hot biosphere. Proc Natl Acad Sci 89:6045–6049

Harms U, Koeberl C, Zoback MD (eds) (2007) Continental scientific drilling, a decade of progress, and challenges for the future. Springer, Berlin

Head IM, Jones DM, Larter SR (2003) Biological activity in the deep subsurface and the origin of heavy oil. Nature 426:344–349

Huber JA, Mark Welch D, Morrison HG, Huse SM, Neal PR, Butterfield DA, Sogin ML (2007) Microbial population structures in the deep marine biosphere. Science 318:97–100

Itavaara M, Nyyssonen M, Kapanen A, Nousiainen A, Ahonen L, Kukkonen I (2011) Characterization of bacterial diversity to a depth of 1500 m in the Outokumpu deep borehole, Fennoscandian Shield. FEMS Microbiol Lett 77:295–309

Kallmeyer J, Pockalny R, Adhikari RR, Smith DC, D'Hondt S (2012) Global distribution of microbial abundance and biomass in subseafloor sediment. Proc Natl Acad Sci 109:16213–16216

Kashefi K, Lovley DR (2003) Extending the upper temperature limit for life. Science 301:934

Kelley DS, Karson JA, Fru GL, Yoerger DR, Shank TM, Butterfield DA, Hayes JM, Schrenk MO, Olson EJ, Proskurowski G, Jakuba M, Bradley A, Larson B, Ludwig K, Glickson D, Buckman K, Bradley AS, Brazelton WJ, Roe K, Bernasconi SM, Elend MJ, Lilley MD, Baross JA, Summons RE, Sylva SP (2005) A serpentinite-hosted ecosystem: the Lost City hydrothermal field. Science 307:1428–1434

Kieft TL (2010) Sampling the deep sub-surface using drilling and coring techniques. In: Timmis KN (ed) Microbiology of hydrocarbons and lipids. Springer, Berlin, pp 3427–3441

Kieft TL, Fredrickson JK, Onstott TC, Gorby YA, Kostandarithes HM, Bailey TJ, Kennedy DW, Li SW, Plymale A, Spadoni CM, Gray MS (1999) Dissimilatory reduction of Fe(III) and other electron acceptors by a *Thermus* isolate. Appl Environ Microbiol 65:1214–1221

Kieft TL, McCuddy SM, Onstott TC, Davidson M, Lin L-H, Mislowac B, Pratt L, Boice E, Sherwood Lollar B, Lippmann-Pipke J, Pfiffner SM, Phelps TJ, Gihring T, Moser D, van Heerden E (2005) Geochemically generated, energy-rich substrates and indigenous microorganisms in deep, ancient groundwater. Geomicrobiol J 22:325–335

Kieft TL, Phelps TJ (1997) Life in the slow lane: Activities of microorganisms in the subsurface. In: Amy PS, Haldeman DL (eds) The microbiology of the terrestrial subsurface. CRC, Boca Raton, FL, pp 137–163

Kieft TL, Phelps TJ, Fredrickson JK (2007) Drilling, coring, and sampling subsurface environments. In: Hurst CJ (ed) Manual of environmental microbiology, 3rd edn. ASM, Washington, DC, pp 799–817

Kyle JE, Eydal HSC, Ferris FG, Pedersen K (2008) Viruses in granitic groundwater from 69 to 450 m depth of the Äspö hard rock laboratory, Sweden. ISME J 2:571–574

Klouche N, Basso O, Lascourrèges JF, Cayol JL, Thmas P, Fauque G, Fardeau ML, Magot M (2009) *Desulfocurvus vexinensis* gen. nov. sp. nov., a sulfate-reducing bacterium isolated from a deep subsurface aquifer. Int J Syst Evol Microbiol 59:3100–3104

Labonté JM, Field EK, Lau M, Chivian D, Van Heerden E, Wommack KE, Kieft TL, Onstott TC, Stepanauskas R (2015) Single cell genomics indicates horizontal gene transfer and viral infections in a deep subsurface Firmicutes population. Front Microbiol 6:349

Lauro FM, Bartlett DH (2008) Prokaryotic lifestyles in deep sea habitats. Extremophiles 12:15–25

Lau MCY, Magnabosco C, Brown CT, Grim S, Lacrampe-Couloume G, Wilkie K, Sherwood Lollar B, Simkus DN, Slater GF, Hendrickson S, Pullin M, Kieft TL, Li L, Snyder L, Kuloyo O, Linage B, Borgonie G, Vermeulen J, Maleke M, Tlalajoe N, Moloantoa KM, van Heerden E, Vermeulen F, Pienaar M, Munro A, Joubert L, Ackerman J, van Jaarsveld C, Onstott TC (2013) Continental subsurface waters support unique but diverse C-acquisition strategies. AGU Fall Meeting, San Francisco, California, US, 9th–13th December 2013

Lehman RM, Colwell FS, Ringelberg DB, White DC (1995) Combined microbial community-level analyses for quality assurance of terrestrial subsurface cores. J Microbiol Methods 22:263–281

Lin LH, Wang P-L, Rumble D, Lippmann-Pipke J, Boice E, Pratt LM, Sherwood Lollar B, Brodie E, Hazen T, Andersen G, DeSantis T, Moser DP, Kershaw D, Onstott TC (2006) Long term biosustainability in a high energy, low diversity crustal biome. Science 314:479–482

Lin X, Kennedy D, Fredrickson J, Bjornstad B, Konopka A (2012a) Vertical stratification of subsurface microbial community composition across geological formations at the Hanford Site. Environ Microbiol 14:414–425

Lin X, McKinley J, Resch CT, Kaluzny R, Lauber CL, Fredrickson J, Knight R, Konopka A (2012b) Spatial and temporal dynamics of the microbial community in the Hanford unconfined aquifer. ISME J 6:1665–1676

Lippmann-Pipke J, Erzinger J, Zimmer M, Kujawa C, Boettcher M, van Heerden E, Bester A, Moller H, Stroncik NA, Reches Z (2011a) Geogas transport in fractured hard rock—Correlations with mining seismicity at 3.54 km depth, TauTona gold mine, South Africa. Appl Geochem 26:2134–2146

Lippmann-Pipke J, Sherwood Lollar B, Neidermann S, Stroncik N, Naumann R, VanHeerden E, Onstott TC (2011b) Neon identifies two billion year old fluid component in Kaapvaal Craton. Chem Geol 282:287–296

Lippmann-Pipke J, Stute M, Torgersen T, Moser DP, Hall J, Lin L, Borcsik M, Bellamy RES, Onstott TC (2003) Dating ultra-deep mine waters with noble gases and ^{36}Cl, Witwatersrand, South Africa. Geochimica Cosmoshimica Acta 67:4597–4619

Liu YT, Karnauchow TM, Jarrell KF, Balkwill DL, Drake GR, Ringelberg D, Clarno R, Boone DR (1997) Description of two new thermophilic *Desulfotomaculum* spp., *Desulfotomaculum putei* sp. nov, from a deep terrestrial subsurface, and *Desulfotomaculum luciae* sp. nov, from a hot spring. Int J System Bacteriol 47:615–621

Lovley DR, Chapelle FH, Woodward JC (1994) Use of dissolved H_2 concentration to determine distribution of microbially catalyzed redox reactions in anoxic groundwater. Environ Sci Technol 28:1205–1210

Madden AS, Swidle AL, Beazley MJ, Moon JW, Ravel B, Phelps TJ (2012) Long-term solid-phase fate of co-precipitated U(VI)-Fe(III) following biological iron reduction by *Thermoanaerobacter*. Am Miner 97:1641–1652

Madigan MT, Martinko JM, Stahl DA, Clark DP (2013) Brock biology of microorganisms, 13th edn. Benjamin Cummings, Boston, MA

Moon J-W, Roh Y, Lucas W, Yeary LW, Lau RJ, Rawn CJ, Love LJ, Phelps TJ (2007) Microbial formation of lanthanide-substituted magnetites by *Thermoanaerobacter* sp. TOR-39. Extremophiles 11:859–867

Moser DP (2012) Deep microbial ecosystems in the U.S. Great Basin: a second home for *Desulforudis audaxviator*? Abstract B41F-08 presented at 2012 Fall Meeting, AGU, San Francisco, CA, 3–7

Moser D, Boston PJ, Martin H (2001) Sampling in caves and mines. In: Bitton GE (ed) Encyclopedia of environmental microbiology. Wiley, New York, pp 821–835

Moser DP, Gihring T, Fredrickson JK, Brockman FJ, Balkwill D, Dollhopf ME, Sherwood-Lollar B, Pratt LM, Boice E, Southam G, Wanger G, Welty AT, Baker BJ, Onstott TC (2005) *Desulfotomaculum* spp. and *Methanobacterium* spp. dominate a 4– to 5-kilometer deep fault. Appl Environ Microbiol 71:8773–8783

Moser DP, Onstott TC, Fredrickson JK, Brockman FJ, Balkwill DL, Drake GR, Pfiffner SM, White DC, Takai K, Pratt LM, Fong J, Sherwood-Lollar B, Slater G, Phelps TJ, Spoelstra N, DeFlaun M, Southam G, Welty AT, Baker BJ, Hoek J (2003) Temporal shifts in microbial community structure and geochemistry of an ultradeep South African gold mine borehole. Geomicrobiol J 20:517–548

Motamedi M, Pedersen K (1998) *Desulfovibrio aespoeensis* sp. nov., a mesophilic sulfate-reducing bacterium from deep groundwater at Äspö hard rock labioratory. Sweden Int J Syst Bacteriol 48:311–315

Murphy EM, Schramke JA, Fredrickson JK, Bledsoe HW, Francis AJ, Sklarew DS, Linehand JC (1992) The influence of microbial activity and sedimentary organic carbon on the isotope geochemistry of the Middendorf aquifer. Water Resour Res 28:723–740

Nagata T, Tamburini C, Arístegui J, Baltar F, Bochdansky A, Fonda-Umani S, Fukuda H, Gogou A, Hansell DA, Hansman RJ, Herndl GJ, Panagiotopoulos C, Reinthaler T, Sohrin R, Verdugo P, Yamada N, Yamashita Y, Yokokawa T, Bartlett DH (2010) Emerging concepts on microbial processes in the bathypelagic ocean – ecology, biogeochemistry, and genomics. Deep-Sea Res II 57:1519–1536

National Science Foundation (2003) EarthLab, NSF-sponsored report of underground opportunities in GeoSciences and GeoEngineering. National Science Foundation, Washington, DC

Nealson KH, Inagaki F, Takai K (2005) Hydrogen-driven subsurface lithoautotrophic microbial ecosystems (SLiMEs): do they exist and why should we care? Trends Microbiol 13:405–410

Nyssönen M, Bomberg M, Kapanen A, Nousiainen A, Pitkänen P, Itävaara M (2012) Methanogenic and sulphate-reducing microbial communities in deep groundwater of crystalline rock fractures in Oliluoto, Finland. Geomicrobiol J 29:863–878

Omar G, Onstott TC, Hoek J (2003) The origin of deep subsurface microbial communities in the Witwatersrand Basin, South Africa as deduced from apatite fission track analyses. Geofluids 3:69–80

Onstott TC, Colwell FS, Kieft TL, Murdoch L (2009) New horizons for deep subsurface microbiology. Microbe 4:499–505

Onstott TC, Lin LH, Davidson M, Mislowac B, Borcsik M, Hall J, Slater G, Ward J, Sherwood Lollar B, Lippmann-Pipke J, Boice E, Pratt L, Pfiffner BS, Moser D, Gihring T, Kieft TL, Phelps TJ, van Heerden E, Litthauer D, DeFlaun M, Rothmel R (2006) The origin and age of biogeochemical trends in deep fracture water of the Witwatersrand basin, South Africa. Geomicrobiol J 23:369–414

Onstott TC, Magnabosco C, Aubrey AD, Burton AS, Dworkin JP, Elsila JE, Grunsfeld S, Cao BH, Hein JE, Glavin DP, Kieft TL, Silver BJ, Phelps TJ, van Heerden E, Opperman DJ, Bada JL (2014) Does aspartic acid racemization constrain the depth limit of the subsurface biosphere? Geobiology 12:1–19

Onstott TC, Moser DP, Pfiffner SM, Fredrickson JK, Brockman FJ, Phelps TJ, White DC, Peacock A, Balkwill D, Hoover R, Krumholz LR, Borscik M, Kieft TL, Wilson R (2003) Indigenous and contaminant microbes in ultradeep mines. Environ Microbiol 5:1168–1191

Onstott TC, Phelps TJ, Colwell FS, Ringelberg D, White DC, Boone DR, McKinley JP, Stevens TO, Long PE, Balkwill DL, Griffin T, Kieft T (1998) Observations pertaining to the origin and ecology of microorganisms recovered from the deep subsurface of Taylorsville Basin, Virginia. Geomicrobiol J 15:353–385

Onstott TC, Phelps TJ, Kieft TL, Colwell FS, Balkwill DL, Fredrickson JK, Brockman FJ (1999) A global perspective on the microbial abundance and activity in the deep subsurface. In: Seckbach J (ed) Enigmatic microorganisms and life in extreme environments. Kluwer, The Netherlands, pp 489–500

Opperman DJ, Sewell BT, Litthauer D, Isupov MN, Littlechild JA, van Heerden E (2010) Biochem Biophys Res Commun 393:426–431

Orcutt BN, Bach W, Becker K, Fisher AT, Hentscher M, Toner BM, Wheat CG, Edwards KJ (2011) Colonization of subsurface microbial observatories deployed in young ocean crust. ISME J 5:692–703

Orcutt B, Wheat CG, Edwards K (2010) Subseafloor ocean crust microbial observatories: development of FLOCS (flow-through osmo colonization system) and evaluation of borehole construction materials. Geomicrobiol J 27:143–157

Orsi W, Biddle JF, Edgcomb V (2013) Deep sequencing of subseafloor eukaryotic rRNA reveals active fungi across marine subsurface provinces. PLoS One 8:e56335

Parkes RJ, Cragg BA, Bale SJ, Getliff JM, Goodman K, Rochelle PA, Fry JC, Weightman AJ, Harvey SM (1994) Deep bacterial biosphere in Pacific Ocean sediments. Nature 371:410–413

Pedersen K (1993) The deep subterranean biosphere. Earth-Sci Rev 34:243–260

Pedersen K (1997) Microbial life in deep granitic rock. FEMS Microbiol Rev 20:399–414

Pedersen K (2000) Exploration of deep intraterrestrial microbial life: current perspectives. FEMS Microbiol Lett 185:9–16

Phelps TJ, Fliermans CB, Garland TR, Pfiffner SM, White DC (1989) Methods for recovery of deep terrestrial subsurface sediments for microbiological studies. J Microbiol Methods 9:267–279

Phelps TJ, Murphy EM, Pfiffner SM, White DC (1994) Comparisons between geochemical and biological estimates of subsurface microbial activities. Microb Ecol 28:335–349

Roh Y, Liu SV, Li G, Huang H, Phelps TJ, Zhou J (2002) Isolation and characterization of metal-reducing *Theroanaerobacter* strains from deep subsurface environments of the Piceance Basin, Colorado. Appl Environ Microbiol 68:6013–6020

Røy H, Kallmeyer J, Adhikari RR, Pockalny R, Jørgensen BB, D'Hondt S (2012) Aerobic microbial respiration in 86-million-year-old deep-sea red clay. Science 336:922–925

Russell BF, Phelps TJ, Griffin WT, Sargent KA (1992) Procedures for sampling deep subsurface communities in unconsolidated sediments. Groundwater Monit Remediat 12:96–104

Sahl JW, Schmidt R, Swanner ED, Mandernack KW, Templeton AS, Kieft TL, Smith RL, Sanford WE, Callaghan RL, Mitton JB, Spear JR (2008) Subsurface microbial diversity in deep-granitic-fracture water in Colorado. Appl Environ Microbiol 74:143–152

Schrenk MO, Brazelton WJ, Lang SQ (2013) Serpentinization, carbon and deep life. Rev Miner Geochem 75:575–606

Silver BJ, Onstott TC, Rose G, Lin L-H, Ralston C, Sherwood-Lollar B, Pfiffner SM, Kieft TL, McCuddy S (2010) In situ cultivation of subsurface microorganisms in a deep mafic sill: implications for SLiMEs. Geomicrobiology J27:329–348

Sinclair JL, Ghiorse WC (1989) Distribution of aerobic bacteria, protozoa, algae, and fungi in deep subsurface sediments. Geomicrobiol J 7:15–32

Sherwood Lollar B, Voglesonger K, Lin L-H, LaCrampe-Couloume G, Telling J, Abrajano TA, Onstott TC, Pratt LM (2007) Hydrologic controls on episodic H_2 release from Precambrian fractured rocks—energy for deep subsurface life on Earth and Mars. Astrobiology 7:971–986

Sherwood Lollar B, Westgate TD, Ward JA, Slater GF, Lacrampe-Couloume G (2002) Abiogenic formation of alkanes in the Earth's crust as a minor source for global hydrocarbon reservoirs. Nature 416:522–524

Smith RL, Harris SH (2007) Determining the terminal electron-accepting reaction in the saturated subsurface. In: Hurst CJ, Crawford RL, Garland JL, Lipson DA, Mills AL, Stetzenbach LD (eds) Manual of environmental microbiology, 3rd edn. ASM Press, Washington, DC, pp 860–871

Smith DC, Spivack AJ, Fisk MR, Haveman SA, Staudigel H, The Leg 185 Shipboard Scientific Party (2000) Methods for quantifying potential microbial contamination during deep ocean coring. ODP Tech. Note 28

Sogin ML, Morrison HG, Huber JA, Welch DM, Huse SM, Neal PR, Arrieta JM, Herndl GJ (2006) Microbial diversity in the deep sea and the underexplored "rare biosphere". Proc Natl Acad Sci USA 103:12115–12120

Stevens TO, McKinley JP (1995) Lithoautotrophic microbial ecosystems in deep basalt aquifers. Science 270:450–454

Stevens TO, McKinley JP (2000) Abiotic controls on H2 production from basalt-water reactions and implications for aquifer biogeochemistry. Environ Sci Technol 34:826–831

Stober I, Bucher K (2004) Fluid sinks within the earth's crust. Geofluids 4:143–151

Takai K, Moser DP, DeFlaun MF, Onstott TC, Fredrickson JK (2001a) Archaeal diversity in waters from deep South African Gold mines. Appl Environ Microbiol 67:5750–5760

Takai K, Moser DP, Onstott TC, Spoelstra N, Pfiffner SM, Dohnalkova A, Fredrickson JK (2001b) *Alakliphilus transvaalensis* gen. nov., sp. nov., an extremely alkaliphilic bacterium isolated from a deep South African gold mine. Int J System Evol Microbiol 51:1245–1256

Takai K, Nakamura K, Toki T, Tsunagai U, Miyazaki M, Hirayama H, Nakagawa S, Nonoura T, Horikoshi K (2008) Cell proliferation at 122C and isotopically heavy CH_4 production by a hyperthermophilic methanogen under high-pressure cultivation. Proc Natl Acad Sci USA 105:10949–10954

Wang FP, Zhou HY, Meng J, Peng XT, Jiang LJ, Sun P, Zhang CL, Van Nostrand JD, Deng Y, He ZL, Wu LY, Zhou JH, Xiao X (2009) GeoChip-based analysis of metabolic diversity of microbial communities at the Juan de Fuca Ridge hydrothermal vent. Proc Natl Acad Sci USA 106:4840–4845

Wanger G, Onstott TC, Southam G (2008) Stars of the terrestrial deep subsurface: a novel 'star-shaped' bacterial morphotype from a South African platinum mine. Geobiology 6:325–330

Whitman WB, Coleman DC, Wiebe WJ (1998) Prokaryotes: the unseen majority. Proc Natl Acad Sci USA 95:6578–6583

Wilhelms A, Larter SR, Head I, Farrimond P, di-Primio R, Zwach C (2001) Biodegradation of oil in uplifted basins prevented by deep-burial sterilization. Nature 411:1034–1037

Wilson JT, McNabb JF, Balkwill DL, Ghiorse WC (1983) Enumeration and characterization of bacteria indigenous to a shallow water-table aquifer. Ground Water 21:134–142

Wu JY, Gao WM, Johnson RH, Zhang WW, Meldrum DR (2013) Integrated metagenomic and metatranscriptomic analyses of microbial communities in the meso- and bathypelagic realm of North Pacific Ocean. Marine Drugs 11:3777–3801

Yayanos AA (1995) Microbiology to 10,500 meters in the deep sea. Ann Rev Microbiol 49:777–805

Yayanos AA (2001) Deep-sea piezophilic bacteria. Meth Microbiol 30:615–637

Yeary LW, Moon J-W, Rawn CJ, Love LJ, Rondinone AJ, Thompson JR, Chakoumakos BC, Phelps TJ (2011) Magnetic properties of bio-synthesized zinc ferrite nanoparticles. J Magnet Magnetic Mater 323:3043–3048

Chapter 7
Microbiology of the Deep Subsurface Geosphere and Its Implications for Used Nuclear Fuel Repositories

J.R. McKelvie, D.R. Korber, and G.M. Wolfaardt

Abstract A number of countries are actively working toward the siting and development of deep geological repositories (DGR) for used nuclear fuel. Given their ubiquity and metabolic capabilities, it is assumed that with sufficient time and appropriate conditions, microorganisms could alter the geochemistry of the repository. As such, the DGR concept provides an invaluable opportunity to evaluate the evolution of subsurface conditions from "disturbance" back to original state. The design concept involves the use of steel or copper/steel used fuel containers, surrounded by a low-permeability, swelling clay buffer material within a low-permeability, stable host rock environment. Within a newly constructed DGR, conditions would be warm, oxidizing, and dry. With sufficient time, these conditions would gradually revert to the original state of the surrounding geology. This chapter discusses how microbes and their metabolic activity may change over time and discusses the potential effects they may have on the engineered barrier system (EBS) that serves to isolate the used fuel containers and on the used fuel itself. The widespread support for the development of underground facilities as a means to ensure safe, long-term storage of increasing inventory of nuclear waste underscores the pressing need to learn more about the impacts of microbial activity on the performance of such facilities over the long term.

J.R. McKelvie (✉)
Nuclear Waste Management Organization, Toronto, ON, Canada
e-mail: jmckelvie@nwmo.ca

D.R. Korber
University of Saskatchewan, Saskatoon, SK, Canada
e-mail: darren.korber@usask.ca

G.M. Wolfaardt
Ryerson University, Toronto, ON, Canada
e-mail: gmw@sun.ac.za

7.1 Introduction

An active and diverse community of microorganisms is commonly found in deep subsurface crystalline and sedimentary rock environments (Ghiorse and Wilson 1988; Balkwill 1989; Pedersen and Ekendahl 1992a, b; Baker et al. 2003; Newby et al. 2004). In fact, the abundance of organisms (~6.1×10^{30} cells) in the subsurface, or geosphere, has been estimated to be approximately the same as the photosynthetically derived green plant biomass present on the earth's surface (Whitman et al. 1998; Amend and Teske 2005; Harvey et al. 2007). The diversity of microorganisms is now known to be considerably greater than previously thought, providing a possible explanation for the range of biological reactions observed within the geosphere.

Factors influencing numbers, diversity, and activity of subsurface microbes, such as pore size, water activity, temperature, pH, available carbon sources, hydrostatic pressures, and electron acceptors often differ notably from those in the biosphere surface. The prevalent geochemistry largely dictates the physiological potential of the resident microbes, as well as the biogeochemical reactions that they catalyze (Fredrickson and Balkwill 2006). In turn, the response of microbes to their physical-chemical environment will determine how they affect their surrounding macro- and micro-environments. When subsurface environments are disturbed through anthropogenic activities, conditions are changed from the original state: the atmosphere becomes more oxidizing, pressures are reduced, and pore space becomes less restrictive. Overall, these generally trend toward becoming more permissive for microbial growth. Any anthropogenic disturbances would also introduce microorganisms previously considered foreign. Even if the host rock originally contained a relatively inactive and sparse indigenous microbial population, a variety of allochthonous microorganisms with a range of metabolic capabilities could come to inhabit a disturbed underground system.

Internationally, many countries are actively working toward the siting and construction of deep geological repositories (DGR) for used nuclear fuel. The DGR concept provides an invaluable opportunity to evaluate the evolution of subsurface conditions from "disturbance" back to original state. Within a newly constructed DGR, located at a depth of 500 m, conditions would be warm, oxidizing, and dry. With sufficient time, these conditions would gradually revert to the original state of the surrounding geology. Data capturing these changes can provide insight into how microbes and their metabolic activity may change over time with potential effects on the engineered barrier system (EBS) that serves to isolate the used fuel container and the used fuel itself. The widespread support for the development of underground facilities as a means to ensure safe, long-term storage of increasing stockpiles of nuclear waste underscores the pressing need to learn more about the impacts of microbial activity on the performance of such facilities over the long term.

7.2 Microbial Diversity in the Deep Subsurface

Interest in subsurface microbiology has increased over the last few decades, originally through culture-dependent approaches, and more recently through culture-independent analyses that have revealed new information about the diversity and range of metabolic activities in this poorly explored environment (Ghiorse and Wilson 1988; Balkwill 1989; Fredrickson and Balkwill 2006; Biddle et al. 2008).

Microorganisms living in geological formations near the surface are influenced by the activity of higher organisms and their products, as well as by above ground climatic conditions that select for generalist microbes capable of responding to environmental change. Under conditions of increasing isolation, the physical and chemical properties of the geologic matrix tend to be more stable, such that microorganisms found deeper in the subsurface tend to be more selective in terms of their metabolism. In particular, the supply of carbon and energy sources has the potential to be growth limiting due to the absence of primary photosynthetic production. Similarly, oxygen becomes increasingly unavailable with depth, giving rise to a corresponding increasing dependence on alternative (oxygen-independent) metabolic strategies (i.e., anaerobic respiration, fermentation, and chemoautotrophy).

Concentrations of microorganisms found in different deep subsurface environments appear to vary with location, ranging from none detected in high salt, low-permeability rock formations (e.g., 750 m salt deposits at Asse, Germany) to diverse and abundant microbial communities living in fractures (West and McKinley 2002). In the deep subsurface, microorganisms may exist as free-living individuals, as cell aggregates in liquid suspension, or as surface-attached colonies called biofilms (Costerton et al. 1978, 1995). Due to their small size and simple, often specialized, metabolism, many microbes have the potential to survive as free-living individuals within the small matrices of subsurface environments, as evidenced by numerous reports of various members of Bacteria (Gram-positive and Gram-negative bacteria, DeFlaun et al. 2007; Rastogi et al. 2009) and Archaea recovered from subsurface samples.

Yeast and fungi have also been detected in subsurface environments and the existence of a food chain that includes higher level bacterivorous predators is not beyond the realm of possibility, as gene sequences from flagellated and ciliated protozoa have been recovered from the deep subsurface (Pedersen 1999a, b). More recently, Borgonie et al. (2011) reported the discovery of a species of the phylum Nematoda from fracture water 0.9–3.6 km deep in South African mines. Interestingly, the nematodes were not detected in the mines' service waters, but in palaeometeoric fracture water with an age of 3000–12,000 years based on carbon-14 data. The authors suggested that these subsurface nematodes might control microbial population density by grazing on biofilms on fracture surfaces. They also noted that the nematodes were able to enter anabiosis (a state of suspended animation) for extended periods and metabolize aerobically when oxygen partial pressures were as low as 0.4 kPa. Even though these larger organisms are confined to fractures and fissures, their discovery demonstrates that the metazoan biosphere reaches much deeper than previously recognized and, more generally, that deep ecosystems are more complex than formerly supposed. Although many reports suggest that the complexity of subsurface communities rivals that of surface environments, this clearly is not the case for deep subsurface consolidated habitats, especially those where oxygen is not readily available. Under anoxic conditions, the diversity of organisms is much lower and their mode of survival more specialized: i.e., involving (chemolitho-) autotrophy and anaerobic respiration, which tends to revolve around hydrogen, bicarbonate, methane, and acetate. Further, bacteriophages, or bacterial viruses, have been found in deep granitic groundwater in Äspö, Sweden (Kyle et al. 2008). Up to 10^7 virus-like particles were detected per milliliter and the authors suggested that these agents could control microbial populations in the deep subsurface.

As discussed further in Sect. 7.4, the excavation of a repository would perturb the natural system given the likely introduction of microorganisms associated with various construction materials and with worker activities (e.g., species from workers, surface dust and airborne/aerosolized microorganisms, machinery), as well as physical and chemical effects. There would also most certainly be long-term equilibration of the installation with its surrounding environment: biologically, chemically, and physically. It is expected that with the passing of sufficient time after emplacement of used nuclear fuel, the repository would eventually return to the natural conditions where low potentials for microbial growth and activity dominate. As such, characterization of the existing microbial community and its relationship to the hydrogeochemistry of the system is a component of DGR site investigations.

7.2.1 Microbial Growth, Activity, and Survival

Microbial cells are capable of exponential growth. A typical first-order exponential growth equation (Eq. 7.1) can accordingly predict the change in the number of microorganisms over a given period of time (t), providing that the initial number (N_o) of bacteria is known as well as the organism's specific growth rate (μ).

$$N = N_o e^{\mu t} \qquad (7.1)$$

Microbial growth in any habitat is dependent on nutrient availability and is usually controlled by the concentration of the most limiting factor (Kieft and Phelps 1997). Limiting factors could be major macronutrients (e.g., carbon, nitrogen, or phosphorus), but could also be physical and chemical conditions such as osmotic pressure, temperature, or pH. In general, microbial growth rates in most deep subsurface environments have been estimated as quite low, more than 1000s of times lower than rates observed for surface environments (Thorn and Ventullo

1988), with doubling times on the order of centuries or longer (Chapelle and Lovley 1990; Phelps et al. 1994; Fredrickson and Onstott 1996).

It has been speculated that in some deep subsurface environments, microbial metabolism would primarily be directed toward cell maintenance rather than growth (Stevens 1997). Considering the in situ growth rate potential for microbes in low-nutrient, subsurface habitats, the activity and reproductive potential for microbes is expected to be low in the host rocks being considered for DGRs. Within the context of the subsurface, the growth, activity, and survival of microorganisms therefore take on the following specific and concise meanings:

- Growth—indicated by the increase in the number of microbial cells.
- Activity—reflected by the outcome of metabolic processes of living cells.
- Survival—the prolonged persistence of cells in a non-growing state, where with the return of appropriate conditions, cells could return to a growing state.

Figure 7.1 is a generalized representation of the continuum between cells that are alive and dead, as described by Davey (2011), where the exact "point of no return" between alive and dead is difficult to determine. In most laboratory studies, or in clinical or industrial settings where microbial presence is determined, the emphasis is on obtaining data in a relatively short time, seldom longer than a few weeks. Because the microorganisms evaluated in these instances typically would be those which have short generation times, such timescales are realistic; in fact, there are sustained efforts to develop more rapid analyses. There has been relatively little discussion, however, about the effectiveness of commonly used microbiological techniques to account for the remarkably slow rates of microbial growth in the subsurface environment due to factors such as low concentrations of nutrients.

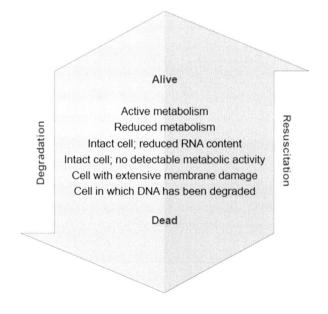

Fig. 7.1 The route from live to dead for a microbial cell, adapted from Davey (2011). Light end of the *arrow* is where the process is less likely to occur

Fig. 7.2 Metabolic state and estimated community doubling times for cells in various environments

	Cell Metabolic State	Typical Environment	Estimated Community Doubling Time (Seconds)
Extended Starvation	Active Growth	Fermenters, Lake Sediments	$10^3 - 10^6$
	Survival	Deep Sea Sediments, Subsurface Aquifers	$10^6 - 10^9$
	Preservation	Confined Clays, Rock Formations	$10^{10} - 10^{13}$

Under these circumstances, the relative position of the perceived point of no return is likely different from that generally applied in everyday (high nutrient, diagnostic) microbiology. For example, cells in consolidated subsurface environments enter a state of preservation in which the calculated doubling time, based on predicted rates of flux of growth requirements, can be on the order of centuries, as summarized in Fig. 7.2.

The increasing number of successful resuscitations of ancient microbial cells sampled from the deep subsurface is a clear indication that a time frame for the point of no return from alive to dead cannot be definitively assigned for microbial cells; rather, environmental conditions have a strong influence on microbial persistence. There is a wide range of conditions under which microorganisms are uncultivable, and there is, thus, a need to expand on the potential offered by genomic and proteomic techniques to delineate the viability of cells from these extreme environments. As with efforts aimed at deriving actual in situ metabolic rates in laboratory columns, there is a justifiable concern that these methods may overestimate rates. This further emphasizes the importance of applying appropriate methods when performing subsurface microbiology and of interacting with specialists from other disciplines.

7.2.1.1 Microbial Energetics

A recurring theme in subsurface microbiology is the source of energy available for sustained microbial growth. A link between the characteristics of a subsurface environment and the presence of specific microorganisms has already been demonstrated (Fredrickson and Balkwill 2006; Lovley 2006). Where subsurface microbial activity is high, an active or recent linkage of the subsurface environment with surface terrestrial processes can usually be shown. For example, subsurface microbial communities may be linked with the aerobic or anaerobic breakdown of dissolved or particulate photosynthetically derived organic matter that either

(1) percolates via surface water into subsurface habitats or (2) was deposited via sedimentary processes (e.g., kerogen, petroliferous deposits, lignin, etc.).

Deeper in the geosphere, chemoautotrophic systems may also become established independent of photosynthesis (Stevens 1997). For such a chemoautotrophic microbial community to develop and be sustained, a sufficient source of energy must be available. Relatively new evidence points to the existence of subsurface ecosystems based on the geochemical production of hydrogen via radiolysis. Primary production supporting this system would be provided by chemoautotrophic homoacetogens, as well as acetoclastic and hydrogenotrophic methanogens, and subsequent heterotrophic organisms based on the autotrophic production of organic carbon (Pedersen 1997, 1999a, b; Stevens and McKinley 1995; Fredrickson and Balkwill 2006). This suggestion is not improbable because conditions in the deep subsurface (high temperatures, reducing gases, etc.) resemble conditions that persisted when life first developed on earth. Still, the energy requirements of microbial cells may be considerable in comparison to the small energy yields and fluxes that dominate in much of the deep subsurface. These differences in energy requirements, along with the maximum potential rates of energy supply, impose a significant constraint on the habitability of consolidated materials, and, as indicated by Hoehler (2004), this important determinant of the presence, distribution, and productivity of life in photosynthesis-independent subsurface environments has seldom been tested. Thus, in the absence of the comparatively high-energy light- and oxygen-based metabolic processes that are the primary drivers of the earth's surface biosphere, life in the deep subsurface is relegated to low-energy anaerobic processes dependent primarily on the mineral geochemistry of the host environment. Jones and Lineweaver (2010) described terrestrial waters that are uninhabited due to limitations in nutrients and energy, and, among others, restrictions on pore space. Whether certain habitats are indeed uninhabited remains a topic for discussion, given our inability to detect very low numbers of organisms with stringent growth requirements.

Hoehler (2004) described the magnitude of the energy required to sustain basic biochemical integrity and function in terms of two concepts: biological energy quantum (BEQ) and maintenance energy (ME). With respect to BEQ, for free energy to be usefully harnessed it must be available at levels equal to or larger than the specific required finite minimum energy needed to drive the synthesis of adenosine triphosphate (ATP) from adenosine diphosphate (ADP): i.e., the minimum free energy that must be available in a given environment to sustain life. Importantly, this free energy change ($\Delta G_{ADP \rightarrow ATP}$) is sensitive to prevailing physical and chemical conditions, and the parameters with specific relevance to the environment of interest in the deep subsurface are (1) magnesium concentration, (2) ionic strength, and (3) pH. Many subsurface environments have high magnesium content, which has a strong affinity for ATP, ADP, and phosphate, and the $\Delta G_{ADP \rightarrow ATP}$ increases with ionic strength. Similarly, $\Delta G_{ADP \rightarrow ATP}$ becomes more positive with increases in pH. It should be noted that in contrast to BEQ requirements described above, Jackson and McInerney (2002) indicated that some fermentative bacteria do have the ability to couple substrate metabolism directly to

ATP synthesis to obtain energy for growth via reactions in which the change in free energy is less than what is needed for ATP synthesis. However, with reference to earlier observations, and based on a combination of theoretical and experimental observations, Hoehler (2004) suggested that the BEQ required for actively growing populations would be twice as much as that required for static populations in a maintenance mode. Further, the emergence and proliferation of a population are prerequisites to its survival in maintenance mode. Given the stability of the deep subsurface geosphere, however, it is not clear how such proliferation could occur in stages other than deposition and early consolidation or some major geologic disturbance.

To preserve life, organisms require a minimum rate of energy intake to maintain molecular and cellular function in addition to integrity. As discussed above, energy is used by microorganisms in one of three broad categories: growth, maintenance of basic metabolism without growth (i.e., metabolic activity), or survival that involves little or no metabolic activity and where energy is primarily used to preserve the integrity of amino acids and nucleic acids (that may be operative in consolidated materials). Ignoring the important constraints on energy flux in the deep subsurface geosphere and failing to carefully consider maintenance requirements of microbial cells in the context of this important control of energy availability could result in a significant overestimation of related microbial activity. Indeed, Morita (1999) suggested that when active bacteria are reported in ancient materials, it is possible that the suspended animation state has been broken either by giving the cells a substrate or electron acceptor or by creating different conditions from the in situ environment or from some other perturbation of the sample.

This concern has been discussed in a number of seminal publications (e.g., Chapelle and Lovley 1990; Phelps et al. 1994; Kieft and Phelps 1997; Krumholz 1998). In brief, the observations are as follows: (1) that the use of inappropriate methodology can lead to a significant difference between potential activity and actual in situ activity of microbes; (2) that laboratory estimates of microbial activity are often orders of magnitude higher than actual in situ rates, potentially by factors up to 10^6 over what geochemical models and knowledge of groundwater flows would substantiate; and (3) that calculated rates of microbial activity are typically averaged over time and distance, while the actual rates of activity are temporally and spatially heterogeneous. Considering the extremely slow rates of energy flux and the levels required for growth, it is probable that the slow rates of metabolism in the deep subsurface will be directed primarily toward *survival* rather than *growth*. The adaptive ability of microorganisms, when stressed by one set of conditions (e.g., starvation) to initiate cross-protection against other stressors, such as osmotic stress, heat stress, or temperature extremes, generally contributes to survival success to the degree that microorganisms in lithotrophic ecosystems have apparently developed mechanisms for survival and extended periods of anabiosis (Krumholz 1998). Another theory now emerging is that individual microorganisms can survive in subsurface environments for millennia while carrying out cellular metabolism at an extremely slow rate (Morita 1999).

The selective pressures on organisms or groups of organisms, as well as how they will survive and potentially proliferate, should also be taken into account (Davey 2011). Baas-Becking's (1934) eloquently stated "everything is everywhere, but the environment selects." For example, organisms that require complex organic materials for their carbon and energy requirements (heterotrophic organisms) cannot grow in the absence of such compounds. However, heterotrophy may occur aerobically (using oxygen as terminal electron acceptor) or anaerobically (by facultative anaerobes using a range of alternate electron acceptors of varying oxidative states). Under saturated conditions, the deep subsurface environment is typically reducing, thus anaerobic processes predominate in these zones (Lovley 2006). Anaerobic chemotrophs can use various chemicals as sources of energy (generally, redox reactions where electrons are shuttled between donor molecules to microbial electron acceptors, with liberation of potential energy) and therefore have the potential for being highly active, depending on the chemistry of the system. The type of chemotrophy that occurs is dependent on the microbes, with some organisms requiring pre-formed organic materials as a source of energy (chemoorganotrophs or chemoheterotrophs), and others requiring or being able to use inorganic materials (chemolithotrophs). In the case of chemolithoautotrophic organisms (e.g., *Acidithiobacillus ferrooxidans* previously known as *Thiobacillus ferrooxidans*, and *Nitrosonomas spp.*), the source of reducing energy comes from the geosphere and includes reduced inorganic electron donors such as H_2, NH_3, NO_2^-, Fe^{2+}, S^o, and H_2S and the utilization of relatively oxidized compounds as electron acceptors (e.g., O_2, NO_3^-, SO_4^{2-}).

To sustain a photosynthesis-independent ecosystem, a primary reductant is necessary; in the subsurface, this reductant could be geologically evolved hydrogen. Stevens and McKinley (1995) offered proof for the existence of such a system and presented evidence where autotrophic organisms outnumbered heterotrophs, and where stable carbon isotope analysis indicated that autotrophic methanogenesis was linked to the disappearance of inorganic carbon. They further demonstrated that production of H_2 from the reaction of basaltic rock with anaerobic water supported microbial growth in laboratory experiments. Pedersen (1999a, b) offered support for a hydrogen-driven microbial ecosystem, as hydrogen (termed by Pedersen as "geogas") can be generated from basaltic rocks by hydrolytic reactions of naturally occurring radioisotopes with water. For example, the decay of alpha-emitting particles, such as radon and radium, causes water hydrolysis, yielding hydrogen. Pedersen's model explains how the hydrogen thus produced could support growth of methanogens and acetogens, as described in Eqs. (7.2 and 7.3), respectively:

$$4H_2 + CO_2 \rightarrow CH_4 + 2H_2O \tag{7.2}$$

$$4H_2 + 2CO_2 \rightarrow CH_3COOH + 2H_2O \tag{7.3}$$

The products of these reactions would then provide the reduced carbon necessary to support other microorganisms in the H_2-driven ecosystem: e.g., production of methane and CO_2 from acetate; production of H_2S, CO_2, and water from acetate and sulfate; and production of water, ferrous iron, and CO_2 from acetate and ferric iron (and acid), as respectively, shown in Eqs. (7.4, 7.5, and 7.6):

$$CH_3COOH \rightarrow CH_4 + CO_2 \qquad (7.4)$$

$$CH_3COOH + SO_4^{2-} + 2H^+ \rightarrow 2CO_2 + 2H_2S + 2H_2O \qquad (7.5)$$

$$CH_3COOH + 8FeOOH + 16H^+ \rightarrow 2CO_2 + 8Fe^{2+} + 14H_2O \qquad (7.6)$$

Fermentation reactions are also common under anaerobic conditions, but rely on the presence of organic compounds that function as either electron donors or electron acceptors. Typical fermentative end products include CO_2, H_2, acetate, propionate, butyrate, ethanol, lactate, and formate, with typical substrates including cellulosic and other carbon-rich materials (e.g., carbohydrates). Fermentation often occurs in syntrophic association with organisms that use H_2 and CO_2 for methane generation (in the rumen, for example), as well as those acetoclastic organisms that strictly use acetate for methanogenesis. Thus, where sufficient organic material is found, these organisms will play a functional metabolic role.

In summary, the deep subsurface imposes significant physical (e.g., extremely small void space) and chemical (e.g., low available energy and energy flux) constraints on microorganisms, and as discussed by McCollom and Amend (2005), a rigorous accounting of controls, such as energy flow, is needed to improve our understanding of the potential biological productivity of chemolithoautotrophic communities and to better describe limits to habitability in subsurface environments. Such information is needed to derive more accurate estimates of microbial contribution to geochemical evolution.

7.2.1.2 Spores, Cell Dormancy, and Death

Microorganisms are known to survive under extreme conditions, including cells entering either an inactive state (also known as moribund, latency, dormancy, and cryptostatic) or a resting state (sporulation) involving chemical-morphological adaptations (Davey 2011). If such cells are present in the surrounding matrix of a DGR, the possibility that bacteria may survive must be considered. Repositories positioned in clay deposits create conditions likely conducive for cell entry into dormancy because of accumulation of endogenous waste and low availability of water. The best known "resting-state" mechanism is the formation of an endospore (spore): that is, heat- and desiccation-resistant, non-vegetative structures formed by members of genera such as *Bacillus* and *Clostridia* that have undergone stress or change in their growth conditions. Spores have a complex, layered wall containing high amounts of calcium, dipicolinic acid, and peptidoglycan, as well as a full copy of the cell's genome, and are much more capable of surviving stress conditions than are their vegetative counterparts. Accordingly, spores undergo extended periods of dormancy until conditions are once again appropriate for vegetative growth.

Reportedly, spores have been recovered from archeological samples as old as 15,000–40,000 years (Grant et al. 2000). Reports of million year-old spores revived from insects entrapped in amber (Cano and Borucki 1995; Greenblatt et al. 1999), though highly controversial, have also been published.

7.2.2 Biofilms

A generic description of a biofilm is the aggregation of microbial cells and their EPS (extracellular polymeric substances) on a surface. Numerous research articles and reviews describe biofilm formation as a sequence of events. Most of these descriptions have the common features described by Busscher and Van der Mei (2006): microbial adhesion to surfaces is the onset of biofilm development, which is typically preceded by formation of a conditioning film of macromolecular components that enables initial microbial adhesion. The initial stages of biofilm formation are typically described as reversible, but formation becomes less reversible once the cells anchor themselves through EPS matrix production. Within this matrix, the cells start to grow and form microcolonies that ultimately develop into a mature biofilm from which viable cells are released back to the environment. Microbial transport to the substratum is typically enabled by microbial motility, as well as by mass transfer processes such as convection, diffusion, or sedimentation.

In the deep subsurface, biofilm research to date has primarily focused on groundwater within fractures of crystalline rock. Anderson et al. (2007) pointed out that in the case of nonporous crystalline rock, an influencing factor is groundwater flow through fractures, which could support biological growth on fracture walls. Earlier studies examining the potential for biofilm formation in granitic systems have employed recirculating flow cells and biofilm samplers connected to boreholes with outlet valves set to maintain pressures close to those found in the boreholes (Vandergraaf et al. 1997; Stroes-Gascoyne et al. 2000; Anderson et al. 2006). In their study, Anderson et al. (2006) described observations on biofilms cultivated in flow cells connected in a closed loop with a borehole that intersected a hydraulically conductive fracture zone. They showed significant biofilm formation (up to 1.86×10^4 cells/mm^2) accompanied with the production of EPS. The EPS had functional groups involved in adsorption, which led the authors to suggest that biosorption was mediated by functional groups on the microbial cell wall and the EPS matrix, without the need for energy from metabolically active cells. A subsequent publication by Jägevall et al. (2011) reported significant numbers of microorganisms on fracture surfaces of a natural hard rock aquifer, as measured by quantitative polymerase chain reaction analyses and clone libraries.

An interesting question is whether the advantages frequently attributed to the biofilm mode of growth will be of relevance in view of the very scarce nutrient supply and limited liquid flow expected in low-permeability host rock environments under consideration for DGRs. For example, increased EPS production by biofilm cells has been described as a mechanism to expand microbial habitat range by trapping nutrients and preventing desiccation; however, the energy requirements for EPS production are poorly described, especially in environments with limited water supply. Among the few articles on energy expenditure related to EPS production, Harder and Dijkhuizen (1983) calculated that in aqueous environments bacteria may invest more than 70 % of their carbon and energy for production of EPS. Delineating the survival mechanisms of the surface-associated cells in oligotrophic, dense matrices could indeed reveal information relevant to biofilms as a survival mechanism, which may provide more credible information to our understanding of survival in the deep subsurface environment.

7.3 Deep Geological Repositories

When solid used nuclear fuel bundles are removed from reactors, they are highly radioactive. Although the radioactivity of used nuclear fuel will decrease with time, its chemical toxicity will persist such that it will remain a potential health risk essentially indefinitely. Sources of radiation include reaction products that form upon splitting of the fissionable U^{235} uranium atoms: radioactive isotopes, including iodine, molybdenum, cesium, technitium, palladium, etc., and those elements that are formed when the uranium atoms that absorb bombarded neutrons do not split—elements with atomic numbers of 93 or greater such as neptunium and plutonium—known as transuranic actinides. The radioactive material is unstable and undergoes decay or breakdown over time, during which radiation is emitted. Some of this radiation is gamma radiation, which is highly energetic and penetrating, like X-rays, and thus damaging to biological (living) material due to direct and indirect (formation of damaging radicals) effects. Those fission products with very short half-lives tend to rapidly emit large amounts of radiation energy, and thus are typically very hazardous, but do not require long-term management. However, due to the presence of significant amounts of radionuclides with long half-lives in used nuclear fuel, safety assessments for deep geological repositories consider a time frame on the order of one million years.

On initial emplacement of used nuclear fuel, the repository environment would experience a sharp increase in temperature because of the release of energy from fission products with short half-lives. Based on known decay rates and proposed density of used nuclear materials, container surface temperatures of 100 °C (Maak et al. 2010), or possibly higher in some designs (120 °C; Jolley et al. 2003), would occur approximately 10 years after placement and then begin to drop. Peak temperatures and time frames will be design and site specific. However, temperatures of the host rock in a DGR would likely range from 55 to 75 °C, and these elevated temperatures would persist for hundreds to a few thousands of years.

Nuclear waste management agencies have considered a number of concepts for the disposal of used nuclear fuel that employ a system of engineered barriers: clay, seals, and metal containers. For example, the design concept of SKB and POSIVA, the Swedish and Finnish nuclear waste management agencies, respectively, uses copper-iron containers, consisting of an outer copper shell and a cast iron insert, for encapsulation of used nuclear fuel prior to emplacement in their deep geological repository. Similarly, the Canadian program design concept employs a steel inner vessel surrounded by copper. In contrast, the engineered barrier system design of NAGRA, the Swiss nuclear waste management agency, presently uses a steel container within a sedimentary host rock design. Corrosion of DGR-relevant materials has undergone considerable study (King 2007) and shows the potential for corrosion will change as repository conditions evolve from relatively warm, dry, and aerobic (in the initial phase) to cooler, wet, and anaerobic (over the longer term). Despite different container designs, both concepts surround the used nuclear fuel containers with bentonite clay to protect the containers and limit groundwater access and potential transport of microbial metabolites, such as sulfide, via groundwater, to the container surfaces (Bennett and Gens 2008; Pedersen 1996, 1999b; Smart et al. 2011). Bennett and Gens (2008) have reviewed repository concepts and EBS designs for high-level waste and used nuclear fuel disposal in European countries.

There are several emplacement options being considered internationally for used fuel containers, including vertical and horizontal boreholes, and horizontal tunnels (Maak et al. 2010). Used nuclear fuel containers would be placed in rooms or tunnels constructed from main access tunnel leads or in boreholes. A clay-based buffer (100 % compacted bentonite rings) would surround containers. The facility would be tailored to the geological matrix in which the DGR is situated, with horizontal and vertical boreholes generally adopted for crystalline rock repositories (as well as hard sedimentary rock), and the horizontal tunnel option preferred for a soft sedimentary rock repository (e.g., analogous to the Opalinus Clay Formation in Switzerland).

The deep geological repository concept for used nuclear fuel involves multiple barriers to ensure long-term safety of humans and the environment. Figure 7.3 shows an example of a multibarrier system conceptual design. Considerable research and development has been undertaken to investigate the potential for microbial activity within each region of the repository. Specifically, work has been undertaken to (1) characterize the microbial communities present in low-permeability, stable geologic formations under consideration as host rocks for a DGR; (2) minimize microbiologically influenced corrosion of the used fuel containers; (3) inhibit microbial growth and activity in the high-density clay barrier that surrounds the used nuclear fuel containers; and (4) create low-permeability seals and fill to backfill the repository. Hence, the multiple protective barriers in the DGR concept include both naturally occurring, low-permeability, stable geologic formations and the engineered barrier system (EBS), which consists of the used nuclear fuel containers and surrounding clay buffer, seal, and backfill materials.

Fig. 7.3 Regions in a repository where microbial activity is evaluated

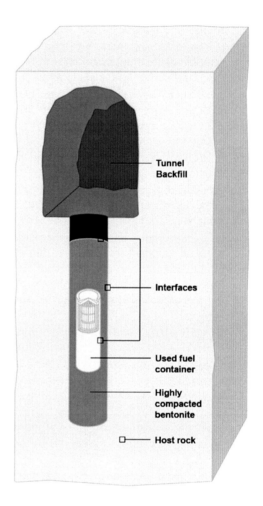

7.3.1 The Host Rock Environment

The host rock is the natural barrier component of the repository. International siting activities are focused on finding low-permeability host rocks that can isolate and contain used nuclear fuel. The type of host rock influences whether in-floor boreholes, horizontal boreholes, or tunnels are most suitable as a waste container emplacement method (Maak et al. 2010). The character of the host rock also determines conditions for microbial growth and represents a possible source of microorganisms. Thus, the potential for microorganisms to affect the performance of the EBS located in either host rock type warrants discussion.

In design concepts where the used nuclear fuel would be deposited within fractured igneous/crystalline rock, more stringent performance criteria of the EBS (particularly with respect to container life expectancy) would be required versus emplacements in intact rock. Groundwater movement through the fractured rock matrix could transport viable microbes and nutrients into and out of the proximity of the EBS where they could exert effects, including (1) affecting waste container integrity (e.g., via microbiologically influenced corrosion), (2) altering radionuclide mobility (in the event of a container failure) by complexation with various microbial components, such as biofilms and their extracellular products (Pedersen and Albinsson 1992), or (3) alteration of the redox state, thereby influencing radionuclide solubility (Pedersen 1996; Anderson and Lovley 2002). These effects could occur in the DGR (e.g., between the buffer material or backfill and the host rock) or within the cracks and fissures in the surrounding geologic matrix.

Repository concepts where the host rock is sedimentary rock or clay typically place greater emphasis on the host rock as a low-permeability barrier against radionuclide movement, and thus the required container lifetimes are relatively shorter. Such sedimentary rock typically does not support extensive microbial growth; evidence to support the existence of bacteria in these formations that would have been emplaced millions of years before the present time is still a matter of conjecture. For example, a preliminary study of microbial communities in potential Canadian sedimentary host rock types with low water activity suggested low to negligible microbial biomass (Stroes-Gascoyne and Hamon 2008a), which in turn suggests that a large contribution of microorganisms from sedimentary host rock types to the EBS is not likely. Others predicted that the small pore space (e.g., <0.02 μm diameter) and low water availability within clay matrices would significantly restrict the number and activity of microorganisms (Stroes-Gascoyne et al. 2007a, 2010b; Poulain et al. 2008). In such an environment, movement of nutrients to, and metabolic wastes from, microbial cells would similarly be hindered, causing nutrient limitation and end-product inhibition, and resulting in a low frequency of cellular division. Small pore throats would further prevent any significant bacterial translocation. Possible mechanisms by which bacteria would survive under the above conditions are as yet unknown. It follows that the potential impact of these scarcely distributed microorganisms within a clay formation would likely be negligible relative to the potential effects of introduced organisms at the time of DGR construction.

The hydrogeochemistry associated with the deep subsurface is site specific; however, there are identifiable trends. Overall, the potential influence of the chemical and physical conditions, and related microbial activity, in the host rock (as discussed by Sherwood Lollar 2011) on the biotic reactions in the repository appears to be minimal, which indeed is the objective of EBS (i.e., to isolate these two zones). Nevertheless, the numbers, distribution, and activities of microorganisms in the host rock provide a realistic scenario of the endpoint toward which these characteristics may evolve in the repository.

Overall, current knowledge of microbial distribution and activity in previously undisturbed deep subsurface environments that were subsequently disturbed, for example, by mining activities, provides indications as to what may happen in a deep geological repository. It is noteworthy that gradients of water activity, pressure, pore space, and temperature will occur locally within the EBS as well as over the larger repository (e.g., it is expected that temperatures in the center of the repository will be higher than those attained at the periphery of the installation). Realistically, we can assume that there will be (1) indigenous microorganisms that may be stimulated by the disturbance and (2) microorganisms introduced during EBS installation. Given this, the impact of microbes on repository performance requires evaluation. Specifically, what are the potential outcomes with regard to microbial behavior and how much of an effect could microorganisms have considering the in situ physical-chemical properties and geochemical evolution? Over what timescale are microorganisms active before transforming from active metabolism—to survival—to preservation, or the reverse scenario? Establishing the boundaries of these behaviors and their effects remains a goal of research on potential microbial outcomes on a DGR.

7.3.2 The Engineered Barrier System

Many countries have adopted clay-based materials as essential components of various EBS designs in both sedimentary and crystalline rock. Bentonite is expected to serve as a buffer between the used nuclear fuel containers and host rock, where it will influence hydraulic, mechanical, thermal, and chemical processes, as well as radionuclide diffusion (Stroes-Gascoyne 2005). Most designs will have used nuclear fuel containers encased in 100 % high-density compacted bentonite when placed within a hole or a tunnel within the host rock. Subsequently, the bentonite clay will swell following saturation by groundwater. Microorganisms indigenous to the water and bentonite will be present in the zones spanning from the container surface to the surrounding host rock. Redox conditions of the infiltrating groundwater may be oxic initially, with oxygen being introduced in the repository atmosphere during installation, but will return to anoxic conditions due to minor container corrosion, mineral oxidation, or microbial respiration. However, once all the oxygen has been consumed, the next most oxidized, relatively abundant alternate electron acceptor (based on the chemistry of the geologic system) would then be used (e.g., NO_3^-). After all the NO_3^- has been consumed, the electron acceptor would switch to NO_2^- followed by species such as manganese (IV), ferric iron, sulfate, and lastly, CO_2. Organisms generally use electron acceptors that yield the greatest amount of energy (Amend and Teske 2005) providing that these acceptors are sufficiently abundant.

7.3.2.1 Metal Containers

It is highly unlikely that biological processes will have an impact on the used nuclear fuel itself, or the inside of the containers holding the used fuel, due to the combination of high temperature, high radiation, absence of water, and lack of nutrients (Meike and Stroes-Gascoyne 2000). Nuclear waste management agencies around the world have considered a variety of container designs and components as part of their engineered barrier systems. For example, Canada is examining two used fuel container designs. One design, which similar to the 4 m long, cylindrical Swedish and Finnish designs, consists of a 10 cm thick carbon-steel, structural inner vessel within an outer copper corrosion-resistant shell approximately 5 cm thick. The other design involves a smaller container, approximately 2 m in length, consisting of a 4 cm thick carbon-steel vessel directly coated with 3 mm of copper. Other international containers, such as the Swiss and French designs, involve carbon-steel only. All of the containers, when located within the anaerobic environment of the deep geological repository, are designed to withstand corrosion from a variety of mechanisms, including microbiologically influenced corrosion (MIC), as discussed further in Sect. 7.4.2.3.

7.3.2.2 Bentonite

Since the activity of microorganisms within the repository has the potential to cause microbiologically influenced corrosion, and may also contribute to the generation of gases, much attention has been focused on the ability of the clay buffer to inhibit microbes and their activity. Microorganisms can grow over a large range of water activities (a_w of 0.75–0.999), but most favor a_w values of 0.98 or above (corresponding to a salinity of water of ~3.6 % or less) (Jay et al. 2005). Studies have shown that pure bentonite that has been compacted to 2 Mg/m^3 (corresponding to a dry density of about 1.6 Mg/m^3) and is water saturated (approximately 26 % v/w) has an a_w of 0.96, which is sufficient to inhibit the activity of the large majority of bacteria likely to be problematic in a repository (Stroes-Gascoyne et al. 2011). Further, water-expanded bentonite clay physically restricts the movement of water and whether emplaced or naturally occurring, hydrated clay matrices are known to result in a low-permeability environment, with hydraulic conductivities in the range of 10^{-12}–10^{-14} m/s (Pusch and Weston 2003). This would serve to directly limit the diffusion of radionuclides, as well as that of nutrients needed by microorganisms for metabolic activity and growth. Given that the average pore size of the clay matrix is on the order of hundreds of times smaller than a bacterial cell, cell growth and movement would also be physically restricted.

In the current Canadian EBS reference design, highly compacted 100 % bentonite buffer material surrounding the waste containers is proposed to prevent or minimize potential negative consequences of microbial activity, such as damage to the container or barrier integrity. Early Canadian EBS design options were based on a blended mix of sand and clay; however, findings reported by Stroes-Gascoyne (2010) related to microbial behavior recommended the use of 100 % bentonite. For the highly compacted 100 % bentonite to inhibit the activity of bacteria and germination of bacterial spores, it has been established that the bentonite would need to meet one or both the following criteria (Stroes-Gascoyne et al. 2006, 2007a, b): (1) have a water activity of less than or equal to 0.96, resulting from either a bentonite dry density of at least 1.6 Mg/m^3 or a porewater salinity of greater than 60 g NaCl/L; or (2) yield a swelling pressure of at least 2 MPa.

7.3.2.3 Seals and Tunnel Backfill

Repository seals isolate the disposal room from the repository access shaft and prevent the backfill mass comprised of either clay alone or clay and rock from expanding into the access tunnels. The repository seal enables the swelling clay surrounding the used fuel containers to meet density and water activity specification targets. Overall, clay-based backfill and sealants play a number of important functional roles (Stroes-Gascoyne 2005), including (1) limiting the rate of liquid movement by diffusion, (2) providing mechanical support to the container, thereby protecting it from movement or shifting of the surrounding host rock, (3) retention of radionuclides in the event of container failure, (4) provision of a thermally conductive medium to transmit heat to the surrounding host rock, and (5) limiting numbers, activities, and transport of microorganisms near the container. Options for seals include bulkheads of cement or compacted clay plugs placed at the entrance to the waste repository rooms.

Clay-based sealants have been the focus of intense research as related to EBS performance, including comprehensive research programs by Canada and Sweden over more than two decades. Clay-based seal and backfill materials have been shown to contain indigenous aerobic and anaerobic microflora, including sulfate reducing bacteria (SRB). For example, *Desulfovibrio africanus* has been found in MX-80 bentonite (Masurat et al. 2010a, b). Many studies have focused on showing how buffer and backfill parameters are likely to affect survival and activity of these and other organisms under relevant EBS conditions (see e.g., Stroes-Gascoyne 2010; Stroes-Gascoyne et al. 1997b, 2007a,b, 2010b; Pedersen 1993a; Pedersen et al. 2000a,b).

Dense backfill that consists of crushed rock, non-swelling clay, and bentonite (dry bulk density of 2.1 Mg/m^3) would be used to fill remaining excavation spaces around the bentonite buffer. Pellets consisting of 50:50 bentonite/crushed rock or 100 % bentonite will be blown into areas where the dense backfill cannot be emplaced. Backfill material serves to (1) support the containers and the 100 % compacted bentonite buffer, (2) aid in creation of anaerobic conditions, and (3) slow saturation of the repository. Upon saturation, the hydraulic conductivities of these buffer and backfill would be low, typically 10^{-12}–10^{-14} m/s for 100 % bentonite buffer and 10^{-10}–10^{-11} m/s for backfill (Pusch and Weston 2003). Seals or bulkheads constructed out of concrete or expandable clays (or both) would be used to seal emplacement rooms and repository openings.

High-performance cement (Portland cement) is an option for seals, grouts, tunnel liners, bulkheads, and floors. Repository applications rely on cement primarily for its mechanical support and sealant properties in the repository, rather than the direct containment of used nuclear fuel containers. Concrete forms the backbone of the building industry and its use goes back for centuries. Thus, there is a relatively large body of data available on the susceptibility of concrete to microbial attack. Pedersen (1999b) has questioned whether "relevant microorganisms survive at pH equivalent to that of repository concrete and can they possibly influence repository performance by concrete degrading activities such as acid production." Evidence from surface (e.g., roads, bridges) and underground cement structures (e.g., sewer pipes) has shown that the integrity of concrete over extended timescales can indeed be influenced by microorganisms, particularly the well-known sulfur-oxidizing bacteria such as *Acidithiobacillus thiooxidans* (previously *Thiobacillus thiooxidans*), which produces sulfuric acid under aerobic conditions through the oxidation of reduced sulfate, sulfide, and thiosulfate compounds. Sulfuric acid contributes to the degradation of concrete by dissolving the calcium silicate hydrate and calcium hydroxide cement matrix constituents. Nitric acid, produced by the combined action of nitrosifying and nitrifying bacteria that use inorganic nitrogen compounds (i.e., ammonium, nitrite), may similarly lead to acid-mediated concrete deterioration. It is noteworthy that within a DGR, oxygen would be in finite supply, in sharp contrast to the situation of roads and bridges; thus, once utilized during corrosion, mineral dissolution, and microbial redox reactions, oxygen would no longer be able to contribute to the formation of sulfuric acid.

The detrimental effects of microbial activity on concrete have been described in a number of reviews. However, the emphasis in most literature on the biodeterioration of concrete and other cement-based products relates to conditions conducive to microbial proliferation. In contrast, the repository environment poses a number of challenges to microbial activity, including the extremely low porosity that will exclude all but those with an ultrasmall cell size. It is recognized though that the study of microbes in the deep subsurface and in highly consolidated materials is at a relatively early stage, and there are likely microbes to be discovered in addition to ultrasmall bacteria adapted for persistence in these extreme environments.

Cement has a high pH (~13) and contains relatively low concentrations of carbon-containing nutrients to support microbial growth. Cement compounds for repository applications typically use 0.5–1 % of a superplasticizer (abbreviated SP and typically made of sulfonated naphthalene condensates) which functions to reduce the amount of water required for cement mixing and to improve the workability and strength of the cement for repository applications. These plasticizers are carbon-containing compounds and thus they may be subject to microbial breakdown as a possible carbon source. The study of plasticizer leaching from concrete still remains incomplete, and despite the small quantities of SP added, there is evidence to suggest that some microorganisms use these compounds as a carbon source (Haveman et al. 1996; Stroes-Gascoyne 1997).

Microbial activity can accelerate other processes in the cement components and potentially affect other repository component requirements (e.g., the ability to maintain swelling pressure of the buffer, and limiting water activity in the clay matrix). It is also known that disturbance of an environment (e.g., introduction of air and nutrients) results in a temporary increase in microbial activity; therefore, the period during emplacement and soon after may be critical to the potential impact of microbial activity and may warrant consideration (Wersin et al. 2011).

7.3.2.4 Repository Interfaces

Interfaces and transition zones between the repository components may support increases in microbial diversity and metabolic activity (Fig. 7.3). Such areas pertain to (1) the area between the used fuel container and the highly compacted bentonite, (2) the area between the highly compacted bentonite and the host rock, and (3) areas where the backfill comes into contact with the bentonite buffer and host rock.

There is a compelling body of research (Stroes-Gascoyne 1997; Stroes-Gascoyne and West 1997; Pedersen et al. 2000a, b), indicating that microbial growth and activity is not expected on the container surface or in the bentonite close to the used fuel container, primarily due to heat and radiation, which will inhibit microbes in this area. Furthermore, the low permeability of the bentonite is expected to impede repopulation of the region near the container as long as the clay buffer remains undamaged.

Of more concern is the area between the bentonite and the host rock, between the backfill and bentonite, or between the backfill and host rock. As discussed above, the low water activity and swelling pressure of sealing materials is integral to impeding microbial growth. However, it is possible in some very low-permeability host rocks that a very long time will be needed to saturate the bentonite materials, offering areas where localized microbial activity may occur. However, the generalized biofilm model, sometimes described as a community of cells, is not likely relevant to this environment due to limitations in water and nutrients. Nonetheless, nuclear waste management agencies are investigating the potential for microbial growth and activity in (1) fractures in the host rock due to excavation damage, (2) areas of incomplete sealing at interfaces between different zones, or (3) fissures that form as a result of gas formation or uneven drying in the bentonite buffer.

If biofilms were to form, they could significantly affect subsurface geochemical interactions. However, it is also possible that biofilms located at interfaces between the host rock and the buffer matrix, or the container and buffer, could function to effectively plug these regions and decrease fluid transport. Because biofilm formation can have a significant impact on the porosity and permeability of fractures and porous media (Coombs et al. 2010 and references therein), it is also important to consider the potential impacts of attachment–detachment dynamics on fluid flow. Biofilms can reduce fluid flow by constricting pore throats and increasing tortuosity of pore flow paths and can also alter pH, redox, as well as groundwater and rock chemistry (Coombs et al. 2010). Porosity and permeability can be reduced by microbiologically mediated precipitation, i.e., biomineralization, resulting in possible plugging or cementation of pore spaces (Coombs et al. 2010). However, biofilm growth that covers mineral surfaces can also potentially block access to sorption sites and decrease sorption (Bass et al. 2002). The relationship between planktonic cells (which could sorb and then transport radionuclides) and attached microbial communities in the subsurface is complex and poorly understood (Onstott et al. 2009). Overall, the assessment for the potential for biofilm formation,

7.3.3 Environmental Factors

Tolerance of microorganisms to nearly every example of extreme conditions is well documented. West and McKinley (2002) summarized the specialized microbes, which have evolved within various extreme habitats of relevance to deep geological repositories for used nuclear fuel (Table 7.1), revealing that there is hardly any foreseeable condition under which microorganisms might not be expected to survive. Thus, the potential for resistant microbes to become established within certain regions of a DGR is not only possible but likely. However, the potential for microorganisms to develop a poly-extremophilic phenotype (e.g., to become multiply resistant to heat, pressure, low water activity, and radiation) so as to survive and proliferate in the immediate proximity of the used nuclear fuel containers is considered most improbable. The extreme conditions would occur almost instantaneously on placement of the used nuclear fuel, thus not allowing microorganisms the time necessary for evolutionary or adaptive change.

A growing body of research (Stroes-Gascoyne et al. 2002; Stroes-Gascoyne 2010) suggests that conditions within the EBS will result in a significant reduction in the number of surviving, viable microorganisms or cause cells to enter a dormant, relatively inactive state. The range of conditions expected to exist in the repository environment after emplacement of used nuclear fuel (e.g., after saturation of the excavated zone) would include (1) extreme antimicrobial conditions at or near the surface of used nuclear fuel containers due to the combined effects of radiation (especially with thin-walled containers), heat, and desiccation; (2) low water activity, heat, and restricted pore space within the compacted bentonite clay material; and iii) higher water activity, pore space, and some nutrients (carbon, nitrogen, and oxygen) within the backfill, as well as the EBS-host rock interface regions (Stroes-Gascoyne 2010; Pedersen 2010). For microbes to have any impact on the emplaced barriers or fuel, they will first need to survive long-term exposure to these harsh conditions. The survival and potential activity of bacterial cells will be dependent upon their distance from the waste container and local conditions of

Table 7.1 Range of tolerances of bacteria to a variety of subsurface conditions

Condition	Range of tolerance
Temperature	−20–113 °C
a_w	Minimum a_w of 0.62
Radiation	Dosages of 17–30 kGy
Salinity	Up to 50 % w/w
Pressure	180 MPa
Pore size	0.2 μm

Adapted from West and McKinley (2002)

moisture, temperature, nutrients, pore space, and time. From Stroes-Gascoyne (2010), it can be concluded that heterogeneities existing at boundaries in the EBS will likely be critical in determining microbial survival; results from this in situ emplacement repository study revealed that interfaces between the bentonite-based buffer, backfill, and sealing materials contained higher numbers of culturable heterotrophic aerobic bacteria and anaerobic bacteria—including SRB—than did the bulk materials themselves. Fractures and incomplete seals between adjacent blocks of expanded bentonite, the expanded bentonite and the host rock walls of the repository, and the expanded bentonite and the surface of the waste container could all therefore offer refuges to microbial cells.

7.3.3.1 Temperature and Water Activity

In a large-scale underground experiment, an electric heater was installed in 50 % sand and 50 % bentonite buffer material at 240 m depth for 2.5 years as part of the Buffer Container Experiment (BCE) at the Atomic Energy of Canada Limited's Underground Research Laboratory (AECL URL) to simulate container emplacement (Stroes-Gascoyne et al. 1997b). Heat surrounding the experimental container resulted in creation of a gradient of moisture within the buffer matrix, ranging from 24 % at the host rock wall to 13 % at the container surface, where temperatures of 50–60 °C were maintained. Following decommissioning of the experiment, it was found that viable microbes, including heterotrophs and specialized organisms, could only be recovered from heated buffer matrix materials where the moisture was >15 % (an $a_w \geq 0.96$), suggesting that the buffer and backfill materials could be populated or repopulated when higher moisture levels prevailed.

A limiting water activity value of 0.96 on SRB survival in bentonite was also reported by Motamedi et al. (1996) using *Desulfomicrobium baculatum* and *Desulfovibrio* sp. over a 60-day incubation at 30 °C. However, the presence of an indigenous SRB (e.g., *Desulfovibrio africanus*, a common corrosion-causing microorganism) isolated from dry Wyoming bentonite MX-80 powder suggests that strategies employed by SRB include the formation of a "dormant" state (Masurat et al. 2010a). In this case, organisms became active upon addition of growth medium containing 4 % salt and temperatures of 40 °C. Notably, viable *D. africanus* cells could still be recovered from dry bentonite powder even after heat treatment of 100 °C for 20 h (but not after treatment of 120 °C for 20 h), providing evidence of considerable tolerance to heat, and suggesting that viable SRB are likely to be present in the emplaced buffer material and can be expected to be active if other controls on their activity are not imposed.

Enhanced survival of desiccated cells is a well-known phenomenon and is likely linked, in part, to the increased efficacy and penetration of moist heat compared to dry heat on various cell components (e.g., protein denaturation effects versus oxidation and dehydration) (Fine and Gervais 2005; Jay et al. 2005). The potential for spores to survive under conditions of elevated temperature and low moisture, which may occur close to the used fuel container, must also be considered in light of evidence of organism survival at temperatures of 120 °C in Yucca mountain tuff (Horn et al. 1998). However, given that the surface temperatures of a used nuclear fuel container would not drop to <60 °C for ~10,000 years, in combination with increasing swelling pressure and low water activity within the compacted bentonite, the potential for microbial activity within the zone most affected by temperature is considered extremely remote (King et al. 2010).

7.3.3.2 Radiation

A series of experiments were conducted by AECL from 1991 to 1997 to evaluate the extent to which radiation emitted from used nuclear fuel would impact microorganisms in situ (Stroes-Gascoyne 1997). Several scenarios were explored with respect to microbial survival in the container-buffer zone, including (1) effects of radiation under near-saturated conditions, (2) effects of different radiation dosages, (3) effect of radiation at three temperature regimes (30, 60, and 90 °C), and (4) effect of radiation under various moisture conditions (0, 23, and 47 % saturation). The studies revealed that radiation and desiccation effects within the bentonite buffer used to surround nuclear fuel waste containers would essentially create a sterilized zone extending a few centimeters and a microbe-depleted zone extending tens of centimeters (Stroes Gascoyne and West 1997). After ~40 cm, radiation levels would decrease by several orders of magnitude, no longer inhibiting organisms solely based on radiation effects (Stroes-Gascoyne et al. 1995). However, there is also evidence suggesting that microbes could survive at elevated cumulative dosages of radiation. For example, radiation-tolerant bacteria have been found in the reactor core at Three Mile Island, where 10 Gy/h radiation was received by the organisms which were also regularly exposed to biocides and hydrogen peroxide in the system (Booth 1987).

Other reports have shown that specialized organisms have the capacity to survive normally lethal conditions of radiation exposure. For example, *Deinococcus* species isolated from Sonoran Desert soils were demonstrated to survive up to 30 kGy radiation (Rainey et al. 2005) and survival of chronic IR (irradiation) of up to 60 Gy/h has also been reported (Daly 2000). Work by Fredrickson et al. (2008), and further reviewed by Meike and Stroes-Gascoyne (2000), described a link between protein oxidation (desiccation) resistance and IR resistance, posing interesting questions about the evolution of IR resistance in bacteria beyond that of genome copy number and DNA repair. In a DGR, those regions closest to the used nuclear fuel container will be lethal given the combined effects of temperature and radiation; however, microbes located further from the containers will be subjected to less extreme conditions. Due to shielding provided by the used nuclear fuel container and surrounding buffer material, the effects of radiation will not, in fact, extend very far from the actual container surface. Thus, other controls in the EBS (i.e., clay swelling pressure, water availability, temperature) will primarily be responsible for limiting the activity and potential effects of microorganisms as the DGR evolves.

7.3.3.3 Salinity, Pressure, and Pore Size

Due to the water adsorption capacity of expandable clays, such as the bentonite used as buffer material, there is less available water for microbial growth. There is also a positive relationship between clay swelling pressure and buffer compaction density; the more compacted the clay (i.e., the more expandable clay per unit volume), the greater the resultant swelling pressure (Craig 1987). The swelling pressure exerted by water-saturated (26 % v/w) 100 % bentonite compacted to a density of 2 Mg/m^3 (dry density of Mg/m^3) is approximately 5 MPa (Pedersen et al. 2000a). It has been demonstrated that SRB isolates *Desulfovibrio aespoeensis* and *Desulfomicrobium bacalatum* become non-cultivable at clay densities higher than 1.8 Mg/m^3. Similarly, results from Hedin (2006) indicated that swelling pressures of 2 MPa or greater were sufficient to eliminate microbial activity within compacted bentonite.

Increasing salinity results in a decrease in water availability due to interaction of water molecules with solute ions. While microorganisms have been documented to survive a_w ranges of 0.74–0.99, most bacteria prefer an a_w of 0.98 or higher (Jay et al. 2005). Several studies have assessed the effect of salinity (NaCl and CaCl) on microbial survival in commercial Wyoming MX-80 bentonite clay (75 % montmorillonite) (Stroes-Gascoyne et al. 2010a, b; Stroes-Gascoyne and Hamon 2008b). Stroes-Gascoyne et al. (2010b) examined the effect of salinity (0, 50, 100, 150, or 200 g/L NaCl) on microbial survival over a 40–90-day period in saturated MX-80 bentonite clay over a range of dry densities (0.8, 1.3, 1.6, 1.8, or 2.0 Mg/m^3). When porewater salinity was $\leq$50 g/L NaCl, a relatively high bentonite dry density of 1.6 Mg/m^3 was required to achieve an a_w of 0.96 and swelling pressure of 2 MPa in order to suppress microbial growth (i.e., no increase in culturable cell numbers compared to dry uncompacted bentonite with ~200 CFU/g). Higher porewater salinities were more effective at inhibiting microbial culturability. For example, when $\geq$60 g/L NaCl porewater was used, a water activity of less than 0.96 was achieved regardless of the bentonite dry density used (Stroes-Gascoyne and Hamon 2008b). Studies evaluating $CaCl_2$ salinity on culturability of aerobic bacteria up to 100 g/L yielded results similar to the NaCl studies (Stroes-Gascoyne et al. 2010a).

A similar limiting water activity value was obtained by Motamedi et al. (1996) for some species of SRB in compacted pure bentonite. Pedersen et al. (2000a, b) examined the effect of MX-80 bentonite buffer density on various microorganisms, including SRB, and their ability to produce sulfide over a 28-week incubation period, and determined that the bulk density of buffer started to exert an inhibitory effect on sulfide generation at a bulk density of ~1.5 Mg/m^3. The authors postulated that microbes would be present to within a few centimeters of the container, either due to spores originally in the clay or subsequently transported by groundwater during saturation, but that the conditions within the compacted bentonite clay would be inhibitory—possibly against all microorganisms—after full swelling of the clay had been achieved. Ultimately, the key microbial control parameters of water activity and swelling pressure are determined by the dry density of the compacted bentonite and the porewater salinity (Stroes-Gascoyne et al. 2010b).

From the porewater salinity effects (Stroes-Gascoyne et al. 2010a) described above, it follows that the performance of clay barriers would be influenced by the ambient solution chemistry; for example, as the buffer and backfill zones saturate, there will be an increase in salinity over time (Dixon et al. 2002). The concentration of total dissolved solids (TDS) within the groundwater will vary depending on the surrounding environment, but will affect the swelling ability of the montmorillonite clay in the bentonite, thereby influencing the hydraulic conductivity of the matrix. Using a maximum TDS of 100 g/L salts, the highly compacted 100 % bentonite matrix would be minimally affected, with less than a one-log effect on hydraulic conductivity (Dixon 2000). Within the backfill, however, there is the possibility of microbial movement during saturation of this zone, but the swelling properties in the compacted bentonite would remain, more or less, unchanged.

It is possible that during the saturation of the EBS buffer and backfill with groundwater, microbes could be transported to the fuel container if heat and desiccation cause radial cracking or other physical damage to the buffer integrity. Stroes-Gascoyne and West (1997) considered cell repopulation events in a series of studies involving penetration of viable *Pseudomonas stutzeri* cells into compacted bentonite buffer plugs (50:50 bentonite:sand; dry densities of 1.2–1.8 Mg/m^3) pre-saturated with sterile water. Their results revealed that cell mobility was restricted to less than 5 mm in all cases, the smallest sampling interval used, but that rapid movement did occur along the metallic holder–buffer interface. Their results suggested that any interfaces or zones with reduced density could provide preferential pathways for cell migration.

In a similar study, Fukunaga et al. (2001) determined that *Escherichia coli* suspensions only penetrated <5 mm into compacted 70 % bentonite 30 % sand buffer over a three-week period. Together, these studies provide evidence that as long as the buffer matrix remains intact, microbial regrowth or colonization of the used nuclear fuel container surface and surrounding regions will be very slow or that these areas will remain devoid of active microorganisms.

Given the small clay particle size, the availability of pore space in water-saturated clays is also highly limiting and appears to have a direct effect on microbial activity. For instance, Fredrickson et al. (1997) found no evidence of metabolic activity, as determined by anaerobic mineralization of [^{14}C]-acetate and [^{14}C]-glucose, and $^{35}SO_4^{2-}$ reduction, in intact shale cores with pore throats <0.2 μm in diameter collected in northwestern New Mexico. Subsequent enrichments revealed the presence of SRB and $^{35}SO_4^{2-}$ reduction in the shale materials after 14 days of incubation. Comparatively rapid rates of metabolic activity were found in sandstone core samples with a large percentage of pore throats >0.2 μm in diameter. From these results, the authors concluded that while viable bacteria can be maintained and stimulated in materials such as shales with pore throats smaller than the size of known bacteria, subsurface bacteria require interconnected pore throats greater than 0.2 μm diameter for sustained activity. A further observation was that the detrital organic matter in the small-pore-diameter shales is not subject to direct microbial attack. In contrast, bacteria in adjacent sandstones with a more open pore structure are likely sustained by endogenous nutrients that are slowly released from the shale. Extrapolating these observations to the DGR environment, it may be possible that similar nutrient exchange takes place between dense and more porous materials, emphasizing the need to avoid the inclusion of components that may serve as nutrients, even in dense EBS materials.

In general, there is a growing body of research (e.g., the Boom Clay at Mol, Belgium, and the Opalinus Clay at Mont Terri, Switzerland) that suggests while argillaceous matrices can host microbial populations (Stroes-Gascoyne et al. 2007c; Mauclaire et al. 2007), their activity has been limited over a much longer time frame than the times currently under consideration for a DGR. Thus, the engineered barrier system is anticipated to function in restricting microbial growth and metabolism in this application.

7.3.3.4 Carbon and Energy Sources

Sources of nutrients within a DGR will vary depending on whether the repository is located within a granitic or sedimentary rock formation. For example, Canadian granitic rock has been categorized as "nutrient poor" (Stroes-Gascoyne 1997); deep groundwaters in the Canadian Shield are similarly limited by availability of organic carbon, which is typically <2 mg/L (Loewen and Flett 1984; Stroes-Gascoyne 1989). Inorganic carbon, in the form of carbonates, would range from 0.01 to 0.1 g/L and concentrations of sulfate could be as high as 3 g/L (McMurry et al. 2003) in these groundwaters.

It is inevitable that DGR construction will introduce a variety of materials that could be used by microorganisms for growth, including those associated with intentionally placed repository materials (e.g., the used nuclear fuel and container, buffer, and backfill), as well as those materials which are inadvertently placed during construction activities, including fuels, detergents, lubricants, wastes from human activities, etc. (Hallbeck 2010). Stroes-Gascoyne and West (1996) performed an analysis of the nutrients that would be available to heterotrophic and chemolithotrophic microorganisms in intentionally emplaced EBS materials and determined that the major nutrients, N and P, would be growth limiting, even using the unlikely scenario of all of the carbon being available for growth.

Bentonite clay proposed for use as buffer and backfill does contain small quantities (up to 1.5 %; Sheppard et al. 1997) of carbon-containing compounds (e.g., humic and fulvic acids), which overall would not be readily available to microorganisms but which may be extractable (Lucht et al. 1997). It has been proposed that heat, along with effects of radiation, may cause the degradation of clay-bound organic compounds rendering them more bioavailable (Stroes-Gascoyne et al. 1997a). In an experiment designed to explain this, heating (60 and 90 °C) and irradiation (25 and 50 kGy) of a 50:50 Avonlea bentonite:silica sand buffer preparation, followed by extraction with a 3:1 water:buffer ratio, resulted in an approximate two-log greater enhancement of cell growth (relative to control extracts not exposed to heat and irradiation) when supplemented with granitic groundwater. While this suggests the potential for stimulation of microbes within a repository exists, any microbial utilization of extracted carbon within these low hydraulic conductivity ($\sim 10^{-12}$ to 10^{-14} m/s) environments would occur far more quickly than the substrate could be replenished by diffusion, inhibiting any sustained effect on microbial growth. Subsequent work has examined other possible nutrient additions to the EBS, with particular focus on the blasting components as potential sources of N and C (e.g., ammonia nitrate fuel oil; ANFO) (Stroes-Gascoyne et al. 1996; Stroes-Gascoyne 1997). Within blast rubble, a significant amount of nitrogen and carbon could remain as a result of incomplete combustion or detonation failure of explosive materials, ranging from 4 to 5 % of the total ANFO (Stroes-Gascoyne et al. 1996) to as much as 10–20 % (Forsyth et al. 1995). Excavated rock from repository construction may therefore constitute a significant potential nutrient source for microorganisms if incorporated into EBS backfill (Stroes-Gascoyne et al. 1996; Stroes-Gascoyne and Gascoyne 1998). Factors affecting nutrient concentration in blast-rubble rock include surface area of rock exposed to the blast and the amount of explosives used, as well as the amount of blasting materials and gases entering the rock via fissures. Direct measurements on leachates from freshly broken rock at the AECL URL indicated that the amount of N in freshly excavated rock varied from 6.5 to 47 % of the total N present in the initial blasting materials (Stroes-Gascoyne et al. 1996; Stroes-Gascoyne and Gascoyne 1998) and could potentially contribute to a two-log increase in microbial numbers.

Concentrations of organic carbon, as high as 120 µg/g of rock matrix, in old and fresh broken rocks could originate from a variety of anthropogenic sources, including leaked oil, greases, and paints. Surface waters used during drilling operations similarly present a potential source of carbon (Stroes-Gascoyne and Gascoyne 1998). Because the net effect of stimulation of microbes within the EBS is unknown, it has been suggested that measures (e.g., washing, leaching) should be taken to reduce the potential for nutrient addition via excavated rock or surface-processed waters (Stroes-Gascoyne and Gascoyne 1998).

Lastly, superplasticizer (SP) ingredients in high-strength cement used in bulkheads, or as grouts to seal groundwater leaks within the repository, are not thought to likely play a significant role as a source of carbon due to the low relative abundance compared to the major components of the EBS, along with the low concentrations and leachability (10^{-16} kg/m^3) of the SP (Onofrei et al. 1991).

It is expected that any labile carbon source introduced in the bentonite buffer or backfill material, due to construction activities, or transported into the repository with the groundwater would be consumed during either aerobic or anaerobic respiration (Wang and Francis 2005). The largest contribution of organic matter in the EBS is likely to be from the buffer and backfill. As such, once materials for buffer and backfill have been chosen, their organic matter content and potential bioavailability for microbial processes should be further assessed (OECD 2012).

7.4 Perturbation and Deep Subsurface Successional Change: Deep Geological Repositories for Used Nuclear Fuel as a Perturbation Model

Discussion of potential impacts of microbial activity in the context of the EBS should include two broad questions: (1) can microbial metabolism realistically have an impact on repository function; and (2) if yes, what can be done to mitigate the impact, or control microbial activity? The specific environment associated with each design and placement method, together with the conditions of the site where they will be applied, can potentially present different outcomes in terms of microbial persistence and activity. This information should allow the development of a "microbial potential" index for different materials under conditions typical of the area at and near the used-fuel containers. The DGR concept provides an invaluable opportunity to evaluate the evolution of subsurface conditions from "disturbed" back to original state. This section discusses the expected evolution of the repository that can be expected, as well as the potential impacts that microbial activity could have on a DGR.

7.4.1 Sources of Microbes and Their Activity

A relatively large variety of microbial functional groups that may affect the overall performance of deep geologic storage facilities for used nuclear fuel has been described, including (1) those with the potential for directly damaging storage containers, (2) those organisms with potential for creating corrosion-aggressive environments where the diffusion of metabolic end products to the container could result in indirect container damage, (3) those that produce metabolites that could lead to the deterioration of EBS components such as concrete and seals, and (4) those that may impact the mobility of released radionuclides.

Specific bacterial groups have received much attention in the literature for their known involvement in metal deterioration, such as the SRB and metal reducing bacteria. The potential for microbiologically influenced corrosion of used nuclear fuel containers is a significant consideration in safety assessments for DGR performance. However, to obtain realistic data, it is important that these metabolic reactions be considered in the context of the EBS environment (and also from a microbial ecology perspective). For example, as pointed out by Chen et al. (2011), sulfide and bisulfide (formed by either mineral dissolution or produced by SRB) presumably are the most likely corrosive agents in the groundwater to which containers will be exposed. In their experiments performed in solution, Chen and coauthors (2011) showed that once HS^- became depleted in solution, its diffusion in the bulk solution became rate determining. Considering the extremely slow rate of microbial metabolism that prevails in low-permeability environments, it is highly improbable that the source values for microbiologically produced sulfide would be sufficient to create a notable concentration gradient (Chen et al. 2011). While the SRB may derive some benefit from the presence of other microorganisms (e.g., specifically those that lower the redox potential to levels conducive for SRB growth), they would also compete with these cells for nutrients in this oligotrophic environment, further limiting the source rates.

As well as physical and chemical controls on microbial activity, microbial metabolic activity could also create conditions conducive to growth of other specialized functional microbial groups, or even organisms that might compete with them for scarce resources. Typically, these microorganisms would be adapted for purposes other than just survival in oligotrophic environments and would span a variety of functional groups. One example is halophilic and halotolerant microorganisms. The potential for halophiles to be present within subsurface microbial communities is not surprising, as halophiles have long been isolated from most environments (particularly within soils where salts periodically become concentrated during dry periods) (Stewart 1938). Extremely halophilic organisms have also been detected, albeit in low numbers, in subsurface environments associated with highly saline environments including the Waste Isolation Pilot Plant located 650 m below ground surface in a bedded salt formation located in Calsbad, New Mexico (Vreeland et al. 1998).

In a long-term test experiment of buffer performance, the moderately halophilic bacterium, *Desulfovibrio salexigens,* isolated from the deep groundwater of the Äspö hard rock laboratory (HRL), was used to evaluate bacterial survival under different clay swelling pressures (Pedersen 2000). This organism did not survive as well, or penetrate as deeply into the clay (~6 mm), as did a *Bacillus subtilis* test strain. It should be pointed out, however, that it is almost impossible to realistically simulate in situ conditions, including the role of microbial species interaction, when performing pure culture studies. A guide for future considerations in this respect is the comprehensive study by Stroes-Gascoyne et al. (2007b) that examined the effect of salinity on the fate of microorganisms existing within bentonite buffers of differing dry densities relevant to the EBS environment.

While there is little evidence to suggest a strong presence of eukaryotes in the deep subsurface, there is some indication that eukaryotes are present at depth. For example, investigations of groundwater from the Äspö HRL have suggested that small fungi may inhabit Fennoscandian Shield igneous rock aquifers, although this would not likely have any direct relevance to a DGR (Pedersen 2000). Given the unique metabolic processes in which eukaryotes participate, fungal chelating agents could offer possible mechanisms for radionuclide transport, and more generally, could affect the activity of those functional microbial groups of interest in the EBS environment.

Concentrations of microbes in the deep subsurface vary widely, ranging up to 10^8 CFU per mL of water or gram of sediment (Fredrickson and Onstott 1996; Balkwill 1989; Ghiorse and Wilson 1988) and comprise a diverse variety of predominantly Gram-negative microbes and Archaea, including SRB, methanogens, and acetogens (Kotelnikova and Pedersen 1997). Backfill and buffer materials undergo handling and processing and therefore will not be sterile. These materials have been reported to contain various organisms (up to 10^4 CFU/mL in as-bought bentonite; Haveman et al. 1995), including those with potential to impact the EBS, such as SRB (Masurat et al. 2010a, b). The intrusion and saturation of backfill by ground or service waters could serve to both inoculate and activate organisms present in, or introduced into, the porous matrix. It is not anticipated that these microorganisms will penetrate through the compacted bentonite buffer, as the average pore throat diameter in clays would be too small for bacterial cells to pass through (Chapelle 1993); however, groundwater, and associated microorganisms, will inevitably reach the backfill and the backfill-host rock interface.

Given the presence of viable microorganisms within the EBS, the only questions remaining are how active these organisms will be, and what effects their activity will have on the DGR. Following closure of the DGR, there would likely be a strong chemical gradient between the repository and host rock because they would be oxidizing and relatively reducing (-225–0 mV; McMurry et al. 2003), respectively. Redox fronts will likely play a key role in defining the types of reactions that organisms would mediate (McKinley et al. 1997) and would create suitable conditions for lithoautotrophic (chemoautotrophic) organisms. Chemoautotrophic organisms are predominantly aerobes and thus would benefit by being positioned at the aerobic–anaerobic interface where the conversion of Fe(II) to Fe(III) by *Acidithiobacillus ferrooxidans*, for example, would yield energy for the fixation of CO_2. Alternative oxidizing electron acceptors (e.g., NO_3^-) would similarly function within a repository environment as conditions evolve from the initial oxidizing state.

A variety of microbial end products could have an impact on a DGR, including production of organic acids and other metabolites (e.g., formate, acetate, lactate, butyrate, nitrite, ammonia), as well as gases such as sulfide, CO_2, and methane, and an assemblage of various extracellular products that might influence both the repository or radioelements. Hydrogen may also exist or be produced by microbes, but in any environment where it is produced, its evolution tends to be coupled with consumption by other organisms (i.e., SRB) so that free microbiologically produced H_2 gas is not common.

It has been speculated that hydrogen-driven ecosystems could exist in the deep subsurface (Pedersen 1993b, 1996, 1999a; Stevens and McKinley 1995; Kotelnikova and Pedersen 1997). If this is the case, the presence of high amounts of sulfate would most certainly drive sulfidogenesis rather than methanogenesis, due to the higher substrate affinity of SRB for hydrogen relative to methanogens (Uberoi and Bhattacharya 1995). However, it is not thought likely that the activity of SRB and other microorganisms would cause corrosion to occur within the buffer region near the used fuel container. Accordingly, sulfide gas or other corrosive metabolites would need to diffuse through the EBS buffer and backfill in order to reach the copper container surface, where possible stress corrosion cracking (SCC) could occur. King and Kolář (2006) previously determined through modeling that the likelihood of significant SCC occurring is very low.

In the absence of an abundance of organic materials in the repository, it is not expected that microbial fermentation will be a dominant process. Only minor quantities of fermentation end products (e.g., acetate, formate, propionate, ethanol, butyrate, lactate) are expected in a used nuclear fuel repository. Because the 100 % highly compacted bentonite buffer is expected to be strongly inhibitory to microbial growth and metabolism, reduced end products resulting from any microbial fermentation that did occur would be produced mainly in the backfill regions.

While there will be microorganisms in both the backfill and compacted bentonite, the backfill zone is significantly more likely to undergo microbiologically mediated changes. To date, relatively few studies have specifically targeted backfill, which will contain a mixture of bentonite and crushed rock, but it is evident that the material will contain a diversity of microorganisms which may be active on saturation with water, with possible short-term benefits to the repository redox conditions by using up the O_2 introduced into the repository (Stroes-Gascoyne et al. 1997a; Stroes-Gascoyne and West 1996).

7.4.2 Impact of Microbes on Introduced Materials

Complex carbon sources are likely of limited availability in the subsurface (Loewen and Flett 1984; Stroes-Gascoyne and West 1996). However, it is likely that construction of an engineered barrier system (EBS) would introduce some sources of carbon. Thus, there may be an enhanced capacity for metabolic activity of heterotrophic organisms in the EBS versus that occurring within the unperturbed host rock. The DGR environment is expected to gradually become more reducing as organisms present in this zone consume O_2 as a terminal electron acceptor and produce CO_2. Development of reducing conditions could enable anaerobic organisms like SRB to become active, providing that sufficient sulfate is present, as well as appropriate electron donors such as hydrogen or organic carbon. This may lead to the production of bisulfide/HS^-, which could diffuse through the clay buffer and contribute to corrosion of metal containers.

7.4.2.1 Microbial Metabolites and Extracellular Materials

Gasses could be generated as a consequence of (1) microbial degradation of organic materials present in the backfill and buffer (e.g., yielding CO_2, hydrogen sulfide, or methane), (2) anaerobic corrosion of the metal container (evolving copper sulfides and hydrogen gas), or (3) abiotic radiolysis of water, yielding hydrogen gas. In general, production of significant quantities of microbial gas within the compacted bentonite buffer of an EBS is thought unlikely for two main reasons: (1) minimal organic material and (2) low numbers of microorganisms and the controls on their activity within the buffer region (i.e., bentonite swelling pressure and water activity). The majority of studies on gas generation, as related to nuclear waste repositories, have been centered on disposal of carbon-rich (often cellulose-rich) intermediate- and low-level wastes, as summarized by Humphreys et al. (2010).

The backfill region and associated interfaces have higher hydraulic conductivities than the host rock and can potentially support notable microbial activity. As discussed by Stroes-Gascoyne (1997), the availability of organic materials contained in bentonite clays is expected to be very low, as evidenced by their extreme stability since the time of their deposition (approximately 75–85 Ma for Wyoming and Saskatchewan-Avonlea bentonites). The effects of exposure of clays in backfill to aerobic conditions are not well understood; hence, any contribution to in situ gas production is speculative (Lucht et al. 1997; Stroes-Gascoyne and West 1997). Organic materials arising from EBS construction (e.g., blasting residues, drilling fluids, machine exhaust, etc.) in the repository could also represent sources of organic carbon for supporting microbial growth.

Microbial activity within the backfill zone following closure of the repository will drive the gas phase from oxidizing to reducing at some point post-closure. Data from the long-term tunnel sealing experiment (TSX) at the AECL URL (Stroes-Gascoyne et al. 2007a) offers indirect evidence of this; samples retrieved from the 10 % bentonite/90 % sand backfill region of the tunnel seal contained higher numbers of SRB and a notable decrease in the number of heterotrophic aerobic, nitrate respirers and nitrate reducing bacteria, as compared to the 70 % bentonite/30 % sand highly compacted buffer material. It was proposed that the high moisture, nutrients, and space associated with this region of the backfill could stimulate aerobes, thereby depleting oxygen and creating conditions conducive for the strict anaerobes. An alternate explanation for the increase in anaerobe numbers is that the elevated temperatures obtained during the second phase of the TSX could have inhibited the aerobes and facultative anaerobes more than the strict anaerobes.

Gas analysis of samples from the isothermal test (ITT), part of a long-term (6.5 year) study on the evolution of buffer gas and microbiological parameters (Stroes-Gascoyne et al. 2002), implied that sulfide production may have just been initiated (only about 0.02–0.5 % of total available bulk sulfate was converted to sulfide) and that evolution of reducing conditions was not yet significant. Analysis of gas from the buffer material showed that gas composition remained relatively unchanged from initial conditions (there was some slight evidence of O_2 reduction and small increases in H_2 and CH_4 levels). This may reflect a loss of viability within the resident microorganisms, as a comparison of culture-based and biochemical analysis of the buffer samples indicated that the viable population of cells was approximately two logs higher than what could be cultured. In general, development of a reducing anaerobic backfill zone after repository closure will likely retard oxidative copper corrosion, thereby serving to stabilize oxidative copper corrosion of the fuel containers over the short term (Pedersen 2000).

The most common of the microbiologically derived gasses likely to be found in the EBS would include CO_2, hydrogen, hydrogen sulfide, and methane. While initially most abundant, CO_2 will be reduced in subsequent microbiologically mediated reactions. Hydrogen and methane are thought to have the greatest potential for having impacts on the EBS environment, since these are reduced compounds that may function as microbiological energy sources (and in the case of methane, a carbon source also). It is likely that SRB would effectively outcompete methanogens at higher sulfate concentrations, given their greater affinity for hydrogen (Lovley and Klug 1983; Uberoi and Bhattacharya 1995). Indeed, Sheppard et al. (1997) showed that methanogenesis would not start until sulfate levels had dropped considerably. Other organisms may oxidize H_2 using nitrate, iron, manganese, and CO_2 as electron acceptors in the absence of methanogens and SRB (Schwartz and Friedrich 2006).

The Belgium and Swiss designs may incorporate the use of stainless steel (e.g., AISI 316 L) for high-level waste disposal, and it has been estimated that this would represent the largest gas generation source term within their EBS (Ortiz et al. 2002). The authors proposed that the hydrogen evolved due to anaerobic corrosion of the steel container surfaces would be microbiologically transformed and potentially reduced into either methane or sulfide/thiosulfate. Modeling results suggested that diffusion would not be sufficient to offset gas generation, which could create preferential pathways in the EBS. However, this is not a likely outcome due to the plasticity of clay buffer materials that would likely self-seal. In general, the nature of the gas phase will depend on the kinetics of its production, with rapid gas evolution leading to formation of bubbles or a separate gas phase, whereas slow evolution of gasses may simply diffuse away or be metabolized by other microorganisms at the time of production.

7.4.2.2 Effect of Microbes on Permeability in the EBS

Following repository closure, the presence of oxygen could stimulate aerobic metabolic processes and drive oxidative precipitation of dissolved metals such as Fe and Mn (Tufenkji et al. 2002; Haveman et al. 2005), which could lead to plugging of flow paths and pore spaces (Howsam 1987; Goldschneider et al. 2007). Ultimately, heterotrophic metabolism would deplete the available O_2 requiring use of alternate electron acceptors. Under anaerobic conditions, reductive dissolution of Fe and Mn oxides in the repository environment could then occur (Lovley 1987, 2006). These reactions tend to occur at interfaces between aerobic and anaerobic environments, or during transition from aerobic to anaerobic conditions. The resulting geochemical changes have the potential to affect backfill permeability.

It is well known that biofilms contribute to the plugging or fouling of flow paths in fractures (Wolfaardt et al. 2007; Characklis 1990; Sharp et al. 1999). Accordingly, it has been hypothesized that biofilms could clog or plug larger pore spaces within the EBS backfill. Evidence to support this is lacking, however. A study by Lucht et al. (1997) at the Whiteshell Laboratories (Pinawa, MB) infused low-nutrient groundwater into columns containing different backfill preparations for a 180-day period and found no evidence to support pore clogging, based on permeameter data, even though there were 10^6–10^7 CFU/mL of viable microbes and up to 30 mg/L of DOC present in the column effluent. Studies demonstrating microbiologically mediated (biofilm) plugging of crushed rock and sediment/sand matrices have been conducted, although it must be emphasized that these simulated highly permeable environments involving the continuous flow of nutrient-rich liquid to support sufficient growth to cause plugging (Hama et al. 2001; Brydie et al. 2005; Coombs et al. 2010). Liquid flow, generally considered a stimulatory factor for biofilm formation, along with sufficient nutrient concentrations, would not be present within the backfill and buffer region of a repository. At the host rock–EBS interface, a potential limiting role could be played by biofilms in terms of scavenging nutrients being transported from the host rock into the EBS, as well as through the direct complexation of biofilms with radionuclides that might be migrating from the repository (via adsorption, uptake, or precipitation reactions; see below) (Anderson et al. 2007).

Due to its importance to the EBS, the integrity of the bentonite clay buffer over time is of primary significance; any functional change in the bentonite could potentially affect its ability to undergo swelling upon saturation and subsequently could affect its capacity to control microbial growth and migration of radionuclides. The transformation of montmorillonite to illite (Eq. 7.7) has been studied primarily from an abiotic point of view (Huang et al. 1993; Wersin et al. 2007); high temperatures (~150 °C), elevated pressure (100 MPa), and time are required.

$$Ca^{2+}/Na^+ - \text{montmorillonite} + K^+ + \left(Al^{3+}\right)$$
$$\rightarrow \text{illite} + \text{silica} + Ca^{2+}/Na^+ \tag{7.7}$$

In general, the scarcity of illite in naturally occurring bentonite deposits suggests that this conversion is not common (McMurry et al. 2003). Biologically mediated changes (biotransformation) of montmorillonite (smectite) to illite have been shown to occur also in the absence of extremes of temperature and pressure, instead by relying upon microbial reduction of Fe(III). In this case, the authors suggested that microbes dissolved smectite at room temperatures, at 101 kPa, over 14 days (Kim et al. 2004). In a more recent study by Jaisi et al. (2011) using X-ray diffraction and high-resolution transmission electron microscopy, it was demonstrated that formation of illite from nontronite by the mesophile *Shewanella putrefaciens* and the thermophile *Thermus scotoductus* occurred at basic pH (8.4) and high temperature (65 °C). While the conversion of Fe(III) to Fe(II) is known to be mediated by various iron-reducing bacteria, as well as by SRB, it is notable that conditions for microbial growth and activity within the compacted bentonite matrix will be severely inhibitory, as indicated elsewhere in this review. The montmorillonite to illite bioconversion is not thought, therefore, to be of potential significance to the functioning of the clay barrier.

A biodeterioration process of potential significance to repository functioning relates to the integrity of cement-containing materials, particularly use of cement bulkheads for sealing tunnels. Within the repository cement, bulkheads play two key roles. The first is to provide a barrier between the used nuclear fuel vault and the main tunnel system. Secondly, but perhaps more importantly, is a role in which the cement bulkhead provides a brace against which expandable clays present in the buffer and backfill may exert swelling pressure. Evidence from the Tunnel Sealing Experiment (Stroes-Gascoyne et al. 2007a) suggests that the interface between the cement bulkheads and the backfill zone is a likely region where the preferential conditions for microbial growth would exist, even possibly supporting a biofilm community of attached microorganisms. Sources for the growth of the biofilm would comprise microbes and materials contained in the backfill and groundwater, as well as those carbon compounds present in the superplasticizers that leach out of the cement. Metabolites of these biofilms would include organic acids, which could accelerate cement weathering and biodeterioration.

7.4.2.3 Microbiologically Influenced Corrosion of the Used Fuel Container

Conditions under which microbiologically influenced corrosion (MIC) of copper-steel or steel containers may occur are obviously a consideration for emplacement options. MIC of metals can occur through direct interaction of microbes on the surface of the container, or indirectly, where microbes growing in another place produce chemicals (i.e., acetate, ammonia, or sulfide, as above) that diffuse to the container. Given the compelling body of research (Stroes-Gascoyne 1997; Stroes-Gascoyne and West 1997; Pedersen et al. 2000a, b) indicating that the container surface and high-density clay buffer surrounding the container would likely be biologically inactive (due primarily to heat, radiation, and water activity effects) for a period of hundreds to thousands of years or longer, there is little evidence to suggest that a biofilm of SRB could develop on a used fuel container, provided that the EBS remained intact. Further, transport studies, such as that conducted by Stroes-Gascoyne and West (1997) using *Pseudomonas stutzeri*, suggest that repopulation of the region near the container is highly unlikely as long as the clay buffer remains undamaged. The rate of SRB-induced corrosion would therefore be limited by the rate of diffusion of sulfide (HS^-) produced allosterically outside of the container zone to the container surface (Kwong 2011). The kinetics of this diffusion would need to be sufficient for continued MIC to be sustained. Given this, MIC rates expected in a DGR are notably slower than when microbes are directly associated with metal surfaces (Sheng et al. 2007; Xu et al. 1999). In addition, they are dependent on the type of metal being corroded, as described below.

Corrosion of Iron and Steel

MIC processes are facilitated by electrochemical reactions on the metal surface. Electrons produced in the oxidation reaction at the anode are consumed in the reduction reaction at the cathode because the two reactions occur at equal rates. Examples of anodic and cathodic reactions (Eqs. 7.8 and 7.9) on ferrous metals in aerobic environments are as follows:

$$\text{Fe} \rightarrow \text{Fe}^{2+} + 2e^- \qquad (7.8)$$

$$2\text{H}_2\text{O} + \text{O}_2 + 4e^- \rightarrow 4\text{OH}^- \qquad (7.9)$$

Products of these reactions (i.e., Fe^{2+}, OH^-) will further react to produce, for example, ferrous hydroxides, in the presence of water, and ferric hydroxides in the presence of oxygen and water (Eqs. 7.10 and 7.11, respectively):

$$\text{Fe}^{2+} + 2(\text{OH}^-) \rightarrow \text{Fe}(\text{OH})_2 \qquad (7.10)$$

$$\text{Fe}(\text{OH})_2 + \tfrac{1}{2}\text{H}_2\text{O} + \tfrac{1}{4}\text{O}_2 \rightarrow \text{Fe}(\text{OH})_3 \qquad (7.11)$$

When dehydration or partial dehydration reactions are considered, many oxides or oxyhydroxides can also be formed, including from Eq. (7.10), FeO, and from Eq. (7.11), Fe_2O_3 and FeOOH, as well as the mixed-species Fe_3O_4 magnetite. Different atomic arrangements of the ferric species (Eq. 7.11) include hematite (α-Fe_2O_3), maghemite (γ-Fe_2O_3), to a lesser extent β- and δ-Fe_2O_3, goethite (α-FeOOH), and lepidocrocite (γ-FeOOH), among others. In addition, when oxygen is present, the solution near the steel surface will contain quantities of Fe^{2+} and Fe^{3+}; the specific solubilities of these components are highly dependent on pH.

Following the consumption of oxygen by corrosion reactions (Eqs. 7.9 and 7.11) and microbial processes, steel corrosion can continue anaerobically in the presence of water, which reduces according to the half reaction shown in Eq. (7.12), to produce hydrogen gas.

$$2\text{H}_2\text{O} + 2e^- \rightarrow 2\text{OH}^- + \text{H}_2 \qquad (7.12)$$

When coupled with iron oxidation (Eq. 7.8), ferrous hydroxides will be produced (as per Eq. 7.10); however, the lower oxidizing power of water compared to hydrogen mitigates formation of ferric hydroxides or ferric oxides listed above (or solution-based Fe^{3+}). Only the mixed-oxide species magnetite (Fe_3O_4) will be formed that contains Fe^{3+}, owing to the high lattice energy of this oxide as a surface species. Thus, the steel surface will consist of significant quantities of magnetite and ferrous oxides or hydroxides, and an absence of ferric oxides, while the solution near the corroding surface will contain only Fe^{2+}.

While the combination of Eqs. (7.8 and 7.9) does constitute corrosion reactions, the iron oxides and hydroxides possibly affect the actual corrosion rates, depending on how they deposit on the surface.

Corrosion of Copper

During the early phase of the DGR life cycle, when conditions are warm and oxidizing, roughening in the form of under-deposit corrosion and fast uniform corrosion of the copper containers will occur only as long as sufficient oxygen is available (Stroes-Gascoyne and West 1996; Kwong 2011). While corrosion is a chemical process, microbes may enhance the process by altering the local chemical environment (e.g., creation of differential aeration cells), in particular pH and Eh, and by production of corrosive metabolites, such as sulfide that would diffuse from a remote location and may impact the durability of the metal waste containers in a repository. It was suggested by Akid et al. (2008) that up to one-third (~33 %) of material loss that arises from copper corrosion may be attributed to microbial activity. In contrast, Kwong (2011) reported that after 1 million years MIC would conservatively contribute up to 1 mm of total predicted copper container wall loss (1.27 mm; i.e., MIC would contribute ~80 %). This value was determined using an extreme-value calculation, based on a HS^- concentration of 3 ppm continuously being supplied by SRB to the container surface in a crystalline rock repository environment, corresponding to an estimated corrosion rate of 1 nm/year (King and Stroes-Gascoyne 1995; King 1996). In summary, the predicted extents of MIC are both design and site specific and require appropriate study accordingly.

King et al. (2001, 2010) noted that under anaerobic conditions, it is expected that copper corrosion, along with H_2 evolution (Eq. 7.13), would occur in the presence of sulfide in solution.

$$2Cu + H^+ + HS^- \rightarrow Cu_2S + H_2 \quad (7.13)$$

After examining HS^- effects on copper corrosion up to 167 days under anaerobic conditions using dilute 5×10^{-5} mol/L Na_2S in 0.1 mol/L NaCl, Chen et al. (2011) concluded that the concentration of HS^- at the copper–water interface would ultimately limit corrosion, given that the diffusion coefficient in compacted clay buffer would be on the order of 10^{-7} cm^2/s.

Overall, conclusions from laboratory experiments suggest that the main MIC effects on copper corrosion would occur indirectly from the diffusion of sulfides produced in regions of the repository (backfill and interface) where high temperature and other conditions would not result in microbial death or inactivation (e.g., Stroes-Gascoyne and King 2002). Diffusion of the sulfide will be limited through the compacted buffer, however, making this source of corrosion a minor component (Wersin et al. 1994; Masurat et al. 2010a; Pedersen 2010; Kwong 2011).

Corrosion of Copper-Iron

Smart et al. (2011) installed miniature copper–cast iron used nuclear fuel containers with 1 mm diameter "defects" in the outer container shell and exposed these to ambient granitic groundwater in SKB's Äspö Hard Rock Laboratory in Sweden. The main aim of the study was to determine how the container would perform over time should the outer copper barrier fail, as part of a worst-case scenario study on galvanic coupling between steel and copper components. The authors came to three general conclusions: (1) water analysis showed that there were compositional differences between water from inside the support cages compared with the external borehole water, and these differences were attributed to enhanced corrosion of iron components in their experimental system, as well as increased microbial activity inside the cages that surrounded their experimental container assemblies; (2) microbiological analysis showed that SRB were active in the boreholes and the support cages, with the microbial activity higher inside the support cages compared with the boreholes outside the support cages; and (3) over time, electrochemically measured corrosion rates for both iron and copper increased in the experimental systems containing low-density bentonite, while increased rates were observed only for iron in the absence of bentonite. It was suggested that microbiologically produced sulfides were responsible for the increased rates. However, the authors noted that this effect has not yet been confirmed in fully compacted bentonite and that the increased microbial activity by SRB could have been stimulated by the corroding iron. Furthermore, because hydrogen may be generated by the anaerobic corrosion of the used fuel containers, as well as by the radiolysis of water, there would seem to exist the potential for these processes to fuel further SRB growth and sulfide generation, thereby accelerating overall corrosion.

7.4.2.4 Impact of Microbes on Radionuclide Speciation, Fate, and Movement

In the early years following a DGR closure, it is expected that any viable microorganisms within the clay buffer immediately surrounding the used nuclear fuel containers would be almost completely inactivated as a consequence of the combined effects of heat, low water activity, limited pore space, and radiation (Motamedi et al. 1996; Stroes-Gascoyne et al. 2010b). On cooling, the region around the containers would eventually become saturated with groundwater, at which time conditions might enable surviving cells to once again become active. However, the swelling pressure exerted by the bentonite buffer is expected to inhibit bacterial growth.

In the event of container failure, microbes can interact with both toxic and nontoxic radionuclides dissolved from the used nuclear fuel. The reaction of microorganisms with inorganic metals is well understood and includes metal oxidation/reduction (redox) reactions, biosorption of metals to cell surfaces and extracellular components, intracellular accumulation of metals, and extracellular precipitation (Vieira and Volesky 2000; Suzuki and Banfield 2004; Frazier et al. 2005; Merroun and Selenska-Pobell 2008). In addition to generating energy, the microbially mediated reactions and microbial products may detoxify compounds. Consequently, microbe–radionuclide interactions may affect the mobility of the radionuclide by mediating its immobilization either onto or within the cell or by altering the radionuclides solubility.

A variety of microorganisms are known for their abilities to involve metal elements, including redox-sensitive radionuclides (i.e., uranium, plutonium, neptunium, technetium), in a wide range of redox reactions. Much of the available literature centers on application of microbial biotechnology for the remediation of uranium-contaminated sites associated with the shallow subsurface (see review by

Anderson and Lovley 2002). Because toxic metals cannot be degraded, remediation strategies have focused on stabilizing these elements so they become less mobile and thus less biologically available. These include using microorganisms to reduce soluble-phase actinides (e.g., U(VI)) to their insoluble (i.e., U(IV)) state, as proposed by Lovley et al. (1991) as a strategy for preventing movement of soluble U into groundwater systems. The initial phase of this approach depends upon creation of anaerobic environmental conditions through stimulation of microbial metabolism, which itself requires having an adequate supply of either organic carbon or hydrogen to serve as electron donors. A variety of organisms that mediate similar metal reduction reactions includes SRB (*Desulfovibrio* spp.) and Fe-reducing bacteria (*Geobacter and Shewanella* spp.). In typical reactions of uranium, U (VI) behaves as an electron acceptor (dissimilatory U reduction) with the production of low-solubility uranium minerals, thereby decreasing the concentration of soluble U(VI).

It is noteworthy that this reaction may proceed in the opposite direction upon interaction of reduced uranium compounds under oxidizing environments (Merroun and Selenska-Pobell 2008), a microbiologically mediated process that has been exploited to recover uranium via bioleaching from low-grade ore deposits (Brierley 1978). In these reactions, addition of Fe(III), under acidic conditions, functions as a U(IV) oxidant and yields the more soluble U(VI). Re-oxidation of the reduced Fe(II) is performed by *Thiobacillus ferrooxidans* (reclassified as *Acidithiobacillus ferrooxidans* in 2000), which use reduced iron as an electron donor for energy production during CO_2 autotrophy.

More relevant to DGR environments, naturally occurring redox cycling by anaerobic microbes has been demonstrated by Wilkins et al. (2006, 2007) to influence the speciation and mobility of radionuclides associated with low-level radioactive waste at the Drigg low-level storage site in the UK. In that system, the removal of both soluble U(VI) and Tc(VII) from groundwaters [via reduction to their respective insoluble forms, U(IV) and Tc(IV)] was correlated with microbial Fe(III) reduction in microcosms constructed with sediment. Wilkins et al. (2007) further confirmed that the potential for nitrite reduction to re-oxidize and hence re-mobilize these elements was radionuclide specific (i.e., uranium underwent oxidative re-solubilization, whereas technetium did not).

The high sorption capacity of clay-based buffer and backfill materials is key to retardation of radionuclide migration. The conditions in compacted bentonite (radiation, water activity, swelling pressure, and temperature) make it unlikely that microbes would be involved in the migration of radionuclides from the zone nearest the used nuclear fuel container. However, there is an increased possibility that fractures within the outer regions of the backfill zone, or at the interface between the backfill and the host rock, could be colonized by microorganisms, which could then potentially affect actinide mobility. One consideration is that soluble radionuclides or radionuclides sorbed to bentonite colloids could diffuse out of the repository and reach attached biofilm communities (West et al. 2002), which could function to retard colloid-facilitated nuclide diffusion by sorbing either the clay–radionuclide complex or free radionuclides. Subsequently, biofilm cells bearing clay-adsorbed radionuclides could themselves be transported through fissure conduits along the backfill-host rock interface, mobilizing radioactive materials from repository into the host rock (Kurosawa and Ueda 2001).

Ohnuki et al. (2010) examined U(VI) and Pu(VI) with *Bacillus subtilis* in kaolinite clay. The authors determined that both U(VI) and Pu(VI) sorbed to bacterial cells in the presence of kaolinite, but that U(VI) became directly sorbed to the cells, whereas Pu(VI) underwent sequential reduction to Pu(VI), Pu(V), and then Pu(IV) before sorbing to cells. The microbial cells were determined to compete directly with clay colloids for radionuclide binding, and similar mechanisms utilized by attached biofilms on rocks in situ may be expected to retard radionuclide migration.

Such a mechanism could explain the stability of uranium species (i.e., no chemical evidence suggesting a conduit for radionuclide migration) associated with the Cigar Lake (Canada) UO_2 deposit, despite the existence of large fractures spanning the clay dome overlying the UO_2 deposit (Brown and Sherriff 1999; Smellie et al. 1997). While there is abundant evidence that microbes sorb radionuclides along with a variety of other metal elements (Merroun and Selenska-Pobell 2008), there are also reports (Anderson et al. 2007) that biofilms in fractured subsurface (granite surfaces) decrease the capacity for adsorption of mobile radionuclide compounds. Using neutral pH anaerobic groundwater conditions in a microcosm, Anderson et al. (2007) determined that biofilms suppress the capacity for the fluid-rock interfaces to act as barriers (i.e., to sorb) to specific nuclear material migration offsite (e.g., ^{60}Co(II), ^{241}Am(III), ^{234}Th(IV), but not trivalent species such as Pm and Am).

Evidence for the indirect facilitated transport of radionuclides via the metabolic products of microbial cells is also available. Frazier et al. (2005) reported that bacterial siderophores promoted dissolution of UO_2 at fivefold greater net dissolution rates than by simple proton-promoted dissolution. Siderophores normally function to sequester iron under limiting conditions, where low-solubility iron exists as iron oxides (Boukhalfa and Crumbliss 2002). The fact that normally stable tetravalent actinides (e.g., U(IV)) held under anaerobic conditions may in fact be mobilized by soluble microbial organic ligands offers an additional route for radionuclide transport. Siderophore-coupled accumulation of radionuclides (Pu (IV)) was also observed for *Microbacterium flavescens* (JG-9), which produces the siderophore desferrioxamine-B (DFOB) (John et al. 2001). In that study, it was shown that only living and metabolically active *M. flavescens* were observed to take up Fe(III)-DFOB and Pu(IV)-DFOB complexes.

Overall, it is evident that the potential for microbes to influence radionuclide migration is complex and will be site specific since it is dependent on the microorganisms present in the repository and host rock, as well as the geochemical conditions of the system (e.g., pH and Eh).

7.5 Concluding Remarks

International nuclear research programs have provided a considerable body of evidence that a diverse community of microorganisms can be expected in the deep geological repository (DGR) environment. Within the engineered barrier system (EBS), there will be a substantial gradient effect with respect to zones of inhibition, from most inhibitory at or near the container surface, decreasing out toward the bentonite clay into the host rock. The primary controls on microbial growth in the bentonite buffer will be water activity, clay swelling pressure, temperature, and radiation. Available evidence to date suggests that these controls will create a biologically inactive zone extending from the container to tens of centimeters into the compacted clay buffer. Microbes positioned further from the container, particularly in the backfill region of lower density clay, would not be subjected to the same inhibitory factors, and hence, could remain viable. The extent of their activity will be governed, in part, by temperature and availability of nutrients, as well as by mass transport conditions dominant within the backfill and surrounding geologic formations. The activity of viable organisms within the nonlethal regions of the EBS would serve to consume nutrients using oxygen as a terminal electron acceptor and help drive redox conditions from oxidizing to reducing within the repository. Ultimately, a combination of biological and chemical reactions would be expected to create a stable anaerobic environment. These effects would lead to the gradual return of the repository conditions to that of the surrounding host rock. However, the time periods required for this to be achieved remain unclear.

Interfacial regions between the various EBS components, primarily the buffer- and backfill-host rock interfaces, as well as interfaces between cement components and the backfill, may support greater microbial activity and hence larger populations than the highly compacted bentonite buffer. Water movement along these interfaces will be governed by saturation rates imposed by the enclosing low-permeability host rock. Microbes positioned at interfaces would be expected to produce greater amounts of soluble and gaseous end products than elsewhere in the repository. If gases are produced in sufficient quantity, so as to accumulate more rapidly than they dissolve, diffuse, or are consumed, the integrity of the EBS could potentially be affected by creation of preferential pathways for groundwater, and hence microbial transport.

While SRB-mediated MIC will probably not occur in the oxidizing phase of repository evolution, allochthonously produced HS^- could migrate to the container surface once conditions become anoxic, after which indirect corrosion could proceed at a rate limited by sulfide diffusion through, and precipitation (as FeS) in, the buffer. In the event of failure of the used fuel container, other potential microbiologically mediated processes would similarly rely on the integrity of the compacted bentonite surrounding the used fuel containers, including microbial movement to and from the container, as well as transport of microbe–radionuclide complexes or radionuclide–siderophore complexes. Water activity within the compacted bentonite is expected to remain sufficiently limiting so as to prevent significant microbial activity.

The overall potential for microbial activity in low-permeability crystalline and sedimentary rock formations considered for deep geological repositories is low compared with soil, surface, and aquatic environments. In a review, Fredrickson and Onstott (1996) stated that average population doubling times within subsurface environments may be so slow (e.g., 100 years or more; see Fig. 7.2) that, realistically, metabolic activity would scarcely meet the requirements of cell preservation, let alone proliferation. For some subsurface environments, it is not yet clear whether viable cells may even be recoverable, and that perhaps cells present in those environments, while metabolically functional, may have lost the ability to replicate. One would expect that with the passing of sufficient time after nuclear waste emplacement, the repository would eventually return to natural host rock conditions where low potentials for microbial growth and activity dominate. Given this, questions that require consideration are (1) to what extent will biotic reaction rates increase as a consequence of the disturbances that would accompany construction of a DGR and (2) how long will it take for microbial rates to return to original levels? Seeking answers to these important questions provides a unique opportunity to advance environmental microbiology and to expand our understanding of microbial behavior under extremely restrictive conditions.

References

Akid R, Wang H, Smith TJ, Greenfield D, Earthman JC (2008) Biological functionalisation of a sol–gel coating for the mitigation of microbial-induced corrosion. Adv Funct Mater 18:203–211
Amend JP, Teske A (2005) Expanding frontiers in deep subsurface microbiology. Palaeogeogr Palaeoclimatol Palaeoecol 219:131–155
Anderson RT, Lovley DR (2002) Microbial redox interactions with uranium: an environmental perspective. In: Keith-Roach MJ, Livens FR (eds) Interactions of microorganisms with radionuclides. Elsevier, Oxford, pp 205–223
Anderson CR, Jakobsson A-M, Pedersen K (2006) Autoradiographic comparisons of radionuclide adsorption between subsurface anaerobic biofilms and granitic host rocks. Geomicrobiol J 23:15–29
Anderson CR, Jakobsson A-M, Pedersen K (2007) Influence of in situ biofilm coverage on the radionuclide adsorption capacity of subsurface granite. Environ Sci Technol 41:830–836
Baas-Becking LGM (1934) Geobiologie of inleiding tot de milieukunde. W.P. Van Stockum and Zoon, The Hague, The Netherlands
Bass CJ, Holtom GJ, Jackson CP, Lappin-Scott H (2002) The potential impact of micro-organisms in the geosphere on radionuclide migration. In: Report # AEAT/ERRA-0239. Nirex Limited, UK
Baker BJ, Moser DB, MacGregor BJ, Fishbain S, Wagner M, Fry NK, Jackson B, Speolstra N, Loos S, Takai K, Sherwood Lollar B, Fredrickson J, Balkwill D, Onstott TC, Wimpee CF, Stahl DA (2003) Related assemblages of sulphate-reducing bacteria associated with ultradeep gold mines of South Africa and deep basalt aquifers of Washington State. Environ Microbiol 5:267–277

Balkwill DL (1989) Numbers, diversity, and morphological characteristics of aerobic, chemoheterotrophic bacteria in deep subsurface sediments from a site in South Carolina. Geomicrobiol J 7:33–52

Bennett DG, Gens R (2008) Overview of European concepts for high-level waste and spent fuel disposal with special reference waste container corrosion. J Nucl Mater 379:1–8

Biddle JF, Fitz-Bibbon S, Schuster SC, Brenchley JE, House CH (2008) Metagenomic signatures of the Peru Margin subseafloor biosphere show a genetically distinct environment. Proc Natl Acad Sci U S A 105:10583–10588

Booth W (1987) Post-mortem on Three Mile Island. Science 238:1342–1345

Borgonie G, García-Moyano A, Litthauer D, Bert W, Bester A, van Heerden E, Möller C, Erasmus M, Onstott TC (2011) Nematoda from the terrestrial deep subsurface of South Africa. Nature 474:79–82

Boukhalfa H, Crumbliss AL (2002) Chemical aspects of siderophore mediated iron transport. Biometals 15:325–339

Brierley CL (1978) Bacterial leaching. Crit Rev Microbiol 6:207–262

Brown DA, Sherriff BL (1999) Evaluation of the effect of microbial subsurface ecosystems on spent nuclear fuel repositories. Environ Technol 20:469–477

Brydie JR, Wogelius RA, Merrifield CM, Boult S, Gilbert P, Allison D, Vaughan DJ (2005) The μ2M project on quantifying the effects of biofilm growth on hydraulic properties of natural porous media and on sorption equilibria: an overview. In: Shaw RA (ed) Understanding the micro to macro behaviour of rock-fluid systems, Special Publication 249. Geological Society, London, pp 131–144

Busscher HJ, van der Mei HC (2006) Microbial adhesion in flow displacement systems. Clin Microbiol Rev 19:127–141

Cano RJ, Borucki MK (1995) Revival and identification of bacterial spores in 25- to 40-million-year-old Dominican amber. Science 268:1060–1064

Chapelle FH (1993) Ground-water microbiology and geochemistry. Wiley, New York

Chapelle FH, Lovley DR (1990) Rates of microbial metabolism in deep coastal plain aquifers. Appl Environ Microbiol 156:1865–1874

Characklis WG (1990) Microbial fouling. In: Characklis WG, Marshall KC (eds) Biofilms. Wiley, New York, pp 523–584

Chen J, Qin Z, Shoesmith DW (2011) Long-term corrosion of copper in a dilute anaerobic sulphide solution. Electrochim Acta 56:7854–7861

Coombs P, Wagner D, Bateman K, Harrison H, Milodowski AE, Noy D, West JM (2010) The role of biofilms in subsurface transport processes. Q J Eng Geol 43:131–139

Costerton JW, Geesey GG, Cheng K-J (1978) How bacteria stick. Sci Am 238:86–95

Costerton JW, Lewandowski Z, Caldwell DE, Korber DR, Lappin-Scott HM (1995) Microbial biofilms. Ann Rev Microbiol 49:711–745

Craig RF (1987) Soil mechanics, 4th edn. Chapman and Hall, London, UK

Daly MJ (2000) Engineering radiation-resistant bacteria for environmental biotechnology. Curr Opin Biotechnol 11:280–285

Davey HH (2011) Life, death and in-between: meanings and methods in microbiology. Appl Environ Microbiol 77:5571–5576

DeFlaun MF, Fredrickson JK, Dong H, Pfiffner SM, Onstott TC, Balkwill DL, Streger SH, Stackenbrandt E, Knoessen S, van Heerden F (2007) Isolation and characterization of a *Geobacillus thermoleovorans* strain from an ultra-deep South African gold mine. Syst Appl Microbiol 30:152–164

Dixon DA (2000) Porewater salinity and the development of swelling pressure in bentonite based buffer and backfill materials. Posiva, Report 2000–04. Helsinki, Finland

Dixon DA, Chandler NA, Baumgartner P (2002) The influence of groundwater salinity and interfaces on the performance of potential backfilling materials. Proc. 6th International Workshop on Design and Construction of Final Repositories, Brussels, Belgium

Fine F, Gervais P (2005) Thermal destruction of dried vegetative yeast cells and dried bacterial spores in a convective hot air flow: strong influence of initial water activity. Environ Microbiol 7:40–46

Forsyth B, Cameron A, Miller A (1995) Explosives and water quality. In: Hynes TP, Blanchette MC (eds) Proceedings of Sudbury '95 mining and the environment, vol II, Ground and surface water. Canmet, Ottawa, pp 795–803

Frazier W, Kretzschmar R, Kraemer SM (2005) Bacterial siderophores promote dissolution of UO_2 under reducing conditions. Environ Sci Technol 39:5709–5715

Fredrickson JK, Balkwill DL (2006) Geomicrobial processes and biodiversity in the deep terrestrial subsurface. Geomicrobiol J 23:345–356

Fredrickson JK, Onstott TC (1996) Microbes deep inside the Earth. Sci Am 275:68–73

Fredrickson JK, McKinley JP, Bjornstad BN, Long PE, Ringelberg DB, White DC, Krumholz LR, Suflita JM, Colwell FS, Lehman RM, Phelps TJ, Onstott TC (1997) Pore-size constraints on the activity and survival of subsurface bacteria in a late cretaceous shale-sandstone sequence, northwestern New Mexico. Geomicrobiol J 14:183–202

Fredrickson JK, Li S-MW, Gaidamakova EK, Matrosova VY, Zhai M, Sulloway HM, Scholten JC, Brown MG, Balkwill DL, Daly MJ (2008) Protein oxidation: key to bacterial desiccation resistance? ISME J 2:393–403

Fukunaga S, Honya M, Yokoyama E, Arai K, Mine T, Mihara M, Senju T (2001) A study on conditions for microbial transport through compacted buffer material. Mater Res Soc Symp Proc 663:675–682

Ghiorse WC, Wilson JT (1988) Microbial ecology of the terrestrial subsurface. Adv Appl Microbiol 33:107–172

Goldschneider AA, Haralampides KA, MacQuarrie KTB (2007) River sediment and flow characteristics near a bank filtration water supply: implications for riverbed clogging. J Hydrol 344:55–69

Grant WD, Holtom GJ, Rosevear A, Widdowson D (2000) A review of environmental microbiology relevant to the disposal of radioactive waste in a deep geological repository. Nirex Report NSS/R329

Greenblatt CL, Davis A, Clement BG, Kitts CL, Cox T, Cano RJ (1999) Diversity of microorganisms isolated from amber. Microb Ecol 38:58–68

Hallbeck L (2010) Principal organic materials in a repository for spent nuclear fuel. SKB Technical Report, TR-2010-19

Hama K, Bateman K, Coombs P, Hards VL, Milodowski AE, West JM, Wetton PD, Yoshida H, Aoki K (2001) Influence of bacteria on rock-water interaction and clay mineral formation in subsurface granitic environments. Clay Miner 36:599–613

Harder W, Dijkhuizen L (1983) Physiological responses to nutrient limitation. Annu Rev Microbiol 37:1–23

Harvey RW, Suflita JM, McInerney MJ, Mills AL (2007) Overview of issues in subsurface and landfill microbiology. In: Hurst CJ, Crawford RL, Garland JL, Lipson DA, Mills AL, Stetzenback LD (eds) Manual of environmental microbiology, 3rd edn. ASM Press, Washington, DC, pp 795–798

Haveman SA, Stroes-Gascoyne S, Hamon CJ (1995) The microbial population of buffer materials. Atomic Energy of Canada Limited Technical Record, TR-654, COG-94-488

Haveman SA, Stroes-Gascoyne S, Hamon CJ (1996) Biodegradation of a sodium sulphonated naphthalene formaldehyde condensate by bacteria naturally present in granitic groundwater. Atomic Energy of Canada Limited Technical Record, TR-72 1, COG-95-547

Haveman SA, Swanson EWA, Voordouw G, Al TA (2005) Microbial populations of the river-recharged Fredericton aquifer. Geomicrobiol J 22:311–324

Hedin A (2006) Safety function indicators in SKB's safety assessments of a KBS-3 repository. In: Proceeding International High-Level Radioactive Waste Management Conference, Las Vegas, NV, April 30–May 04, 2006

Hoehler TM (2004) Biological energy requirements as quantitative boundary conditions for life in the subsurface. Geobiology 2:205–215

Horn JM, Davies M, Martin S, Lian T, Jones D (1998) Assessing microbiologically induced corrosion of waste package materials in the Yucca Mountain repository. Presented at ICONE-6, May 10–15. Available at: www.osti.gov/bridge/servlets/purl/289885-UhgQya/webviewable/289885.pdf

Howsam P (1987) Biofouling in wells and aquifers. Water Environ J 2:209–215

Huang W-L, Longo JM, Pevear DR (1993) An experimentally derived kinetic model for smectite-to-illite conversion and its use as a geothermometer. Clays Clay Miner 41:162–177

Humphreys PN, West JN, Metcalfe R (2010) Microbial effects on repository performance. NDA report QRS-1378Q-1. February 2010

Jackson BE, McInerney MJ (2002) Anaerobic microbial metabolism can proceed close to thermodynamic limits. Nature 415:454–456

Jägevall S, Rabe L, Pedersen K (2011) Abundance and diversity of biofilms in natural and artificial aquifers of the Äspö Hard Rock Laboratory. Sweden Microb Ecol 61:410–422

Jaisi DP, Eberl DD, Dong H, Kim J (2011) The formation of illite from nontronite by mesophilic and thermophilic bacterial reaction. Clays Clay Miner 59:21–33

Jay JM, Loessner MJ, Golden DA (2005) Modern food microbiology (7th Edn). Springer, New York, 810

John SG, Ruggiero CE, Hersman LE, Tung CS, Neu MP (2001) Siderophore mediated plutonium accumulation by *Microbacterium flavescens* (JG-9). Environ Sci Technol 35:2942–2948

Jolley DM, Ehrhorn TF, Horn J (2003) Microbial impacts to the near-field environment geochemistry: a model for estimating microbial communities in repository drifts at Yucca Mountain. J Contam Hydrol 62–63:553–575

Jones EG, Lineweaver CH (2010) To what extent does terrestrial life "Follow the water"? Astrobiology 10:349–361

Kieft TL, Phelps TJ (1997) Life in the slow lane: activities of microorganisms in the subsurface. In: Amy P, Haldeman D (eds) The microbiology of the terrestrial subsurface. CRC, Boca Raton, FL, pp 135–161

Kim J, Dong H, Seabaugh J, Newell SW, Ebert DD (2004) Role of microbes in the smectite-to-illite reaction. Science 303:830–832

King F (1996) A copper container corrosion model for the in-room emplacement of used CANDU fuel. AECL-11552, COG-96-105. Atomic Energy of Canada Limited Report

King F (2007) Status of the understanding of used fuel container corrosion processes—summary of current knowledge and gap analysis. Nuclear Waste Management Organization Report. NWMO-TR-2007-09. Toronto, ON

King F, Kolář M (2006) Consequences of microbial activity for corrosion of copper used fuel containers—Analyses using the CCM – MIC.0.1 Code. Ontario Power Generation Report 06819-REP-01300-10120-R00. Toronto, Canada

King F, Stroes-Gascoyne S (1995) Microbiologically influenced corrosion of nuclear fuel waste disposal containers. In: Angel P et al. (Eds) Proceedings of the 1995 International Conference of Microbial Influenced Corrosion. 35/1-35/14. Nace International

King F, Ahonen L, Taxen C, Vuorinen U, Werme L (2001) Copper corrosion under expected conditions in a deep geologic repository. Swedish Nuclear Fuel and Waste Management Company Report, SKB TR 01–23

King F, Lilja C, Pedersen K, Pitkänen P, Vähänen M (2010) An update of the state-of the- art report on the corrosion of copper under expected conditions in a deep geologic repository. Swedish Nuclear Fuel and Waste Management Company Report, SKB TR-10-67

Kotelnikova S, Pedersen K (1997) Evidence for methanogenic archaea and homoacetogenic bacteria in deep granitic rock aquifers. FEMS Microbiol Rev 20:339–349

Krumholz LR (1998) Microbial ecosystems in the Earth's subsurface. ASM News 64:197–202

Kurosawa S, Ueda S (2001) Effect of colloids on radionuclide migration for performance assessment of HLW disposal in Japan. Pure Appl Chem 73:2027–2037

Kwong GM (2011) Status of corrosion studies for copper used fuel containers under low salinity conditions. Nuclear Waste Management Organization Report, NWMO TR-2011-14. Toronto, Ontario

Kyle JE, Eydal HSC, Ferris FG, Pedersen K (2008) Viruses in granitic groundwater from 69 to 450m depth of the Äspö hard rock laboratory. Sweden ISME J 2:571–574

Loewen NR, Flett RJ (1984) The possible effects of microorganisms upon the mobility of radionuclides in the groundwaters of the Precambrian shield. Atomic Energy of Canada Limited Technical Report, TR-217

Lovley DR (1987) Anaerobic production of magnetite by a dissimilatory iron-reducing microorganism. Nature 30:252–254

Lovley DR (2006) Dissimilatory Fe (III)- and Mn (IV)-reducing prokaryotes. Prokaryotes 2:635–658

Lovley DR, Klug MJ (1983) Sulphate reducers can outcompete methanogens at freshwater sulphate concentrations. Appl Environ Microbiol 45:187–192

Lovley DR, Phillips EJP, Gorby YA, Landa ER (1991) Microbial reduction of uranium. Nature 350:413–416

Lucht LM, Stroes-Gascoyne S, Miller SH, Hamon CJ, Dixon DA (1997) Colonization of compacted backfill materials by microorganisms. Atomic Energy of Canada Limited Report, AECL-11832, COG-97-321-I

Maak P, Birch K, Simmons GR (2010) Evaluation of container placement methods for the conceptual design of a deep geological repository. Nuclear Waste Management Organization Report, NWMO TR-2010-20, Toronto, ON

Masurat P, Eriksson S, Pedersen K (2010a) Evidence of indigenous sulphate-reducing bacteria in commercial Wyoming bentonite MX-80. Appl Clay Sci 47:51–57

Masurat P, Eriksson S, Pedersen K (2010b) Microbial sulphide production in a compacted Wyoming bentonite MX-80 under in situ conditions relevant to a repository for high-level radioactive waste. Appl Clay Sci 47:58–64

Mauclaire L, McKenzie JA, Schwyn B, Bossart P (2007) Detection and cultivation of indigenous microorganisms in Mesozoic claystone core samples from the Opalinus Clay Formation (Mont Terri Rock Laboratory). Phys Chem Earth 32:232–240

McCollom TM, Amend JP (2005) A thermodynamic assessment of energy requirements for biomass synthesis by chemolithoautotrophic micro-organisms in oxic and anoxic environments. Geobiology 3:135–144

McKinley IG, Hagenlocher I, Alexander WR, Schwyn B (1997) Microbiology in nuclear waste disposal: interfaces and reaction fronts. FEMS Microbiol Rev 20:545–556

McMurry J, Dixon DA, Garroni JD, Ikeda BM, Stroes-Gascoyne S, Baumgartner P, Melnyk TW (2003) Evolution of a Canadian deep geologic repository: base scenario. AECL Report No: 06819-REP-01200-10092-R00

Meike A, Stroes-Gascoyne S (2000) Review of microbial responses to abiotic environmental factors in the context of the proposed Yucca Mountain repository. Atomic Energy of Canada Limited Report AECL-12101. Pinawa, Canada

Merroun JL, Selenska-Pobell S (2008) Bacterial interactions with uranium: an environmental perspective. J Contam Hydrol 102:285–295

Morita RY (1999) Is H_2 the universal energy source for long-term survival? Microb Ecol 38:307–320

Motamedi M, Karland O, Pedersen K (1996) Survival of sulphate reducing bacteria at different water activities in compacted bentonite. FEMS Microbiol Lett 141:83–87

Newby DT, Reed DW, Petzke LM, Igoe AL, Delwiche ME, Roberto FF, McKinley JP, Whiticar MJ, Colwell FS (2004) Diversity of methanotroph communities in a basalt aquifer. FEMS Microbiol Ecol 48:333–344

OECD (2012) The post-closure radiological safety case for a spent fuel repository in Sweden: An International Peer Review of the SKB Licence-application Study of March 2011. Nuclear

Energy Agency, Organisation for Economic Co-operation and Development. NEA Report No. 7084. ISBN 978-92-64-99191-0

Ohnuki T, Kozai N, Sakamoto F, Ozaki T, Nankawa T, Suzuki Y, Francis AJ (2010) Association of actinides with microorganisms and clay: Implications for radionuclide migration from waste-repository sites. Geomicrobiol J 27:225–230

Onofrei M, Gray MN, Roe LH (1991) Superplasticizer function and sorption in high performance cement-based grouts. Swedish Nuclear Fuel and Waste Management Company Stripa Project Report, SKB-TR-91-21. Also Atomic Energy of Canada Limited Report, AECL-10141, COG-91-293, 1992

Onstott TC, Colwell FS, Kieft TL, Murdoch L, Phelps TJ (2009) New horizons for deep subsurface microbiology. Microbe 4:499–505

Ortiz L, Volckaert G, Mallants D (2002) Gas generation and migration in Boom Clay, a potential host rock formation for nuclear waste storage. Eng Geol 64:287–296

Pedersen K (1993a) Bacterial processes in nuclear waste disposal. Microbiol Eur 1:18–23

Pedersen K (1993b) The deep subterranean biosphere. Earth Sci Rev 34:42–47

Pedersen K (1996) Investigations of subterranean bacteria in deep crystalline bedrock and their importance for the disposal of nuclear waste. Can J Microbiol 42:382–391

Pedersen K (1997) Microbial life in deep granitic rock. FEMS Microbiol Rev 20:399–414

Pedersen KA (1999a) Evidence for a hydrogen-driven, intro-terrestrial biosphere in deep granitic rock aquifers. Microbial biosystems: New Frontiers. In: Bell CR, Brylinski M, Johnson-Green P (eds.) Proceedings of the 8th annual symposium of microbial ecology. Atlantic Society for Microbial Ecology, Halifax, NS, Canada

Pedersen KA (1999b) Subterranean microorganisms and radioactive waste disposal in Sweden. Eng Geol 52:163–176

Pedersen K (2000) Microbial processes in radioactive waste disposal. SKB technical report TR-00-04. April 2000

Pedersen KA (2010) Analysis of copper corrosion in compacted bentonite clay as a function of clay density and growth conditions for sulphate-reducing bacteria. J Appl Microbiol 108:1094–1104

Pedersen K, Albinsson Y (1992) Possible effects of bacteria on trace element migration in crystalline bed-rock. Radiochim Acta 58(59):365–369

Pedersen K, Ekendahl S (1992a) Assimilation of CO_2 and introduced organic compounds by bacterial communities in groundwater from southeastern Sweden deep crystalline bedrock. Microb Ecol 23:1–14

Pedersen K, Ekendahl S (1992b) Distribution and activity of bacteria in deep granitic groundwaters of south-eastern Sweden. Microb Ecol 20:37–52

Pedersen K, Motamedi M, Karnland O, Sanden T (2000a) Cultivability of microorganisms introduced into a compacted bentonite clay buffer under high-level radioactive waste repository conditions. Eng Geol 58:149–161

Pedersen K, Motamedi M, Karnland O, Sanden T (2000b) Mixing and sulphate-reducing activity of bacteria in swelling, compacted bentonite clay under high-level radioactive waste repository conditions. J Appl Microbiol 89:1038–1047

Phelps TJ, Murphy EM, Pfiffner SM, White DC (1994) Comparison between geochemical and biological estimates of subsurface microbial activities. Microb Ecol 28:335–349

Poulain S, Le Marrec C, Altmann S (2008) Microbial investigations in Opalinus clay, an argillaceous formation under evaluation as a potential host rock for a radioactive waste repository. Geomicrobiol J 25:240–249

Pusch R, Weston R (2003) Microstructural stability controls the hydraulic conductivity of smectitic buffer clay. Appl Clay Sci 23:35–41

Rainey FA, Ray K, Ferreira M, Gatz BZ, Fernanda Nobre M, Gagaley D, Rash BA, Park M-J, Earl AM, Shank NC, Small AM, Henk MC, Battista JR, Kämpfer PK, Costa MS (2005) Extensive diversity of ionizing-radiation-resistant bacteria recovered from Sonoran Desert soil and description of nine new species of the genus *Deinococcus* obtained from a single soil sample. Appl Environ Microbiol 71:5225–5235

Rastogi G, Stetler LD, Peyton BM, Sani RK (2009) Molecular analysis of prokaryotic diversity in the deep subsurface of the former Homestake Gold Mine, South Dakota. USA J Microbiol 47:371–384

Schwartz E, Friedrich B (2006) The H_2-metabolizing prokaryotes. In: Dworkin M, Falkow S, Rosenberg E, Schleifer K-H, Stackebrandt (Eds) The prokaryotes, Vol 2, Ecophysiology and biochemistry. Chapter 1.17. Springer, New York

Sharp AA, Cunningham AB, Komlos J, Billmayer J (1999) Observation of thick biofilm accumulation and structure in porous media and corresponding hydrodynamic and mass transfer effects. Water Sci Technol 3:1195–1201

Sheng XX, Ting UP, Pehkonen SA (2007) The influence of sulphate-reducing bacteria bioflim on the corrosion of stainless steel AISI 316. Corrosion Sci 49:2159–2176

Sheppard MI, Stroes-Gascoyne S, Motycka M, Haveman SA (1997) The influence of the presence of sulphate on methanogenesis in the backfill of a Canadian nuclear fuel waste disposal vault; A laboratory study. Atomic Energy of Canada Limited Report, AECL-11764, COG-97-21-I

Sherwood Lollar B (2011) Far-field microbiology considerations relevant to a deep geological repository—State of Science review. Nuclear Waste Management Organization, Techncial Report NWMO TR-2011-09. Toronto, ON

Smart N, Rance A, Reddy B, Lydmark S, Pedersen K, Lilja C (2011) Further studies of in situ corrosion testing of miniature copper–cast iron nuclear waste canisters. Corrosion Eng Sci Technol 46:142–147

Smellie JAT, Karlsson F, Alexander WR (1997) Natural analogue studies: present status and performance assessment implications. J Contam Hydrol 26:3–17

Stevens T (1997) Lithoautotrophy in the subsurface. FEMS Microbiol Rev 20:327–337

Stevens TO, McKinley JP (1995) Lithoautotrophic microbial ecosystems in deep basalt aquifers. Science 270:450–453

Stewart LS (1938) Isolation of halophilic bacteria from soil, water, and dung. J Food Sci 3:417–420

Stroes-Gascoyne, S. 1989. The Potential for Microbial Life in a Canadian High-Level Nuclear Fuel Waste Disposal Vault: A Nutrient and Energy Source Analysis. Atomic Energy of Canada Limited Report, AECL-9574

Stroes-Gascoyne S (1997) Microbial aspects of the Canadian used fuel disposal concept—Status of current knowledge from applied experiments. Atomic Energy of Canada Limited Report, 06819-REP-01200-0026-R00

Stroes-Gascoyne S (2005) A review of international experience with microbial activity in bentonite-based sealing materials and argillacous host rocks. Atomic Energy of Canada Limited. Report No: 06819-REP-01300-10109-R00

Stroes-Gascoyne S (2010) Microbial occurrence in bentonite-based buffer, backfill and sealing materials from large-scale experiments at AECL's Underground Research Laboratory. Appl Clay Sci 47:36–42

Stroes-Gascoyne S, Gascoyne M (1998) The introduction of microbial nutrients into a nuclear waste disposal vault during excavation and operation. Environ Sci Technol 32:317–326

Stroes-Gascoyne S, Hamon CJ (2008) Preliminary microbial analysis of limestone and shale rock samples. Nuclear Waste Management Organization, NWMO TR-2008-09, Toronto, ON

Stroes-Gascoyne S, Hamon CJ (2008b) The effect of intermediate dry densities (1.1–1.5 g/cm^3) and intermediate porewater salinities (60–90 g NaCl/L) on the culturability of heterotrophic aerobic bacteria in compacted 100 % bentonite. Nuclear Waste Management Organization, NWMO TR-2008-11, Toronto, ON

Stroes-Gascoyne S, King F (2002) Microbiologically influenced corrosion issues in high-level nuclear waste repositories. In: Little B (Eds.) Proceedings of CORROSION/ 2002 Research Topical Symposium Microbiologically Influenced Corrosion. NACE International, Houston, TX, p. 79

Stroes-Gascoyne S, West JM (1996) An overview of microbial research related to high-level nuclear waste disposal with emphasis on the Canadian concept for the disposal of nuclear fuel waste. Can J Microbiol 42:349–366

Stroes-Gascoyne S, West JM (1997) Microbial studies in the Canadian nuclear fuel waste management program. FEMS Microbiol Rev 20:573–590

Stroes-Gascoyne S, Lucht LM, Borsa J, Delaney TL, Haveman SA, Hamon CJ (1995) Radiation resistance of the natural microbial population in buffer materials. Mater Res Soc Symp Proc 353:345–352

Stroes-Gascoyne S, Gascoyne M, Onagi D, Thomas DA, Hamon CJ, Watson R, Porth RJ (1996) Introduction of microbial nutrients in a nuclear fuel waste disposal vault as a result of excavation and operation activities. Atomic Energy of Canada Limited Report, AECL-11532, COG-96-14

Stroes-Gascoyne S, Haveman SA, Vilks P (1997a) The change in bioavailability of organic matter associated with clay-based buffer material as a result of heat and radiation treatment. Mater Res Soc Symp Proc 465:987–994

Stroes-Gascoyne S, Pedersen K, Haveman SA, Dekeyser K, Arlinger J, Daumas S, Ekendahl S, Hallbeck L, Hamon CJ, Jahromi N, Delaney T-L (1997b) Occurrence and identification of organisms in compacted clay-based buffer material designed for use in nuclear fuel waste disposal vault. Can J Microbiol 43:1133–1146

Stroes-Gascoyne S, Haveman SA, Hamon CJ, Ticknor KV (2000) Analysis of biofilms grown in situ at AECL's Underground Research Laboratory on Granite, Titanium and Copper Coupons. Atomic Energy of Canada Limited Report, AECL-12098

Stroes-Gascoyne S, Hamon CJ, Vilks P, Gierszewski P (2002) Microbial, redox and organic characteristics of compacted clay-based buffer after 6.5 years of burial at AECL's Underground Research Laboratory. Appl Geochem 17:1287–1303

Stroes-Gascoyne S, Hamon CJ, Kohle C, Dixon DA (2006). The effects of dry density and porewater salinity on the physical and microbiological characteristics of highly compacted bentonite. Ontario Power Generation, Nuclear Waste Management Division Report 06819-REP-01200-10016-R00

Stroes-Gascoyne S, Hamon CJ, Dixon DA, Martino JB (2007a) Microbial analysis of samples from the tunnel sealing experiments at AECL's Underground Research Laboratory. Phys Chem Earth 32:219–231

Stroes-Gascoyne S, Maak P, Hamon CJ, Kohle C. (2007b). Potential implications of microbes and salinity on the design of repository sealing system components. NWMO TR-2007-10

Stroes-Gascoyne S, Schippers A, Schwyn B, Poulain S, Sergeant C, Simanoff M, Le Marrec C, Altmann S, Nagaoka T, Mauclaire L, McKenzie J, Daumas S, Vinsot A, Beaucaire C, Matray S-M (2007c) Microbial community analysis of Opalinus Clay drill core samples from the Mont Terri Underground Research Laboratory. Switzerland Geomicrobiol J 24:1–17

Stroes-Gascoyne S, Hamon CJ, Dixon DA, Priyanto DG (2010a) The effect of $CaCl_2$ Porewater Salinity (50–100 g/L) on the culturability of heterotrophic aerobic bacteria in compacted 100 % bentonite with dry densities of 0.8 and 1.3 g/cm^3. Nuclear Waste Management Organization, NWMO TR-2010-06, Toronto, ON

Stroes-Gascoyne S, Hamon CJ, Maak P, Russell S (2010b) The effects of the physical properties of highly compacted smectitic clay (bentonite) on the culturability of indigenous microorganisms. Appl Clay Sci 47:155–162

Stroes-Gascoyne S, Sergeant C, Schippers A, Hamon CJ, Nèble S, Vesvres M-H, Barsotti V, Poulain S, Le Marrec C (2011) Biogeochemical processes in a clay formation in situ experiment: Part D—Microbial analyses—synthesis of results. Appl Geochem 26:980–989

Suzuki Y, Banfield J (2004) Resistance to, and accumulation of, uranium by bacteria from a uranium-contaminated site. Geomicrobiol J 21:113–121

Thorn PM, Ventullo RM (1988) Measurement of bacterial growth rates in subsurface sediments using the incorporation of tritiated thymidine into DNA. Microb Ecol 16:3–16

Tufenkji N, Ryan JN, Elimelech M (2002) The promise of bank filtration. Environ Sci Technol 36:422A–428A

Uberoi V, Bhattacharya SK (1995) Interactions among sulphate reducers, acetogens, and methanogens in anaerobic propionate systems. Water Environ Res 67:330–339

Vandergraaf TT, Miller HG, Jain DK, Hamon CJ, Stroes-Gascoyne S (1997) The effect of biofilms on radionuclide transport in the geosphere: Results from an initial investigation. Atomic Energy of Canada Limited Technical Record, TR-774, COG-96-635-I

Vieira R, Volesky B (2000) Biosorption: a solution to pollution? Int Microbiology 3:17–24

Vreeland RH, Piselli AF Jr, McDonnough S, Meyers SS (1998) Distribution and diversity of halophilic bacteria in a subsurface salt formation. Extremophiles 2:321–331

Wang Y, Francis AJ (2005) Evaluation of microbial activity for long-term performance assessments of deep geologic nuclear waste repositories. J Nucl Radiochem Proc 6:43–50

Wersin P, Spahiu K, Bruno J (1994) Time evolution of dissolved oxygen and redox conditions in a HLW repository. Swedish Nuclear Fuel and Waste Management Company Technical Report, TR 94–02

Wersin P, Johnson LH, McKinley IG (2007) Performance of the bentonite barrier at temperatures beyond 100 °C: a critical review. Phys Chem Earth 32:780–788

Wersin P, Stroes-Gascoyne S, Pearson FJ, Tournassat C, Leupin OX, Schwyn B (2011) Biogeochemical processes in a clay formation in-situ experiment: Part G—key interpretations & conclusions. Implications for repository safety. Appl Geochem 26(6):1023–1034

West JM, McKinley IG (2002) The geomicrobiology of radioactive waste disposal. In: Bitton G (ed) The encyclopaedia of environmental microbiology. Wiley, New York, pp 2661–2674

West JM, McKinley IG, Stroes-Gascoyne S (2002) Microbial effects on waste repository materials. In: Keith-Roach M, Livens F (eds) Interactions of microorganisms with radionuclides. Elsevier Sciences, Oxford, UK, pp 255–277

Whitman WB, Coleman DC, Wiebe WJ (1998) Prokaryotes: the unseen majority. Proc Natl Acad Sci USA 95:6578–6583

Wilkins MJ, Livens FR, Vaughan DJ, Lloyd JR (2006) The impact of Fe(III) reducing bacteria on uranium mobility. Biogeochemistry 78:125–150

Wilkins MJ, Livens FR, Vaughan DJ, Beadle I, Lloyd JR (2007) The influence of microbial redox cycling on radionuclide mobility in the subsurface at a low-level radioactive waste storage site. Geobiology 5:293–301

Wolfaardt GM, Lawrence JR, Korber DR (2007) Cultivation of microbial communities. In: Hurst CJ, Crawford RL, Garland JL, Lipson DA, Mills AL, Stetzenback LD (eds) Manual of environmental microbiology, 3rd edn. ASM Press, Washington, DC, pp 101–111

Xu LC, Fang HHP, Chan KY (1999) Atomic force micrsoscopy study of microbiologically influenced corrosion of mild steel. J Electrochem Soc 146:4455–5560

Chapter 8
Life in Hypersaline Environments

Aharon Oren

Abstract Many microorganisms are adapted to life at high-salt concentrations. Halophilic representatives are found in each of the three domains of life: Archaea, Bacteria, and Eukarya. Halophilic viruses exist as well. In NaCl-saturated brines such as found in the northern part of Great Salt Lake, Utah, in a few other natural salt lakes, and in saltern crystallizer ponds for the production of salt, we find members of all groups. Blooms of microorganisms have occasionally been observed in the magnesium- and calcium-rich waters of the Dead Sea. Dense communities of extremely halophilic Archaea (family Halobacteriaceae) and of the alga *Dunaliella salina* often impart a red color to salt-saturated brines. There are different strategies that enable halophilic or halotolerant microorganisms to grow in the presence of high-salt concentrations. A few groups (Archaea of the family Halobacteriaceae; the red extremely halophilic bacterium *Salinibacter*) maintain molar concentrations of salts (K^+, Cl^-) intracellularly, and their proteins are functional in a high-salt environment. Other groups (most salt-adapted members of the Bacteria, halophilic algae, and fungi) accumulate organic solutes to provide osmotic balance of their cytoplasm with the hypersaline medium.

8.1 Introduction

One day I rode to a large salt lake, or Salina, which is distant fifteen miles from the town [of El Carmen]. During the winter it consists of a shallow lake of brine, which in summer is converted into a field of snow-white salt. ... One of these brilliantly-white and level expanses, in the midst of the brown and desolate plain, offers an extraordinary spectacle. A large quantity of salt is annually drawn from the salina; and great piles, some hundred tons in weight, were lying ready for exportation. ... The border of the lake is formed of mud: and in this numerous large crystals of gypsum, some of which are three inches long, lie embedded; whilst on the surface, others of sulphate of magnesia lie scattered about. ... The mud is black, and has a fetid odour. I could not, at first, imagine the cause of this, but I

A. Oren (✉)
Department of Plant and Environmental Sciences, The Institute of Life Sciences, The Hebrew University of Jerusalem, Edmond J. Safra Campus, Givat Ram, Jerusalem 91904, Israel
e-mail: aharon.oren@mail.huji.ac.il

afterwards perceived that the froth, which the wind drifted on shore was coloured green, as if by confervæ: I attempted to carry home some of this green matter, but from an accident failed. Parts of the lake seen from a short distance appeared of a reddish colour, and this, perhaps, was owing to some infusorial animalcula. The mud in many places was thrown up by numbers of some kind of annelidous animal. How surprising it is that any creatures should be able to exist in a fluid, saturated with brine, and that they should be crawling among crystals of sulphate of soda and lime! ... Flamingoes in considerable numbers inhabit this lake; they breed here, and their bodies are sometimes found by the workmen, preserved in the salt. I saw several wading about in search of food, - probably for the worms which burrow in the mud; and these latter, perhaps, feed on infusoria or confervæ. Thus we have a little world within itself, adapted to these little inland seas of brine.

Thus, Charles Darwin described his 1833 visit to a coastal hypersaline lake in Patagonia (Darwin 1839). A reddish color is often associated with hypersaline brines as many types of halophilic microorganisms are pigmented pink, orange, or purple. The waters of the northern part of Great Salt Lake, Utah, are colored red, and so are the brines of crystallizer ponds of salterns where salt is produced by evaporation of seawater (Fig. 8.1). Such red brines typically contain between 10^7 and 10^8 prokaryote cells per milliliter, numbers two to three orders of magnitude higher than the numbers of bacteria normally found in ocean water. Beautiful displays of colored microorganisms, including green "confervæ," can be seen among the "crystals of sulphate of ... lime" shown in Fig. 8.2: masses of gypsum ($CaSO_4 \cdot 2H_2O$) deposited on the bottom of a saltern evaporation pond.

Fig. 8.1 A saltern crystallizer pond of Salt of the Earth Ltd., Eilat, Israel: NaCl-saturated brine (total dissolved salt concentration ~340 g/l) colored *red* due to dense communities of halophilic Archaea (*Haloquadratum walsbyi* and others), the unicellular β-carotene-rich alga *Dunaliella salina*, and possibly red-pigmented members of the Bacteria as well

8 Life in Hypersaline Environments

Fig. 8.2 Gypsum crusts on the bottom of an evaporation pond (total dissolved salt concentration of the brine ~200 g/l) of Salt of the Earth Ltd., Eilat, Israel. The *upper panel* shows the surface of the crust, colored *orange brown* by carotenoid-rich unicellular cyanobacteria (*Aphanothece*-type); the *lower panel* shows the stratification of the microbial communities within a piece of gypsum crust (diameter ~30 cm). A layer of dark-green filamentous cyanobacteria (*Phormidium*-type) is found below the *orange* surface layer. Then follows a *red-purple* layer that contains anoxygenic photosynthetic sulfur bacteria (*Halochromatium*-type; Gammaproteobacteria) using sulfide as electron donor for phototrophic growth. The sulfide is supplied by sulfate-reducing bacteria in the black mud underlying the gypsum layer

Most water bodies on our planet are saline. The oceans contain on average 35 g/l dissolved salts, 86 % of which is NaCl. Far higher salt concentrations are found in natural hypersaline lakes such as Great Salt Lake, the Dead Sea, and other inland lakes, evaporation ponds constructed in coastal areas for the production of salt from seawater, brine pools found at several sites on the bottom of the oceans, and others. All these environments are inhabited by diverse communities of microorganisms "thriving in salt" (Boetius and Joye 2009). Halophilic (salt-loving) and halotolerant microorganisms can even be found within salt crystals, and they may survive there for long periods, possibly even for millions of years.

This chapter provides an overview of the biology of hypersaline environments and of the properties of the microorganisms inhabiting them. It will first discuss the phylogenetic diversity within the world of the halophiles: we find halophiles within each of the three domains of life: Archaea, Bacteria, and Eukarya. Halophilic viruses exist as well, and we have learned much about these in recent years. Then it will explore the mechanisms used by the microorganisms to withstand the osmotic pressure of their highly saline habitats, and show how different groups of halophiles have solved the problem of osmotic adaptation in different ways, and it will examine the upper limits of salinity at which certain microbial processes were found to proceed. Although hypersaline ecosystems can be quite diverse, not all physiological types of microorganisms known from freshwater or marine environments are active at the highest salt concentrations. The chapter deals with "thalassohaline" environments, i.e., hypersaline environments formed by evaporation of seawater to form sodium- and chloride-dominated brines of near-neutral pH, as well as "athalassohaline environments" whose ionic composition is highly different from that of ocean water, and sometimes may show extremes of pH as well. As examples of athalassohaline environments the microbiology of the Dead Sea (high magnesium, high calcium, relatively low sodium, and a relatively low pH) and of the Wadi an Natrun lakes in Egypt (extremely high pH, very low magnesium and calcium) will be presented. Some halophilic microorganisms inhabiting such environments are true "polyextremophiles" that can tolerate not only the highest salt concentrations but extremes of temperature and pH as well.

The extensive literature about hypersaline environments and their biota has been reviewed by many authors in the past. The book "Hypersaline environments. Microbiology and biogeochemistry" by Barbara Javor (1989) remains a rich source of information, and the older reviews by Kushner (1978) and Borowitzka (1981) are still relevant today. Many aspects of the biology of halophilic microorganisms and their habitats have been discussed by the author of this chapter in earlier review papers and book chapters (McGenity and Oren 2012; Oren 2002a, b, 2006a, 2007, 2008, 2009, 2011b, c).

8.2 Diversity of Halophilic and Halotolerant Microorganisms

Some of the diversity of halophilic microbial life can be observed even without a microscope. The bright colors of the saltern crystallizer brine shown in Fig. 8.1 and the differently pigmented zones within the benthic gypsum crust illustrated in Fig. 8.2 show that microorganisms with different pigmentations may develop in certain hypersaline environments to densities sufficiently high to be seen with the naked eye. Microscopic examination of the salt-saturated brine of the crystallizer pond shows that in the numerically dominant organism there is an unusual type of flat square- or rectangular-shaped cell. This organism, first recognized as a prokaryote by Walsby thanks to the presence of gas vesicles within the cells (Walsby 1980, 2005) and now known as *Haloquadratum walsbyi* (Bolhuis et al. 2004, 2006; Burns et al. 2007; Legault et al. 2006), is now known to be a major contributor to the biota of salt lakes worldwide, natural as well as artifical such as saltern ponds, when NaCl concentration approaches saturation. *Haloquadratum* is a representative of the domain Archaea. In addition, a pigmented eukaryote, the unicellular flagellated green alga *Dunaliella salina*, is present in the saltern brine at numbers up to a few thousands per milliliter. It is colored orange red due to massive accumulation of β-carotene granules within its chloroplast. The orange-brown, the green, and the purple layers within the gypsum crust shown in Fig. 8.2 are colored due to the growth of cyanobacteria and purple sulfur bacteria, organisms representing different branches of the domain Bacteria. All three domains of life are thus represented within the microbial communities adapted to life at high-salt concentrations.

8.2.1 Archaea

Halophilic Archaea (order Halobacteriales, family Halobacteriaceae) are the extreme halophiles par excellence (Oren 1994b, 2006b). Nearly without exceptions they are pigmented pink red due to the presence of bacterioruberin derivatives, 50-carbon carotenoid pigments that provide protection to photooxidative damage to the cells. Today more than 40 genera and more than 150 species of Halobacteriaceae have been described. Most members of the family require at least 100–150 g/l salts for growth. When suspended in solutions containing less salt, the cells are irreversibly damaged and many species lyse when exposed to freshwater. Only a few members of the family can grow and survive at salt concentrations below 50 g/l.

Since representatives of the genera *Halobacterium* and *Halococcus* were first isolated from salted food products more than a century ago, these red extremely halophilic prokaryotes have fascinated scientists because of their unusual properties, so different from all other known forms of life, as shown for example by the title of Helge Larsen's classic essay on "the halobacteria's confusion to biology" (Larsen 1973). The Halobacteriaceae are typically aerobic organisms that use organic substrates as carbon and energy source. Their metabolic potential was in the past considered to be limited, amino acids and small organic acids being the preferred carbon and energy sources for most species; many types also use simple sugars, as well as polymers such as starch, proteins, and lipids. Recent studies have shown that the metabolic diversity of the group was underestimated in the past. Some representatives can even metabolize hydrocarbons and aromatic compounds (Andrei et al. 2012; Falb et al. 2008). Members of the genus *Halobacterium* can also grow in the absence of molecular oxygen by fermentation of arginine (Hartmann et al. 1980), and *Halorhabdus tiamatea*, a non-pigmented isolate from a deep hypersaline anoxic basin near the bottom of the Red Sea (see Sect. 8.6.5), only grows on complex substrates by fermentation (Antunes et al. 2008a). Representatives of several genera can grow anaerobically with nitrate as the electron acceptor and some can use other electron acceptors such as dimethylsulfoxide, trimethyl-N-oxide, or fumarate. Another unusual mode of life displayed by some members of the Halobacteriaceae is photoheterotrophy: *Halobacterium* spp. and some other related organisms can produce bacteriorhodopsin, a membrane-bound purple protein that carries a retinal moiety and that serves as a light-driven proton pump.

Some species of Halobacteriaceae are polyextremophiles that combine the ability to grow at high-salt concentrations and the requirement of high salt for growth and structural stability with the ability to grow at extremes of pH and sometimes of temperature as well. Many genera of the family are haloalkaliphiles that do not grow at neutral pH but require pH 9–11 for growth (Bowers and Wiegel 2011; Soliman and Trüper 1982). Such organisms are adapted to life in hypersaline soda lakes such as Lake Magadi in Kenya and the lakes of the Wadi an Natrun in Egypt described below in further depth (see Sect. 8.6.4). Such haloalkaliphilic Archaea can color the brines of soda lakes red purple (Imhoff et al. 1979). A single acidophilic member has been described: *Halarchaeum acidiphilum*. It grows only within the narrow pH range between 4.1 and 4.8. Surprisingly, it was isolated from solar salt that showed an alkaline reaction (Minegishi et al. 2010), and its true ecological niche is still unknown. *Halorubrum lacusprofundi* isolated from Deep Lake, Antarctica (Franzmann et al. 1988), is a facultative psychrophile that can grow at temperatures as low as 4 °C, even if its temperature optimum is above 30 °C. *Halorubrum sodomense* from the Dead Sea is extremely tolerant to the high concentrations of magnesium found in its habitat (Oren 1983).

Not all halophilic members of the Archaea belong to the Halobacteriales order. There also are a few genera of obligatory halophilic methanogens (*Methanohalophilus*, *Methanohalobium*) that can grow up to very high-salt concentrations. Culture-independent studies of the archaeal communities in saltern crystallizer ponds in Spain and in Lake Tyrrell, NW Australia, recently led to the recognition that there is a group of extremely small (~0.6 μm) spherical shaped Archaea present in high numbers. These "Nanohaloarchaea" are not closely related to the Halobacteriaceae. The genome sequence for a representative of this group could be reconstructed from metagenomic data (Ghai et al. 2011; Narasingarao et al. 2012), but all attempts to grow these intriguing organisms have failed thus far.

8.2.2 Bacteria

In the past salt-saturated brines were considered the habitat of halophilic Archaea, while hypersaline environments with salt concentrations up to ~200–250 g/l are the habitat for a great variety of halophilic or very halotolerant members of the domain Bacteria (Oren 2011b; Ventosa et al. 1998). And indeed, many groups of Bacteria, belonging to diverse phylogenetic branches and showing different types of metabolism, can grow at elevated salinities. Most are moderate halophiles that grow optimally at salt concentrations between 50 and 100 g/l but are able to grow also at higher salinities, albeit at lower rates. Among these moderately halophilic Bacteria are representatives of the Cyanobacteria (Fig. 8.2) (Oren 2012c), different branches of the Proteobacteria including aerobic heterotrophs and anoxygenic phototrophic members of the Gammaproteobacteria (Fig. 8.2), sulfate-reducing anaerobes belonging to the Deltaproteobacteria, different types of Firmicutes, members of the Actinobacteria, the Spirochaetes, and the Bacteroidetes. Among the aerobic heterotrophic Proteobacteria the best-known group is the family *Halomonadaceae* (Gammaproteobacteria) whose members of the genera *Halomonas* and *Chromohalobacter* are extremely versatile prokaryotes with respect to their adaptability to a wide range of salt concentrations (Ventosa et al. 1998).

An interesting, phylogenetically coherent group of Bacteria that require high-salt concentrations for growth is the order Halanaerobiales with two families: the Halanaerobiaceae and the Halobacteroidaceae (Kivisto and Karp 2011; Oren 2006c; Rainey et al. 1995). Phylogenetically the group is affiliated with the Low G+C Firmicutes. The ~30 described species are all obligate anaerobes, and most obtain their energy from fermentation of simple sugars or amino acids. Some representatives (e.g., *Selenihalanaerobacter shriftii*) also grow by anaerobic respiration using selenate or nitrate as the electron acceptors; others (*Acetohalobium*, *Natroniella*) are homoacetogens. The group also contains the polyextremophilic *Halothermothrix orenii*, recovered from sediment of a Tunisian salt lake, and growing up to 200 g/l and up to 68 °C (optimum at 100 g/l salt and 60 °C) (Cayol et al. 1994). Other anaerobes, belonging to a separate phylogenetic lineage that like it not only hot and salty but alkaline as well, are the halophilic alkalithermophilic *Natronovirga wadinatrunensis*, *Natranaerobius trueperi*, and *Natranaerobius thermophilus* isolated from the soda lakes of the Wadi an Natrun, Egypt (Bowers et al. 2009; Mesbah and Wiegel 2009; Mesbah et al. 2007a, b; see Sect. 8.6.4). Another example of an unusual anaerobic moderate halophile is *Haloplasma contractile* found in the brine-filled Shaban Deep in the Red Sea. It requires between 15 and 180 g/l salt and lives by fermentation or denitrification. It shows an unusual mode of motility by contraction of tentacle-like protrusions. It is unrelated to any major branch of the Bacteria, and its phylogenetic position is between the Firmicutes and the Mollicutes (Antunes et al. 2008b; see Sect. 8.6.5). The order Haloplasmatales was established for this organism, and thus far it is the only member of this order.

A relatively recent addition to the list of halophilic members of the Bacteria is the genus *Salinibacter* with type species *Salinibacter ruber*, the first real non-archaeal extreme halophile known. The existence of a rod-shaped extremely halophilic prokaryote phylogenetically affiliated with the Bacteroidetes was first inferred from culture-independent studies: fluorescence in situ hybridization studies of prokaryote communities in saltern crystallizer brines in Spain showed an abundance (~15–20 % of the total cell numbers, and sometimes even more) of slightly curved rods that stained with fluorescent 16S rRNA probes targeting Bacteria (Antón et al. 1999). The organism was soon brought into culture (Antón et al. 2002, 2008). *Salinibacter* readily develops colonies on agar media with 200–250 g/l salt and low concentrations of yeast extract. Cells are pigmented orange red due to the presence of a novel C_{40}-carotenoid acyl glycoside compound named salinixanthin (Lutnæs et al. 2002). One of the novel features of *Salinibacter* is the presence of unusual acylhalocapnine lipids in its membrane (Baronio et al. 2010; Corcelli et al. 2004). The organism is an aerobic chemoorganotroph, but it can also use light as an energy source thanks to the presence of a bacteriorhodopsin-like light-driven proton pump (xanthorhodopsin) that can accept photons absorbed by the salinixanthin carotenoid (Lanyi 2005). *Salinibacter* does not grow below 150 g/l salt and is thus no less halophilic and salt requiring than the most halophilic representatives of the Halobacteriaceae of the archaeal domain. It is easy to grow, and it can be selectively enriched by exploiting differences in sensitivity toward antibiotics as compared to the Archaea (Elevi Bardavid et al. 2007). Physiological experiments as well as genomic analysis (Mongodin et al. 2005) has shown that *Salinibacter* resembles the aerobic halophilic Archaea in many aspects, in spite of its different phylogenetic affiliation (Oren 2013a; Oren et al. 2004; Peña et al. 2010, 2011).

8.2.3 Eukarya

Microbial communities at high-salt concentrations are generally dominated by prokaryotes (Bacteria and/or Archaea, depending on the salinity). Eukaryotes generally play minor roles, but still different groups of eukaryotic microorganisms may occur up to the highest salinities: unicellular algae of the genus *Dunaliella*, salt-adapted fungi and yeast, and different types of protozoa. There even is a macroorganism genus, the brine shrimp *Artemia* ("sea monkey") which thrives in thalassohaline hypersaline habitats.

Species of the green algae genus *Dunaliella* are the main primary producers in saltern crystallizer ponds, in the Dead Sea, in the northern arm of Great Salt Lake, and in many other hypersaline water bodies (Elevi Bardavid et al. 2008; Javor 1989; Oren 2001, 2005; Oren et al. 1995c; Post 1977; Stephens and Gillespie 1976). *Dunaliella* is a halotolerant rather than a truly halophilic alga, but some species, especially *D. salina*, can still grow at salinities too high for the most salt-tolerating cyanobacteria. *Dunaliella* has a single chloroplast that occupies about half of its cell volume. Some species become orange red colored when grown under stress conditions (high light intensity, nutrient limitation, supra-optimal salinity) due to the massive accumulation of granules of β-carotene between the thylakoids of the chloroplast. Thus, *Dunaliella* can also contribute to the red coloration of hypersaline brines (see above). Other eukaryotic algal groups such as diatoms do not contain representatives that are as well adapted to life at high-salt concentrations as *Dunaliella* spp.

The contribution of salt-loving and salt-tolerant fungi to the biota of hypersaline environments became recognized only recently. Species of black yeasts, notably

Hortaea werneckii and *Trimmatostroma salinum*, are commonly found in marine salterns (Butinar et al. 2005a, b; Gunde-Cimerman et al. 2000, 2009; Zajc et al. 2011; Zalar et al. 2005), and they are well adapted to life at high-salt concentrations (Gostinčar et al. 2009, 2011; Lenassi et al. 2013). The most halophilic fungus known is *Wallemia ichthyophaga*, which can grow up to 250 g/l salt (Zajc et al. 2013). Many other genera of fungi and yeasts have been found in salterns and natural salt lakes, including in the athalassohaline waters of the Dead Sea (Buchalo et al. 1998; Butinar et al. 2005a). However, little is yet known about the quantitative contribution of fungi to the heterotrophic activity in hypersaline ecosystems.

Amoeboid, ciliate, and flagellate protozoa are often observed in the waters and in the sediments of hypersaline water bodies up to NaCl saturation (Cho 2005; Hauer and Rogerson 2005; Heidelberg et al. 2013; Park et al. 2003; Triado-Margarit and Casamayor 2013). Among the isolated halophilic heterotrophic flagellates are *Pleurostomum flabellatum* (optimum growth at 300 g/l salt, requiring at least 150–200 g/l) (Park et al. 2007) and *Halocafeteria seosinensis* (no growth below 75 g/l salt, optimum at 150 g/l) (Park et al. 2006). Analysis of the microbial food web along the salinity gradient in saltern evaporation ponds in Spain showed that the activity of eukaryotic predators controls the density of the heterotrophic prokaryotic community in ponds of intermediate salinity (100–150 g/l salt). Protozoa were occasionally seen in the crystallizer ponds as well, but there viruses were much more important than protozoa as factors regulating the community density and activity (Pedrós-Alió et al. 2000a, b).

Eukaryote protists are also abundantly found in the deep anoxic brines in the eastern Mediterranean. Fluorescence in situ hybridization and analysis of small-subunit rRNA gene libraries showed fungi to be the most diverse eukaryote group in the ecosystem, followed by ciliates and stramenopiles (heterokonts). More than 40 protist phylotypes were found in a sample from the lower halocline in the anoxic brine pool of L'Atalante basin in the Eastern Mediterranean (depth 3.5 km) (Alexander et al. 2009; Filker et al. 2012; Stock et al. 2012, 2013; see also Sect. 8.6.5).

8.2.4 Viruses

In nearly all aquatic ecosystems, viruses or virus-like particles outnumber prokaryotic microorganisms by at least one order of magnitude. Hypersaline environments are no exception, as shown in electron microscopy studies of saltern ponds, the Dead Sea, and the alkaline hypersaline Mono Lake in California (Guixa-Boixareu et al. 1996; Jiang et al. 2004; Oren et al. 1997b). As stated above, viral lysis is probably the most important loss factor to prokaryotic communities at the highest salinities (Guixa-Boixareu et al. 1996; Pedrós-Alió et al. 2000a, b). In recent years, the search for viruses attacking halophilic Archaea in salterns and in natural hypersaline lakes such as Lake Retba in Senegal has yielded a wealth of novel types, including head-and-tail, spherical, and spindle-shaped phages, lytic as well as lysogenic types, viruses with double-stranded DNA, and single-stranded DNA viruses with a membrane envelope (Atanasova et al. 2012; Pietilä et al. 2009; Porter et al. 2007b; Sabet 2012; Sime-Ngando et al. 2011). Culture-independent, metagenomic approaches were also used to characterize the viral communities in saltern ponds (Santos et al. 2007, 2011, 2012; Rodriguez-Brito et al. 2010).

8.3 Strategies of Osmotic Adaptation in the Microbial World

Biological membranes are permeable to water. This basic feature implies that cells, including cells inhabiting high-salt environments, must maintain a cytoplasm that is at least isosmotic with their outside medium (Brown 1976, 1990). Having a cytoplasm more diluted than the environment would result in rapid loss of water. Active pumping of water to prevent drying out of the cell is not an option as the energy cost will be far too high.

Most cells even maintain a cytoplasm with a somewhat higher concentration of osmotically active molecules and ions than the concentration in their surroundings. This leads to the buildup of a turgor pressure. A positive turgor pressure is essential for the expansion of the growing cell leading to its division into two cells. The halophilic Archaea of the family Halobacteriaceae present a rare case of cells that lack a significant turgor pressure (Walsby 1971). The absence of a turgor pressure in this group enables the cells of *Haloquadratum walsbyi* to maintain their unusual flat, square to rectangular shape, a shape not compatible with a cytoplasm that exerts pressure on the cell wall as a result of its turgor (Walsby 2005).

The problem of how to achieve the necessary osmotic balance between the cells' cytoplasm and the high osmotic pressure of the hypersaline environment in which halophilic microorganisms thrive was solved in different ways by different groups of halophilic and halotolerant organisms. One common principle, however, applies: intracellular sodium concentrations are always maintained at levels much below the concentration in the medium. Cells of all domains of life pump sodium out, using Na^+/H^+ antiporters, Na^+/K^+ antiporters, or other mechanisms such as primary sodium pumps.

Two basically different strategies are found in the microbial world to provide the necessary osmotic stabilization of the cells:

1. Use of inorganic ions to balance the salts in the outside medium. Potassium rather than sodium is accumulated in molar concentrations, and chloride is the counterion. This strategy requires adaptations of all the intracellular processes which must proceed in the presence of high-salt concentrations. These adaptations, as explained below, are generally so far going that growth, and even survival, at low-salt concentrations is no longer possible. Organisms using this "high salt-in" strategy are therefore obligate halophiles that need the constant presence of high-salt concentrations.

2. Use of organic solutes to provide osmotic balance of the cytoplasm. Organic solutes used for this purpose need to be small molecules, highly soluble in water, that do not negatively interfere with the activity of intracellular enzymes. Such "compatible" solutes are generally uncharged or zwitterionic molecules. As the intracellular concentrations of these osmotic solutes can be regulated in accordance with the outside salt concentrations, this "low salt-in" strategy generally allows for a high degree of flexibility, so that cells can rapidly adapt to changes in the salinity of their medium (Galinski 1993, 1995; Grant 2004; Oren 2006a).

Combinations of the two strategies are seldom found in nature.

The best-known group in which the "high salt-in" strategy is found is the archaeal family Halobacteriaceae. *Halobacterium* and related genera maintain molar concentrations of KCl within their cytoplasm. Na^+ is removed from the cells by Na^+/H^+ antiporter activity; K^+ in part enters the cells in an energy-independent manner, but can also be accumulated by energy-dependent carriers. Chloride is accumulated against the (inside-negative) membrane potential by co-transport with Na^+ and/or by the light-driven primary chloride pump halorhodopsin (Oren 2006b). Another organism that uses KCl for osmotic equilibrium of its cytoplasm with the environment is *Salinibacter ruber*, the extremely halophilic representative of the Bacteroidetes (Bacteria). Analysis of the *Salinibacter* genome showed all the elements for the establishment of high KCl concentrations inside the cells as in *Halobacterium* to be present, including a halorhodopsin-like light-driven chloride pump (Mongodin et al. 2005). A third, phylogenetically unrelated group in which use of the "high salt-in" mode of life was indicated is the anaerobic fermentative Halanaerobiaceae (Firmicutes) (Rainey et al. 1995). *Halanaerobium praevalens*, *Haloanaerobium acetethylicum*, and *Halobacteroides halobius* cells were found to contain K^+, Na^+, and Cl^- at concentrations high enough to be at least isotonic with the medium (Oren 1986, 2006c; Oren et al. 1997a; Rengpipat et al. 1988).

A common property of most enzymes and other proteins synthesized by members of the Halobacteriaceae—a property shared with many proteins of *Salinibacter*—is their highly acidic nature. They have a high excess of negatively charged amino acids (glutamate, aspartate) over basic amino acids (lysine, arginine) (Lanyi 1974; Reistad 1970). These proteins typically require molar concentrations of salt to stabilize their native conformation. When suspended in distilled water or in low-salt solutions most proteins of *Halobacterium* unfold, and enzymes lose their activity. This is the reason why the "high salt-in" strategy allows little flexibility to adapt to changing salt concentrations. As the cell wall of *Halobacterium*, *Haloquadratum*, and most other members of the family consists of highly acidic glycoprotein subunits that also require salt for correct folding, removal of the salt causes instantaneous dissolution of the cell wall and lysis of the cell. Different explanations have been brought forward for the salt dependence of such halophilic proteins, including the need for high concentrations of cations to shield the excess of negative charges on the protein surface and the need for salt to bind water and thus stabilize the relatively weak hydrophobic forces that must keep the protein in proper shape; other explanations may apply as well (Eisenberg and Wachtel 1987; Eisenberg et al. 1992; Kushner 1978; Lanyi 1974; Madern et al. 2000; Mevarech et al. 2000).

The proteome of *Salinibacter ruber* is nearly as acidic as that of *Halobacterium* and relatives (Mongodin et al. 2005; Oren and Mana 2002). However, analysis of genome sequences of members of the Halanaerobiaceae did not show the expected excess in acidic residues in the proteins (Elevi Bardavid and Oren 2012b). The correlation between an acidic proteome, the salt requirement of enzymes, and accumulation of ions or organic osmotic solutes thus needs to be reevaluated. The recent finding that one species of halophilic and alkaliphilic anoxygenic phototrophs of the genus *Halorhodospira* has a highly acidic proteome while a phylogenetically closely related and physiologically similar species does not show a significant excess of acidic residues in its proteins (Deole et al. 2013) proves that our understanding of the topic is still incomplete.

Much more widespread is the "low salt-in" strategy in which organic solutes, generally small uncharged or zwitterionic compounds, are accumulated intracellularly, either by de novo biosynthesis or by uptake of such solutes from the medium when available. Concentrations of Na^+, Cl^-, and other inorganic ions are kept low, and no special modification of the enzymes and other components of the intracellular machinery is necessary when using this strategy.

One of the best-known cases of use of an organic compatible solute is the accumulation of molar concentrations of glycerol within the cells of the unicellular alga *Dunaliella* (Ben-Amotz and Avron 1973; Oren 2005). Glycerol is produced photosynthetically from CO_2, it is the smallest organic solute known, and it can be mixed with water in any ratio. *Dunaliella* regulates its glycerol content according to the salt concentration outside the cells. Also some halophilic and halotolerant fungi produce glycerol for the purpose of osmotic stabilization.

Use of glycerol as a compatible solute depends on the presence of a cytoplasmic membrane with a low permeability to the compound. Prokaryote cell membranes are highly permeable to glycerol. Accordingly, other organic compounds are used for the same purpose in the prokaryotic world. These include simple sugars such as sucrose and trehalose, different amino acid derivatives such as glycine betaine (*N,N,N*-trimethylglycine), ectoine (1,4,5,6-tetrahydro-2-methyl-pyrimidine-4-carboxylic acid), and others (Galinski 1993, 1995; Roberts 2005). Glycine betaine is produced as an osmotic solute by many halophilic and halotolerant cyanobacteria, anoxygenic phototrophs (*Ectothiorhodospira*, *Halorhodospira*), and a few other types of prokaryotes. Most heterotrophic members of the Bacteria domain are unable to synthesize glycine betaine, but they rapidly accumulate the compound when supplied in their medium. If it is not available, most halophilic and halotolerant bacteria produce ectoine and its 5-hydroxy derivative, hydroxyectoine, as their osmotic solute (Galinski 1995; Ventosa et al. 1998; Wohlfarth et al. 1990). Cyanobacteria living in moderately salty media accumulate sugars (sucrose, trehalose), many species adapted to more elevated salinities produce glucosylglycerol, and the most salt-tolerant ones generally synthesize glycine betaine (Hagemann 2011, 2013; Mackay et al. 1984; Oren 2012c). The above list of osmotic solutes is only a partial one. For example, ε-acetyl-α-lysine and Nδ-acetylornithine were detected in aerobic members of the Firmicutes. Many halophilic prokaryote species contain cocktails of compatible solutes rather than relying on a single compound. Organic osmotic solutes such as ectoine and hydroxyectoine are of considerable interest for biotechnological applications: they are known to stabilize proteins and other biomolecules, and they recently found medical applications as well (Lentzen and Schwarz 2006).

Organic osmotic solutes are also found in the domain Archaea. The halophilic methanogens produce glycine betaine, β-glutamine, β-glutamate, and N-ε-acetyl-β-lysine (Lai et al. 1991). And some alkaliphilic members of the Halobacteriaceae, a group that, as shown above, primarily uses KCl for osmotic stabilization, can also produce 2-sulfotrehalose as an additional osmotically active compound (Desmarais et al. 1997).

8.4 Trophic Interrelationships Between Different Types of Halophiles Based on Organic Osmotic Solutes

The fact that molecules such as glycerol and glycine betaine are synthesized in large amount by key components of many hypersaline ecosystems to serve as osmotic solutes implies that such molecules may become available to other partners inhabiting the same environment. Therefore, it is relevant to investigate the fate of such compounds along the food chain as these molecules are important links in the carbon cycle in hypersaline ecosystems and (in the case of glycine betaine) in the nitrogen cycle as well (Elevi Bardavid et al. 2008; Borowitzka 1981; Welsh 2000).

As explained above, glycerol is the osmotic solute of *Dunaliella*, found worldwide as the main or even sole primary producer at the highest salinities. It can accumulate intracellularly at concentrations as high as 5–6 M (i.e., >50 % weight/volume) (Ben-Amotz and Avron 1973). Many heterotrophic members of the halophilic microbial communities, and notably members of the Halobacteriaceae, can use glycerol as a carbon and energy source. When incubated in the presence of glycerol, members of the archaeal genera *Haloferax*, *Haloarcula*, and others convert part of the glycerol to acids: D-lactate, acetate, and pyruvate. Following incubation of Dead Sea water with micromolar concentrations of ^{14}C-glycerol, part of the added label was converted to lactate, acetate, and pyruvate (Oren and Gurevich 1994). In similar experiments with saltern crystallizer brine formation of lactate and acetate was observed, but not of pyruvate. It is interesting to note that pyruvate is a favorite nutrient of *Haloquadratum* present in massive numbers in these brines. Glycerol turnover in the saltern brine was found to be a very rapid process (Oren 1995).

Attempts were made to show uptake and incorporation of glycerol by the members of the heterotrophic community in saltern crystallizer ponds in Spain, using fluorescence in situ hybridization combined with microautoradiography following incubation with radiolabeled glycerol. Surprisingly, neither the flat square cells of *Haloquadratum walsbyi* nor the rod-shaped *Salinibacter ruber* were observed to incorporate glycerol. Labeled amino acids and acetate were readily incorporated by both types of cells in these experiments (Rosselló-Mora et al. 2003). However, in pure culture *Salinibacter* readily take up glycerol, enabling higher growth yields (Sher et al. 2004). Part of the glycerol taken up by *Salinibacter* is oxidized to dihydroxyacetone and excreted to the medium (Elevi Bardavid and Oren 2008). As *Haloquadratum* efficiently takes up dihydroxyacetone (a property first predicted from the genome sequence), this compound may link the metabolism of the two organisms which occur together at high densities in the saltern crystallizer ecosystem. Interesting interrelationships are thus possible between members of the community based on the metabolism of the osmotic solute glycerol (Orellana et al. 2013; Oren 2011c, 2012a).

Glycerol can also be transformed in hypersaline ecosystems under anaerobic conditions. *Halanaerobium* spp. may slowly convert glycerol to acetate, hydrogen, and CO_2. In combination with the sulfate-reducing halophilic bacterium *Desulfohalobium retbaense* glycerol consumption rates increased, probably due to interspecies hydrogen transfer processes (Cayol et al. 2002).

Glycine betaine, produced as an osmotic solute by halophilic cyanobacteria and by some anoxygenic phototrophs that grow at high-salt concentrations (e.g., the genus *Halorhodospira*), is readily taken up by many other microorganisms that cannot produce the compound themselves. Thus, they save energy that must otherwise be spent to synthesize other compatible solutes such as ectoine. Therefore, it is not surprising to find glycine betaine as the major or even as the sole osmotic solute in complex communities that include phototrophic members that produce the compound. The saltern gypsum crust illustrated in Fig. 8.2 is an example of such an ecosystem: no other osmotic solutes than glycine betaine could be detected (Oren et al. 2013). Under aerobic conditions, glycine betaine is degraded to CO_2 and NH_4^+; anaerobic breakdown leads to the formation of trimethylamine and other methylated amines (Oren 1990). Such methylated amines are the major substrates for methanogenic Archaea that live at high-salt concentrations (Oremland and King 1989), as methanogenesis by reduction of CO_2 with H_2 or by the splitting of acetate does not function in hypersaline environments (see Sect. 8.5).

8.5 Microbial Processes at High Salt: Possibilities and Limitations

As described above, hypersaline environments are inhabited by a great diversity of microorganisms. Still, not all microbial processes known from freshwater and marine environments were documented to occur at the highest salinities.

Comparative studies of salt lakes and studies of the microbial activities along the salinity gradient in solar saltern evaporation ponds show that there is a distinct upper salt concentration limit for different microbial processes. Examples of processes never found at salt concentrations above 150–200 g/l are autotrophic nitrification (i.e., the oxidation of ammonium ions to nitrate via nitrite under aerobic conditions), methanogenesis by reduction of CO_2 with H_2 or by splitting of acetate, or the oxidation of acetate by sulfate-reducing bacteria. In contrast, autotrophic sulfide oxidation was documented at salt concentrations above 250 g/l, and the same is true for sulfate reduction with lactate as the electron donor or for methanogenesis using methylated amines as the energy source.

To explain such observations, a theory was proposed (Oren 1999a, b, 2001, 2011a) based on the following assumptions:

1. Osmotic adaptation requires a large expenditure of energy. Pumping sodium ions out of the cells against a concentration gradient is energetically expensive, and so are active transport of other ions and the biosynthesis of organic osmotic solutes.
2. The possibility for a certain type of metabolism to occur at the highest salt concentrations therefore depends on the amount of energy that can be gained and on the energy required for the regulation of intracellular salt and solute concentrations.
3. Production of organic osmotic solutes, in addition to the need for removal of Na^+ from the cells, requires more energy than the "high salt-in" strategy where cells use KCl for osmotic stabilization.

It must be realized that the way nature works is more complex than what can be predicted in a simple theory. Thus, it must be taken into account that processes that provide little energy based on thermodynamical calculations in terms of free energy change per mol of substrate transformed may still proceed when substrate concentrations are high and turnover is fast. And a number of cases were identified that cannot be easily explained according to the simple model (Oren 2011a). Still, most observations fit well with the predictions, as shown by the examples below. The examples below show that in most cases the theory explains the observations, both from field data and from laboratory studies of pure cultures. Literature references to document the different observations are given in Oren (1999a, b, 2011a).

- Light energy is generally abundantly available in shallow hypersaline lakes and in saltern ponds. Oxygenic photosynthesis by *Dunaliella* and (to a lesser extent) by cyanobacteria proceeds up to the highest salt concentrations, and so does anoxygenic photosynthesis by *Halorhodospira*. Light energy absorbed by bacteriorhodopsin and similar light-driven proton pumps can also help heterotrophic aerobes (*Halobacterium*, *Haloquadratum* and relatives, *Salinibacter*) to grow or at least to survive in saturated brines when the supply of organic substrates for respiration is limiting.
- Aerobic respiration and dissimilatory nitrate reduction (denitrification) are processes yielding much energy for each electron transferred to oxygen or to nitrate.

Accordingly there is a large diversity of heterotrophic prokaryotes that obtain their energy from these processes. These include organisms that use the "high salt-in" strategy (Halobacteriaceae, *Salinibacter*) as well as Bacteria that synthesize organic osmotic solutes such as ectoine (e.g., members of the Halomonadaceae) (Ventosa et al. 1998). It may be noted that the first group performs better in salt-saturated brines than the second.
- Fermentation processes yield little energy only. Still, fermentative bacteria of the order Halanaerobiales can function up to very high-salt concentrations (200–250 g/l). The fact that this group of fermentative organisms it not known to synthesize organic compatible solutes but instead accumulates inorganic ions (Oren 1986) allows fermentation to proceed without the need to spend excessive amounts of energy for the biosynthesis of organic osmotic solutes.
- Dissimilatory sulfate reduction can proceed up to salt concentrations approaching saturation (Porter et al. 2007a; Roychoudhury et al. 2013). Halophilic and halotolerant sulfate reducers characterized belong to the Deltaproteobacteria or to the Firmicutes. These groups are known to use organic solutes for osmotic adaptation. Different electron donors support sulfate reduction in nature, including lactate (which is converted to acetate and CO_2 by "incomplete oxidizers"), H_2, and acetate (converted to CO_2 by a group of sulfate reducers designated "complete oxidizers"). Oxidation of lactate and of molecular hydrogen with sulfate as the electron acceptor yields relatively much energy and these reactions can proceed up to high salinities (Brandt et al. 2001; Kjeldsen et al. 2006). *Desulfohalobium retbaense*, an isolate from Lake Retba in Senegal, is the most halotolerant species in culture, tolerating up to 240 g/l salt and growing optimally at 100 g/l (Ollivier et al. 1991). However, only little energy is released when acetate serves as the electron donor. The most halotolerant acetate-oxidizing sulfate reducer characterized, *Desulfobacter halotolerans* from the sediments of Great Salt Lake (Utah) (see Sect. 8.6.2) thrives optimally at 10–20 g/l salt only, and growth ceases above 130 g/l (Brandt and Ingvorsen 1997).
- The two principal reactions yielding methane in nature, i.e., the reduction of carbon dioxide with hydrogen and the aceticlastic split, were never shown to occur at high-salt concentrations. No methane formation from $CO_2 + H_2$ was documented above 100–120 g/l, and no methanogens were ever isolated that split acetate to $CH_4 + CO_2$ above 40–50 g/l salt. These two reactions yield small amounts of energy only, probably insufficient to support growth at higher salt concentrations by methanogenic Archaea that use different organic osmotic solutes to provide osmotic balance. However, methanogenesis based on trimethylamine and other methylated compounds proceeds nearly up to NaCl saturation. Here a good correlation is found between the upper salinity limit of the organisms and the amount of free energy released in the different reactions.
- Although ammonium ions are abundantly present in many hypersaline environments, including the waters of the Dead Sea, occurrence of autotrophic oxidation of ammonium to nitrite and further to nitrate (nitrification) was never shown to occur at high salinities. Salt concentrations of 100–150 g/l are probably the

upper limit for the first step, and the most salt-tolerant autotrophic ammonia oxidizer in culture, known as "*Nitrosococcus halophilus*," does not grow above 94 g/l salt with an optimum at 40 g/l. The upper salt limit for the oxidation of nitrite to nitrate is probably even lower. The sensitivity of autotrophic nitrification to high salt is in full agreement with the very low amounts of energy gained in this process.

– The amount of energy gained per electron transferred to oxygen is much larger for the autotrophic oxidation of sulfide and other reduced sulfur compounds than for the oxidation of ammonia or nitrite. Accordingly, sulfur compounds can be used as electron donors by chemoautotrophs up to much higher salt concentrations. *Thiobacillus halophilus*, isolated from an Australian hypersaline lake, grows up to about 240 g/l salt, and *Thiohalorhabdus denitrificans* found both in salt lakes in Siberia and in Mediterranean salterns is even more halophilic: it grows optimally at 175 g/l NaCl and tolerates 290 g/l. There are more examples of highly halotolerant sulfur autotrophs (Oren 2011a).

However, the simple theory does not explain all observations. One notable case is the apparent absence of aerobic methane oxidation in high-salt environments. Methane is formed in hypersaline sediments mainly from trimethylamine, a breakdown product of glycine betaine. But in spite of the large amount of energy released during the aerobic oxidation of methane, no truly halophilic methanotrophs are known, and there are no clear indications that methane is oxidized at significant rates in brines with more than 100–150 g/l salt (Conrad et al. 1995).

8.6 Hypersaline Ecosystems: Case Studies

It is impossible to provide here an in-depth discussion of the microbiology of all hypersaline environments that have been studied. Therefore, only a few selected biotopes will be briefly presented here: solar salterns, Great Salt Lake, the Dead Sea, the alkaline lakes of the Wadi an Natrun, Egypt, deep-sea brines, and the interesting "ecosystem" of brine inclusions within salt crystals. These topics were selected as recent research has provided much new and interesting information.

8.6.1 Solar Saltern Evaporation Ponds

Solar salterns have always been popular ecosystems for the study of halophilic microbial life: they are found in many coastal areas all over the world (and sometimes inland as well), they are easily accessible, and they provide a gradient of salt concentrations from seawater salinity to NaCl saturation and beyond ("bitterns"). Along the gradient, sequential precipitation occurs of calcium carbonate, calcium sulfate (gypsum), and finally halite when the total salt concentration exceeds ~320 g/l (Javor 1989; Oren 2002a). The ponds are shallow, so that phototrophic communities can also develop on the bottom sediments (Fig. 8.2). The microorganisms present in the evaporation ponds and in the crystallizer ponds influence the salt production process, generally in a positive way by enhancing evaporation, but sometimes negatively when cyanobacteria excrete large amounts of viscous polysaccharides that cause the formation of poor quality salt (Javor 2002).

The high levels of nutrients generally present enable an abundant development of photosynthetic communities. In the evaporation ponds of lower salinities, most of the primary production is performed by benthic mats dominated by cyanobacteria, and the same is true for those ponds where gypsum precipitates. Both unicellular cyanobacteria (*Aphanothece halophytica* and similar forms) and filamentous types (e.g., *Coleofasciculus chthonoplastes*, *Halospirulina tapeticola*, *Phormidium* spp., and others) are common (Oren 2012a). Diatoms and other types of eukaryotic algae (other than *Dunaliella*) occur up to ~100–150 g/l salt. The upper layer of the gypsum crust illustrated in Fig. 8.2 is colored brown orange due to *Aphanothece*—*Halothece*-like unicellular types; the dark green layer below contains *Phormidium*-type filamentous cyanobacteria. Similar gypsum crusts were documented from salterns elsewhere (Caumette 1993; Caumette et al. 1994). These benthic cyanobacterial mats and gypsum crusts with their colored layered communities of phototrophic microorganisms, oxygenic as well as anoxygenic, have been popular objects for multidisciplinary studies that included culture-dependent approaches, culture-independent 16S rRNA-based characterization of the communities (Mouné et al. 2002; Sørensen et al. 2005, 2009), pigment analysis and light penetration (Oren et al. 1995c), microelectrode and optode studies to estimate rates of photosynthesis and other processes at different depths within the sediment (Canfield et al. 2004; Woelfel et al. 2009), advanced spectroscopic methods to study photosynthesis in situ (Prášil et al. 2009), presence of different fatty acids as biomarkers for certain groups of cyanobacteria (Ionescu et al. 2007), analysis of the presence of organic osmotic solutes within the community (Oren et al. 2013), and others (Oren et al. 2009a). Metaproteomic analysis of a layered microbial mat in the evaporation lagoons at Guerrero Negro, Baja California, Mexico (~90 g/l salt), showed a marked acid-shifted isoelectric point profile (Kunin et al. 2008). Whether or not this pattern is correlated with salt adaptation remains to be confirmed, as many marine bacteria show a similar excess of acidic amino acids in their proteins (Elevi Bardavid and Oren 2012a).

Saltern crystallizer ponds where halite accumulates on the bottom until the salt is harvested are generally colored pink red as a result of dense communities of halophilic Archaea, *Dunaliella salina*, and *Salinibacter*. Generally, prokaryotes are present in such brines at densities of 10^7–10^8/ml, but higher numbers have also been reported (Borowitzka 1981; Javor 1983, 1989; Oren 2002a) (Fig. 8.1). The square flat *Haloquadratum* is often a prominent member to the community. Salterns thus provide a unique opportunity to study the behavior of *Haloquadratum* in its natural environment, such as the possible function of its gas vesicles; in

8 Life in Hypersaline Environments

contrast with the expectations, no high degree of buoyancy is bestowed on the cells by the gas vesicles (Oren et al. 2006).

Saltern crystallizer ponds worldwide have been intensively studied using a variety of approaches:

- Culture-dependent approaches. *Haloquadratum walsbyi* is difficult to grow (Bolhuis et al. 2004; Burns et al. 2004a), but with some skill and with much patience most groups of halophilic Archaea present in saltern crystallizer ponds can be cultured (Burns et al. 2004b; Sabet et al. 2009). Long incubation times are needed (8–12 weeks), and low-nutrient media give the best results. Most types of prokaryotes identified in a culture-independent study of an Australian saltern crystallizer pond were obtained as colonies on agar plates (Burns et al. 2004b). Novel types of halophilic fungi were also obtained from saltern ponds (Butinar et al. 2005b; Gunde-Cimerman et al. 2000).
- Culture-independent, small-subunit rRNA gene-based methods. Such studies first identified the importance of *Haloquadratum* and *Salinibacter* to the biota of the salterns (Antón et al. 2000; Benlloch et al. 1995, 1996, 2002; Pašić et al. 2005, 2007; Manikandan et al. 2009; Oh et al. 2009; Oren 2002c; Oren et al. 2009b, and others).
- Fluorescence in situ hybridization studies. Adaptation of FISH methodology to the study of hypersaline environment led to the recognition of the morphology of *Salinibacter* and its possible modes of metabolism (Antón et al. 2000; Maturrano et al. 2006; Rosselló-Mora et al. 2003).
- Culture-independent studies based on other genes. Ten different phylotypes of the *bop* gene, coding for the protein moiety of the light-driven proton pump bacteriorhodopsin, were obtained from a solar saltern on the Adriatic coast (Pašić et al. 2005).
- Use of DNA melting profiles and reassociation kinetics to obtain information on the complexity of the microbial community and the nature of its dominant members (Øvreås et al. 2003).
- Metagenomic studies. The metagenomic approach led to understanding of the extent of variability between *Haloquadratum* genomes within a single population in a saltern and enabled the discovery of a large pool of accessory genes, including phage-related genes (Dyall-Smith et al. 2011; Fernandez et al. 2013; Legault et al. 2006; Oren 2012b; Pašić et al. 2005, 2007). Metagenomics also led to the discovery of a new type of extremely halophilic Archaea with very small cells not closely related to the Halobacteriaceae, the "Nanohaloarchaea" (Ghai et al. 2011; Narasingarao et al. 2012). No members of this group have yet been cultured. These "Nanohaloarchaeota" were also detected using by flow cytometry and cell sorting (Zhaxybayeva et al. 2013).
- Pigment studies. The different pigments present (β-carotene, chlorophyll *a*, and chlorophyll *b* of *Dunaliella*, bacterioruberins of the Halobacteriaceae, salinixanthin of *Salinibacter*) are excellent biomarkers for the characterization of the microbial communities in the salterns, qualitative as well as quantitative

(Litchfield and Oren 2001; Oren 2009; Oren and Dubinsky 1994; Oren and Rodríguez-Valera 2001).
- Lipid studies. The presence of specific polar lipids, especially of glycolipids, in different genera of halophilic Archaea, and the presence of unique sulfonolipids in *Salinibacter* have triggered "lipidomic" studies of saltern communities, using electrospray ionization mass spectrometry or MALDI-TOF (Litchfield and Oren 2001; Lopalco et al. 2011, Oren 1994a; Oren et al. 1996).
- In situ activity measurements and assessment of the interrelationships between different members of the microbial community (Elevi Bardavid et al. 2008; Pedrós-Alió et al. 2000a, b; Warkentin et al. 2009).
- Virus studies. Virus-like particles are abundantly found in saltern crystallizer brines (Guixa-Boixareu et al. 1996), and different viruses attacking halophilic Archaea of the family Halobacteriaceae were obtained from such ponds (Atanasova et al. 2012; Sabet 2012).

8.6.2 Great Salt Lake

Although not connected to the oceans, the waters of Great Salt Lake have a thalassohaline composition. The lake is a remnant of the ice-age saline Lake Bonneville that has largely dried out. Since the lake was divided into two parts by a railway causeway in the 1950s, the north arm has increased in salinity and is nearly saturated with respect to NaCl; the south arm, where rivers enter the lake, decreased in salt concentration (currently up to ~170 g/l). Climatic changes in the past decades have greatly influenced the water level and the salinity.

Brines colored red as a result of dense blooms of halophilic Archaea were reported in Great Salt Lake since the pioneering studies of Fred Post (1977); *Dunaliella* spp. were the main primary producers at the time (Stephens and Gillespie 1976). Today Halobacteriaceae and *Dunaliella* are still the main microorganisms inhabiting the north arm, while the biota of the less saline southern part is much more diverse.

Surprisingly, relatively few studies were devoted to the characterization of the microbial communities in Great Salt Lake and the processes they perform until quite recently. A few novel organisms were isolated from its waters (e.g., the archaeon *Halorhabdus utahensis*) and from its sediments (e.g., the already mentioned acetate oxidizer *Desulfobacter halotolerans* and *Desulfosalsimonas propionicica* (Brandt and Ingvorsen 1997; Kjeldsen et al. 2010), and some effort was dedicated to the characterization of sulfate reduction in the lake's sediments at different salinities (Brandt et al. 2001). Using *drsAB*, coding for the dissimilatory sulfite reductase, as a marker, it was shown that in the north arm (270 g/l salt) *Desulfohalobium* phylotypes dominated, with smaller contributions from the Desulfobacteraceae and from sulfate-reducing Firmicutes (Kjeldsen et al. 2006).

The potential of culture-independent 16S rRNA gene-based methods for the exploration of the microbial biodiversity in Great Salt Lake was realized only

8 Life in Hypersaline Environments

recently, and results of such studies are now becoming available (Baxter et al. 2005; Meuser et al. 2013; Oren 2012b; Parnell et al. 2011; Weimer et al. 2009). The "GeoChip," a microarray with multiple functional probes, was also applied to obtain information on the presence of different functional groups of microorganisms in the lake (Parnell et al. 2010).

8.6.3 The Dead Sea

The Dead Sea is an example of an athalassohaline hypersaline water body: the ionic composition of its waters differs greatly from that of seawater. Divalent cations are present in much higher concentrations than monovalent cations (~2 M Mg^{2+} and ~0.5 M Ca^{2+}, as compared to ~1.5 M Na^+ and ~0.2 M K^+). Chloride (99 % of the anion sum) and bromide (1 %) are the main anions, and sulfate is present in very low concentrations. The pH is around 6. The lake is saturated with respect to NaCl. As the water level is decreasing at a rate of more than a meter each year, massive amounts of halite precipitate to the bottom, and the ratio between Mg^{2+} and Na^+ is continuously increasing. As high concentrations of divalent cations are toxic even to the most salt-tolerant and salt-loving microorganisms, the conditions for life in the Dead Sea are becoming ever more extreme, to the extent that today it is very difficult to show that anything is alive in the lake. The exceedingly high concentrations of "chaotropic" ions that destabilize biological structures (Baldwin 1996; Hofmeister 1888) inhibit growth of even the most halophilic organisms.

In earlier times, the lake was permanently stratified, the upper 40 m of the water column being less dense (~300 g/l total dissolved salts) than the lower water mass (~340 g/l salts down to the maximum depth of about 330 m at the time of the 1959–1960 survey of the lake). The negative water balance caused a steady increase of the salinity of the surface waters, and in 1979 complete mixing of two layers occurred. This also ended the existence of an anaerobic hypolimnion. Today the lake is well mixed during the winter months and oxygen penetrates down to the bottom. In summer, the water column becomes temporarily stratified due to the formation of a thermocline.

Different salt- and magnesium-tolerant microorganisms have been isolated from the Dead Sea. Examples are the Archaea *Haloferax volcanii, Haloarcula marismortui, Halorubrum sodomense,* and *Halobaculum gomorrense* (Mullakhanbhai and Larsen 1975; Oren 1983, 1988; Oren et al. 1995b). Such archaea prefer to live in NaCl brines, but they can grow at suboptimal rates at high magnesium concentrations. *Dunaliella* sp. (a green variety that does not accumulate massive amounts of β-carotene) is the sole primary producer in the Dead Sea. Anaerobic fermentative bacteria that are members of the Halanaerobiales were recovered from the sediments: *Halobacteroides halobius, Sporohalobacter lortetii, Orenia marismortui,* and *Selenihalanaerobacter shriftii*. Fungi were also isolated from the Dead Sea (Buchalo et al. 1998; Butinar et al. 2005a). Virus-like particles were abundantly found in a period following a dense bloom of Archaea in the lake (Oren et al. 1997b).

Hardly any quantitative information is available about the densities of the microbial communities in the Dead Sea prior to 1980. Since that year two events of massive blooming of *Dunaliella* and halophilic Archaea were observed. The first event started in the summer of 1980 and lasted until the end of 1982; the second episode started in the spring of 1992 and lasted until the end of 1995. Both microbial bloom events were triggered by massive inflow of freshwater during unusually rainy winters. On both occasions, the lake became colored red by the bacterioruberin pigments of the Archaea which reached densities of up to 2×10^7 and 3.5×10^7, respectively—densities similar to those found in saltern crystallizer ponds such as shown in Fig. 8.1. In both cases, the blooms were limited to a few upper meters of the water column, where the dense Dead Sea brines had become diluted by the fresh flood waters (maximum dilution ~10–15 % in 1980, ~30 % in 1992 (Oren 1988, 1993, 1999b; Oren and Gurevich 1993, 1995; Oren and Shilo 1981; Oren et al. 1995a). No massive inflow of freshwater occurred in the Dead Sea in the past two decades, the water level steadily deceased, causing halite precipitation, and this resulted in a steady increase in the concentrations of chaotropic cations, hostile to life (Oren 2010).

Metagenomic studies were peformed using both material collected in recent years when the densities of microorganisms were very small, as well as material preserved from the 1992 surface bloom. The small halophilic archaeal community left in 2007 was quite diverse, but during the bloom a single phylotype dominated the community (Bodaker et al. 2010; Rhodes et al. 2012). The metaproteomes were highly acidic as expected for an environment dominated by the most extremeophilic types of halophilic Archaea (Rhodes et al. 2010). Freshwater springs on the bottom of the lake display a large diversity of microorganisms as shown by metagenomic analysis, and such springs continuously inoculate the lake with different types of prokaryotic microorganisms (Ionescu et al. 2012).

8.6.4 The Hypersaline Alkaline Lakes of Wadi an Natrun, Egypt

There are other athalassohaline environments where concentrations of divalent cations such as Mg^{2+} and Ca^{2+} are extremely low and the pH is very high. In hypersaline soda lakes, where the pH is around 10–11 or even higher, Mg^{2+} and Ca^{2+} are very little soluble. Carbonate and bicarbonate ions, in addition to chloride and sulfate, are the main anions. Such soda lakes are found in many geographical locations. Well-known examples are Mono Lake, California, and Lake Magadi and other similar lakes in Kenya and Tanganyika. There also are hypersaline soda lakes in Tibet and in India.

The soda lakes of the Wadi an Natrun, Egypt, deserve a special discussion as a novel and very intriguing group of anaerobic haloalkalithermophiles was recently isolated from this environment (Mesbah and Wiegel 2012; Mesbah et al. 2007a, b). The long history of the microbiological exploration of these lakes was recently reviewed (Oren 2013b). The red coloration of the brines of some of these very shallow lakes has been attributed in the past to massive presence of anoxygenic phototrophic sulfur bacteria (Jannasch 1957). And indeed a number of such phototrophs were isolated from the Wadi an Natrun lakes such as the red-purple *Halorhodospira abdelmalekii* and the green-colored *Halorhodospira halochloris*. These are haloalkaliphiles that require both high salt and high pH for optimal growth. However, red haloalkaliphilic Archaea belonging to the Halobacteriaceae also exists. The first haloalkaliphilic member of the family, *Natronomonas pharaonis*, was isolated from the Wadi an Natrun lakes (Soliman and Trüper 1982). Analysis of the pigments extracted from the brines showed that archaeal bacterioruberin pigments were present in much larger quantities than the carotenoids and bacteriochlorophyll pigments of the anoxygenic phototrophs (Imhoff et al. 1979). Additional species of haloalkaliphilic prokaryotes were isolated from these Egyptian soda lakes such as *Alkalibacillus haloalkaliphilus*.

A most interesting group of organisms, recovered thus far only from the Wadi an Natrun lakes, are the anaerobic halophilic alkalithermophiles (Bacteria, order Natranaerobiales, phylogenetically affiliated with the clostridia). Three species were described: *Natranaerobius thermophilus* (optimal growth at 54 °C (up to 56 °C) at a total Na^+ concentration between 3.3 and 3.9 M (range: 3.1–4.9 M) and pH 9.5) (Mesbah and Wiegel 2008; Mesbah et al. 2007a, b), *Natranaerobius trueperi* (optimum growth at 52 °C, 3.7–5.4 M Na^+, and pH 9.5) (Mesbah and Wiegel 2009), and *Natronovirga wadinatrunensis* (growing optimally at 51 °C, 3.7–3.9 M Na^+, and pH 9.9) (Mesbah and Wiegel 2009). These bacteria ferment simple sugars or pyruvate mainly to lactate, acetate, and formate. The existence of such organisms that have to withstand three forms of stress simultaneously while obtaining only little energy from fermentation processes raises fundamental questions about the bioenergetic basis of life under some of the harshest possible environmental conditions (Bowers et al. 2009).

Culture-independent 16S rRNA gene sequence-based methods have also been applied to samples from the Wadi an Natrun lakes to obtain more in-depth information on the diversity of life forms present in this intriguing environment (Mesbah et al. 2007a, b).

8.6.5 Deep-Sea Brines

One of the most unusual but still little explored habitats for halophilic microorganisms is the anoxic brines found in a number of locations near the bottom of the oceans at several kilometers depth below the sea surface. Examples of such deep-sea brine lakes were found in the Eastern Mediterranean Sea, the Red Sea, and the

Gulf of Mexico. The ionic composition of these brines varies greatly: some are NaCl dominated, but Discovery Basin, located at a depth of 3.5 km on the Mediterranean Sea floor off the coast of Crete, has a gradient from 0.05 to 5.05 M $MgCl_2$ (Hallsworth et al. 2007). The brine lakes in the Eastern Mediterranean Sea were formed at sites where tectonic processes brought seawater in contact with evaporites that were deposited during the Messinian salinity crisis (5.96–5.33 million years ago).

Both culture-dependent and culture-independent approaches have been used in the study of the deep-sea brine pools (Antunes et al. 2011). Two highly unusual prokaryotes were isolated from the Shaban Deep on the bottom of the Red Sea at a depth of 1447 m. One is *Halorhabdus tiamatea*. In contrast to nearly all other members of the Halobacteriaceae which are red-pink pigmented aerobes, *H. tiamatea* is colorless and prefers an anaerobic lifestyle (Antunes et al. 2008a). *Haloplasma contractile* is a contractile bacterium, thus far the only known representative of a new lineage remotely affiliated with the Firmicutes and the Mollicutes. It can grow by fermentation or by anaerobic respiration with nitrate as the electron acceptor at salt concentrations between 15 and 180 g/l (Antunes et al. 2008b).

Culture-independent, 16S rRNA gene-based approaches were used to characterize the prokaryote communities at a number of sites: Kebrit Deep and Shaban Deep in the Red Sea (Eder et al. 1999, 2001, 2002), and Lake Thetis, L'Atalante, Urania, and Discovery Brine in the Eastern Mediterranean (Borin et al. 2009; La Cono et al. 2011; Mapelli et al. 2012; van der Wielen et al. 2005). Of special interest are the changes that occur in the microbial communities along the very steep (a few meters only) haloclines and chemoclines that separate the dense anaerobic brines from the oxygenated seawater above (Sass et al. 2001). Convergence of carbon and energy sources, inorganic nutrients, and potential electron acceptors from above and from below makes these transition zones hotspots of life, as shown also by the numbers of prokaryotes present (more than an order of magnitude higher than the overlying seawater) (Daffonchio et al. 2006). Novel phylogenetic lineages of Bacteria and Archaea were found, but the function of the organisms remains unknown. A wealth of novel protist 18S rRNA phylotypes were retrieved from the L'Atalante and Thetis basins (Alexander et al. 2009; Filker et al. 2012; Stock et al. 2012, 2013). Functional genes such as *dsrAB* for sulfate-reducing bacteria and *mrcA* for methyl coenzyme M reductase of methanogens were also targeted to probe the microbial diversity in L'Atalante and Urania Basins and in a brine pool at a depth of 650 m in the Gulf of Mexico (Joye et al. 2009; van der Wielen and Heijs 2007).

The most intriguing site to explore the limits of life at high-salt concentrations is Discovery Basin. Its deepest parts contain 5.05 M magnesium chloride brine and hardly any other ions. Thus, this is a brine of "chaotropic" cations only, with stabilizing "kosmotropic" (anti-chaotropic) ions such as Na^+ being absent. To estimate the upper $MgCl_2$ concentration for life, presence of mRNA rather than rRNA was assessed along the salinity gradient in the halocline, as presence of ribosomal RNA does not prove that the cells are alive and active. In fact, rRNA and

DNA can be preserved in concentrated brines long after the death of a cell. Based on the distribution of *dsrAB* and *mrcA* mRNA, a concentration of 2.3 M $MgCl_2$ was estimated to be the upper limit to support life (Hallsworth et al. 2007).

8.6.6 Life Within Brine Inclusions in Salt Crystals

When halite crystallizes, brine inclusions are often left within the crystals, and microorganisms present in the brine can become entrapped within these inclusions. Brine inclusions can also be found in rock salt deposited hundreds of millions of years ago. The question can therefore be asked how long different types of halophilic microorganisms can survive within such salt crystals.

When Vreeland et al. (2000) claimed to have cultured a *Bacillus*-type bacterium from a crystal of Permian rock salt that most probably had remained unaltered since it was deposited 250 million years ago, the finding met with much skepticism (Graur and Pupko 2001; Nickle et al. 2002). Even if it can be proven that the crystal did not undergo any alterations during such a long period, and even if the procedures of surface sterilization of the sample completely ruled out contamination by recently grown microorganisms (Sankaranarayanan et al. 2011), the properties of the organism, very similar to *Virgibacillus marismortui*, were strikingly "modern." Still, different types of halophiles, notably Archaea of the family Halobacteriaceae, were isolated from rock salt of different geological age (Denner et al. 1994; Grant et al. 1998; McGenity et al. 2000; Norton et al. 1993; Vreeland et al. 1998, 2007), and laboratory simulations showed that different halophiles can survive for prolonged periods within salt (Norton and Grant 1988).

Recent studies of different rock salts confirmed the presence of a variety of halophilic microorganisms within brine inclusions in salt. Techniques employed in such studies include a combination of microscopy, culture-dependent approaches, 16S rRNA gene sequencing, and pigment analysis using Raman spectroscopy (Fendrihan et al. 2009; Lowenstein et al. 2011; Osterrothová and Jehlička 2011; Winters et al. 2013). A heterogeneous distribution of cells was documented in brine inclusions along salt cores from beneath Death Valley (Schubert et al. 2009a, b, 2010a) and from Salar Grande, Chile (Gramain et al. 2011). Presence of remains of *Dunaliella* as well as of halophilic Archaea in the crystals (Schubert et al. 2010b) is of special interest as the glycerol leaking out of the dying algal cells may have supplied a source of carbon and energy for long-term survival of the prokaryotes. Evidence is thus accumulating that indeed prokaryotes, and notably members of the Halobacteriaceae, may remain viable in rock salt for millions of years (McGenity et al. 2000). Endolithic microbial communities within halite evaporites in the Atacama Desert present another interesting model system for the study of survival of different types of microorganisms on and within halite crystals (de los Ríos et al. 2010, Vítek et al. 2012; Wierzchos et al. 2006).

8.7 Final Comments

The examples discussed above show many of the fascinating aspects of the diverse world of the halophilic microorganisms and their interactions with the environments in which they live. The discussions did not cover all the environments in which halophiles are found. For example, communities of halophilic and halotolerant microorganisms inhabit the leaves of certain plants such as the tamarix tree and *Atriplex* spp. that grow in arid areas and excrete salt from salt glands on their leaves (Simon et al. 1994; Qvit-Raz et al. 2008). Halophiles are found in the nasal cavities of certain seabirds (Brito-Echeverría et al. 2009) and desert iguanas (Deutch 1994). Hypersaline environments with unusual chemical properties may be inhabited by unusual microorganisms. One example is the brines of the alkaline salt-saturated Searles Lake, California, which contain ~3.9 mM arsenic. The lake supports a biogeochemical arsenic cycle: As(V) is reduced to As(III) in anaerobic respiration processes, and As(III) is oxidized to As(V) by chemoautotrophs (Oremland et al. 2005). Another example is the subglacial "ferrous ocean" beneath Taylor Glacier in Antarctica, where Fe(III) serves as the terminal electron acceptor linked to the oxidation of sulfur compounds by the biota in the cold (~−5 °C) and hypersaline (>80 g/l total salts) brine (Mikucki et al. 2009). Based on these examples, it is clear that future exploration of unusual high-salt environments may lead to more exciting discoveries showing the diversity of life adapted to the many hypersaline environments on our planet.

References

Alexander E, Stock A, Breiner H-W, Behnke A, Bunge J, Yakimov MM, Stoeck T (2009) Microbial eukaryotes in the hypersaline anoxic L'Atalante deep-sea basin. Environ Microbiol 11:360–381

Andrei A-Ş, Banciu HL, Oren A (2012) Metabolic diversity in Archaea living in saline ecosystems. FEMS Microbiol Lett 330:1–9

Antón J, Llobet-Brossa E, Rodríguez-Valera F, Amann R (1999) Fluorescence *in situ* hybridization analysis of the prokaryotic community inhabiting crystallizer ponds. Environ Microbiol 1:517–523

Antón J, Rosselló-Mora R, Rodríguez-Valera F, Amann R (2000) Extremely halophilic *Bacteria* in crystallizer ponds from solar salterns. Appl Environ Microbiol 66:3052–3057

Antón J, Oren A, Benlloch S, Rodríguez-Valera F, Amann R, Rosselló-Mora R (2002) *Salinibacter ruber* gen. nov., sp. nov., a novel extreme halophilic member of the *Bacteria* from saltern crystallizer ponds. Int J Syst Evol Microbiol 52:485–491

Antón J, Peña A, Santos F, Martínez-García M, Schmitt-Kopplin P, Rosselló-Mora R (2008) Distribution, abundance and diversity of the extremely halophilic bacterium *Salinibacter ruber*. Saline Syst 4:15

Antunes A, Taborda M, Huber R, Moissl C, Nobre MF, da Costa MS (2008a) *Halorhabdus tiamatea* sp. nov., a non-pigmented, extremely halophilic archaeon from a deep-sea, hypersaline anoxic basin of the Red Sea, and emended description of the genus *Halorhabdus*. Int J Syst Evol Microbiol 58:215–220

Antunes A, Rainey FA, Wanner G, Taborda M, Pätzold J, Nobre MF, da Costa MS, Huber R (2008b) A new lineage of halophilic, wall-less, contractile bacteria from a brine-filled deep on the Red Sea. J Bacteriol 190:3580–3587

Antunes A, Kamanda Ngugu D, Stingl U (2011) Microbiology of the Red Sea (and other) deep-sea anoxic brine lakes. Environ Microbiol Rep 3:416–433

Atanasova NS, Roine E, Oren A, Bamford DH, Oksanen HM (2012) Global network of specific virus-host interactions in hypersaline environments. Environ Microbiol 14:426–440

Baldwin RL (1996) How Hofmeister ion interactions affect protein stability. Biophys J 71:2056–2063

Baronio M, Lattanzio VMT, Vaisman N, Oren A, Corcelli A (2010) The acylhalocapnines of halophilic bacteria: structural details of unusual sulfonate sphingoids. J Lipid Res 51:1878–1885

Baxter BK, Litchfield CD, Sowers K, Griffith J, Arora DasSarma R, DasSarma S (2005) Microbial diversity of Great Salt Lake. In: Gunde-Cimerman N, Oren A, Plemenitaš A (eds) Adaptation to life at high salt concentrations in Archaea, Bacteria, and Eukarya. Springer, Dordrecht, pp 11–25

Ben-Amotz A, Avron M (1973) The role of glycerol in the osmotic regulation of the halophilic alga *Dunaliella parva*. Plant Physiol 51:875–878

Benlloch S, Martínez-Murcia AJ, Rodríguez-Valera F (1995) Sequencing of bacterial and archaeal 16S rRNA genes directly amplified from a hypersaline environment. Syst Appl Microbiol 18:574–581

Benlloch S, Acinas SG, Martínez-Murcia AJ, Rodríguez-Valera F (1996) Description of prokaryotic biodiversity along the salinity gradient of a multipond saltern by direct PCR amplification of 16S rDNA. Hydrobiologia 329:19–31

Benlloch S, López-López A, Casamayor EO, Øvreås L, Goddard V, Dane FL, Smerdon G, Massana R, Joint I, Thingstad F, Pedrós-Alió C, Rodríguez-Valera F (2002) Prokaryotic genetic diversity throughout the salinity gradient of a coastal solar saltern. Environ Microbiol 4:349–360

Bodaker I, Sharon I, Suzuki MT, Reingersch R, Shmoish M, Andreishcheva F, Sogin ML, Rosenberg M, Belkin S, Oren A, Béjà O (2010) Comparative community genomics in the Dead Sea: an increasingly extreme environment. ISME J 4:399–407

Boetius A, Joye S (2009) Thriving in salt. Science 324:1523–1525

Bolhuis H, te Poele EM, Rodríguez-Valera F (2004) Isolation and cultivation of Walsby's square archaeon. Environ Microbiol 6:1287–1291

Bolhuis H, Palm P, Wende A, Falb M, Rampp M, Rodriguez-Valera F, Pfeiffer F, Oesterhelt D (2006) The genome of the square archaeon *Haloquadratum walsbyi*: life at the limits of water activity. BMC Genomics 7:169

Borin S, Brusetti L, Mapelli F, D'Auria G, Brusa T, Marzorati M, Rizzi A, Yakimov M, Marty D, de Lange GJ, van der Wielen P, Bolhuis H, McGenity TJ, Polymenakou PN, Malinverno E, Giuliano L, Corselli C, Daffonchio D (2009) Sulfur cycling and methanogenesis primarily drive microbial colonization of the highly sulfidic Urania deep hypersaline basin. Proc Natl Acad Sci USA 106:9151–9156

Borowitzka LJ (1981) The microflora. Adaptations to life in extremely saline lakes. Hydrobiologia 81:33–46

Bowers KJ, Wiegel J (2011) Temperature and pH optima of extremely halophilic archaea: a mini-review. Extremophiles 15:119–128

Bowers KJ, Mesbah NM, Wiegel J (2009) Biodiversity of polyextremophilic *Bacteria*: does combining the extremes of high salt, alkaline pH and elevated temperature approach a physico-chemical boundary for life? Saline Syst 5:9

Brandt KK, Ingvorsen K (1997) *Desulfobacter halotolerans* sp. nov., a halotolerant acetate-oxidizing sulfate-reducing bacterium isolated from sediments of Great Salt Lake, Utah. Syst Appl Microbiol 20:366–373

Brandt KK, Vester F, Jensen AN, Ingvorsen K (2001) Sulfate reduction dynamics and enumeration of sulfate-reducing bacteria in hypersaline sediments of the Great Salt Lake (Utah, USA). Microb Ecol 41:1–11

Brito-Echeverría J, López-López A, Yarza P, Antón J, Rosselló-Móra R (2009) Occurrence of *Halococcus* spp. in the nostrils salt glands of the seabird *Calonextris diomedea*. Extremophiles 13:557–565

Brown AD (1976) Microbial water stress. Bacteriol Rev 40:803–846

Brown AD (1990) Microbial water stress physiology. Principles and perspectives. Wiley, Chichester

Buchalo AS, Nevo E, Wasser SP, Oren A, Molitoris HP (1998) Fungal life in the extremely hypersaline water of the Dead Sea: first records. Proc R Soc Lond B 265:1461–1465

Burns DG, Camakaris HM, Janssen PH, Dyall-Smith ML (2004a) Cultivation of Walsby's square haloarchaeon. FEMS Microbiol Lett 238:469–473

Burns DG, Camakaris HM, Janssen PH, Dyall-Smith ML (2004b) Combined use of cultivation-dependent and cultivation-independent methods indicates that members of most haloarchaeal groups in an Australian crystallizer pond are cultivable. Appl Environ Microbiol 70:5258–5265

Burns DG, Janssen PH, Itoh T, Kamekura M, Li Z, Jensen G, Rodríguez-Valera F, Bolhuis H, Dyall-Smith ML (2007) *Haloquadratum walsbyi* gen. nov., sp. nov., the square haloarchaeon of Walsby, isolated from saltern crystallizers in Australia and Spain. Int J Syst Evol Microbiol 57:387–392

Butinar L, Santos S, Spencer-Martins I, Oren A, Gunde-Cimerman N (2005a) Yeast diversity in hypersaline habitats. FEMS Microbiol Lett 244:229–234

Butinar L, Sonjak S, Zalar P, Plemenitaš A, Gunde-Cimerman N (2005b) Melanized halophilic fungi are eukaryotic members of microbial communities in hypersaline waters of solar salterns. Bot Mar 48:73–79

Canfield DE, Sørensen KB, Oren A (2004) Biogeochemistry of a gypsum-encrusted microbial ecosystem. Geobiology 2:133–150

Caumette P (1993) Ecology and physiology of phototrophic bacteria and sulfate-reducing bacteria in marine salterns. Experientia 49:473–481

Caumette P, Matheron R, Raymond N, Relexans J-C (1994) Microbial mats in the hypersaline ponds of Mediterranean salterns (Salins-de-Giraud, France). FEMS Microbiol Ecol 13:273–286

Cayol J-L, Ollivier B, Patel BKC, Prensier G, Guezennec J, Garcia J-L (1994) Isolation and characterization of *Halothermothrix orenii* gen. nov., sp. nov., a halophilic, thermophilic, fermentative, strictly anaerobic bacterium. Int J Syst Bacteriol 44:534–540

Cayol J-L, Fardeau M-L, Garcia J-L, Ollivier B (2002) Evidence of interspecies hydrogen transfer from glycerol in saline environments. Extremophiles 6:131–134

Cho BC (2005) Heterotrophic flagellates in hypersaline waters. In: Gunde-Cimerman N, Oren A, Plemenitaš A (eds) Adaptation to life at high salt concentrations in Archaea, Bacteria, and Eukarya. Springer, Dordrecht, pp 543–549

Conrad R, Frenzel P, Cohen Y (1995) Methane emission from hypersaline microbial mats: lack of aerobic methane oxidation activity. FEMS Microbiol Ecol 16:297–305

Corcelli A, Lattanzio VMT, Mascolo G, Babudri F, Oren A, Kates M (2004) Novel sulfonolipid in the extremely halophilic bacterium *Salinibacter ruber*. Appl Environ Microbiol 70:6678–6685

Daffonchio D, Borin S, Brusa T, Brusetti L, van der Wielen PWJJ, Bolhuis H, Yakimov MM, D'Auria G, Giuliano L, Marty D, Tamburini C, McGenity TJ, Hallsworth JE, Sass AM, Timmis KN, Tselepides A, de Lange GJ, Hübner A, Thomson J, Varnavas SP, Gasparoni F, Gerber HW, Malinverno E, Corselli C, Biodeep Scientific Party (2006) Stratified prokaryote network in the oxic-anoxic transition of a deep-sea halocline. Nature 440:203–207

Darwin C (1839) Journal of researches into the geology and natural history of the various countries visited by H.M.S. Beagle, under the command of Captain Fitzroy, R.N. from 1832 to 1836. Henry Colburn, London de los Ríos A, Valea S, Ascaso C, Davila A, Kastovsky J, Mckay CP, Gómez-Silva B, Wierzchos J (2010) Comparative analysis of the microbial communities inhabiting halite evaporates of the Atacama Desert. Int Microbiol 13:79–89

Denner EBM, McGenity TJ, Busse H-J, Grant WD, Wanner G, Stan-Lotter H (1994) *Halococcus salifodinae* sp. nov., an archaeal isolate from an Austrian salt mine. Int J Syst Bacteriol 44:774–780

Deole R, Challacombe J, Raiford DW, Hoff WD (2013) An extremely halophilic proteobacterium combines a highly acidic proteome with a low cytoplasmic potassium content. J Biol Chem 288:581–588

Desmarais D, Jablonski PE, Fedarko NS, Roberts MF (1997) 2-Sulfotrehalose, a novel osmolyte in haloalkaliphilic Archaea. J Bacteriol 179:3146–3153

Deutch CE (1994) Characterization of a novel salt-tolerant *Bacillus* sp. from the nasal cavities of desert iguanas. FEMS Microbiol Lett 121:55–60

Dyall-Smith ML, Pfeiffer F, Klee K, Palm P, Gross K, Schuster SC, Rampp M, Oesterhelt D (2011) *Haloquadratum walsbyi*: limited diversity in a global pond. PLoS One 6(6):e20968

Eder W, Ludwig W, Huber R (1999) Novel 16S rRNA gene sequences retrieved from highly saline brine sediments of Kebrit Deep, Red Sea. Arch Microbiol 172:213–218

Eder W, Jahnke LL, Schmidt M, Huber R (2001) Microbial diversity of the brine-seawater interface of the Kebrit Deep, Red Sea, studied via 16S rRNA gene sequences and cultivation methods. Appl Environ Microbiol 67:3077–3085

Eder W, Schmidt M, Koch M, Garbe-Schönberg D, Huber R (2002) Prokaryotic phylogenetic diversity and corresponding geochemical data of the brine-seawater interface of the Shaban Deep, Red Sea. Environ Microbiol 4:758–763

Eisenberg H, Wachtel EJ (1987) Structural studies of halophilic proteins, ribosomes, and organelles of bacteria adapted to extreme salt concentrations. Annu Rev Biophys Biophys Chem 16:69–92

Eisenberg H, Mevarech M, Zaccai G (1992) Biochemical, structural, and molecular genetic aspects of halophilism. Adv Protein Chem 43:1–62

Elevi Bardavid R, Oren A (2008) Dihydroxyacetone metabolism in *Salinibacter ruber* and in *Haloquadratum walsbyi*. Extremophiles 12:125–131

Elevi Bardavid R, Oren A (2012a) Acid-shifted isoelectric point profiles of the proteins in a hypersaline microbial mat – an adaptation to life at high salt concentrations? Extremophiles 16:787–792

Elevi Bardavid R, Oren A (2012b) The amino acid composition of proteins from anaerobic halophilic bacteria of the order Halanaerobiales. Extremophiles 16:567–572

Elevi Bardavid R, Ionescu D, Oren A, Rainey FA, Hollen BJ, Bagaley DR, Small AM, McKay CM (2007) Selective enrichment, isolation and molecular detection of *Salinibacter* and related extremely halophilic *Bacteria* from hypersaline environments. Hydrobiologia 576:3–13

Elevi Bardavid R, Khristo P, Oren A (2008) Interrelationships between *Dunaliella* and halophilic prokaryotes in saltern crystallizer ponds. Extremophiles 12:5–14

Falb M, Müller K, Königsmaier L, Oberwinkler T, Horn P, von Gronau S, Gonzalez O, Pfeiffer F, Bornberg-Bauer E, Oesterhelt D (2008) Metabolism of halophilic archaea. Extremophiles 12:177–196

Fendrihan S, Musso M, Stan-Lotter H (2009) Raman spectroscopy as a potential method for the detection of extremely halophilic archaea embedded in halite in terrestrial and possibly extraterrestrial samples. J Raman Spectrosc 40:1996–2003

Fernandez AB, Ghai R, Martin-Cuadrado AB, Sanchez-Porro C, Rodriguez-Valera F, Ventosa A (2013) Metagenome sequencing of prokaryotic microbiota from two hypersaline ponds of a marine saltern in Santa Pola, Spain. Genome Announce 1:e00933

Filker S, Stock A, Breiner H-W, Edgcomb V, Orsi W, Yakimov MM, Stoeck T (2012) Environmental selection of protistan plankton communities in hypersaline anoxic deep-sea basins, Eastern Mediterranean Sea. Microbiol Open 2:54–63

Franzmann PD, Stackebrandt E, Sanderson K, Volkman JK, Cameron DE, Stevenson PL, McMeekin TA, Burton HR (1988) *Halobacterium lacusprofundi* sp. nov., a halophilic bacterium isolated from Deep Lake, Antarctica. Syst Appl Microbiol 11:20–27

Galinski EA (1993) Compatible solutes of halophilic eubacteria: molecular principles, water-solute interaction, stress protection. Experientia 49:487–496

Galinski EA (1995) Osmoadaptation in bacteria. Adv Microb Physiol 37:273–328

Ghai R, Fernández AB, Martin-Cuadrado A-B, Megumi Mizuno C, McMahon KD, Papke RT, Stepanauskas R, Rodriguez-Brito B, Rohwer F, Sánchez-Porro C, Ventosa A, Rodríguez-Valera F (2011) New abundant microbial groups in aquatic hypersaline environments. Sci Rep 1:135

Gostinčar C, Grube M, de Hoog GS, Zalar P, Gunde-Cimerman N (2009) Extremotolerance in fungi: evolution on the edge. FEMS Microbiol Ecol 71:2–11

Gostinčar G, Lenassi M, Gunde-Cimerman N, Plemenitaš A (2011) Fungal adaptation to extremely high salt concentrations. Adv Appl Microbiol 77:71–107

Gramain A, Chong Diaz G, Demergasso C, Lowenstein TK, McGenity TJ (2011) Archaeal diversity along a subterranean salt core from the Salar Grande (Chile). Environ Microbiol 13:2105–2121

Grant WD (2004) Life at low water activity. Philos Trans R Soc London B 359:1249–1267

Grant WD, Gemmell RT, McGenity TJ (1998) Halobacteria: the evidence for longevity. Extremophiles 2:279–287

Graur D, Pupko T (2001) The Permian bacterium that isn't. Mol Biol Evol 18:1143–1146

Guixa-Boixareu N, Calderón-Paz JI, Heldal M, Bratbak G, Pedrós-Alió C (1996) Viral lysis and bacterivory as prokaryotic loss factors along a salinity gradient. Aquat Microb Ecol 11:215–227

Gunde-Cimerman N, Zalar P, de Hoog GS, Plemenitaš A (2000) Hypersaline water in salterns – natural ecological niches for halophilic black yeasts. FEMS Microbiol Ecol 32:235–240

Gunde-Cimerman N, Ramos J, Plemenitaš A (2009) Halotolerant and halophilic fungi. Mycol Res 113:1231–1241

Hagemann M (2011) Molecular biology of cyanobacterial salt acclimation. FEMS Microbiol Rev 35:87–123

Hagemann M (2013) Genomics of salt acclimation: synthesis of compatible solutes among cyanobacteria. Adv Bot Res 65:27–55

Hallsworth JE, Yakimov MM, Golyshin PN, Gillion JLM, D'Auria G, de Lima AF, La Cono V, Genovese M, McKew BA, Hayes SL, Harris G, Giuliano L, Timmis KN, McGenity TJ (2007) Limits of life in $MgCl_2$-containing environments: chaotropicity defines the window. Environ Microbiol 9:801–813

Hartmann R, Sickinger H-D, Oesterhelt D (1980) Anaerobic growth of halobacteria. Proc Natl Acad Sci USA 77:3821–3825

Hauer G, Rogerson A (2005) Heterotrophic protozoa from hypersaline environments. In: Gunde-Cimerman N, Oren A, Plemenitaš A (eds) Adaptation to life at high salt concentrations in Archaea, Bacteria, and Eukarya. Springer, Dordrecht, pp 522–539

Heidelberg KB, Nelson WC, Holm JB, Eisenkolb N, Andrade K, Emerson JB (2013) Characterization of eukaryotic microbial diversity in hypersaline Lake Tyrrell, Australia. Front Microbiol 4:115

Hofmeister F (1888) Zur Lehre von der Wirkung der Salze. Zweite Mittheilung. Arch Exp Pathol Pharmakol 24:247–260

Imhoff JF, Sahl HG, Soliman GHS, Trüper HG (1979) The Wadi Natrun: chemical composition and microbial mass development in alkaline brines of eutrophic desert lakes. Geomicrobiol J 1:219–234

Ionescu D, Lipski A, Altendorf K, Oren A (2007) Characterization of the endoevaporitic microbial communities in a hypersaline gypsum crust by fatty acid analysis. Hydrobiologia 576:15–26

Ionescu D, Siebert C, Polerecky L, Munwes YY, Lott C, Häusler S, Bižić-Ionescu M, Quast C, Peplies J, Glöckner FO, Ramette A, Rödiger T, Dittmar T, Oren A, Geyer S, Stärk H-J,

Sauter M, Licha T, Laronne JB, de Beer D (2012) Microbial and chemical characterization of submarine freshwater springs in the Dead Sea, harboring rich microbial communities. PLoS One 7:e38319

Jannasch HW (1957) Die bakterielle Rotfärbung der Salzseen des Wadi Natrun (Ägypten). Arch Hydrobiol 53:425–433

Javor BJ (1983) Planktonic standing crop and nutrients in a saltern ecosystem. Limnol Oceanogr 28:153–159

Javor B (1989) Hypersaline environments. Microbiology and biogeochemistry. Springer, Berlin

Javor BJ (2002) Industrial microbiology of solar salt production. J Ind Microbiol Biotechnol 28:42–47

Jiang S, Steward G, Jellison R, Chu W, Choi S (2004) Abundance, distribution, and diversity of viruses in alkaline, hypersaline Mono Lake, California. Microb Ecol 47:9–17

Joye SB, Samarkin VA, Orcutt BM, MacDonald IR, Hinrichs K-U, Elvert M, Teske AP, Lloyd KG, Lever MA, Montoya JP, Meile CD (2009) Metabolic variability in seafloor brines revealed by carbon and sulphur dynamics. Nat Geosci 2:349–354

Kivisto AT, Karp MT (2011) Halophilic anaerobic fermentative bacteria. J Biotechnol 152:114–124

Kjeldsen KU, Loy A, Jakobsen TF, Thomsen TR, Wagner M, Ingvorsen K (2006) Diversity of sulfate-reducing bacteria from an extreme hypersaline sediment, Great Salt Lake (Utah). FEMS Microbiol Ecol 60:287–298

Kjeldsen KU, Jakobsen TF, Glastrup J, Ingvorsen K (2010) *Desulfosalsimonas propionicica* gen. nov., sp. nov., a halophilic, sulfate-reducing member of the family *Desulfobacteraceae* isolated from a salt-lake sediment. Int J Syst Evol Microbiol 60:1060–1065

Kunin V, Raes J, Harris JK, Spear JR, Walker JJ, Ivanova N, von Mering C, Bebout BM, Pace NR, Bork P, Hugenholtz P (2008) Millimeter scale genetic gradients and community-level molecular convergence in a hypersaline microbial mat. Mol Syst Biol 4:198

Kushner DJ (1978) Life in high salt and solute concentrations: halophilic bacteria. In: Kushner DJ (ed) Microbial life in extreme environments. Academic Press, London, pp 317–368

La Cono V, Smedile F, Bortoluzzi G, Arcadi E, Maimone G, Messina E, Borghini M, Oliveri E, Mazzola S, L'Haridon S, Toffin L, Genovese L, Ferrer M, Giuliano L, Golyshin PN, Yakimov MM (2011) Unveiling microbial life in new deep-sea hypersaline Lake Thetis. Part I: Prokaryotes and environmental settings. Environ Microbiol 13:2250–2268

Lai M-C, Sowers KR, Robertson DE, Roberts MF, Gunsalus RP (1991) Distribution of compatible solutes in the halophilic methanogenic archaebacteria. J Bacteriol 173:5352–5358

Lanyi JK (1974) Salt-dependent properties of proteins from extremely halophilic bacteria. Bacteriol Rev 38:272–290

Lanyi JK (2005) Xanthorhodopsin: a proton pump with a light-harvesting carotenoid antenna. Science 309:2061–2064

Larsen H (1973) The halobacteria's confusion to biology. Antonie van Leeuwenhoek 39:383–396

Legault BA, Lopez-Lopez A, Alba-Casado JC, Doolittle WF, Bolhuis H, Rodríguez-Valera F, Papke RT (2006) Environmental genomics of "*Haloquadratum walsbyi*" in a saltern crystallizer indicates a large pool of accessory genes in an otherwise coherent species. BMC Genomics 7:171

Lenassi M, Gostinčar C, Jackman S, Turk M, Sadowski I, Nislow C, Jones S, Birol I, Gunde-Cimerman N, Plemenitaš A (2013) Whole genome duplication and enrichment of metal cation transporters revealed by de novo genome sequencing of extremely halotolerant black yeast *Hortaea werneckii*. PLoS One 8:e71328

Lentzen G, Schwarz T (2006) Extremolytes: natural compounds from extremophiles for versatile applications. Appl Microbiol Biotechnol 72:623–634

Litchfield CD, Oren A (2001) Polar lipids and pigments as biomarkers for the study of the microbial community structure of solar salterns. Hydrobiologia 466:81–89

Lopalco P, Lobasso S, Baronio M, Angelini R, Corcelli A (2011) Impact of lipidomics on the microbial world of hypersaline environments. In: Ventosa A, Oren A, Ma Y (eds) Halophiles and hypersaline environments. Springer, Berlin, pp 123–135

Lowenstein TK, Schubert BA, Timofeeff MN (2011) Microbial communities in fluid inclusions and long-term survival in halite. GSA Today Jan:4–9

Lutnæs BF, Oren A, Liaaen-Jensen S (2002) New C_{40}-carotenoid acyl glycoside as principal carotenoid of *Salinibacter ruber*, an extremely halophilic eubacterium. J Nat Prod 65:1340–1343

Mackay MA, Norton RS, Borowitzka LJ (1984) Organic osmoregulatory solutes in cyanobacteria. J Gen Microbiol 130:2177–2191

Madern D, Ebel C, Zaccai G (2000) Halophilic adaptation of enzymes. Extremophiles 4:91–98

Manikandan M, Kannan V, Pašić L (2009) Diversity of microorganisms in solar salterns of Tamil Nadu, India. World J Microbiol Biotechnol 25:1007–1017

Mapelli F, Borin S, Daffoncio D (2012) Microbial diversity in deep hypersaline anoxic basins. In: Stan-Lotter H, Fendrihan S (eds) Adaptation of microbial life to environmental extremes: novel research results and application. Springer, New York, pp 21–36

Maturrano L, Santos F, Rosselló-Mora R, Antón J (2006) Microbial diversity in Maras salterns, a hypersaline environment in the Peruvian Andes. Appl Environ Microbiol 72:3887–3895

McGenity TJ, Oren A (2012) Hypersaline environments. In: Bell EM (ed) Life at extremes. Environments, organisms and strategies for survival. CABI International, London, pp 402–437

McGenity TJ, Gemmell RT, Grant WD, Stan-Lotter H (2000) Origins of halophilic microorganisms in ancient salt deposits. Environ Microbiol 2:243–250

Mesbah NM, Wiegel J (2008) Life at extreme limits. The anaerobic halophilic alkalithermophiles. Ann NY Acad Sci 1125:44–57

Mesbah NM, Wiegel J (2009) *Natronovirga wadinatrunensis* gen. nov., sp. nov. and *Natranaerobius trueperi* sp. nov., halophilic alkalithermophilic micro-organisms from soda lakes of the Wadi An Natrun, Egypt. Int J Syst Evol Microbiol 59:2042–2048

Mesbah NM, Wiegel J (2012) Life under multiple extreme conditions: diversity and physiology of the halophilic alkalithermophiles. Appl Environ Microbiol 78:4074–4082

Mesbah NM, Abou-El-Ela SH, Wiegel J (2007a) Novel and unexpected prokaryotic diversity in water and sediments of the alkaline, hypersaline lakes of the Wadi An Natrun, Egypt. Microb Ecol 54:598–617

Mesbah NM, Hedrick DB, Peacock AD, Rohde M, Wiegel J (2007b) *Natranaerobius thermophilus* gen. nov., sp. nov., a halophilic alkalithermophilic bacterium from soda lakes of the Wadi An Natrun, Egypt, and proposal of *Natranaerobiaceae* fam. nov. and *Natranaerobiales* ord. nov. Int J Syst Evol Microbiol 57:2507–2512

Meuser JE, Baxter BK, Spear JR, Peters JW, Posewitz MC, Boyd ES (2013) Contrasting patterns of community assembly in the stratified water column of Great Salt Lake, Utah. Microb Ecol 66:268–280

Mevarech M, Frolow F, Gloss LM (2000) Halophilic enzymes: proteins with a grain of salt. Biophys Chem 86:155–164

Mikucki JA, Pearson A, Johnston DT, Turchyn AV, Farquhar J, Schrag DP, Anbar AA, Priscu JC, Lee PA (2009) A contemporary microbially maintained subglacial ferrous "ocean". Science 324:397–400

Minegishi H, Echigo A, Nagaoka S, Kamekura M, Usami R (2010) *Halarchaeum acidiphilum* gen. nov., sp. nov., a moderately acidophilic haloarchaeon isolated from commercial solar salt. Int J Syst Evol Microbiol 60:2513–2516

Mongodin MEF, Nelson KE, Duagherty S, DeBoy RT, Wister J, Khouri H, Weidman J, Balsh DA, Papke RT, Sanchez Perez G, Sharma AK, Nesbo CL, MacLeod D, Bapteste E, Doolittle WF, Charlebois RL, Legault B, Rodríguez-Valera F (2005) The genome of *Salinibacter ruber*: convergence and gene exchange among hyperhalophilic bacteria and archaea. Proc Natl Acad Sci USA 102:18147–18152

Mouné S, Caumette P, Matheron R, Willison JC (2002) Molecular sequence analysis of prokaryotic diversity in the anoxic sediments underlying cyanobacterial mats of two hypersaline ponds in Mediterranean salterns. FEMS Microbiol Ecol 44:117–130

Mullakhanbhai MF, Larsen H (1975) *Halobacterium volcanii*, spec. nov., a Dead Sea halobacterium with a moderate salt requirement. Arch Microbiol 104:207–214

Narasingarao P, Podell S, Ugalde JA, Brochier-Armanet C, Emerson JB, Brocks JJ, Heidelberg KB, Banfield JF, Allen EE (2012) *De novo* assembly reveals abundant novel major lineage of Archaea in hypersaline microbial communities. ISME J 6:81–93

Nickle DC, Learn GH, Rain MW, Mullins JI, Miller JE (2002) Curiously modern DNA for a "250 million-year-old" bacterium. J Mol Evol 54:134–137

Norton CF, Grant WD (1988) Survival of halobacteria within fluid inclusions in salt crystals. J Gen Microbiol 134:1365–1373

Norton CF, McGenity TJ, Grant WD (1993) Archaeal halophiles (halobacteria) from two British salt mines. J Gen Microbiol 139:1077–1081

Oh D, Porter K, Russ B, Burns D, Dyall-Smith M (2009) Diversity of *Haloquadratum* and other haloarchaea in three, geographically distant, Australian saltern crystallizer ponds. Extremophiles 14:161–169

Ollivier B, Hatchikian CE, Prensier G, Guezennec J, Garcia J-L (1991) *Desulfohalobium retbaense* gen. nov. sp. nov., a halophilic sulfate-reducing bacterium from sediments of a hypersaline lake in Senegal. Int J Syst Bacteriol 41:74–81

Orellana MV, Pang WL, Durand PM, Whitehead M, Baliga MS (2013) A role for programmed cell death in the microbial loop. PLoS One 8:e62595

Oremland RS, King GM (1989) Methanogenesis in hypersaline environments. In: Cohen Y, Rosenberg E (eds) Microbial mats. Physiological ecology of benthic microbial communities. American Society for Microbiology, Washington, DC, pp 180–190

Oremland RS, Kulp TR, Switzer Blum J, Hoeft SE, Baesman S, Miller LG, Stolz JF (2005) A microbial arsenic cycle in a salt-saturated extreme environment. Science 308:1305–1308

Oren A (1983) *Halobacterium sodomense* sp. nov., a Dead Sea halobacterium with an extremely high magnesium requirement. Int J Syst Bacteriol 33:381–386

Oren A (1986) Intracellular salt concentrations of the anaerobic halophilic eubacteria *Haloanaerobium praevalens* and *Halobacteroides halobius*. Can J Microbiol 32:4–9

Oren A (1988) The microbial ecology of the Dead Sea. In: Marshall KC (ed) Advances in microbial ecology, vol 10. Plenum, New York, pp 193–229

Oren A (1990) Formation and breakdown of glycine betaine and trimethylamine in hypersaline environments. Antonie van Leeuwenhoek 58:291–298

Oren A (1993) The Dead Sea – alive again. Experientia 49:518–522

Oren A (1994a) Characterization of the halophilic archaeal community in saltern crystallizer ponds by means of polar lipid analysis. Int J Salt Lake Res 3:15–29

Oren A (1994b) The ecology of the extremely halophilic archaea. FEMS Microbiol Rev 13:415–440

Oren A (1995) The role of glycerol in the nutrition of halophilic archaeal communities: a study of respiratory electron transport. FEMS Microbiol Ecol 16:281–290

Oren A (1999a) Bioenergetic aspects of halophilism. Microbiol Mol Biol Rev 63:334–348

Oren A (1999b) Microbiological studies in the Dead Sea: future challenges toward the understanding of life at the limit of salt concentrations. Hydrobiologia 405:1–9

Oren A (2001) The bioenergetic basis for the decrease in metabolic diversity at increasing salt concentrations: implications for the functioning of salt lake ecosystem. Hydrobiologia 466:61–72

Oren A (2002a) Halophilic microorganisms and their environments. Kluwer Scientific, Dordrecht

Oren A (2002b) Diversity of halophilic microorganisms: environments, phylogeny, physiology, and applications. J Ind Microbiol Biotechnol 28:56–63

Oren A (2002c) Molecular ecology of extremely halophilic Archaea and Bacteria. FEMS Microbiol Ecol 39:1–7

Oren A (2005) A hundred years of *Dunaliella* research – 1905-2005. Saline Syst 1:2

Oren A (2006a) Life at high salt concentrations. In: Dworkin M, Falkow S, Rosenberg E, Schleifer K-H, Stackebrandt E (eds) The prokaryotes. A handbook on the biology of bacteria, vol 2, 3rd edn. Springer, New York, pp 263–282

Oren A (2006b) The order *Halobacteriales*. In: Dworkin M, Falkow S, Rosenberg E, Schleifer K-H, Stackebrandt E (eds) The prokaryotes. A handbook on the biology of bacteria, vol 3, 3rd edn. Springer, New York, pp 113–164

Oren A (2006c) The order *Haloanaerobiales*. In: Dworkin M, Falkow S, Rosenberg E, Schleifer K-H, Stackebrandt E (eds) The prokaryotes. A handbook on the biology of bacteria, vol 4, 3rd edn. Springer, New York, pp 804–817

Oren A (2007) Biodiversity in highly saline environments. In: Gerdes C, Glansdorff N (eds) Physiology and biochemistry of extremophiles. ASM Press, Washington, DC, pp 223–231

Oren A (2008) Microbial life at high salt concentrations: phylogenetic and metabolic diversity. Saline Syst 4:2

Oren A (2009) Microbial diversity and microbial abundance in salt-saturated brines: why are the waters of hypersaline lakes red? In: Oren A, Naftz DL, Palacios P, Wurtsbaugh WA (eds) Saline lakes around the world: unique systems with unique values. The SJ and Jessie E Quinney Natural Resources Research Library, College of Natural Resources, Utah State University, Salt Lake City, UT, pp 247–255 (open access at http://www.cnr.usu.edu/quinney/htm/publications/nrei)

Oren A (2010) The dying Dead Sea: the microbiology of an increasingly extreme environment. Lakes Reservoirs Res Manage 15:215–222

Oren A (2011a) Thermodynamic limits to microbial life at high salt concentrations. Environ Microbiol 13:1908–1923

Oren A (2011b) Diversity of halophiles. In: Horikoshi K (ed) Extremophiles handbook. Springer, Tokyo, pp 309–325

Oren A (2011c) Ecology of halophiles. In: Horikoshi K (ed) Extremophiles handbook. Springer, Tokyo, pp 343–361

Oren A (2012a) Approaches toward the study of halophilic microorganisms in their natural environments: who are they and what are they doing. In: Vreeland RH (ed) Advances in the understanding of halophilic microorganisms. Springer, Dordrecht, pp 1–33

Oren A (2012b) Metagenomics of salt lakes. In: Nelson K (ed) Encyclopedia of metagenomics. Springer, New York. http://www.springerreference.com/docs/edit/chapterdbid/303297.html

Oren A (2012c) Salts and brines. In: Whitton BA (ed) Ecology of cyanobacteria II. Their diversity in time and space, 2nd edn. Springer, Dordrecht, pp 401–426

Oren A (2013a) *Salinibacter*: an extremely halophilic bacterium with archaeal properties. FEMS Microbiol Lett 342:1–9

Oren A (2013b) Two centuries of microbiological research in the Wadi Natrun, Egypt: a model system for the study of the ecology, physiology, and taxonomy of haloalkaliphilic microorganisms. In: Seckbach J, Oren A, Stan-Lotter H (eds) Polyextremophiles – organisms living under multiple forms of stress. Springer, Dordrecht, pp 103–119

Oren A, Dubinsky Z (1994) On the red coloration of saltern crystallizer ponds. II. Additional evidence for the contribution of halobacterial pigments. Int J Salt Lake Res 3:9–13

Oren A, Gurevich P (1993) Characterization of the dominant halophilic archaea in a bacterial bloom in the Dead Sea. FEMS Microbiol Ecol 12:249–256

Oren A, Gurevich P (1994) Production of D-lactate, acetate, and pyruvate from glycerol in communities of halophilic archaea in the Dead Sea and in saltern crystallizer ponds. FEMS Microbiol Ecol 14:147–156

Oren A, Gurevich P (1995) Dynamics of a bloom of halophilic archaea in the Dead Sea. Hydrobiologia 315:149–158

Oren A, Mana L (2002) Amino acid composition of bulk protein and salt relationships of selected enzymes of *Salinibacter ruber*, an extremely halophilic bacterium. Extremophiles 6:217–223

Oren A, Rodríguez-Valera F (2001) The contribution of *Salinibacter* species to the red coloration of saltern crystallizer ponds. FEMS Microbiol Ecol 36:123–130

Oren A, Shilo M (1981) Bacteriorhodopsin in a bloom of halobacteria in the Dead Sea. Arch Microbiol 130:185–187

Oren A, Gurevich P, Anati DA, Barkan E, Luz B (1995a) A bloom of *Dunaliella parva* in the Dead Sea in 1992: biological and biogeochemical aspects. Hydrobiologia 297:173–185

Oren A, Gurevich P, Gemmell RT, Teske A (1995b) *Halobaculum gomorrense* gen. nov., sp. nov., a novel extremely halophilic archaeon from the Dead Sea. Int J Syst Bacteriol 45:747–754

Oren A, Kühl M, Karsten U (1995c) An endoevaporitic microbial mat within a gypsum crust: zonation of phototrophs, photopigments, and light penetration. Mar Ecol Prog Ser 128:151–159

Oren A, Duker S, Ritter S (1996) The polar lipid composition of Walsby's square bacterium. FEMS Microbiol Lett 138:135–140

Oren A, Heldal M, Norland S (1997a) X-ray microanalysis of intracellular ions in the anaerobic halophilic eubacterium *Haloanaerobium praevalens*. Can J Microbiol 43:588–592

Oren A, Bratbak G, Heldal M (1997b) Occurrence of virus-like particles in the Dead Sea. Extremophiles 1:143–149

Oren A, Rodríguez-Valera F, Antón J, Benlloch S, Rosselló-Mora R, Amann R, Coleman J, Russell NJ (2004) Red, extremely halophilic, but not archaeal: the physiology and ecology of *Salinibacter ruber*, a bacterium isolated from saltern crystallizer ponds. In: Ventosa A (ed) Halophilic microorganisms. Springer, Berlin, pp 63–76

Oren A, Pri-El N, Shapiro O, Siboni N (2006) Buoyancy studies in natural communities of square gas-vacuolate archaea in saltern crystallizer ponds. Saline Syst 2:4

Oren A, Sørensen KB, Canfield DE, Teske AP, Ionescu D, Lipski A, Altendorf K (2009a) Microbial communities and processes within a hypersaline gypsum crust in a saltern evaporation pond (Eilat, Israel). Hydrobiologia 626:15–26

Oren A, Baxter BK, Weimer BC (2009b) Microbial communities in salt lakes: phylogenetic diversity, metabolic diversity, and *in situ* activities. In: Oren A, Naftz DL, Palacios P, Wurtsbaugh WA (eds) Saline lakes around the world: unique systems with unique values. The SJ and Jessie E Quinney Natural Resources Research Library, College of Natural Resources, Utah State University, Salt Lake City, UT, pp 257–263

Oren A, Elevi Bardavid R, Kandel N, Aizenshtat Z, Jehlicka J (2013) Glycine betaine is the main organic osmotic solute in a stratified microbial community in a hypersaline evaporitic gypsum crust. Extremophiles 17:445–451

Osterrothová K, Jehlička J (2011) Investigation of biomolecules trapped in fluid inclusions inside halite crystals by Raman spectroscopy. Spectrochim Acta A 83:288–296

Øvreås L, Daae FL, Torsvik V, Rodríguez-Valera F (2003) Characterization of microbial diversity in hypersaline environments by melting profiles and reassociation kinetics in combination with terminal restriction fragment length polymorphism (T-RFLP). Microb Ecol 46:291–301

Park JS, Kim H, Choi DH, Cho BC (2003) Active flagellates grazing on prokaryotes in high salinity waters of a solar saltern. Aquat Microb Ecol 33:173–179

Park S-J, Kang C-H, Rhee S-K (2006) Characterization of the microbial diversity in a Korean solar saltern by 16S rRNA gene analysis. J Microbiol Biotechnol 16:1640–1645

Park JS, Simpson AGB, Lee WJ, Cho BC (2007) Ultrastructure and phylogenetic placement within Heterolobosea of the previously unclassified, extremely halophilic heterotrophic flagellate *Pleurostomum flabellatum* (Ruinen 1938). Protist 158:397–413

Parnell JJ, Rompato G, Latta LC IV, Pfender ME, van Nostrand JD, He Z, Zhou J, Andersen G, Champine P, Ganesan B, Weimer BC (2010) Functional biogeography as evidence of gene transfer in hypersaline microbial communities. PLoS One 5:e12919

Parnell JJ, Rompato G, Crowl TA, Weimer BC, Pfrender ME (2011) Phylogenetic distance in Great Salt Lake microbial communities. Aquat Microb Ecol 64:267–273

Pašić L, Galán Bartual S, Poklar Ulrih N, Grabnar M, Herzog Velikonja B (2005) Diversity of halophilic archaea in the crystallizers of an Adriatic solar saltern. FEMS Microbiol Ecol 54:491–498

Pašić L, Poklar Ulrih N, Črnigoj M, Grabnar M, Herzog Velikonja B (2007) Haloarchaeal communities in the crystallizers of two Adriatic solar salterns. Can J Microbiol 53:8–18

Pedrós-Alió C, Calderón-Paz JI, MacLean MH, Medina G, Marassé C, Gasol JM, Guixa-Boixereu N (2000a) The microbial food web along salinity gradients. FEMS Microbiol Ecol 32:143–155

Pedrós-Alió C, Calderón-Paz JI, Gasol JM (2000b) Comparative analysis shows that bacterivory, not viral lysis, controls the abundance of heterotrophic prokaryotic plankton. FEMS Microbiol Ecol 32:157–165

Peña A, Teeling H, Huerta-Cepas J, Santos F, Yarza P, Brito-Echeverría J, Lucio M, Schmitt-Kopplin P, Meseguer I, Schenowitz C, Dossat C, Barbe V, Dopazo J, Rosselló-Mora R, Schüler M, Oliver Glöckner M, Amann R, Gabaldón T, Antón J (2010) Fine-scale evolution: genomic, phenotypic and ecological differentiation in two coexisting *Salinibacter ruber* strains. ISME J 4:882–895

Peña A, Teeling H, Huerta-Cepas J, Santos F, Meseguer I, Lucio M, Schmitt-Kopplin P, Dopazo J, Rosselló-Móra R, Schuler M, Oliver Glöckner F, Amann R, Gabaldón T, Antón J (2011) From genomics to microevolution and ecology: the case of *Salinibacter ruber*. In: Ventosa A, Oren A, Ma Y (eds) Halophiles and hypersaline environments. Springer, Berlin, pp 109–122

Pietilä MK, Roine E, Paulin L, Kalkkinen N, Bamford DH (2009) An ssDNA virus infecting archaea: a new lineage of viruses with a membrane envelope. Mol Microbiol 72:307–319

Porter D, Roychoudhury AN, Cowan D (2007a) Dissimilatory sulfate reduction in hypersaline coastal pans: activity across a salinity gradient. Geochim Cosmochim Acta 71:5102–5116

Porter K, Russ BE, Dyall-Smith ML (2007b) Virus-host interactions in salt lakes. Curr Opin Microbiol 10:418–424

Post FJ (1977) The microbial ecology of the Great Salt Lake. Microb Ecol 3:143–165

Prášil O, Bína D, Medová H, Řeháková K, Zapomělová E, Veselá J, Oren A (2009) Emission spectroscopy and kinetic fluorometry studies of phototrophic microbial communities along a salinity gradient in solar saltern evaporation ponds of Eilat, Israel. Aquat Microb Ecol 56:285–296

Qvit-Raz N, Jurkevitch E, Belkin S (2008) Drop-size soda lakes: transient microbial habitats on a salt-secreting desert tree. Genetics 178:1615–1622

Rainey FA, Zhilina TN, Boulygina ES, Stackebrandt E, Tourova TP, Zavarzin GA (1995) The taxonomic status of the fermentative halophilic anaerobic bacteria: description of Haloanaerobiales ord. nov., *Halobacteroidaceae* fam. nov., *Orenia* gen. nov. and further taxonomic rearrangements at the genus and species level. Anaerobe 1:185–199

Reistad R (1970) On the composition and nature of the bulk protein of extremely halophilic bacteria. Arch Microbiol 71:353–360

Rengpipat S, Lowe SE, Zeikus JG (1988) Effect of extreme salt concentrations on the physiology and biochemistry of *Halobacteroides acetoethylicus*. J Bacteriol 170:3065–3071

Rhodes ME, Fitz-Gibbon ST, Oren A, House CH (2010) Amino acid signatures of salinity on an environmental scale with a focus on the Dead Sea. Environ Microbiol 12:2613–2623

Rhodes ME, Oren A, House CH (2012) Dynamics and persistence of Dead Sea microbial populations as shown by high throughput sequencing of rRNA. Appl Environ Microbiol 78:2489–2492

Roberts MF (2005) Organic compatible solutes of halotolerant and halophilic microorganisms. Saline Syst 1:5

Rodriguez-Brito B, Li L, Wegley L, Furlam M, Angly F, Breitbart M, Buchanan J, Desnues C, Dinsdale E, Edwards R, Felts B, Haynes M, Liu H, Lipson D, Mahaffy J, Martin-Cuadrado AB, Mira A, Nulton J, Pašić L, Rayhawk S, Rodriguez-Mueller J, Rodriguez-Valera F, Salamon P, Srinagesh S, Thingstad TF, Tran T, Thurber RV, Willner D, Youle M, Rohwer F (2010) Viral and microbial community dynamics in four aquatic environments. ISME J 4:739–751

Roselló-Mora R, Lee N, Antón J, Wagner M (2003) Substrate uptake in extremely halophilic microbial communities revealed by microautoradiography and fluorescence in situ hybridization. Extremophiles 7:409–413

Roychoudhury AN, Cowan D, Porter D, Valverde A (2013) Dissimilatory sulphate reduction in hypersaline coastal pans: an integrated microbiological and geochemical study. Geobiology 11:224–233

Sabet S (2012) Halophilic viruses. In: Vreeland RH (ed) Advances in understanding of the biology of halophilic microorganisms. Springer, Dordrecht, pp 81–116

Sabet S, Diallo L, Hays L, Jung W, Dillon JG (2009) Characterization of halophiles isolated from solar salterns in Baja California, Mexico. Extremophiles 13:643–656

Sankaranarayanan K, Timofeeff MN, Spathis R, Lowenstein TK, Koji Lum J (2011) Ancient microbes from halite fluid inclusions: optimized surface sterilization and DNA extraction. PloS One 6:e20683

Santos F, Meyerdierks A, Peña A, Rosselló-Mora R, Amann R, Antón J (2007) Metagenomic approach to the study of halophages: the environmental halophage 1. Environ Microbiol 9:1711–1723

Santos F, Moreno-Paz M, Meseguer I, López C, Rosselló-Mora R, Parro V, Antón J (2011) Metatranscriptomic analysis of extremely halophilic viral communities. ISME J 5:1621–1633

Santos F, Yarza P, Parro V, Meseguer I, Rosselló-Móra R, Antón J (2012) Viruses from hypersaline environments: a culture-independent approach. Appl Environ Microbiol 78:1635–1643

Sass AM, Sass H, Coolen MJL, Cypionka H, Overmann J (2001) Microbial communities in the chemocline of a hypersaline deep-sea basin (Urania Basin, Mediterranean Sea). Appl Environ Microbiol 67:5392–5402

Schubert BA, Lowenstein TK, Timofeeff MN (2009a) Microscopic identification of prokaryotes in modern and ancient halite, Saline Valley and Death Valley, California. Astrobiology 9:467–482

Schubert BA, Lowenstein TK, Timofeeff MN, Parker MA (2009b) How do prokaryotes survive in fluid inclusions in halite for 30 k.y.? Geology 37:1059–1062

Schubert BA, Lowenstein TK, Timofeeff MN, Parker MA (2010a) Halophilic Archaea cultured from ancient halite, Death Valley, California. Environ Microbiol 12:440–454

Schubert BA, Timofeeff MN, Polle JEW, Lowenstein TK (2010b) *Dunaliella* cells in fluid inclusions in halite: significance for long-term survival of prokaryotes. Geomicrobiol J 27:61–75

Sher J, Elevi R, Mana L, Oren A (2004) Glycerol metabolism in the extremely halophilic bacterium *Salinibacter ruber*. FEMS Microbiol Lett 232:211–215

Sime-Ngando T, Lucas S, Robin A, Pause Tucker K, Colombet J, Forterre P, Breitbart M, Prangishvili D (2011) Diversity of virus-host systems in hypersaline Lake Retba, Senegal. Environ Microbiol 13:1956–1972

Simon RD, Abeliovich A, Belkin S (1994) A novel terrestrial halophilic environment: the phylloplane of *Atriplex halimus*, a salt-excreting plant. FEMS Microbiol Ecol 14:99–110

Soliman GSH, Trüper HG (1982) *Halobacterium pharaonis* sp. nov., a new, extremely haloalkaliphilic archaebacterium with low magnesium requirement. Zbl Bakt Hyg I Abt Orig C 3:318–329

Sørensen KB, Canfield DE, Teske AP, Oren A (2005) Community composition of a hypersaline endoevaporitic microbial mat. Appl Environ Microbiol 71:7352–7365

Sørensen K, Řeháková K, Zapomělová E, Oren A (2009) Distribution of benthic phototrophs, sulfate reducers, and methanogens in two adjacent salt ponds in Eilat, Israel. Aquat Microb Ecol 56:275–284

Stephens DW, Gillespie DM (1976) Phytoplankton production in the Great Salt Lake, Utah, and a laboratory study of algal response to enrichment. Limnol Oceanogr 21:74–87

Stock A, Breiner H-W, Pachiadaki M, Edgcomb V, Filker S, La Cono V, Yakimov MM, Stoeck T (2012) Microbial eukaryote life in the new hypersaline deep-sea basin Thetis. Extremophiles 16:21–34

Stock A, Edgcomb V, Orsi W, Filker S, Breiner H-W, Yakimov MM, Stoeck T (2013) Evidence for isolated evolution of deep-sea ciliate communities through geological separation and environmental selection. BMC Microbiol 13:150

Triado-Margarit X, Casamayor EO (2013) High genetic diversity and novelty in planktonic protists inhabiting inland and coastal high salinity water bodies. FEMS Microbiol Ecol 85:27–36

van der Wielen PWJJ, Heijs SK (2007) Sulfate-reducing prokaryotic communities in two deep hypersaline anoxic basins in the Eastern Mediterranean deep sea. Environ Microbiol 9:1335–1340

van der Wielen PWJJ, Bolhuis H, Borin S, Daffonchio D, Corselli C, Giuliano L, D'Auria G, de Lange GJ, Huebner A, Varnavas SP, Thomson J, Tamburini C, Marty D, McGenity TJ, Timmis KN, BioDeep Scientific Party (2005) The enigma of prokaryotic life in deep hypersaline anoxic basins. Science 307:121–123

Ventosa A, Nieto JJ, Oren A (1998) Biology of moderately aerobic bacteria. Microbiol Mol Biol Rev 62:504–544

Vítek P, Jehlička J, Edwards HGM, Hutchinson I, Ascaso C, Wierzchos J (2012) The miniaturized Raman system and detection of traces of life in halite from the Atacama desert: some considerations for the search for life signatures on Mars. Astrobiology 12:1095–1099

Vreeland RH, Piselli AF Jr, McDonnough S, Meyers SS (1998) Distribution and diversity of halophilic bacteria in a subsurface salt formation. Extremophiles 2:321–331

Vreeland RH, Rosenzweig WD, Powers DW (2000) Isolation of a 250 million-year-old halotolerant bacterium from a primary salt crystal. Nature 407:897–900

Vreeland RH, Jones J, Monson A, Rosenzweig WD, Lowenstein TK, Timofeeff M, Satterfield C, Cho BC, Park JS, Wallace A, Grant WD (2007) Isolation of live Cretaceous (121–112 million years old) halophilic Archaea from primary salt crystals. Geomicrobiol J 24:275–282

Walsby AE (1971) The pressure relationships of gas vacuoles. Proc R Soc London B 178:301–326

Walsby AE (1980) A square bacterium. Nature 283:69–71

Walsby AE (2005) Archaea with square cells. Trends Microbiol 13:193–195

Warkentin M, Schumann R, Oren A (2009) Community respiration studies in saltern crystallizer ponds. Aquat Microb Ecol 56:255–261

Weimer BC, Rompato G, Parnell J, Gann R, Balasubramanian G, Navas C, Gonzalez M, Clavel M, Albee-Scott S (2009) Microbial biodiversity of Great Salt Lake, Utah. In: Oren A, Naftz DL, Palacios P, Wurtsbaugh WA (eds) Saline lakes around the world: unique systems with unique values. The SJ and Jessie E Quinney Natural Resources Research Library, College of Natural Resources, Utah State University, Salt Lake City, UT, pp 15–22 (open access at http://www.cnr.usu.edu/quinney/htm/publications/nrei)

Welsh DT (2000) Ecological significance of compatible solute accumulation by micro-organisms: from single cells to global climate. FEMS Microbiol Rev 24:263–290

Wierzchos J, Ascaso C, McKay CP (2006) Endolithic cyanobacteria in halite rocks from the hyperarid core of the Atacama Desert. Astrobiology 6:415–423

Winters YD, Timofeeff MK, Lowenstein TK (2013) Identification of carotenoids in ancient salt from Death Valley, Saline Valley, and Searles Lake, California using laser Raman spectroscopy. Astrobiology 13:1065–1080

Woelfel J, Sørensen K, Warkentin M, Forster S, Oren A, Schumann R (2009) Oxygen evolution in a hypersaline crust: photosynthesis quantification by *in situ* microelectrode profiling and planar optodes in incubation chambers. Aquat Microb Ecol 56:263–273

Wohlfarth A, Severin J, Galinski EA (1990) The spectrum of compatible solutes in heterotrophic halophilic eubacteria of the family *Halomonadaceae*. J Gen Microbiol 136:705–712

Zajc J, Zalar P, Plemenitas A, Gunde-Cimerman N (2011) The mycobiota of the salterns. In: Raghukumar C (ed) Biology of marine fungi. Springer, Berlin, pp 133–158

Zajc J, Liu Y, Dai W, Yang Z, Hu J, Gostinčar C, Gunde-Cimerman N (2013) Genome and transcriptome sequencing of the halophilic fungus *Wallemia ichthyophaga*: haloadaptations present and absent. BMC Genomics 14:617

Zalar P, Kocuvan MA, Plemenitaš A, Gunde-Cimerman N (2005) Halophilic black yeasts colonize wood immersed in hypersaline water. Bot Mar 48:323–326

Zhaxybayeva O, Stepanauskas R, Mohan NR, Papke RT (2013) Cell-sorting analysis of geographically separated hypersaline environments. Extremophiles 17:265–275

Chapter 9
Microbes and the Arctic Ocean

Iain Dickinson, Giselle Walker, and David A. Pearce

Abstract It is surprising how little we really know about the microorganisms that live within the Arctic Ocean and in particular their role in the marine food web. A comprehensive food web study is yet to be conducted, and although many studies talk of an Arctic marine food web, publications normally focus on just one aspect of it. In some way, it mirrors our knowledge of the deep sea. We know it exists, we know the main interactions and pathways, but a great deal of the background information and detail is lacking. The single most important part of our research, therefore, is to understand the role and function of microorganisms in the environment. A fitting analogy is perhaps an iceberg. We know what it looks like, we know roughly what it does and how it behaves, we even know that the majority of the iceberg is hidden from view. However, we know very little about the effects of an iceberg on the general environment around it. Further, if we induce change in the behaviour of that iceberg, we have little idea what effect it might have. This is due to a combination of factors; (1) the complexity of the science, (2) the ambition, cost and logistics of conducting experimental work in these systems, (3) the fact that the existing, relatively simple model is sufficient for most purposes, (4) technological and methodological limitations and (5) research funding tends not to favour supporting ambitious long term ecological studies which can be very expensive. However, as with most questions in science, the harder we look, the more there is to

I. Dickinson
Northumbria University, Faculty of Health and Life Sciences, Ellison Building, Newcastle-upon-Tyne NE1 8ST, UK
e-mail: iain.dickinson@northumbria.ac.uk

G. Walker
The University Centre in Svalbard (UNIS), P.O. Box 156, 9171 Longyearbyen, Svalbard, Norway

D.A. Pearce (✉)
Northumbria University, Faculty of Health and Life Sciences, Ellison Building, Newcastle-upon-Tyne NE1 8ST, UK

The University Centre in Svalbard (UNIS), P.O. Box 156, 9171 Longyearbyen, Svalbard, Norway

British Antarctic Survey, Natural Environment Research Council, High Cross, Madingley Road, Cambridge CB3 0ET, UK
e-mail: david.pearce@northumbria.ac.uk

find, and for every question we answer, a number of new questions arise. In this chapter, we attempt to give an overview of what is known about the microbial community in the Arctic marine food web, assess why this knowledge is relatively limited and pose some of the questions that remain to be answered.

9.1 What Is a Food Web?

A food web describes feeding relationships in an ecosystem. A simple food chain describes a linear flow of energy from primary producers to consumers, eventually leading to the apex predator. Decomposers recycle material from the food chain back into the environment. However, in reality there are trophic links between many different producers and consumers of various sizes, so it is more appropriate to consider the whole food web rather than a simple chain.

9.2 What Is Special About the Arctic Ocean?

The Arctic Ocean and its environment encompass parts of Russia, Canada, Greenland, Norway, Alaska and Iceland. The Arctic Ocean is unique, due to its location on the planet and its relationship with the sun. North of the Arctic Circle beginning at 66°33″N, the region experiences 24 h daylight in the summer and 24 h darkness in the winter, and this, perhaps more than any other factor, drives the Arctic Ocean's unique ecology. As well as light, low temperatures may also have cascading effects on the interlinked and delicately balanced food web. The Arctic Ocean is relatively shallow and is mostly surrounded by land, unlike the Southern Ocean. Land masses surrounding the Arctic Ocean drain huge areas of land and contribute vast quantities of nutrients into the coastal ecosystem. As a result of the unique nature of the Arctic Ocean, the nutrients available and the long periods of daylight during the summer, the Arctic Ocean is rich in life. The presence of ice and its attendant snow cover, and total darkness during the winter months, means that this region has a distinctive ecology, in terms of the phytoplankton and zooplankton, the animal populations and their associated environmental factors. The Arctic Ocean ecosystem therefore has a distinct and complex food web, quite unlike those found elsewhere. Arctic marine species have adapted to the cold temperatures to take advantage of the nutrient-rich waters surrounding the ice edges and continental shelves. Indeed, the highest biomass occurs along the coastlines. This observation is consistent with river runoff, nutrient availability and increased stability along coastlines, as well as the convergence of Atlantic water and Arctic water in the Barents Sea (Smetacek and Nicol 2005). Examples of specialist species endemic to the Arctic Ocean are exemplified by the charismatic megafauna: the polar bear, the walrus and the narwhal, all of which depend on sea ice for hunting, reproduction or protection (Stirling 1997; Tynan and DeMaster 1997). However, most of what we know about this ecosystem is based mainly on summer observations as expeditions to the remote Arctic in the winter are both difficult and expensive.

9.3 The Fundamental Importance of Sea Ice

Sea ice still covers significant portions of the Arctic Ocean, particularly in winter, forming at temperatures below the freezing point of seawater. In the polar regions, with an ocean salinity of 35 ppt, the water begins to freeze at -1.8 °C (NSIDC 2015). The formation and structure of sea ice is reviewed in detail by Petrich and Eicken (2009). With its high albedo, the ice and its snow cover reduce the amount of incoming solar radiation absorbed at the ocean surface by reflecting much of that radiation back into space. Changes in the timing of formation and extent of ocean coverage by the sea ice impose temporal and spatial variation in energy requirements and food availability for higher trophic levels. Mismatches in the patterns of climate and biology can lead to decreased reproductive success, lower abundances and changes in the distribution of organisms (Moline et al. 2008). Sea ice is critical for Arctic marine ecosystems in at least two important ways: it provides a habitat for photosynthetic algae and a nursery ground for both invertebrates and fish during times when the water column does not support phytoplankton growth. As the ice melts, releasing organisms into the surface water, a shallow mixed layer forms which fosters large ice-edge blooms important to the overall productivity of Arctic seas (Nielsen et al. 2002). Sea ice is one of the largest habitats in polar oceans and can cover up to 13 % of the world's surface in winter, so it has a massive influence on the Arctic marine food web (Eicken 1992). In winter, sea ice constitutes a thermal barrier against the cold winter atmosphere with the result that the interface between the ice and the seawater remains near the temperature of seawater (Krembs and Deming 2011).

Sea ice imposes a three dimensional structure on the Arctic Ocean microbial community. Sea-ice algal growth is limited by space availability within the sea ice itself rather than by nutrient availability, due to the high retention of nutrients in the sea ice and low abundance of prokaryote grazers (Becquevort et al. 2009). Frost flowers contain high concentrations of bacteria compared to other formations of sea ice; they can mediate long-range wind transportation of microbes, which can themselves influence atmospheric processes by influencing (e.g.) photochemistry (Bowman and Deming 2010). Bacterial–algal relationships in sea ice are also hindered by substrate and temperature limitations, limiting primary production through the microbial loop within the sea ice (Stewart and Fritsen 2004).

The Polar regions have, so far, shown the greatest sensitivity to rising global temperatures. Sea ice has shown a rapid decrease in both area and thickness over the past 10 years (Comiso et al. 2008). Multilayer sea ice occurs when sea ice survives the summer months and refreezes, often increasing in thickness. Bowman et al. (2012) found unique microbes in multiyear ice not seen in first year ice, including *Cyanobacteria*, *Spartobacteria* and *Coraliomargarita* in sufficient numbers to suggest a niche occupation. However, the thickness of the sea ice has rapidly reduced in recent years (Serreze et al. 2007). Ongoing climate warming is causing a dramatic loss of sea ice in the Arctic Ocean, and it is projected that the Arctic Ocean will become seasonally ice-free by 2040 (Kedra et al. 2015). The thickness of sea ice can also drastically alter the activity and composition of marine microorganisms (Mock and Junge 2007). The common theory that the formation of first year sea ice acts as a selective pressure on microbial communities was found to be false by Collins, Rocap and Deming (2010), who sampled sea ice on a weekly basis over winter and only found a drop in microbial abundance rather than a change in community structure. High latitude changes in ice dynamics and their impact on polar marine ecosystems are reviewed by Moline et al. (2008). Whilst the effect of ice dynamics is easily visible when considering the impact on organisms higher in the food chain such as polar bears (through reduction in hunting grounds, etc.), the effect of changing ice dynamics is much less visible on microorganisms at the base of the food chain. Marine microorganisms play a vital role in global biogeochemical cycling (Helmke and Weyland 2004), and are important primary producers in polar environments (Legendre et al. 1992), a decline in microbial activity will have a direct influence on species higher up the food chain such as mammals, fish and birds.

Sea ice can be divided into four distinct parts: ice surface, ice matrix, under the ice and brine channels and pockets. When sea water freezes and sea ice forms, dissolved air, inorganic and organic matter are expelled into concentrated brine which forms pockets and channels in the ice depending on the temperature of the ice (Eicken 1992). Temperature affects the salinity of the pockets, with colder temperatures forcing a higher salinity into a smaller volume of brine. In columnar ice, $-5\ °C$ is the tipping point between permeable and impermeable ice, allowing channels to form between brine pockets. This is because warmer ice has a higher volume of brine in the channels (above 5 %) allowing for more convective flow within the sea ice (Golden et al. 1998). Conditions within these pockets and channels can vary greatly depending on their depth in the sea-ice column. Virus-containing bacteria that were isolated from brine pockets in sea ice had a much greater viral load than those isolated from sea water (Wells and Deming 2006). Typically, the ice closer to the ice–air interface is much colder than the equivalent ice–sea water boundary, reducing surface ice volume and increasing its salt concentration. Conversely, irradiance will be higher at lower depths in the ice since the light has less ice to travel through than to deeper ice layers. When sea ice begins to freeze, diverse mixed populations of microorganisms are forced into a new cold, dark, hypersaline environment and must possess the correct adaptations to survive and remain active. As the season progresses, there is a gradual transition from a microbial population similar to that of the open seawater to a new psychrophilic heterotrophic population (Thomas and Dieckmann 2002) which is fuelled by the organic carbon released from the death and lysis of microorganisms less adapted to survive (Thomas et al. 2001).

The Arctic marine food web relies heavily on the availability of organic carbon within the sea ice. Microbial heterotrophs rely on dissolved or organismal organic carbon, provided by marine autotrophs or indirectly via the microbial loop (Fenchel 1982; Sanders et al. 1992). The melting or thinning of sea ice in spring releases a vast number of microbes into the water column under the sea ice, and coupled with the phytoplankton bloom brought about by increasing light availability, this melting provides a high concentration of organic matter into the underlying seawater (Sherr and Sherr 2007). There are two main beneficiaries of this influx of organic matter. Cells avoiding high-wavelength radiation, and dead cells, quickly sink and provide nutrients to the benthic community such as crustaceans and molluscs. Cells growing in surface waters provide a food source to pelagic zooplankton. The water column directly below the sea ice of both poles is dominated by calanoid copepods, along with pteropod gastropods, siphonophores, appendicularians, chaetognaths, hyperiid amphipods, mysids and benthic invertebrate larvae (Schnack-Schiel 2003). The high concentration of these organisms provides the perfect feeding ground for polar fish including the polar cod, a species that represents the bridge between the upper and lower trophic levels (Gradinger and Bluhm 2004).

Higher trophic level organisms such as seals and birds must rely on polynyas to access the rich pelagic zone below the sea ice. Polynyas are characterised by an open water source surrounded by sea ice (Stringer and Groves 1991), and allow diving mammals and birds to hunt for polar cod and other fish, which in turn feed on pelagic and benthic invertebrates. The sea ice also provides nesting grounds for these birds and mammals, which in turn provides larger predators such as Polar bears and Arctic foxes with an opportunity to hunt for these animals as food. Seasonal sea ice can have a great effect on primary production in an area. However, the decline in primary production associated with sea ice formation may not have as great an effect on species at it does at higher trophic levels, who can change their diet from fresh algal material to detritus when primary production drops (Norkko et al. 2007).

9.4 Structure of the Arctic Marine Food Web

In its simplest form, the Arctic marine food web starts with primary producers, which comprise photosynthetic microbes that use either chlorophyll or bacteriorhodopsins to convert light into carbon (Béjà et al. 2000). Microorganisms exist within the sea ice in pockets and channels containing high concentrations of salt and other elements such as carbon, nitrogen, silica and iron. The cycling of these elements by microorganisms plays a vital role in the initial food chain, altering the abundance of microorganisms available to grazers between summer and winter. Carbon dioxide, ammonium, silicic acid and iron are all mass utilised in summer months by photosynthesising microalgae, which in turn provide organic carbon and oxygen for heterotrophic bacteria during the winter, which provides a large biomass for grazers during spring melts. The two main sources of primary production in Arctic ecosystems are sea-ice algae and phytoplankton in the water column (Søreide et al. 2006). These are then consumed by primary consumers, heterotrophic protists, other small eukaryotes, larvae and also bacteria (Fenchel 1988; Sanders et al. 1992; Sherr and Sherr 2007). Ice algae are a very important part of the marine food web, contributing on average 57 % to the total Arctic marine primary production (Gosselin et al. 1997). The organisms that eat algae, called zooplankton grazers (such as *Gammarus wilkitzkii*), seek not only food in this algal-rich ice but also protection from their own predators. Arctic cod (*Boreogadus saida*), an important food source for many marine mammals and birds, use the same habitat as nursery grounds (Krembs and Deming 2011), and their larvae feed on single-celled protists (i.e. algae and protozoa). The protists and bacteria are also lysed by the virus community, known to be about 10× as numerous as the bacteria. Large phytoplankton cells, such as diatoms and dinoflagellates, are the primary food of the zooplankton. Secondary consumers, largely fish species including polar cod, feed on these zooplankton, acting as a bridge between lower trophic levels and apex predators such as seabirds, whales, arctic foxes and polar bears. Excretions of metabolic products and debris from dying cells contribute to an increasing pool of organic material. As the ice melts in summer, this material releases into the water column, where it contributes to the vertical flux of material that fuels both pelagic and benthic food webs (Krembs and Deming 2011). However, this simplified description does not take into account the huge number of species involved, estimated to be between 9500 and 54,500 taxa (Archambault et al. 2010). In a recent study, Lovejoy and Potvin (2011) estimated that the number of picoplankton (plankton species less than 2 μm in diameter) operational taxonomic units (considered as roughly analogous to species) was ~45,000 (which include the archaea, picoeukaryotes and bacteria) in the Arctic Ocean. The food web is complex: species of different sizes can compete for the same sources of food, and losses of energy (through e.g. death, viral lysis, leakage of nutrients) at each trophic level can be picked up by heterotrophs at lower trophic levels (Azam et al. 1983). This makes the complexity absolutely staggering and our knowledge of the detailed interactions limited. To add to this, at least 10 % of the prokaryote population is as yet unknown. Many studies of distinct aspects of the Arctic food web now exist. Indeed, a number of very good reviews cover what is known of the topic in some detail: Bluhm and Gradinger (2008), Gradinger (2009), Darnis et al. (2012) and Kedra et al. (2015).

9.5 Viruses

Viruses are an important and often overlooked component of the marine food web (Fuhrman 1999; Suttle 2005, 2007), and they have been known to exist in the Polar regions for some time (see review by Pearce and Wilson 2003) where high viral infection rates have been observed (Säwström et al. 2007). Genome size distributions indicate variability and similarities among marine viral assemblages from diverse environments (Stewart and Possingham 2005). Sequence analysis of marine virus communities reveals that groups of related algal viruses are widely distributed in nature (Short and Suttle 2002). Global diversity is very high, presumably encompassing several hundred thousand viral species, and regional richness varies on a North-South latitudinal gradient. The marine regions have been shown to have different assemblages of viruses, e.g. prophage-like sequences are most common in the Arctic (Angly et al. 2006), following similar patterns to other Polar environments (Aguirre de Cárcer et al. 2015). However, ubiquity has also been described. To this end, nearly identical bacteriophage structural gene sequences are widely distributed in both marine and freshwater environments (Short and Suttle 2005), with marine T4-type bacteriophages found to be a ubiquitous component of the dark matter of the biosphere (Filée et al. 2005). Regional differences include the abundance and production of bacteria and viruses in the Bering and Chukchi Seas (Grieg et al. 1996). These communities are not static and can be quite dynamic, over time (Wells and Deming 2006). Studies exist which target specific aspects of the microbial ecology of Arctic marine viruses, such as those that infect psychrophiles (Borriss et al. 2003) and whether there is a cost associated with virus resistance (Lennon et al. 2007). High-throughput sequencing will allow better studies of specific microbiomes, such as the highly divergent picornavirus found in marine mammals (Kapoor et al. 2008), and lead us to better understand the natural role of viruses in gene exchange (Hambly and Suttle 2005). Going forward, we are sure to gain a better understanding of what shapes viral distribution, including how viral activity controls both prokaryotic and eukaryotic populations, their activity and other interactions within the Arctic food web. However, the cultivation of key viral groups and an understanding of the activity of specific virus populations in biogeochemical cycling remain as key challenges in this field.

9.6 Prokaryote Microbial Diversity

Much more is known about the diversity and phylogenetic relationships of the prokaryote community in Arctic marine food webs. In particular, there are several studies dealing with total prokaryote diversity and its limits (Curtis et al. 2002), diversity in permanently cold marine sediments (Ravenschlag et al. 1999) and the water column (Bano and Hollibaugh 2002), the differences between Arctic and Antarctic pack ice (Brinkmeyer et al. 2003), the abundance and production of bacterial groups in the western Arctic Ocean (Rex et al. 2007) and the ecology of the rare microbial biosphere of the Arctic Ocean (Galand et al. 2009). Indeed, global patterns have been recognised in the diversity and community structure in marine bacterioplankton (Pommier et al. 2007), and a latitudinal diversity gradient has been observed in planktonic marine bacteria (Fuhrman et al. 2008). It has recently been suggested that hydrography shapes the bacterial biogeography of the deep Arctic Ocean (Galand et al. 2010), and new insights emerge frequently, into specific bacterioplankton groups, for example, the alpha-Proteobacteria in coastal seawater (González and Moran 1997). Psychrophiles are clearly important in the polar regions (Deming 2002), and a great deal of work has been done on the diversity and association of psychrophilic bacteria in sea ice (Bowman et al. 1997). However, biodiversity of many taxonomic groups remains relatively unknown (Archambault et al. 2010), including areas of the High Arctic where biological data are almost non-existent (Piepenburg et al. 2011).

Unlike the virioplankton, great strides have been made in the cultivation of marine bacteria, including the numerically important Arctic sea-ice bacteria cultured at subzero temperatures (Junge et al. 2002). High-throughput methods for culturing microorganisms in very-low-nutrient media have yielded diverse new marine isolates (Connon and Giovannoni 2002) enabling researchers to focus their attention on specific groups such as the ubiquitous SAR11 marine bacterioplankton clade (Rappe et al. 2002) or the oligotrophic marine gamma-Proteobacteria (Cho and Giovannoni 2004). Such amenability to culture also permits bioprospecting for cold-active enzymes such as the lipases, amylases and proteases, from culturable bacteria, for example from Kongsfjorden near Ny-Ålesund in Svalbard (Srinivas et al. 2009). The cold-active glutaredoxin enzyme was discovered in a sea-ice bacterium *Pseudoalteromonas* sp. AN178 (Wang et al. 2014).

Studies of the prokaryote community in the Arctic Ocean have included activity measurements, for example, the bacterial contribution to respiration in the water column may be substantial: 3–60 % in the Chukchi Sea and Canada Basin and 25 % on average in the Arctic (Kirchman et al. 2009). Bacterial activity has also been studied at -2 to -20 °C in Arctic wintertime sea ice (Junge et al. 2004). Specific functional activity has been studied relating to biogeochemical cycling such as the analysis of the sulphate-reducing bacterial and methanogenic archaeal populations in contrasting Antarctic sediments (Purdy et al. 2003) and the fact that anaerobic ammonium-oxidising bacteria in marine environments are of widespread occurrence but low diversity (Schmid et al. 2007), although progress in this area is far more restricted compared to other bacterial groups.

9.7 Eukaryotic Microbial Diversity

Single-celled eukaryotes, or protists, a category largely comprised of protozoa and micro-algae, make up most of the diversity of eukaryotes. Typically, the first classification applied in microbial ecology is that of size classes. Protists typically range in size from 0.2 to 200 μm (some are larger) and are usually segregated into the pico- (<2 μm), nano- (2–20 μm) and micro-sized fractions (20–200 μm) of the scaling plankton nomenclature (Murphy and Haugen 1985). These size classes may be referred to as "picophytoplankton" or "nanozooplankton", etc. according to whether the groups under study are photosynthetic or heterotrophic, though many taxa include heterotrophs, autotrophs and mixotrophs, due to the complex evolutionary history of gains, transfers and losses of chloroplasts among microbial eukaryotes.

During the classification of the eukaryotes, what has recently emerged is a tree that splits into mostly previously unknown large groups, dubbed "supergroups": Opisthokonts (including animals and fungi), Amoebozoa (amoebae and slime moulds), Excavates (e.g. trypanosomes, amitochondriate parasites), Archaeplastida (green algae and plants, red algae), Stramenopiles (brown algae and kelps), Alveolates (ciliates, apicomplexa, dinoflagellates), Rhizaria (foraminifera, radiolaria, cercozoans) and several smaller groups such as the clade containing the haptophyte algae (all reviewed in Walker et al. 2011). These groups are now widely accepted and are increasingly referred to in ecological literature, meaning that the taxa discussed in recent ecological analyses of particular environments (e.g. Massana et al. 2004) have different affiliations from their former placements as "protozoa" or "algae" in classical literature (Van den Hoek et al. 1996).

In the Arctic, protists have mostly been described from the upper water column in coastal and oceanic areas or in ice-associated (sympagic) communities (Poulin et al. 2011). Arctic marine protists in the water column contribute about 98 % of primary production in coastal regions, while in the mid-Arctic ocean about 57 % of primary production is in ice-associated protist communities (Gosselin et al. 1997). Poulin et al. (2011) estimated that 1874 taxa of phytoplankton organisms and 1027 of sympagic unicellular eukaryotes had been described by microscopical methods from Arctic samples. Protists are important as primary producers and as consumers of prokaryotes such as the cyanobacteria. They are key to the "microbial loop", the set of trophic interactions that operates below the level of classical food chains, from dissolved organic carbon through to unicellular organisms that are eaten by invertebrates (Azam et al. 1983; Fenchel 1988).

The microbial food web sustains the Arctic Ocean under most circumstances, dominating where there is not a large phytoplankton bloom. Energy is released as Dissolved Organic Matter from phytoplankton (either through death of cells, viral lysis or "leaky" photosynthesis) and taken up by bacteria and the smallest eukaryotes, who are then eaten by heterotrophic or mixotrophic protists, who are in turn eaten by larger protists and zooplankton, with energy losses at every stage permitting blooms of smaller taxa and heterotrophy permitting blooms of larger taxa (Sheldon 1972).

9.8 Adaptations

Abiotic extremes characterise the Arctic Ocean. These extremes include the intensity, duration and wavelength of incoming solar radiation, the extent, thickness and duration of ice cover, temperature extremes, stratification of the water column and structures associated with the sea ice (Gosselin et al. 1997). Temperatures within the sea ice form a depth gradient, where temperatures at the surface can be $-20\ °C$

and temperatures within the sea ice itself around −2 °C. There are several mechanisms employed by microorganisms to survive this rapid decline in temperature. Mykytczuk et al. (2013) describes the mechanisms used by *Planococcus halocryophilus* to grow and divide at −15 °C, the lowest temperature recorded to date, suggesting cryoenvironments harbour a more active microbial ecosystem than previously thought. One mechanism sea-ice diatoms employ to thrive in the sea-ice environment is the production of ice binding proteins, which act as a cryoprotectant (Janech et al. 2006). These ice binding proteins have both strong recrystallisation inhibition activity and act to slow the drainage of brine from sea ice, allowing them to maintain a liquid environment around the diatom (Raymond et al. 2009). A method often employed by algae and protozoans in particular is to form a robust stress resistant cyst, triggered by low temperature or a lack of nutrients—depending on the species (Stoecker et al. 1998). Whilst a viable method of survival the cyst is often dormant, not contributing to nutrient cycling or affecting the microbial population. Similarly, some bacteria may become associated with particles such as soil already present in the seawater before freezing occurs (Junge et al. 2004). This allows the bacteria to survive in the slightly less harsh microenvironment of the soil (higher temperature, less salinity, etc.) until the sea ice thaws in spring.

However, in order to continue population growth and metabolic activity, microorganisms need to possess two key adaptations: the ability to maintain membrane fluidity in order to allow nutrients and waste to enter and leave the cell (Thomas and Dieckmann 2002) and cold functioning enzymes. Membrane fluidity at sub-zero temperatures can be achieved in three ways; by an increase in the proportion of unsaturated fatty acids, a decrease in membrane chain length and an increase in polyunsaturated fatty acids (Russell 1997). The increase in polyunsaturated fatty acids is brought about by an increase in the activity of the polyketide synthase group of enzymes (Metz et al. 2001) and as well as being important to the microorganism, it plays an important part in the diet of grazers (Thomas and Dieckmann 2002). Another limiting factor is the reduced affinity for substrates due to enzyme denaturation. Psychrophilic bacteria must maintain high catalytic activity at low temperatures, which is best achieved by either maintaining enzyme structure or increasing enzyme concentration. Psychrophilic bacteria have been found to contain cold-adapted proteases, β-galactosidases, phosphatases and amylases (Pomeroy and Weibe 2001). Maintaining enzyme structure allows for a higher proportion of substrate–active site binding and allows the microorganism to maintain its function. If the microbe does not possess psychrophilic enzymes, it can maintain catalytic activity by increasing the concentration of the enzyme, an option which allows more active sites for substrates to bind to, although with less specificity. This increase in total activity at the cost of specificity can be seen in some microalgae with the photosynthetic enzyme, Ribulose-1,5-biphospate carboxylase/oxygenase (Devos et al. 1998).

Tolerance to salinity is an important characteristic for sea-ice microorganisms. They must be able to survive both the high salinity of the brine pockets and channels and the sudden exposure to hypo saline conditions that occur when the sea ice melts in spring (Thomas and Dieckmann 2002). The hyper-saline conditions of brine pockets can cause severe dehydration stress, as they can contain up to three times the salinity of open seawater (Eicken 1992). In order to cope with this stress, microorganisms must carefully regulate the uptake of osmolytes such as proline, mannitol and inorganic ions to restore osmotic balance (Thomas and Dieckmann 2002). Alongside, salt-tolerant enzymes have been described in psychrophilic isolates (Nichols et al. 2000) as has the regulation of fatty acid proportion for temperature tolerance.

Low light conditions primarily affect photosynthetic algae, and as they are important primary producers, their success affects the whole food chain. In order to continue to carry out photosynthesis within brine pockets, algae must be able to carry out extremely efficient photosynthesis reactions to maximise the little light they receive. The algae may also contain pigments that absorb those wavelengths which can penetrate sea ice. One such pigment, Fucoxanthin, is highly effective at absorbing said wavelengths and shows an increase in concentration during winter months (Lizotte and Priscu 1998). The early bloom depends on light, as is the case for the spike in photosynthetically active radiation which affects the prasinophytes in late January when sun comes back (Terrado et al. 2008).

The analysis of nutrient concentrations and the pulse-amplitude-modulated fluorescence signal of ice algae and phytoplankton suggests that nutrients are the prime limiting factor for sea-ice algal productivity (Gradinger 2008). Over much of the Arctic Ocean's vast shallow continental shelf, brine rejection during sea-ice formation triggers convective mixing that replenishes nutrients into surface waters (Stabeno et al. 2010). The traditional bloom depends on nutrients (Lovejoy et al. 2004), but very early and late seasonal blooms are light-dependent (Massana et al. 2007; Seuthe et al. 2011). Such strong dependence on light, and dominance by successive blooms of photosynthetic taxa (and in turn heterotrophic taxa), also suggests that the Arctic Ocean may have a different ecology from those of other oceans. However, as sampling intensity increases, it may become apparent that the ecology of the Arctic Ocean is similar to that of other oceans (Seuthe et al. 2011).

Marine psychrophiles form the base of the Arctic Ocean food web, and their population and species distribution is affected by the presence of organic and inorganic compounds. The cycling of carbon, nitrogen, silica and iron is responsible for the waxing and waning between population groups that influence species higher up in the food web.

Light availability plays the most important role in the sea-ice carbon cycle, with seasonal changes in light having an effect on autotroph/heterotroph dominance. During the summer, light availability is at its highest and at some latitudes photosynthesis can occur for 24 h a day. This leads to an increase in the autotrophic population both within the sea ice and in sea water (Horner and Schrader 1982). Pre-bloom to post-bloom algal cell concentrations can go from $<10^4$ cells to $>10^9$ cells within some brine pockets (Arrigo et al. 2010). The high populations of autotrophic microorganisms exhaust a lot of CO_2 from the brine pockets, causing a supersaturated O_2 solution of up to 932 µmol kg^{-1} (Skidmore et al. 2012).

During the transition to winter, the population switches to a dominant heterotrophic one. Free sea water autotrophs are forced into brine channels during sea-ice formation and those not adapted die releasing organic carbon into the sea ice. Within the sea ice, autotrophs go into a dormant state, and the high O_2 concentration coupled with the readily available organic carbon allows a heterotrophic population increase. The increase in the heterotrophic population uses up O_2 and replenishes CO_2 inside the ice pockets. During the following spring, the increase in light availability and high CO_2 concentration allow another autotroph bloom. The melting sea ice also releases a high concentration of microbial cells into the water column, providing a food source for grazers.

The uptake of nitrogen happens in two distinct ways; through nitrification (oxidation of ammonium (NH_4^+) to produce NO_2 or NO_3) or assimilation (absorption of nitrites or ammonium, followed by reduction and incorporation into amino acids, etc.). As with carbon, the assimilation or nitrification of nitrogen seems highly dependent on light availability and species distribution, with assimilation at a much higher rate in summer. Experiments have shown that nitrification rates show no difference over a temperature gradient (Horak et al. 2013), so changes in the balance of nitrification/assimilation of NH_4^+ are driven by species abundance and distribution which are in turn a consequence of light and organic carbon availability.

Although bacterial nitrifiers are present during the summer (Hollibaugh et al. 2007), there is a sudden increase in autotrophic algae brought about by increased light availability. Phytoplankton outcompete nitrifiers for NH_4^+ in well-lit ocean layers (Bronk and Ward 2005) which could account for the spike in NH_4^+ assimilation seen in the winter months. When light availability is little to none in winter months, and autotrophic populations are severely reduced, there is little competition for nitrifying bacteria, and nitrification becomes the prominent way in which NH_4^+ is utilised.

Silicon in sea ice is most common in the form of Silicic acid $Si(OH)_4$ and is present in the sea ice through exchange with the surrounding water column (Vancoppenolle et al. 2010). Silicic acid is a vital component of diatom cell walls, and studies have shown that Silicic acid is a major limiting factor in sea-ice autotroph growth (Gosselin et al. 1990). Perhaps the best example is the correlation of $Si(OH)_4$ and chlorophyll concentration in sea ice shown in Cota et al. (1990). Iron is required for phytoplankton growth, photosynthesis and enzyme production (Wang et al. 2014). The concentration of iron present in the ocean is often a limiting factor to primary production, especially in Antarctica (Boyd and Abraham 2001). It has however been shown that sea ice has a much higher concentration of iron (up to two orders of magnitude (Lannuzel et al. 2010) than the surrounding ocean, and so iron is unlikely to have a limiting effect on sea-ice ecosystems.

9.9 Methods Used to Study the Arctic Marine Food Web

Stable isotopes have been used extensively to study the flow of energy and matter between organisms (or food-web function). Earlier studies in this field were based on the natural variability of isotopes and limited to larger organisms that could be physically separated from their environment. For example, the stable carbon and nitrogen isotopic fractionation between diet and tissues of captive seals (Hobson et al. 1996). However, recent methodological developments have allowed isotope ratio measurements of microorganisms, and this in turn allows the measurement of entire food webs, from primary producers at the bottom to apex predators at the top (Middelburg 2014; Søreide et al. 2006). Fatty acid (FA) analysis is a well-established tool for studying trophic interactions in marine habitats (Budge et al. 2008; Dalsgaard et al. 2003). It is now being used for comparatively large-scale studies such as that published by Thiemann et al. (2008) who used quantitative fatty acid signature analysis (QFASA) to examine the diets of 1738 individual polar bears (*Ursus maritimus*) sampled across the Canadian Arctic over a 30-year span. The results of their large-scale study indicated a complex relationship between sea-ice conditions, prey population dynamics and polar bear foraging. The method also allows for experimental manipulations (Fraser et al. 1989) and for specific studies of environmental function (Knoblauch et al. 1999).

A significant issue in the Arctic is the bioaccumulation of toxic substances in the food chain. These include polybrominated diphenyl ethers (PBDEs) (Kelly et al. 2008b), hydroxylated and methoxylated polybrominated diphenyl ethers (Kelly et al. 2008a), organochlorines (Skarphedinsdottir et al. 2010), methyl sulfone PCB, 4,4′-DDE metabolites (Letcher et al. 1998) and mercury (Atwell et al. 1998; Gobeil et al. 1999) among many. One fortunate benefit of such bioaccumulation of toxins is that it can actually be used to trace the route of those chemicals through the food chain.

Most targeted food web studies, by necessity, require some form of environmental manipulation, and this can occur at a number of different spatial scales. From laboratory microcosms, people have been able to examine topics such as seasonal carbon and nutrient mineralisation (Rysgaard et al. 1998), nutrient cycling (Freitag et al. 2006), methane production (Damm et al. 2010) and biodegradation studies (Bagi et al. 2014). Mesocosm experiments have been used to study response to elevated CO_2 levels (Niehoff et al. 2013; Pavlov et al. 2014), nutrient manipulation (Larsen et al. 2015), ocean acidification (Riebesell et al. 2013) and differential transfer of marine bacteria to aerosols (Fahlgren et al. 2015). Then there are the exclosures, where specific components are excluded that can be both biological or chemical, and which have been used to study diet-induced changes in the fatty acid composition of Arctic herbivorous copepods (Graeve et al. 1994) or the effects of crude oil on marine phytoplankton (Harrison et al. 1986).

At the still larger scale, there have been foraging behaviour studies, particularly of seabirds, for example, northern fulmars (Hobson and Welch 1992), thick-billed murres (Falk et al. 2002), little auks (Karnovsky et al. 2011), shallow-diving seabirds and Arctic cod (Matley et al. 2012). In addition, population monitoring has been used for organisms such as marine mammals (Kovacs 2014; Laidre et al. 2008, 2015; Moore and Huntington 2008; Tynan and DeMaster 1997). Ultimately, satellite remote sensing data is also used over very large areas to examine recent trends in sea-ice cover and net primary productivity (NPP) (Brown and Arrigo 2013). So, as food web studies cover an ever wider range of spatial scales (Smith et al. 2005), modelling becomes an essential component in understanding the Arctic marine food web in order to disentangle complex trophic relationships (Legagneux et al. 2012).

Observations of Whole Cells and Pigments Traditionally, protistan diversity in the Arctic has been detected through light microscopy of live samples (Poulin et al. 2011), meaning this seems the ideal method to combine new data with previous observations. Given taxonomically experienced observers, sufficient resolution and magnification on the microscope and the ability to capture uninterpreted records (photographs or videos rather than drawings), light microscopy provides records that are easy to interpret (e.g. Daugbjerg and Moestrup 1993; Vørs 1993; Ikävalko and Gradinger 1997; Lovejoy et al. 2004) in the light of the worldwide taxonomic and natural history literature from the last two centuries (e.g. Larsen and Patterson 1990; Vørs 1992).

The efficiency of microscopy-based sampling can be improved in ecological studies (where species-level identification may not be important) by the use of epifluorescence microscopy, whereby specific cellular compounds such as DNA, chlorophylls a, b and c or phycobiliproteins have fluorescent signals at particular wavelengths. Stains such as DAPI, primulin and lysotracker can be used to show cellular features, e.g. the nucleus, endomembrane system and extracellular scales (Sintes and del Giorgio 2010; Vaulot et al. 2008). Photosynthetic pigments and their cellular distribution in unicellular algae are taxonomically diagnostic (e.g. phycoerythrins in cryptomonad chloroplasts) and can be used to count types of cells very quickly with the aid of image analysis. This technique has been widely used in ecological studies that have established the abundance and importance of bacteria and protists in food webs (Hobbie et al. 1977; Murphy and Haugen 1985).

Samples can be cooled and returned to laboratories for subsequent microscopy (Tong et al. 1997), dilution culturing to estimate abundances (Throndsen 1978; Throndsen and Kristiansen 1991), enrichment culturing to study specific organisms from the culture in higher numbers (e.g. Daugbjerg and Moestrup 1993; Vørs 1993) or isolation of individual species for culture and further study (Andersen and Kawachi 2005). Fine-structural diagnostic features of these organisms can further be studied by electron microscopy, either of filtered environmental samples or of cultures, with numerous rare and novel taxa often described this way (Andersen et al. 1993; Booth and Marchant 1987; Moestrup 1979; Thomsen 1980 *inter alia*).

Where species-level diagnosis is less important than information on abundances of higher-level taxa, the size structure and photosynthetic (fluorescent) characteristics of populations of cells can be studied by flow cytometry (Li and Dickie 2001; Marie et al. 1999; Simon et al. 1994; Vaulot et al. 2008). Pigment signatures can also be analysed in detail by HPLC (e.g. Andersen et al. 1996), which provides relatively taxon-specific information at varying levels of diversity, dependent on the group, and can be analysed using algorithms such as CHEMTAX (Mackey et al. 1996; Wright et al. 2009), along with the analysis of plankton blooms observed in satellite images (Batten et al. 2003; Gons et al. 2002). Interpretation of HPLC signals can be difficult, as there can be huge diversity within a taxon despite all having a common signature (e.g. haptophytes, Liu et al. 2009), some taxa can share pigment signatures (e.g. diatoms and bolidophytes, Guillou et al. 1999) and other taxa can have multiple signatures, even among ecotypes of a single species (Latasa et al. 2004).

Recently, the powerful tools of molecular biology have been applied to the study of biodiversity in Arctic seas (Radulovici et al. 2010). These new approaches have revealed a surprising diversity including new taxa of bacterioplankton and archaeans (Galand et al. 2009; Kirchman et al. 2009), eukaryotic microbes (Lovejoy and Potvin 2011; Lovejoy et al. 2007) for example gregarine parasites (gregarines are apicomplexans) of amphipods (Prokopowicz et al. 2010). The precise role of these recently discovered assemblages in the pelagic marine food webs and in the cycling of organic matter remains obscure. As well, benthic processes and biodiversity are not yet well resolved for the Arctic seafloor although increased attention has been given in recent years to study the coupling between pelagic production and benthic carbon turnover (Morata et al. 2008; Renaud et al. 2007).

9.10 Biogeography

Whilst on the surface, the Arctic Ocean may seem relatively homogeneous, it is important to remember that there are quite a number of distinct biogeographical zones. This includes the different vertical layers of the Ocean from the air–sea interface down to the benthos. There is then the distinction between coastal regions as compared to the open ocean. For example, Arctic shelves comprise roughly 50 % of the Arctic Ocean and are the regions where sea-ice conditions have changed most dramatically over the last decades (Hassol 2004). There are also the transition and brackish zones such as fjord systems and polynyas. The differences between pack or sea ice and fast ice impose different selection pressures, and there is also the issue of special ecosystems such as hydrothermal vents and commensal communities on or inside marine organisms. Niederberger et al. (2010) carried out a microbial characterisation of a terrestrial methane seep into an Arctic hypersaline, subzero perennial spring. It was found to support a viable microbial community that may use methane as an energy and carbon source to sustain anaerobic oxidation. Then there is the issue of locational stability, with microorganisms being moved around by both air and water currents. Picocyanobacteria are wind transported year round from warmer ocean waters and can establish themselves in the warmer summer climate of the Arctic. This can be demonstrated in Polar ice cores, which contain higher concentrations of chlorophyll and tryptophan at depths corresponding to local summers that bore similarities to regional fluorescence patterns seen in the picocyanobacteria *Prochlorococcus* and *Synechococcus* (Price and Bay 2012). Next there is the issue of research effort, for example, the fact that proximity of research facilities and comparative ease of logistic access results in specific Arctic locations such as the Amundsen Gulf, Lancaster Sound, Disko, Zackenberg or Svalbard often being more studied than any others. Within each of the separate Arctic zones, there will be also differences in coverage, connectivity and spatial patchiness.

The total number of microbial species is estimated to be between 10^3 and 10^9, and the bacteria are responsible for around half of photosynthesis on Earth (Pedrós-Alió 2006). Marine microorganisms, collectively present at billions of cells per litre, grow at rates of around one doubling per day in surface waters and are consumed at the same rate (Whitman et al. 1998). There could therefore be huge untapped potential in Arctic marine food web biodiversity. It was previously believed that due to small body sizes and huge population sizes, few geographical barriers and mixing of waters due to wind, waves and currents (Collins 2001), marine planktonic microorganisms should not exhibit biogeographical patterns across latitudinal gradients like those seen in many macroscopic animals and plants (Hillebrand 2004). Furthermore, marine planktonic microorganisms should be cosmopolitan, endemic species should be rare and their global diversity low (Fenchel and Finlay 2004; Finlay 2002). However, Pommier et al. (2007) have shown that compositions of marine free-living microbial taxa differ in different locations and this difference may correlate with environmental factors (Giovannoni and Stingl 2005; Martiny et al. 2006) such as salinity gradients (Bouvier and del Giorgio 2002; Crump et al. 2004), depth stratification (Garcia-Martinez and Rodriguez-Valera 2000; Giovannoni et al. 1996; Gordon and Giovannoni 1996; Riemann et al. 1999) and Ocean fronts (Pinhassi and Berman 2003). Arctic biodiversity is therefore much more extensive, ecologically diverse and biogeographically structured than previously thought. Understanding how this diversity is distributed in marine ecosystems, the mechanisms underlying its spatial variation and the significance of the microbiota concerned is growing rapidly (Chown et al. 2015). To date, there have been a variety of specific studies of bacterial biodiversity in Arctic waters. The distribution (Ferrari and Hollibaugh 1999) and phylogenetic composition of bacterioplankton assemblages from the Arctic Ocean (Bano and Hollibaugh 2002) were the first to be addressed, but now we know that hydrography may shape bacterial biogeography of the deep Arctic Ocean (Galand et al. 2010) and that pole-to-pole biogeography of surface and deep marine bacterial communities may exist (Ghiglione et al. 2012).

9.11 DNA Sequencing

A partial, but very powerful solution to the limitations of microscopy is provided by sequencing DNA directly from environmental samples, without first observing or culturing the organisms in the environment (Giovannoni et al. 1990; Pace 1997). For Arctic microbial samples, this specifically removes both the logistical and time limitations associated with light microscopy and eliminates the need to identify organisms by morphology; hence, considerably more organisms can be detected (e.g. Lovejoy et al. 2002, 2006).

Initially, sequencing methods relied on PCR amplification of marker genes from environmental genomic DNA or RNA samples and cloning and sequencing of these PCR products for phylogenetic identification (Giovannoni et al. 1990). By comparing libraries from genomic DNA with ones based on RNA, it is possible to distinguish between DNA samples from cells that are active in the environment, and ones that are from resting stages or dead cells (Stoeck et al. 2007), though the comparison is made complex by the copy number of rDNA molecules contained in the genomes of some taxa (Not et al. 2009). Typically, the marker used to identify protists has been SSU rRNA (small subunit ribosomal RNA), which is highly abundant within cells, evolves slowly enough to provide consistent relationships within distantly related groups and has no evidence of lateral gene transfer. Fragments of this gene may be used for broad phylogenetic identification, although using either the whole gene or both SSU and large subunit ribosomal RNA (LSU rRNA) subunits produces better phylogenetic resolution (Marande et al. 2009; Vaulot et al. 2008). Either universal or specific primers for particular taxa can be used (Bass and Cavalier-Smith 2004), and multiple PCR-primer approaches have considerably improved detection of diversity (Stoeck et al. 2006). Specific "barcoding" markers such as ITS, psbA or RuBisCO have been developed to easily identify diversity at lower levels in particular eukaryotic groups (Frezal and Leblois 2008; Goldstein and DeSalle 2011), including among the protists (e.g. Nassonova et al. 2010).

The clone library approach has revealed significant novel diversity, particularly of very small eukaryotes (Massana and Pedrós-Alió 2008; Vaulot et al. 2008) such as prasinophytes (Guillou et al. 2004) and haptophytes (Liu et al. 2009); and including whole clades that had previously not been recorded, such as the MAST (marine stramenopile) and MALV (marine alveolate) groups (Lopez-Garcia et al. 2001; Moon-van der Staay et al. 2001). Some of the groups found by clone library sequencing methods, such as picozoa (Seenivasan et al. 2013 http://www.ncbi.nlm.nih.gov/pubmed/23555709), were initially deemed unaffiliated with any other eukaryotic taxa (Dawson and Pace 2002; Not et al. 2007), although most of the diversity detected was related to already-known groups (Berney et al. 2004).

However, clone libraries are subject to inherent biases of their own, such as taxon-specific PCR amplification biases, errors in amplification, ligation biases towards short sequences, chimaeric sequences, the expense of Sanger sequencing, the occurrence of rDNA genes in multiple copies within a single cell and the capacity of DNA to persist as extracellular material (Berney et al. 2004; Foerstner et al. 2006; Stoeck et al. 2010; Vaulot et al. 2008). Libraries also do not show sampling saturation, so even the most intensive library analyses have only recovered a relatively small proportion of the diversity detectable by metagenomic approaches (Not et al. 2009; Stoeck et al. 2010).

Metagenomic sequencing, the direct cloning and sequencing of fragments of environmental DNA (DeLong et al. 2006; Venter et al. 2004) have been developed on the basis of recent advances in high-throughput sequencing technology, based on micro-fabricated high-density picolitre reactors (Margulies et al. 2005). By removing the ligation and transformation steps, it removes some of the larger biases of the clone library approach and considerably more sequencing can be done as hundreds of thousands of sequences are produced in a single sequencing run. The tagging of sequences also allows direct quantification of their abundance (Tedersoo et al. 2010). In prokaryotic microbial communities, metagenomic sequencing has permitted the identification of two orders of magnitude of further novel diversity, particularly of very rare taxa (Sogin et al. 2006) that are impossible to detect by clone-library techniques (Pedrós-Alió 2006). Not et al. (2009) and Stoeck et al. (2010) applied these methods to eukaryotic microbes and discovered that their samples were similarly dominated by very rare sequences. While high-throughput sequencing methods are relatively new, there is increasing application of metagenomic sequencing methods in microbial ecology, allowing the examination of questions regarding diversity, abundance, correlation with abiotic factors (e.g. Comeau et al. 2011) and biogeography.

9.12 Linking Microscopy and Sequence Data

9.12.1 *Quantitative Approaches: FISH, FACS and qPCR*

Whole-cell microscopical approaches can be linked to environmental sequence data of cultured or uncultured protists through the use of specific oligonucleotide probes targeting taxonomically specific genes such as 18S rRNA (DeLong et al. 1989; Lim et al. 1999). This allows environmental sequences to be tied to their "owners", as seen by microscopy using fluorescent in-situ hybridisation (FISH), and can thus be used to determine both biological identity and abundance (Not et al. 2004), complex morphology (in combination with other stains as examined by Hirst et al. 2011), enabling studies of ecological succession (Larsen et al. 2004), and the ecological functions of unknown protists through stained, live feeding experiments (Massana et al. 2009). Such probes can also be hybridised to gold nanoparticles, for performing simultaneous FISH correlated with immunogold electron microscopy, allowing taxonomically diagnostic morphological details to be observed in previously unknown, uncultured taxa (Gerard et al. 2005; Kolodziej and Stoeck 2007; Stoeck et al. 2003). The same fluorescent oligonucleotide probes can also be used in flow cytometry (FACS), allowing isolation of individuals (Seenivasan et al. 2013 http://www.ncbi.nlm.nih.gov/pubmed/23555709), quantification of the abundance of particular taxa in a sample, as well as sorting of the sample into pure populations from which large numbers of cells can be obtained (Li 1994; Simon et al. 1994). Quantification and spatiotemporal mapping of protists can also be done using quantitative PCR (qPCR), where oligonucleotide probes are hybridised to PCR-amplified products from environmental samples (Marie et al. 2006; Suzuki et al. 2000; Zhu et al. 2005). This allows a complex picture of distribution and abundance to be built up quickly, even for uncultured organisms of relatively unknown biology, such as the MAST stramenopiles (Rodriguez-Martinez et al. 2009).

9.12.2 Qualitative Approaches: Detecting Species

Sequencing of isolated or cultured protists has linked ecological data based on microscopy and sequence identity, permitting studies on the distribution of particular morphological species or ecotypes (Koch and Ekelund 2005). However, the correspondence between identity determined from isolated organisms, "operational taxonomic units" (OTUs) in environmental sequencing, population genetic structure and morphospecies or morphological ecotypes is highly variable and complex (Schlegel and Meisterfeld 2003), and cases where the alpha-taxonomy of a group is well-known both from sequences and from morphology are rare (Alverson 2008). There is also considerable sequence diversity within many morphospecies (e.g. *Micromonas pusilla*, c.f. Throndsen and Kristiansen 1991; Slapeta et al. 2006), meaning that the degree of correspondence between uncultured environmental sequencing records and natural history records from the last two centuries is unclear (Lopez-Garcia and Moreira 2008). Some progress has been made in distinguishing species from OTUs, using algorithmic methods to distinguish species-level clades (Leliaert et al. 2009) and ribosomal RNA internal transcribed spacer (ITS) sequences (Coleman 2009), which can predict interbreeding in diatoms (Poulickova et al. 2010; Sorhannus et al. 2010) and can potentially be used to detect clades to species level in uncultured stramenopiles (Rodriguez-Martinez et al. 2012).

9.13 Methodological Limitations

The overall picture of ecological diversity of protists is currently changing, due to major improvements in methods for detecting diversity, from light-microscopy-based to sequencing-based approaches. The taxonomy of protists has also recently changed significantly, formerly being based on light microscopy, but more recently reflecting results from molecular phylogenetics (Walker et al. 2011). It is now clear that ecological studies of Arctic protists have only begun to scratch the surface in understanding population diversity (Archambault et al. 2010; Poulin et al. 2011), despite protists make up a large proportion of the diversity present in Arctic food chains. As the Arctic climate changes rapidly in the next few decades, changes in the communities of protists may have significant effects on food webs generally.

However, while whole-cell methods maintain the link between "real biology" in the field and laboratory-based observations, there are disadvantages and biases inherent in most methods. The length of time required to make taxonomically useful observations using either a light or fluorescence microscope often severely limits the amount of work that can be done on fresh material during time-constrained fieldwork in remote locations such as the Arctic. Culturing-based methods show strong biases, excluding "unculturable" organisms from study. For ecological studies where cells are being counted sometime after initial collection, samples are often preserved in fixatives such as Lugol's iodine, formalin or glutaraldehyde, which can significantly change estimates of diversity and abundance (Epstein 1995; Montagnes et al. 1994). Similar fixatives are also used for electron microscopy of ecological samples and can similarly change estimates of diversity and cell volumes (Vaulot et al. 2008). Sampling, observation time and preservation aspects of microscopy can lead to biases in observations such as missing the early "spike" in prasinophyte green algae during yearly succession in marine communities (Terrado et al. 2008), or very strong biases towards observation of easily-preserved diatoms and dinoflagellates, over unarmoured taxa that become unrecognisable after preservation (Poulin et al. 2011). There are also strong biases in sampling effort, particularly due to the highly variable numbers of taxa observed across the Arctic. For example, the number of taxa observed in Hudson Bay and the Barents Sea tends to be larger than that from Alaska; however, this reported variation is likely to be related to an unintentional bias in the number of field stations present in these regions, the number of microbial researchers often visiting them (Poulin et al. 2011) and the research funding directed to each field station (Archambault et al. 2010). The "taxonomic impediment", linked to the continuous declining numbers of specialist taxonomists with skills in identifying organisms, also affects the documentation of Arctic microbial diversity (Archambault et al. 2010). The unfortunate flip side to that impediment is the fact that many individual taxonomists are able to recognise only certain taxa, often specialising in particular groups and consequently under-reporting other taxa, leading to strong biases in detecting diversity. These factors are compounded by other extrinsic factors such as nomenclatural instability (Lee and Patterson 1998). However, it is worth remembering that despite these biases and limitations, microscopy is the key to both understanding the biological interactions among diverse microbial eukaryotes and linking new data to past records.

9.14 Arctic Microbial Diversity and Endemism

The potential for direct impacts of sea-ice loss begins at the base of the food web, where significant numbers of obligate low-temperature, shade-adapted algal species (Lovejoy et al. 2006, 2007) and radiolarian species endemic to the Arctic Ocean Basin have been found (Lovejoy et al. 2007). The ecological consequences of understanding species-level sequence variation are in detecting diversity from environmental sequence data and assessing whether there is endemism in particular environments (Caron 2009). With this in mind, environmental sequence data can be used to address the ubiquity hypothesis in microbial ecology, which states that "everything is everywhere, but the environment selects" (De Wit and Bouvier 2006). This suggests that latitudinal gradients do not exist in microbial ecology (when ecological factors are accounted for) and, consequently, microbial ecology is fundamentally different from that of macroorganisms (Fenchel and Finlay 2003). This has long been a prevailing paradigm in protistan ecology, though it has generated considerable debate (Caron 2009). The pattern of endemism in protists that emerges from morphological data is mixed (Schlegel and Meisterfeld 2003), with some observations suggesting overwhelming cosmopolitanism of morphospecies (Larsen and Patterson 1990) and that apparent endemism is strongly dependent on extrinsic factors such as the distribution of taxonomically experienced microbial ecologists (Lee and Patterson 2000; Poulin et al. 2011). Other observers have emphasised the hidden diversity in morphospecies (e.g. Mann 1999), and studies of haptophytes, diatoms and green algae in the Antarctic point to strong endemism and geographical isolation with strong latitudinal gradients (Gravalosa et al. 2008; Vyverman et al. 2010). Sampling in polar regions, particularly, is often incomplete and inconsistent, making it difficult to determine whether data supports endemism or cosmopolitanism (Chown and Convey 2007). Diversity and endemism detected by sequence data provide a mixed picture, suggesting some cosmopolitanism (Cermeno et al. 2010; Finlay et al. 2006; Koch and Ekelund 2005) and some endemism (Bass and Cavalier-Smith 2004; Bass et al. 2007), and the assessment may also strongly depend on the relevance of the data to the question. Trivially, the level of sequence identity that is considered an "operational taxonomic unit" determines what diversity is detected, and different taxa may vary at different levels of identity for particular sequences, giving very different overall pictures of diversity between, for example, 95 % and 99 % sequence identity or the V4 and V9 regions of SSU rRNA (Stoeck et al. 2010). Both alpha- and beta-diversity, and endemism, also depend strongly on the genetic marker used (Bass et al. 2007), with rDNA copy number in genomes potentially confounding some results (Not et al. 2009). Considerable morphological diversity presently is known from microscopy of live samples (e.g. Vørs 1993; Ikävalko and Gradinger 1997; Daugbjerg and Moestrup 1993); however, most of the data gathered suggest that a large proportion of this diversity is composed of cosmopolitan species, rather than endemic (e.g. Larsen and Patterson 1990; Vørs 1992; Lee and

Patterson 2000), possibly as a consequence of still relying upon traditional microscopy-based species concepts.

9.15 Yearly Dynamics

To superimpose another layer of complexity onto understanding the biodiversity of different geographic regions, there is also the issue of yearly dynamics (Giovannoni and Vergin 2012; Seuthe et al. 2011; Terrado et al. 2008). The *winter* Ocean is dark, ice-covered, cold, with fresh water at the surface beneath first-year ice, and warmer water below. While autotrophs from previous seasons persist (Niemi et al. 2011), alveolate heterotrophs dominate in this environment (Terrado et al. 2008, 2009), and more heterotrophs are present for longer in high snow regions (Riedel et al. 2008). In the *winter*–spring *transition*, there is a rapid increase in solar radiation, and although the ocean is still ice-covered, cold (with temp gradient below) and with freshwater at the surface; multiyear ice contains old communities (Terrado et al. 2009). A Prasinophyte spike occurs in late January, while diatoms dominate at least around Svalbard. In spring, there is a rapid increase in solar radiation, retreating ice, upwelling of nutrients and runoff from rivers providing silicates and iron (Andreassen and Wassmann 1998), during which time pennate diatoms dominate in the retreating ice, while haptophytes and centric diatoms start to appear in the now open water. Terrado et al. (2008) have suggested that photosynthetically active radiation drives such changes. Although alveolate diversity still dominates during the spring, prasinophytes bloom early in the season. There occurs a second bloom during which diatoms and cryptophytes subsequently dominate the total activity (Terrado et al. 2011), while ciliates and dinoflagellates come to dominate the diversity (Lovejoy et al. 2002). The Arctic microbial marine food web is dynamic at this point in the yearly cycle with high levels of nutrients and unlimited growth. During spring, and throughout the summer, when light becomes available for photosynthesis, a large biomass of unicellular photosynthetic ice algae develops within the lowermost sections of the ice (Krembs and Deming 2011). In *summer*, very high solar radiation penetrates into the surface layers, and no ice occurs in almost all areas. The microbial loop becomes important at this part of the annual cycle, with characteristically low levels of nutrients and high dissolved organic carbon. July sees the highest biomass for heterotrophs and large autotrophs (Seuthe et al. 2011), and while alveolates still dominate the genomic DNA libraries, there is a radiolatian spike and stramenopile diversity decreases (Terrado et al. 2009). Alveolates also dominate sediments, although they do also appear in surface layers, and stramenopiles dominate between the depths of 20–30 m down from the surface (Luo et al. 2009). In the *autumn*, solar radiation now rapidly starts to decrease, sea ice forms and upwelling returns. In Kongsfjorden on Svalbard, all biomass starts to decrease by July, although a later bloom during August and September does occur in the far north (Seuthe et al. 2011). As for the *autumn*, alveolates strongly dominate and heterotrophs increase throughout the season(Terrado et al. 2009). Sanders and

Gast (2012) found that in the Canada Basin, around the autumn equinox, mixotrophs dominate a low biomass level. Some algae remain dormant in the ice throughout the winter, bloom when the spring sunshine kick-starts a growth cycle and, eventually, migrate down to the bottom of the ice and enter the water column where they provide a nutritious dietary food source to many marine organisms.

9.16 Environmental Change

Environmental changes are already observed at a pan-Arctic scale and include a decline in the volume and extent of the sea-ice cover (Comiso et al. 2008), an advance in the melt period (Comiso 2006; Markus et al. 2009), and an increase in river discharge to the Arctic Ocean (McClelland et al. 2006; Peterson et al. 2002) due to increasing precipitation and terrestrial ice melt (Peterson et al. 2006). Changes in the microbial community composition can have a direct effect on environmental function. The degree of ice nucleation activity of microbes, either limited ice nucleation activity (as seen by sea-ice bacteria) or high nucleation activity (as seen by *Pseudomonas antarctica* and *Pseudomonas syringae*) will depend on the bacterial diversity and activity in the atmosphere (Junge and Swanson 2008). Human activity in the area is also on the increase. Gerdes et al. (2005) incubated sea ice with crude oil, in order to assess how an oil spill would affect a sea-ice community and saw a shift in community structure to a *Marinobacter-*, *Shewanella-* and *Pseudomonas*-dominated population. The sea-ice community was also shown to degrade [^{14}C] hexadecane at 2–50 % the efficiency of a mesophilic *Marinobacter*.

9.17 Warming Ocean

Global sea surface temperature is approximately 1 °C higher now than 140 years ago and is one of the primary physical impacts of climate change. Sea surface temperature is increasing more rapidly in some areas than others. Projections show the temperature increases will persist throughout this century. Ice-free summers are expected in the Arctic by the end of this century, perhaps even as early as 2040. Already, there is evidence that many marine ecosystems are affected by rising sea temperature. Over the past 25 years, the rate of increase in sea surface temperature has been about 10 times faster than the average rate of increase during the previous century. The microbial communities in the Arctic Ocean are changing due to rising sea temperatures and the dramatic loss of sea ice. In particular, the archaea Marine group I of Thaumarchaeota went from being 60 % of the archaeal community in 2003 to less than 10 % in 2010, and Bacteriodetes populations dropped dramatically greatly affecting bacterial community diversity (Comeau et al. 2011).

9.18 Loss of Sea Ice

Since the mid-twentieth century, the Arctic Ocean has experienced unprecedented sea-ice loss that has accelerated in recent years (Stroeve et al. 2007; Walsh and Chapman 2001). Future sea-ice loss will likely stimulate additional net primary production over the productive Bering Sea shelves, potentially reducing nutrient flux to the downstream western Arctic Ocean (Brown and Arrigo 2013). Increased ice-free conditions may also favour and extend northward the intrusion of Atlantic phytoplankton species (Poulin et al. 2011). On the other hand, sea-ice loss creates additional open-water habitat for phytoplankton, whose growth is traditionally thought to be light-limited under the sea-ice cover (Hill and Cota 2005; Loeng et al. 2005; Smetacek and Nicol 2005). Area-normalised rates of CO_2 fixation in the ice-free zone are generally far higher than in adjacent sea-ice habitats (Arrigo and van Dijken 2003), so sea-ice loss potentially leads to a more productive Arctic Ocean (Brown and Arrigo 2013). A freshening of surface waters may significantly change mineralised eukaryote ecology and distributions; though this may also affect the climate via dimethyl sulphide (DMS) production. Sea-ice loss is also accelerating (Comiso 2006; Comiso et al. 2008; Stroeve et al. 2007) with less multiyear ice, increases in autumn, and winter temp, more wave actions and coastal erosion. The salinity of the Arctic Ocean is also decreasing (McClelland et al. 2006; Peterson et al. 2002, 2006) due to a changing balance between increased river flow from a rainier Eurasia concurrent with decreased inflow originating from a warming and drying North America. We might expect a northwards movement of taxa as polar conditions become more favourable (Hegseth and Sundfjord 2008; Poulin et al. 2011).

9.19 Productivity

Annual mean open water area correlates with primary production (Pabi et al. 2008), and marine phytoplankton accounts for ca. 45 % of the net annual Arctic primary production (Falkowski et al. 2004). Ice algae contribute 57 % of total primary production in Arctic Ocean (Gosselin et al. 1997) and 3–25 % in the Arctic shelf regions (Legendre et al. 1992). Satellite data of phytoplankton concentrations since 1979 suggest decadal-scale fluctuations linked to climate forcing. Data from ocean transparency measurements and chlorophyll suggest global phytoplankton declines by approximately 1 % per year overall, with local fluctuations linked to climate (Boyce et al. 2010). Primary production increases as ice retreats (Frey et al. 2014); encouraged by an earlier melt and later freeze particularly in the Siberian and Svalbard Arctic (Markus et al. 2009).

9.20 Toxin Accumulation

Organochlorine, total mercury (THg), methylmercury (MeHg) and polybrominated diphenyl ether (PBDEs) contaminants accumulate in Arctic marine food chains (Campbell et al. 2005; Kelly et al. 2008a; Norstrom et al. 1998). Borgå et al. (2010) simulated climate change-induced alterations in bioaccumulation of organic contaminants in an Arctic marine food web. Their study demonstrated that organisms with a limited natural habitat range are likely to suffer the most under changing climatic and oceanic conditions. Organisms with a wide natural range are likely to cope better. Climate change is expected to alter environmental distribution of contaminants and their bioaccumulation due to changes in transport, partitioning, carbon pathways and bioaccumulation process rates. The magnitude and direction of these changes and resulting overall bioaccumulation in food webs are currently unknown.

9.21 Atmospheric CO_2 Concentration and Ocean Acidification

Such increasing CO_2 concentration in the atmosphere, leads to increased concentrations in the Oceans, resulting in a decrease in pH. Haptophytes are considered to be the most productive calcifying organisms on the planet (Fiorini et al. 2011) as they play a crucial role in the marine carbon cycle through calcification and photosynthetic carbon production. More specifically, they note that coccolithophores are responsible for about half of the global surface ocean calcification and contribute significantly to the flux of organic matter from the sea surface to deep waters and sediments (Klaas and Archer 2002). Consistent findings have inferred that coccolithophore mass has increased in relation to rising atmospheric pCO_2 over the past two centuries (Halloran et al. 2008; Iglesias-Rodriguez et al. 2008).

9.22 Questions Raised

So as we start to understand more and more about Arctic marine food webs, a number of lines of enquiry stand out. Much of the current interest in the area concerns the functioning of different parts of the communities and their interactions with each other; nonetheless, perhaps most importantly, we are interested in understanding how these interactions and functions might or would change in the face of a changing environment. As the global temperature warms and the climate changes, we are seeing in some places increased terrestrial run off, which will deliver more nutrients to the marine food webs, along with progressively melting permafrost, which releases both nutrients and potentially new biodiversity into the Arctic Ocean. We know little of the role of sea ice and the effect of its loss. As the sea ice recedes, the patterns of Ocean circulation may change. These circulation patterns deliver nutrients and biodiversity to different regions of the world. We know very little about keystone microbial species and in particular, why are they "keystone" species. With changing selection pressures, we may see colonisation by "new" species, and we have scant understanding of this displacement process. Adding to this are the unknown rates of evolution of species in a changing environment. Gene exchange and dispersal (and indeed gene flow through the microbial community), particularly from the "rare" diversity, have unknown levels of functional redundancy. Rising sea surface temperature is moving us towards ice-free Arctic summers and a changing marine food chain, with progressive Ocean acidification, deoxygenation, bioaccumulation of toxins and oil exploration in the Arctic. Yet we still do not know the full extent of microbial diversity and biogeography, its extent and importance, interactions and feedback loops. The extent of this unknown diversity was vividly illustrated with the recent description of 35 new bacterial phyla (Brown et al. 2015).

9.23 Conclusions

Arctic marine food webs are important because they regulate productivity, biogeochemical cycling and biodiversity, essentially controlling ecosystem function. Polar regions, and more specifically the Arctic, have received a growing interest over the past decades due to the threatening impact of global warming (e.g. Johannessen et al. 1999; Moritz et al. 2002; Serreze et al. 2007; Thomas and Dieckmann 2010). The extreme climate that has prevailed over the Arctic Ocean for several million years has shaped unique marine ecosystems characterised by organisms that are adapted to frigid temperatures, the alternation between polar night and midnight sun, a perennial or seasonal sea-ice cover, limiting nutrients in the stratified surface layer and an extremely pulsed cycle of primary production (Darnis et al. 2012). Sea-ice loss and enhanced productivity have characterised the Arctic Ocean over recent years, and it has been hypothesised that as the ice recedes human impact on these environments will increase in the form of ship traffic, exploration, industrial activities and fisheries. This increase in pollution and further introduction of foreign contaminants will surely have a negative effect on the native species distribution and populations, further exacerbating the problem (Moline et al. 2008). Many questions still remain, and in particular the gaps in our knowledge of the winter microbial food web, so what is urgently needed is a large coordinated, comprehensive "food web" study, the likes of which occurred in the recent International Polar Year.

References

Aguirre de Cárcer D, López-Bueno A, Pearce DA, Alcamí A (2015) Biodiversity and distribution of polar freshwater DNA viruses. Sci Adv 1(5):e1400127

Alverson AJ (2008) Molecular systematics and the diatom species. Protist 159:339–353. doi:10.1016/j.protis.2008.04.001

Andersen RA, Kawachi M (2005) Traditional microalgae isolation techniques. In: Andersen RA (ed) Algal culturing techniques. Academic Press, London, pp 83–101

Andersen BM, Steigerwalt AG, O'Connor SP, Hollis DG, Weyant RS, Weaver RE, Brenner DJ (1993) *Neisseria weaveri* sp. nov., formerly CDC group M-5, a gram-negative bacterium associated with dog bite wounds. J Clin Microbiol 31:2456–2466

Andersen P, Hald B, Emsholm H (1996) Toxicity of *Dinophysis acuminata* in Danish coastal waters. In: Oshima Y, Fukuyo Y, Yasumoto T (eds) Harmful and toxic blooms. UNESCO, Paris, pp 281–284

Andreassen IJ, Wassmann P (1998) Vertical flux of phytoplankton and particulate biogenic matter in the marginal ice zone of the Barents Sea in May 1993. Mar Ecol Prog Ser 170:1–14

Angly FE et al (2006) The marine viromes of four oceanic regions. PLoS Biol 4:e368. doi:10.1371/journal.pbio.0040368

Archambault P et al (2010) From sea to sea: Canada's three oceans of biodiversity. PLoS One 5:e12182. doi:10.1371/journal.pone.0012182

Arrigo KR, van Dijken GL (2003) Phytoplankton dynamics within 37 Antarctic coastal polynya systems. J Geophys Res Oceans 108. doi:10.1029/2002jc001739

Arrigo KR, Pabi S, van Dijken GL, Maslowski W (2010) Air-sea flux of CO_2 in the Arctic Ocean, 1998–2003. J Geophys Res Biogeosci 115. doi:10.1029/2009JG001224

Atwell L, Hobson KA, Welch HE (1998) Biomagnification and bioaccumulation of mercury in an Arctic marine food web: insights from stable nitrogen isotope analysis. Can J Fish Aquat Sci 55: 1114–1121. doi:10.1139/f98-001

Azam F, Fenchel T, Field JG, Gray JS, Meyer-Reil LA, Thingstad F (1983) The ecological role of water-column microbes in the sea. Mar Ecol Prog Ser 10:257–263. doi:10.3354/meps010257

Bagi A, Pampanin DM, Lanzen A, Bilstad T, Kommedal R (2014) Naphthalene biodegradation in temperate and Arctic marine microcosms. Biodegradation 25:111–125. doi:10.1007/s10532-013-9644-3

Bano N, Hollibaugh JT (2002) Phylogenetic composition of bacterioplankton assemblages from the Arctic Ocean. Appl Environ Microbiol 68:505–518

Bass D, Cavalier-Smith T (2004) Phylum-specific environmental DNA analysis reveals remarkably high global biodiversity of Cercozoa (Protozoa). Int J Syst Evol Microbiol 54:2393–2404. doi:10.1099/ijs.0.63229-0

Bass D, Richards TA, Matthai L, Marsh V, Cavalier-Smith T (2007) DNA evidence for global dispersal and probable endemicity of protozoa. BMC Evol Biol 7:162. doi:10.1186/1471-2148-7-162

Batten SD et al (2003) CPR sampling: the technical background, materials and methods, consistency and comparability. Prog Oceanogr 58:193–215. doi:10.1016/j.pocean.2003.08.004

Becquevort S, Dumont I, Tison JL, Lannuzel D, Sauvée ML, Chou L, Schoemann V (2009) Biogeochemistry and microbial community composition in sea ice and underlying seawater off East Antarctica during early spring. Polar Biol 32:879–895. doi:10.1007/s00300-009-0589-2

Béjà O et al (2000) Bacterial rhodopsin: evidence for a new type of phototrophy in the sea. Science 289:1902–1906

Berney C, Fahrni J, Pawlowski J (2004) How many novel eukaryotic 'kingdoms'? Pitfalls and limitations of environmental DNA surveys. BMC Biol 2:13. doi:10.1186/1741-7007-2-13

Bluhm BA, Gradinger R (2008) Regional variability in food availability for Arctic marine mammals. Ecol Appl 18:S77–S96. doi:10.1890/06-0562.1

Booth BC, Marchant HJ (1987) Parmales, a new order of marine chrysophytes, with descriptions of three new genera and seven new species. J Phycol 23:245–260. doi:10.1111/j.1529-8817. 1987.tb04132.x

Borgå K, Saloranta TM, Ruus A (2010) Simulating climate change-induced alterations in bioaccumulation of organic contaminants in an Arctic marine food web. Environ Toxicol Chem 29:1349–1357. doi:10.1002/etc.159

Borriss M, Helmke E, Hanschke R, Schweder T (2003) Isolation and characterization of marine psychrophilic phage-host systems from Arctic sea ice. Extremophiles 7:377–384. doi:10.1007/s00792-003-0334-7

Bouvier TC, del Giorgio PA (2002) Compositional changes in free-living bacterial communities along a salinity gradient in two temperate estuaries. Limnol Oceanogr 47:453–470. doi:10.4319/lo.2002.47.2.0453

Bowman JS, Deming JW (2010) Elevated bacterial abundance and exopolymers in saline frost flowers and implications for atmospheric chemistry and microbial dispersal. Geophys Res Lett 37. doi:10.1029/2010GL043020

Bowman JP, McCammon SA, Brown MV, Nichols DS, McMeekin TA (1997) Diversity and association of psychrophilic bacteria in Antarctic sea ice. Appl Environ Microbiol 63: 3068–3078

Bowman JS, Rasmussen S, Blom N, Deming JW, Rysgaard S, Sicheritz-Ponten T (2012) Microbial community structure of Arctic multiyear sea ice and surface seawater by 454 sequencing of the 16S RNA gene. ISME J 6:11–20. doi:10.1038/ismej.2011.76

Boyce DG, Lewis MR (2010) Worm B. Global phytoplankton decline over the past century Nature 466:591–596, http://www.nature.com/nature/journal/v466/n7306/abs/nature09268. html#supplementary-information

Boyd PW, Abraham ER (2001) Iron-mediated changes in phytoplankton photosynthetic competence during SOIREE. Deep Sea Res II Top Stud Oceanogr 48:2529–2550. doi:10.1016/S0967-0645(01)00007-8

Brinkmeyer R, Knittel K, Jürgens J, Weyland H, Amann R, Helmke E (2003) Diversity and structure of bacterial communities in Arctic versus Antarctic pack ice. Appl Environ Microbiol 69:6610–6619. doi:10.1128/AEM.69.11.6610-6619.2003

Bronk DA, Ward BB (2005) Inorganic and organic nitrogen cycling in the Southern California Bight. Deep Sea Res I Oceanogr Res Pap 52:2285–2300. doi:10.1016/j.dsr.2005.08.002

Brown ZW, Arrigo KR (2013) Sea ice impacts on spring bloom dynamics and net primary production in the Eastern Bering Sea. J Geophys Res Oceans 118:43–62. doi:10.1029/2012JC008034

Brown CT et al (2015) Unusual biology across a group comprising more than 15% of domain Bacteria. Nature 523:208–211. doi:10.1038/nature14486, http://www.nature.com/nature/journal/v523/n7559/abs/nature14486.html#supplementary-information

Budge SM, Wooller MJ, Springer AM, Iverson SJ, McRoy CP, Divoky GJ (2008) Tracing carbon flow in an Arctic marine food web using fatty acid-stable isotope analysis. Oecologia 157: 117–129. doi:10.1007/s00442-008-1053-7

Campbell LM, Norstrom RJ, Hobson KA, Muir DCG, Backus S, Fisk AT (2005) Mercury and other trace elements in a pelagic Arctic marine food web (Northwater Polynya, Baffin Bay). Sci Total Environ 351–352:247–263. doi:10.1016/j.scitotenv.2005.02.043

Caron DA (2009) Past president's address: Protistan biogeography: why all the fuss? J Eukaryot Microbiol 56:105–112. doi:10.1111/j.1550-7408.2008.00381.x

Cermeno P, de Vargas C, Abrantes F, Falkowski PG (2010) Phytoplankton biogeography and community stability in the ocean. PLoS One 5:e10037. doi:10.1371/journal.pone.0010037

Cho JC, Giovannoni SJ (2004) Cultivation and growth characteristics of a diverse group of oligotrophic marine Gammaproteobacteria. Appl Environ Microbiol 70:432–440

Chown SL, Convey P (2007) Spatial and temporal variability across life's hierarchies in the terrestrial Antarctic. Philos Trans R Soc B Biol Sci 362:2307–2331. doi:10.1098/rstb.2006.1949

Chown SL, Clarke A, Fraser CI, Cary SC, Moon KL, McGeoch MA (2015) The changing form of Antarctic biodiversity. Nature 522:431–438. doi:10.1038/nature14505

Coleman AW (2009) Is there a molecular key to the level of "biological species" in eukaryotes? A DNA guide. Mol Phylogenet Evol 50:197–203. doi:10.1016/j.ympev.2008.10.008

Collins A (2001) Ocean circulation. In: University O (ed) Ocean circulation. Butterworth-Heinemann, Boston, MA, pp 208–209

Collins RE, Rocap G, Deming JW (2010) Persistence of bacterial and archaeal communities in sea ice through an Arctic winter. Environ Microbiol 12:1828–1841. doi:10.1111/j.1462-2920.2010.02179.x

Comeau AM, Li WKW, Tremblay J-É, Carmack EC, Lovejoy C (2011) Arctic Ocean microbial community structure before and after the 2007 record sea ice minimum. PLoS One 6:e27492. doi:10.1371/journal.pone.0027492

Comiso JC (2006) Arctic warming signals from satellite observations. Weather 61:70–76. doi:10.1256/wea.222.05

Comiso JC, Parkinson CL, Gersten R, Stock L (2008) Accelerated decline in the Arctic sea ice cover. Geophys Res Lett 35. doi:10.1029/2007GL031972

Connon SA, Giovannoni SJ (2002) High-throughput methods for culturing microorganisms in very-low-nutrient media yield diverse new marine isolates. Appl Environ Microbiol 68:3878–3885

Cota GF, Kottmeier ST, Robinson DH, Smith WO Jr, Sullivan CW (1990) Bacterioplankton in the marginal ice zone of the Weddell Sea: biomass, production and metabolic activities during austral autumn. Deep Sea Res A Oceanogr Res Pap 37:1145–1167. doi:10.1016/0198-0149(90)90056-2

Crump BC, Hopkinson CS, Sogin ML, Hobbie JE (2004) Microbial biogeography along an estuarine salinity gradient: combined influences of bacterial growth and residence time. Appl Environ Microbiol 70:1494–1505. doi:10.1128/AEM.70.3.1494-1505.2004

Curtis TP, Sloan WT, Scannell JW (2002) Estimating prokaryotic diversity and its limits. Proc Natl Acad Sci USA 99:10494–10499

Dalsgaard J, St John M, Kattner G, Muller-Navarra D, Hagen W (2003) Fatty acid trophic markers in the pelagic marine environment. Adv Mar Biol 46:225–340

Damm E, Helmke E, Thoms S, Schauer U, Nöthig E, Bakker K, Kiene RP (2010) Methane production in aerobic oligotrophic surface water in the central Arctic Ocean. Biogeosciences 7:1099–1108. doi:10.5194/bg-7-1099-2010

Darnis G, Robert D, Pomerleau C, Link H, Archambault P, Nelson RJ, Geoffroy M, Tremblay J-É, Lovejoy C, Ferguson SH, Hunt BPV, Fortier L (2012) Current state and trends in Canadian Arctic marine ecosystems: II. Heterotrophic food web, pelagic-benthic coupling, and biodiversity. Clim Change 115(1):179–205

Daugbjerg N, Moestrup Ø (1993) Four new species of Pyramimonas (Prasinophyceae) from Arctic Canada including a light and electron microscopic description of *Pyramimonas quadrifolia* sp. nov. Eur J Phycol 28:3–16. doi:10.1080/09670269300650021

Dawson SC, Pace NR (2002) Novel kingdom-level eukaryotic diversity in anoxic environments. Proc Natl Acad Sci USA 99:8324–8329. doi:10.1073/pnas.062169599

De Wit R, Bouvier T (2006) 'Everything is everywhere, but, the environment selects'; what did Baas Becking and Beijerinck really say? Environ Microbiol 8:755–758. doi:10.1111/j.1462-2920.2006.01017.x

DeLong EF, Wickham GS, Pace NR (1989) Phylogenetic stains: ribosomal RNA-based probes for the identification of single cells. Science 243:1360–1363

DeLong EF et al (2006) Community genomics among stratified microbial assemblages in the ocean's interior. Science 311:496–503. doi:10.1126/science.1120250

Deming JW (2002) Psychrophiles and polar regions. Curr Opin Microbiol 5:301–309

Devos N, Ingouff M, Loppes R, Matagne RF (1998) Rubisco adaptation to low temperatures: a comparative study in psychrophilic and mesophilic unicellular algae. J Phycol 34:655–660. doi:10.1046/j.1529-8817.1998.340655.x

Eicken H (1992) Salinity profiles of Antarctic sea ice: field data and model results. J Geophys Res Oceans 97:15545–15557. doi:10.1029/92JC01588

Epstein SS (1995) Simultaneous enumeration of protozoa and micrometazoa from marine sandy sediments. Aquat Microb Ecol 9(3):219–227

Fahlgren C et al (2015) Seawater mesocosm experiments in the Arctic uncover differential transfer of marine bacteria to aerosols. Environ Microbiol Rep 7:460–470. doi:10.1111/1758-2229.12273

Falk K, Benvenuti S, Dall Antonia L, Kampp K (2002) Foraging behaviour of thick-billed murres breeding in different sectors of the North Water polynya: an inter-colony comparison. Mar Ecol Prog Ser 231:293–302

Falkowski P, Koblfzek M, Gorbunov M, Kolber Z (2004) Development and application of variable chlorophyll fluorescence techniques in marine ecosystems. In: Papageorgiou G, Govindjee (eds) Chlorophyll a fluorescence, vol 19, Advances in photosynthesis and respiration. Springer, Dordrecht, pp 757–778. doi:10.1007/978-1-4020-3218-9_30

Fenchel T (1982) Ecology of heterotrophic microflagellates. IV. Quantitative occurrence and importance as bacterial consumers. Mar Ecol Prog Ser 9:35–42

Fenchel T (1988) Marine plankton food chains. Annu Rev Ecol Syst 19:19–38. doi:10.1146/annurev.es.19.110188.000315

Fenchel T, Finlay BJ (2003) Is microbial diversity fundamentally different from biodiversity of larger animals and plants? Eur J Protistol 39:486–490. doi:10.1078/0932-4739-00025

Fenchel T, Finlay BJ (2004) The ubiquity of small species: patterns of local and global diversity. BioScience 54(8):777–784

Ferrari VC, Hollibaugh JT (1999) Distribution of microbial assemblages in the Central Arctic Ocean Basin studied by PCR/DGGE: analysis of a large data set. Hydrobiologia 401:55–68. doi:10.1023/A:1003773907789

Filée J, Tétart F, Suttle CA, Krisch HM (2005) Marine T4-type bacteriophages, a ubiquitous component of the dark matter of the biosphere. Proc Natl Acad Sci USA 102:12471–12476. doi:10.1073/pnas.0503404102

Finlay BJ (2002) Global dispersal of free-living microbial eukaryote species. Science 296:1061–1063. doi:10.1126/science.1070710

Finlay BJ, Esteban GF, Brown S, Fenchel T, Hoef-Emden K (2006) Multiple cosmopolitan ecotypes within a microbial eukaryote morphospecies. Protist 157:377–390. doi:10.1016/j.protis.2006.05.012

Fiorini S, Middelburg JJ, Gattuso J-P (2011) Testing the effects of elevated pCo_2 on Coccolithophores (Prymnesiophyceae): comparison between haploid and diploid life stages. J Phycol 47:1281–1291. doi:10.1111/j.1529-8817.2011.01080.x

Foerstner KU, von Mering C, Bork P (2006) Comparative analysis of environmental sequences: potential and challenges. Philos Trans R Soc B Biol Sci 361:519–523. doi:10.1098/rstb.2005.1809

Fraser AJ, Sargent JR, Gamble JC, Seaton DD (1989) Formation and transfer of fatty acids in an enclosed marine food chain comprising phytoplankton, zooplankton and herring (*Clupea harengus* L.) larvae. Mar Chem 27:1–18. doi:10.1016/0304-4203(89)90024-8

Freitag TE, Chang L, Prosser JI (2006) Changes in the community structure and activity of betaproteobacterial ammonia-oxidizing sediment bacteria along a freshwater-marine gradient. Environ Microbiol 8:684–696. doi:10.1111/j.1462-2920.2005.00947.x

Frey KE, Comiso JC, Cooper LW, Gradinger RR, Grebmeier JM, Saitoh S-I, Tremblay J-E (2014) Arctic Ocean primary productivity. National Oceanic and Atmospheric Administration, Washington, DC, http://www.arctic.noaa.gov/reportcard/ocean_primary_productivity.html

Frezal L, Leblois R (2008) Four years of DNA barcoding: current advances and prospects. Infect Genet Evol 8:727–736. doi:10.1016/j.meegid.2008.05.005

Fuhrman JA (1999) Marine viruses and their biogeochemical and ecological effects. Nature 399:541–548

Fuhrman JA, Steele JA, Hewson I, Schwalbach MS, Brown MV, Green JL, Brown JH (2008) A latitudinal diversity gradient in planktonic marine bacteria. Proc Natl Acad Sci USA 105:7774–7778. doi:10.1073/pnas.0803070105

Galand PE, Casamayor EO, Kirchman DL, Lovejoy C (2009) Ecology of the rare microbial biosphere of the Arctic Ocean. Proc Natl Acad Sci USA 106:22427–22432. doi:10.1073/pnas.0908284106

Galand PE, Potvin M, Casamayor EO, Lovejoy C (2010) Hydrography shapes bacterial biogeography of the deep Arctic Ocean. ISME J 4:564–576. doi:10.1038/ismej.2009.134

Garcia-Martinez J, Rodriguez-Valera F (2000) Microdiversity of uncultured marine prokaryotes: the SAR11 cluster and the marine Archaea of Group I. Mol Ecol 9:935–948

Gerard E, Guyot F, Philippot P, Lopez-Garcia P (2005) Fluorescence in situ hybridisation coupled to ultra small immunogold detection to identify prokaryotic cells using transmission and scanning electron microscopy. J Microbiol Methods 63:20–28. doi:10.1016/j.mimet.2005.02.018

Gerdes B, Brinkmeyer R, Dieckmann G, Helmke E (2005) Influence of crude oil on changes of bacterial communities in Arctic sea-ice. FEMS Microbiol Ecol 53:129–139. doi:10.1016/j.femsec.2004.11.010

Ghiglione J-F et al (2012) Pole-to-pole biogeography of surface and deep marine bacterial communities. Proc Natl Acad Sci USA 109:17633–17638. doi:10.1073/pnas.1208160109

Giovannoni SJ, Stingl U (2005) Molecular diversity and ecology of microbial plankton. Nature 437:343–348. doi:10.1038/nature04158

Giovannoni SJ, Vergin KL (2012) Seasonality in ocean microbial communities. Science 335: 671–676. doi:10.1126/science.1198078

Giovannoni SJ, Britschgi TB, Moyer CL, Field KG (1990) Genetic diversity in Sargasso Sea bacterioplankton. Nature 345:60–63. doi:10.1038/345060a0

Giovannoni SJ, Rappe MS, Vergin KL, Adair NL (1996) 16S rRNA genes reveal stratified open ocean bacterioplankton populations related to the Green Non-Sulfur bacteria. Proc Natl Acad Sci USA 93:7979–7984

Gobeil C, Macdonald RW, Smith JN (1999) Mercury profiles in sediments of the Arctic Ocean Basins. Environ Sci Technol 33:4194–4198. doi:10.1021/es990471p

Golden KM, Ackley SF, Lytle VI (1998) The percolation phase transition in sea ice. Science 282: 2238–2241

Goldstein PZ, DeSalle R (2011) Integrating DNA barcode data and taxonomic practice: determination, discovery, and description. BioEssays 33:135–147. doi:10.1002/bies.201000036

Gons HJ, Ebert J, Hoogveld HL, van den Hove L, Pel R, Takkenberg W, Woldringh CJ (2002) Observations on cyanobacterial population collapse in eutrophic lake water. Antonie Van Leeuwenhoek 81:319–326

González JM, Moran MA (1997) Numerical dominance of a group of marine bacteria in the alpha-subclass of the class Proteobacteria in coastal seawater. Appl Environ Microbiol 63:4237–4242

Gordon DA, Giovannoni SJ (1996) Detection of stratified microbial populations related to Chlorobium and Fibrobacter species in the Atlantic and Pacific oceans. Appl Environ Microbiol 62:1171–1177

Gosselin M, Legendre L, Therriault J-C, Demers S (1990) Light and nutrient limitation of sea-ice microalgae (Hudson Bay, Canadian Arctic). J Phycol 26:220–232. doi:10.1111/j.0022-3646.1990.00220.x

Gosselin M, Levasseur M, Wheeler PA, Horner RA, Booth BC (1997) New measurements of phytoplankton and ice algal production in the Arctic Ocean. Deep Sea Res II Top Stud Oceanogr 44:1623–1644. doi:10.1016/S0967-0645(97)00054-4

Gradinger R (2008) Sea ice. Institute of Marine Sciences, University of Alaska, Fairbanks, AK, http://www.arcodiv.org/news/NPRB_report2_final.pdf

Gradinger R (2009) Sea-ice algae: major contributors to primary production and algal biomass in the Chukchi and Beaufort Seas during May/June 2002. Deep Sea Res II Top Stud Oceanogr 56: 1201–1212. doi:10.1016/j.dsr2.2008.10.016

Gradinger R, Bluhm B (2004) In-situ observations on the distribution and behavior of amphipods and Arctic cod (*Boreogadus saida*) under the sea ice of the High Arctic Canada Basin. Polar Biol 27:595–603. doi:10.1007/s00300-004-0630-4

Graeve M, Kattner G, Hagen W (1994) Diet-induced changes in the fatty acid composition of Arctic herbivorous copepods: experimental evidence of trophic markers. J Exp Mar Biol Ecol 182:97–110. doi:10.1016/0022-0981(94)90213-5

Gravalosa JM, Flores J-A, Sierro FJ, Gersonde R (2008) Sea surface distribution of coccolithophores in the eastern Pacific sector of the Southern Ocean (Bellingshausen and Amundsen Seas) during the late austral summer of 2001. Mar Micropaleontol 69:16–25. doi:10.1016/j.marmicro.2007.11.006

Grieg FS, Smith C, Azam F (1996) Abundance and production of bacteria and viruses in the Bering and Chukchi Seas. Mar Ecol Prog Ser 131:287–300

Guillou L, Chrétiennot-Dinet M-J, Medlin LK, Claustre H, Goër SL-d, Vaulot D (1999) Bolidomonas: a new genus with two species belonging to a new algal class, the Bolidophyceae (Heterokonta). J Phycol 35:368–381. doi:10.1046/j.1529-8817.1999.3520368.x

Guillou L et al (2004) Diversity of picoplanktonic prasinophytes assessed by direct nuclear SSU rDNA sequencing of environmental samples and novel isolates retrieved from oceanic and coastal marine ecosystems. Protist 155:193–214. doi:10.1078/143446104774199592

Halloran PR, Hall IR, Colmenero-Hidalgo E, Rickaby REM (2008) Evidence for a multi-species coccolith volume change over the past two centuries: understanding a potential ocean acidification response. Biogeosciences 5:1651–1655. doi:10.5194/bg-5-1651-2008

Hambly E, Suttle CA (2005) The viriosphere, diversity, and genetic exchange within phage communities. Curr Opin Microbiol 8:444–450. doi:10.1016/j.mib.2005.06.005

Harrison PJ, Cochlan WP, Acreman JC, Parsons TR, Thompson PA, Dovey HM, Xiaolin C (1986) The effects of crude oil and Corexit 9527 on marine phytoplankton in an experimental enclosure. Mar Environ Res 18:93–109. doi:10.1016/0141-1136(86)90002-4

Hassol SJ (2004) Impacts of a warming Arctic. Arctic Climate Impact Assessment, Canada

Hegseth EN, Sundfjord A (2008) Intrusion and blooming of Atlantic phytoplankton species in the high Arctic. J Mar Syst 74:108–119. doi:10.1016/j.jmarsys.2007.11.011

Helmke E, Weyland H (2004) Psychrophilic versus psychrotolerant bacteria–occurrence and significance in polar and temperate marine habitats. Cell Mol Biol 50:553–561

Hill V, Cota G (2005) Spatial patterns of primary production on the shelf, slope and basin of the Western Arctic in 2002. Deep Sea Res II Top Stud Oceanogr 52:3344–3354. doi:10.1016/j.dsr2.2005.10.001

Hillebrand H (2004) On the generality of the latitudinal diversity gradient. Am Nat 163:192–211. doi:10.1086/381004

Hirst MB, Kita KN, Dawson SC (2011) Uncultivated microbial eukaryotic diversity: a method to link SSU rRNA gene sequences with morphology. PLoS One 6:e28158. doi:10.1371/journal.pone.0028158

Hobbie JE, Daley RJ, Jasper S (1977) Use of nucleopore filters for counting bacteria by fluorescence microscopy. Appl Environ Microbiol 33:1225–1228

Hobson KA, Welch HE (1992) Observations of foraging northern fulmars (*Fulmarus glacialis*) in the Canadian High Arctic. Arctic 45(2):150–153

Hobson KA, Schell DM, Renouf D, Noseworthy E (1996) Stable carbon and nitrogen isotopic fractionation between diet and tissues of captive seals: implications for dietary reconstructions involving marine mammals. Can J Fish Aquat Sci 53:528–533. doi:10.1139/f95-209

Hollibaugh JT, Lovejoy C, Murray AE (2007) Microbiology in polar oceans. Oceanography 20:140–145

Horak REA, Whitney H, Shull DH, Mordy CW, Devol AH (2013) The role of sediments on the Bering Sea shelf N cycle: insights from measurements of benthic denitrification and benthic DIN fluxes. Deep Sea Res II Top Stud Oceanogr 94:95–105. doi:10.1016/j.dsr2.2013.03.014

Horner R, Schrader GC (1982) Relative contributions of ice algae, phytoplankton, and benthic microalgae to primary production in nearshore regions of the Beaufort Sea. Arctic 35(4): 457–571

Iglesias-Rodriguez MD et al (2008) Phytoplankton calcification in a high-CO_2 world. Science 320: 336–340

Ikävalko J, Gradinger R (1997) Flagellates and heliozoans in the Greenland Sea ice studied alive using light microscopy. Polar Biol 17:473–481. doi:10.1007/s003000050145

Janech MG, Krell A, Mock T, Kang J-S, Raymond JA (2006) Ice-binding proteins from sea ice diatoms (Bacillariophyceae). J Phycol 42:410–416. doi:10.1111/j.1529-8817.2006.00208.x

Johannessen OM, Shalina EV, Miles MW (1999) Satellite evidence for an Arctic sea ice cover in transformation. Science 286:1937–1939

Junge K, Swanson BD (2008) High-resolution ice nucleation spectra of sea-ice bacteria: implications for cloud formation and life in frozen environments. Biogeosciences 5:865–873. doi:10.5194/bg-5-865-2008

Junge K, Imhoff F, Staley T, Deming JW (2002) Phylogenetic diversity of numerically important Arctic sea-ice bacteria cultured at subzero temperature. Microb Ecol 43:315–328. doi:10.1007/s00248-001-1026-4

Junge K, Eicken H, Deming JW (2004) Bacterial activity at -2 to -20 degrees C in Arctic wintertime sea ice. Appl Environ Microbiol 70:550–557

Kapoor A et al (2008) A highly divergent picornavirus in a marine mammal. J Virol 82:311–320. doi:10.1128/jvi.01240-07

Karnovsky NJ et al (2011) Inter-colony comparison of diving behavior of an Arctic top predator: implications for warming in the Greenland Sea. Mar Ecol Prog Ser 440:229–240

Kedra M et al (2015) Status and trends in the structure of Arctic benthic food webs. Polar Res 34 (incl suppl)

Kelly BC, Ikonomou MG, Blair JD, Gobas FA (2008a) Bioaccumulation behaviour of polybrominated diphenyl ethers (PBDEs) in a Canadian Arctic marine food web. Sci Total Environ 401:60–72. doi:10.1016/j.scitotenv.2008.03.045

Kelly BC, Ikonomou MG, Blair JD, Gobas FAPC (2008b) Hydroxylated and methoxylated polybrominated diphenyl ethers in a Canadian Arctic marine food web. Environ Sci Technol 42:7069–7077. doi:10.1021/es801275d

Kirchman DL, Moran XAG, Ducklow H (2009) Microbial growth in the polar oceans - role of temperature and potential impact of climate change. Nat Rev Microbiol 7:451–459. doi:10.1038/nrmicro2115

Klaas C, Archer DE (2002) Association of sinking organic matter with various types of mineral ballast in the deep sea: implications for the rain ratio. Glob Biogeochem Cycles 16(4):1116. doi:10.1029/2001GB001765

Knoblauch C, Jorgensen BB, Harder J (1999) Community size and metabolic rates of psychrophilic sulfate-reducing bacteria in Arctic marine sediments. Appl Environ Microbiol 65: 4230–4233

Koch TA, Ekelund F (2005) Strains of the heterotrophic flagellate *Bodo designis* from different environments vary considerably with respect to salinity preference and SSU rRNA gene composition. Protist 156:97–112. doi:10.1016/j.protis.2004.12.001

Kolodziej K, Stoeck T (2007) Cellular identification of a novel uncultured marine stramenopile (MAST-12 Clade) small-subunit rRNA gene sequence from a Norwegian estuary by use of fluorescence in situ hybridization-scanning electron microscopy. Appl Environ Microbiol 73: 2718–2726. doi:10.1128/aem.02158-06

Kovacs KM (2014) Circumpolar ringed seal (*Pusa hispida*) monitoring: CAFF's ringed seal monitoring network. Norsk Polarinstitutt, Tromsø http://brage.bibsys.no/xmlui/handle/11250/191472

Krembs C, Deming J (2011) Sea ice: a refuge for life in polar seas? National Oceanic and Atmospheric Administration. Accessed 31 Jul 2015

Laidre KL, Stirling I, Lowry LF, Wiig Ø, Heide-Jørgensen MP, Ferguson SH (2008) Quantifying the sensitivity of Arctic marine mammals to climate-induced habitat change. Ecol Appl 18: S97–S125. doi:10.1890/06-0546.1

Laidre KL et al (2015) Arctic marine mammal population status, sea ice habitat loss, and conservation recommendations for the 21st century. Conserv Biol 29:724–737. doi:10.1111/cobi.12474

Lannuzel D et al. (2010) Distribution of dissolved iron in Antarctic sea ice: spatial, seasonal, and inter-annual variability. J Geophys Res Biogeosci 115. doi:10.1029/2009JG001031

Larsen J, Patterson DJ (1990) Some flagellates (Protista) from tropical marine sediments. J Nat Hist 24:801–937. doi:10.1080/00222939000770571

Larsen A et al (2004) Spring phytoplankton bloom dynamics in Norwegian coastal waters: microbial community succession and diversity. Limnol Oceanogr 49:180–190. doi:10.4319/lo.2004.49.1.0180

Larsen A, Egge JK, Nejstgaard JC, Di Capua I, Thyrhaug R, Bratbak G, Thingstad TF (2015) Contrasting response to nutrient manipulation in Arctic mesocosms are reproduced by a minimum microbial food web model. Limnol Oceanogr 60:360–374. doi:10.1002/lno.10025

Latasa M, Scharek R, Gall FL, Guillou L (2004) Pigment suites and taxonomic groups in Prasinophyceae. J Phycol 40:1149–1155. doi:10.1111/j.1529-8817.2004.03136.x

Lee WJ, Patterson DJ (1998) Diversity and geographic distribution of free-living heterotrophic flagellates - analysis by primer. Protist 149:229–244. doi:10.1016/s1434-4610(98)70031-8

Lee WJ, Patterson DJ (2000) Heterotrophic flagellates (Protista) from marine sediments of Botany Bay. Aust J Nat Hist 34:483–562. doi:10.1080/002229300299435

Legagneux P et al (2012) Disentangling trophic relationships in a High Arctic tundra ecosystem through food web modeling. Ecology 93:1707–1716

Legendre M, Teugels GG, Cauty C, Jalabert B (1992) A comparative study on morphology, growth rate and reproduction of *Clarias gariepinus* (Burchell, 1822), *Heterobranchus longifilis* Valenciennes, 1840, and their reciprocal hybrids (Pisces, Clariidae). J Fish Biol 40:59–79. doi:10.1111/j.1095-8649.1992.tb02554.x

Leliaert F, Verbruggen H, Wysor B, De Clerck O (2009) DNA taxonomy in morphologically plastic taxa: algorithmic species delimitation in the Boodlea complex (Chlorophyta: Cladophorales). Mol Phylogenet Evol 53:122–133. doi:10.1016/j.ympev.2009.06.004

Lennon JT, Khatana SA, Marston MF, Martiny JB (2007) Is there a cost of virus resistance in marine cyanobacteria? ISME J 1:300–312. doi:10.1038/ismej.2007.37

Letcher RJ, Norstrom RJ, Muir DCG (1998) Biotransformation versus bioaccumulation: sources of methyl sulfone PCB and 4,4′-DDE metabolites in the polar bear food chain. Environ Sci Technol 32:1656–1661. doi:10.1021/es970886f

Li WKW (1994) Primary production of prochlorophytes, cyanobacteria, and eucaryotic ultra-phytoplankton: measurements from flow cytometric sorting. Limnol Oceanogr 39:169–175. doi:10.4319/lo.1994.39.1.0169

Li WK, Dickie PM (2001) Monitoring phytoplankton, bacterioplankton, and virioplankton in a coastal inlet (Bedford Basin) by flow cytometry. Cytometry 44:236–246

Lim EL, Dennett MR, Caron DA (1999) The ecology of *Paraphysomonas imperforata* based on studies employing oligonucleotide probe identification in coastal water samples and enrichment cultures. Limnol Oceanogr 44:37–51. doi:10.4319/lo.1999.44.1.0037

Liu H et al (2009) Extreme diversity in noncalcifying haptophytes explains a major pigment paradox in open oceans. Proc Natl Acad Sci USA 106:12803–12808. doi:10.1073/pnas.0905841106

Lizotte MP, Priscu JC (1998) Pigment analysis of the distribution, succession, and fate of phytoplankton in the Mcmurdo Dry Valley Lakes of Antarctica. In: Ecosystem dynamics in a polar desert: the Mcmurdo Dry Valleys, Antarctica. American Geophysical Union, Washington, DC, pp 229–239. doi:10.1029/AR072p0229

Loeng H et al (2005) Marine systems. In: Arris L, Heal B, Symon C (eds) Arctic climate impact assessment. Cambridge University Press, Cambridge, pp 453–538

Lopez-Garcia P, Moreira D (2008) Tracking microbial biodiversity through molecular and genomic ecology. Res Microbiol 159:67–73. doi:10.1016/j.resmic.2007.11.019

Lopez-Garcia P, Rodriguez-Valera F, Pedrós-Alió C, Moreira D (2001) Unexpected diversity of small eukaryotes in deep-sea Antarctic plankton. Nature 409:603–607. doi:10.1038/35054537

Lovejoy C, Potvin M (2011) Microbial eukaryotic distribution in a dynamic Beaufort Sea and the Arctic Ocean. J Plankton Res 33:431–444. doi:10.1093/plankt/fbq124

Lovejoy C, Legendre L, Martineau M-J, Bâcle J, von Quillfeldt CH (2002) Distribution of phytoplankton and other protists in the North Water. Deep Sea Res II Top Stud Oceanogr 49:5027–5047. doi:10.1016/S0967-0645(02)00176-5

Lovejoy C, Price NM, Legendre L (2004) Role of nutrient supply and loss in controlling protist species dominance and microbial food-webs during spring blooms. Aquat Microb Ecol 34:79–92

Lovejoy C, Massana R, Pedrós-Alió C (2006) Diversity and distribution of marine microbial eukaryotes in the Arctic Ocean and adjacent seas. Appl Environ Microbiol 72:3085–3095. doi:10.1128/aem.72.5.3085-3095.2006

Lovejoy C et al (2007) Distribution, phylogeny, and growth of cold-adapted Picoprasinophytes in Arctic seas. J Phycol 43:78–89. doi:10.1111/j.1529-8817.2006.00310.x

Luo W, Li H, Cai M, He J (2009) Diversity of microbial eukaryotes in Kongsfjorden. Svalbard Hydrobiologia 636:233–248. doi:10.1007/s10750-009-9953-z

Mackey DJ, Higgins HW, Wright SW (1996) CHEMTAX - a program for estimating class abundances from chemical markers: application to HPLC measurements of phytoplankton. Mar Ecol Prog Ser 144:265–283. doi:10.3354/meps144265

Mann DG (1999) The species concept in diatoms. Phycologia 38:437–495

Marande W, López-García P, Moreira D (2009) Eukaryotic diversity and phylogeny using small- and large-subunit ribosomal RNA genes from environmental samples. Environ Microbiol 11:3179–3188. doi:10.1111/j.1462-2920.2009.02023.x

Margulies M et al (2005) Genome sequencing in microfabricated high-density picolitre reactors. Nature 437:376–380. doi:10.1038/nature03959

Marie D, Brussaard CPD, Thyrhaug R, Bratbak G, Vaulot D (1999) Enumeration of marine viruses in culture and natural samples by flow cytometry. Appl Environ Microbiol 65:45–52

Marie D, Zhu F, Balagué V, Ras J, Vaulot D (2006) Eukaryotic picoplankton communities of the Mediterranean Sea in summer assessed by molecular approaches (DGGE, TTGE, QPCR). FEMS Microbiol Ecol 55:403–415. doi:10.1111/j.1574-6941.2005.00058.x

Markus T, Stroeve JC, Miller J (2009) Recent changes in Arctic sea ice melt onset, freezeup, and melt season length. J Geophys Res Oceans 114. doi:10.1029/2009JC005436

Martiny JB et al (2006) Microbial biogeography: putting microorganisms on the map. Nat Rev Microbiol 4:102–112. doi:10.1038/nrmicro1341

Massana R, Pedrós-Alió C (2008) Unveiling new microbial eukaryotes in the surface ocean. Curr Opin Microbiol 11:213–218. doi:10.1016/j.mib.2008.04.004

Massana R et al (2004) Phylogenetic and ecological analysis of novel marine stramenopiles. Appl Environ Microbiol 70:3528–3534. doi:10.1128/AEM.70.6.3528-3534.2004

Massana R, del Campo J, Dinter C, Sommaruga R (2007) Crash of a population of the marine heterotrophic flagellate *Cafeteria roenbergensis* by viral infection. Environ Microbiol 9:2660–2669. doi:10.1111/j.1462-2920.2007.01378.x

Massana R, Unrein F, Rodriguez-Martinez R, Forn I, Lefort T, Pinhassi J, Not F (2009) Grazing rates and functional diversity of uncultured heterotrophic flagellates. ISME J 3:588–596

Matley JK, Crawford RE, Dick TA (2012) Summer foraging behaviour of shallow-diving seabirds and distribution of their prey, Arctic cod (*Boreogadus saida*), in the Canadian Arctic. Polar Res 31(incl suppl)

McClelland JW, Déry SJ, Peterson BJ, Holmes RM, Wood EF (2006) A pan-Arctic evaluation of changes in river discharge during the latter half of the 20th century. Geophys Res Lett 33. doi:10.1029/2006GL025753

Metz JG et al (2001) Production of polyunsaturated fatty acids by polyketide synthases in both prokaryotes and eukaryotes. Science 293:290–293. doi:10.1126/science.1059593

Middelburg JJ (2014) Stable isotopes dissect aquatic food webs from the top to the bottom. Biogeosciences 11:2357–2371. doi:10.5194/bg-11-2357-2014

Mock T, Junge K (2007) Psychrophilic diatoms: mechanisms for survival in freeze-thaw cycles. In: Seckbach J (ed) Algae and cyanobacteria in extreme environments. Springer, New York

Moestrup Ø (1979) Identification by electron microscopy of marine nanoplankton from New Zealand, including the description of four new species. NZ J Bot 17:61–95. doi:10.1080/0028825X.1979.10425161

Moline MA et al (2008) High latitude changes in ice dynamics and their impact on polar marine ecosystems. Ann NY Acad Sci 1134:267–319. doi:10.1196/annals.1439.010

Montagnes DJS, Berges JA, Harrison PJ, Taylor FJR (1994) Estimating carbon, nitrogen, protein, and chlorophyll a from volume in marine phytoplankton. Limnol Oceanogr 39:1044–1060. doi:10.4319/lo.1994.39.5.1044

Moon-van der Staay SY, De Wachter R, Vaulot D (2001) Oceanic 18S rDNA sequences from picoplankton reveal unsuspected eukaryotic diversity. Nature 409:607–610. doi:10.1038/35054541

Moore SE, Huntington HP (2008) Arctic marine mammals and climate change: impacts and resilience. Ecol Appl 18:S157–S165. doi:10.1890/06-0571.1

Morata N, Renaud PE, Brugel S, Hobson KA, Johnson BJ (2008) Spatial and seasonal variations in the pelagic–benthic coupling of the southeastern Beaufort Sea revealed by sedimentary biomarkers. Mar Ecol Prog Ser 371:47–63

Moritz RE, Bitz CM, Steig EJ (2002) Dynamics of recent climate change in the Arctic. Science 297:1497–1502

Murphy LS, Haugen EM (1985) The distribution and abundance of phototrophic ultraplankton in the North Atlantic. Limnol Oceanogr 30:47–58. doi:10.4319/lo.1985.30.1.0047

Mykytczuk NC, Foote SJ, Omelon CR, Southam G, Greer CW, Whyte LG (2013) Bacterial growth at -15 degrees C; molecular insights from the permafrost bacterium *Planococcus halocryophilus* Or1. ISME J 7:1211–1226. doi:10.1038/ismej.2013.8

Nassonova E, Smirnov A, Fahrni J, Pawlowski J (2010) Barcoding amoebae: comparison of SSU, ITS and COI genes as tools for molecular identification of naked lobose amoebae. Protist 161:102–115. doi:10.1016/j.protis.2009.07.003

Nichols DS, Olley J, Garda H, Brenner RR, McMeekin TA (2000) Effect of temperature and salinity stress on growth and lipid composition of *Shewanella gelidimarina*. Appl Environ Microbiol 66:2422–2429

Niederberger TD et al (2010) Microbial characterization of a subzero, hypersaline methane seep in the Canadian High Arctic. ISME J 4:1326–1339. doi:10.1038/ismej.2010.57

Niehoff B, Schmithüsen T, Knüppel N, Daase M, Czerny J, Boxhammer T (2013) Mesozooplankton community development at elevated CO2 concentrations: results from a mesocosm experiment in an Arctic fjord. Biogeosciences 10:1391–1406. doi:10.5194/bg-10-1391-2013

Nielsen S, Sand-Jensen K, Borum J, Geertz-Hansen O (2002) Depth colonization of eelgrass (*Zostera marina*) and macroalgae as determined by water transparency in Danish coastal waters. Estuaries 25:1025–1032. doi:10.1007/BF02691349

Niemi A, Michel C, Hille K, Poulin M (2011) Protist assemblages in winter sea ice: setting the stage for the spring ice algal bloom. Polar Biol 34:1803–1817. doi:10.1007/s00300-011-1059-1

Norkko A, Thrush SF, Cummings VJ, Gibbs MM, Andrew NL, Norkko J, Schwarz AM (2007) Trophic structure of coastal Antarctic food webs associated with changes in sea ice and food supply. Ecology 88:2810–2820

Norstrom RJ et al (1998) Chlorinated hydrocarbon contaminants in polar bears from eastern Russia, North America, Greenland, and Svalbard: biomonitoring of Arctic pollution. Arch Environ Contam Toxicol 35:354–367

Not F, Latasa M, Marie D, Cariou T, Vaulot D, Simon N (2004) A single species, *Micromonas pusilla* (Prasinophyceae), dominates the eukaryotic picoplankton in the Western English Channel. Appl Environ Microbiol 70:4064–4072. doi:10.1128/aem.70.7.4064-4072.2004

Not F et al (2007) Picobiliphytes: a marine picoplanktonic algal group with unknown affinities to other eukaryotes. Science 315:253–255. doi:10.1126/science.1136264

Not F, del Campo J, Balague V, de Vargas C, Massana R (2009) New insights into the diversity of marine picoeukaryotes. PLoS One 4:e7143. doi:10.1371/journal.pone.0007143

NSIDC (2015) All about sea ice. Accessed 31 Jul 2015

Pabi S, van Dijken GL, Arrigo KR (2008) Primary production in the Arctic Ocean, 1998–2006. J Geophys Res Oceans 113. doi:10.1029/2007JC004578

Pace NR (1997) A molecular view of microbial diversity and the biosphere. Science 276:734–740

Pavlov AK, Silyakova A, Granskog MA, Bellerby RGJ, Engel A, Schulz KG, Brussaard CPD (2014) Marine CDOM accumulation during a coastal Arctic mesocosm experiment: no response to elevated pCO_2 levels. J Geophys Res Biogeosci 119:1216–1230. doi:10.1002/2013JG002587

Pearce DA, Wilson WH (2003) Viruses in Antarctic ecosystems. Antarct Sci 15:319–331

Pedrós-Alió C (2006) Marine microbial diversity: can it be determined? Trends Microbiol 14:257–263. doi:10.1016/j.tim.2006.04.007

Peterson BJ et al (2002) Increasing river discharge to the Arctic. Ocean Sci 298:2171–2173. doi:10.1126/science.1077445

Peterson BJ, McClelland J, Curry R, Holmes RM, Walsh JE, Aagaard K (2006) Trajectory shifts in the Arctic and Subarctic freshwater cycle. Science 313:1061–1066

Petrich C, Eicken H (2009) Growth, structure and properties of sea ice. In: Dieckman GS, Thomas DN (eds) Sea ice. Wiley-Blackwell, Oxford, pp 23–77

Piepenburg D et al (2011) Towards a pan-Arctic inventory of the species diversity of the macro- and megabenthic fauna of the Arctic shelf seas. Mar Biodivers 41:51–70. doi:10.1007/s12526-010-0059-7

Pinhassi J, Berman T (2003) Differential growth response of colony-forming alpha- and gamma-proteobacteria in dilution culture and nutrient addition experiments from Lake Kinneret (Israel), the eastern Mediterranean Sea, and the Gulf of Eilat. Appl Environ Microbiol 69:199–211

Pomeroy LR, Weibe WJ (2001) Temperature and substrates as interactive limiting factors for marine heterotrophic bacteria. Aquat Microb Ecol 23:187–204

Pommier T et al (2007) Global patterns of diversity and community structure in marine bacterioplankton. Mol Ecol 16:867–880. doi:10.1111/j.1365-294X.2006.03189.x

Poulickova A, Vesela J, Neustupa J, Skaloud P (2010) Pseudocryptic diversity versus cosmopolitanism in diatoms: a case study on *Navicula cryptocephala* Kutz. (Bacillariophyceae) and morphologically similar taxa. Protist 161:353–369. doi:10.1016/j.protis.2009.12.003

Poulin M, Daugbjerg N, Gradinger R, Ilyash L, Ratkova T, von Quillfeldt C (2011) The pan-Arctic biodiversity of marine pelagic and sea-ice unicellular eukaryotes: a first-attempt assessment. Mar Biodivers 41:13–28. doi:10.1007/s12526-010-0058-8

Price PB, Bay RC (2012) Marine bacteria in deep Arctic and Antarctic ice cores: a proxy for evolution in oceans over 300 million generations. Biogeosciences 9:3799–3815. doi:10.5194/bg-9-3799-2012

Prokopowicz A, Rueckert S, Leander B, Michaud J, Fortier L (2010) Parasitic infection of the hyperiid amphipod *Themisto libellula* in the Canadian Beaufort Sea (Arctic Ocean), with a description of *Ganymedes themistos* sp. n. (Apicomplexa, Eugregarinorida). Polar Biol 33:1339–1350. doi:10.1007/s00300-010-0821-0

Purdy KJ, Nedwell DB, Embley TM (2003) Analysis of the sulfate-reducing bacterial and methanogenic archaeal populations in contrasting Antarctic sediments. Appl Environ Microbiol 69:3181–3191

Radulovici AE, Archambault P, Dufresne F (2010) DNA barcodes for marine biodiversity: moving fast forward? Diversity 2:450–472

Rappe MS, Connon SA, Vergin KL, Giovannoni SJ (2002) Cultivation of the ubiquitous SAR11 marine bacterioplankton clade. Nature 418:630–633. doi:10.1038/nature00917

Ravenschlag K, Sahm K, Pernthaler J, Amann R (1999) High bacterial diversity in permanently cold marine sediments. Appl Environ Microbiol 65:3982–3989

Raymond JA, Janech MG, Fritsen CH (2009) Novel ice-binding proteins from a psychrophilic Antarctic alga (chlamydomonadaceae, chlorophyceae). J Phycol 45:130–136. doi:10.1111/j.1529-8817.2008.00623.x

Renaud PE, Morata N, Ambrose WG Jr, Bowie JJ, Chiuchiolo A (2007) Carbon cycling by seafloor communities on the eastern Beaufort Sea shelf. J Exp Mar Biol Ecol 349:248–260. doi:10.1016/j.jembe.2007.05.021

Rex RM, Tiffany RAS, Matthew TC, David LK (2007) Diversity, abundance, and biomass production of bacterial groups in the western Arctic Ocean. Aquat Microb Ecol 47:45–55

Riebesell U, Gattuso JP, Thingstad TF, Middelburg JJ (2013) Preface — Arctic Ocean acidification: pelagic ecosystem and biogeochemical responses during a mesocosm study. Biogeosciences 10:5619–5626. doi:10.5194/bg-10-5619-2013

Riedel A, Michel C, Gosselin M, LeBlanc B (2008) Winter–spring dynamics in sea-ice carbon cycling in the coastal Arctic Ocean. J Mar Syst 74:918–932. doi:10.1016/j.jmarsys.2008.01.003

Riemann LF, Steward G, Fandino LB, Campbell L, Landry MR, Azam F (1999) Bacterial community composition during two consecutive NE Monsoon periods in the Arabian Sea studied by denaturing gradient gel electrophoresis (DGGE) of rRNA genes. Deep Sea Res II Top Stud Oceanogr 46:1791–1811. doi:10.1016/S0967-0645(99)00044-2

Rodriguez-Martinez R, Labrenz M, del Campo J, Forn I, Jurgens K, Massana R (2009) Distribution of the uncultured protist MAST-4 in the Indian Ocean, Drake Passage and Mediterranean Sea assessed by real-time quantitative PCR. Environ Microbiol 11:397–408. doi:10.1111/j.1462-2920.2008.01779.x

Rodriguez-Martinez R, Rocap G, Logares R, Romac S, Massana R (2012) Low evolutionary diversification in a widespread and abundant uncultured protist (MAST-4). Mol Biol Evol 29:1393–1406. doi:10.1093/molbev/msr303

Russell NJ (1997) Psychrophilic bacteria–molecular adaptations of membrane lipids. Comp Biochem Physiol A Physiol 118:489–493

Rysgaard SR, Thamdrup B, Risgaard-Petersen N, Fossing H, Berg P, Christensen PB, Dalsgaard T (1998) Seasonal carbon and nutrient mineralization in a high-Arctic coastal marine sediment, Young Sound, Northeast Greenland. Mar Ecol Prog Ser 175:261–276

Sanders RW, Gast RJ (2012) Bacterivory by phototrophic picoplankton and nanoplankton in Arctic waters. FEMS Microbiol Ecol 82:242–253. doi:10.1111/j.1574-6941.2011.01253.x

Sanders RW, Caron DA, Berninger U-G (1992) Relationships between bacteria and heterotrophic nanoplankton in marine and fresh waters: an inter-ecosystem comparison. Mar Ecol Prog Ser 86:1–14

Säwström C, Granéli W, Laybourn-Parry J, Anesio AM (2007) High viral infection rates in Antarctic and Arctic bacterioplankton. Environ Microbiol 9:250–255. doi:10.1111/j.1462-2920.2006.01135.x

Schlegel M, Meisterfeld R (2003) The species problem in protozoa revisited. Eur J Protistol 39:349–355. doi:10.1078/0932-4739-00003

Schmid MC et al (2007) Anaerobic ammonium-oxidizing bacteria in marine environments: widespread occurrence but low diversity. Environ Microbiol 9:1476–1484. doi:10.1111/j.1462-2920.2007.01266.x

Schnack-Schiel SB (2003) The macrobiology of sea ice. In: Thomas DN, Dieckmann GS (eds) Sea ice: an introduction to its physics, chemistry, biology and geology. Wiley-Blackwell, Oxford, pp 211–239

Serreze MC, Holland MM, Stroeve J (2007) Perspectives on the Arctic's shrinking sea-ice cover. Science 315:1533–1536

Seuthe L, Töpper B, Reigstad M, Thyrhaug R, Vaquer-Sunyer R (2011) Microbial communities and processes in ice-covered Arctic waters of the northwestern Fram Strait (75 to 80°N) during the vernal pre-bloom phase. Aquat Microb Ecol 64:253–266

Sheldon RW (1972) Size separation of marine seston by membrane and glass-fiber filters. Limnol Oceanogr 17:494–498. doi:10.4319/lo.1972.17.3.0494

Sherr EB, Sherr BF (2007) Heterotrophic dinoflagellates: a significant component of microzooplankton biomass and major grazers of diatoms in the sea. Mar Ecol Prog Ser 352:187–197

Short SM, Suttle CA (2002) Sequence analysis of marine virus communities reveals that groups of related algal viruses are widely distributed in nature. Appl Environ Microbiol 68:1290–1296. doi:10.1128/AEM.68.3.1290-1296.2002

Short CM, Suttle CA (2005) Nearly identical bacteriophage structural gene sequences are widely distributed in both marine and freshwater environments. Appl Environ Microbiol 71:480–486. doi:10.1128/aem.71.1.480-486.2005

Simon N, Barlow RG, Marie D, Partensky F, Vaulot D (1994) Characterization of oceanic photosynthetic picoeukaryotes by flow cytometry. J Phycol 30:922–935. doi:10.1111/j.0022-3646.1994.00922.x

Sintes E, del Giorgio PA (2010) Community heterogeneity and single-cell digestive activity of estuarine heterotrophic nanoflagellates assessed using lysotracker and flow cytometry. Environ Microbiol 12:1913–1925. doi:10.1111/j.1462-2920.2010.02196.x

Skarphedinsdottir H, Gunnarsson K, Gudmundsson GA, Nfon E (2010) Bioaccumulation and biomagnification of organochlorines in a marine food web at a pristine site in Iceland. Arch Environ Contamin Toxicol 58:800–809. doi:10.1007/s00244-009-9376-x

Skidmore M, Jungblut A, Urschel K, Junge (2012) Cryospheric environments in Polar regions (glaciers and ice sheets, sea ice and ice shelves). In: Whyte L, Miller RV (eds) Polar microbiology: life in a deep freeze. ASM Press, Washington, DC, pp 218–239

Slapeta J, Lopez-Garcia P, Moreira D (2006) Global dispersal and ancient cryptic species in the smallest marine eukaryotes. Mol Biol Evol 23:23–29. doi:10.1093/molbev/msj001

Smetacek V, Nicol S (2005) Polar ocean ecosystems in a changing world. Nature 437:362–368

Smith VH, Foster BL, Grover JP, Holt RD, Leibold MA, Denoyelles F Jr (2005) Phytoplankton species richness scales consistently from laboratory microcosms to the world's oceans. Proc Natl Acad Sci USA 102:4393–4396. doi:10.1073/pnas.0500094102

Sogin ML et al (2006) Microbial diversity in the deep sea and the underexplored "rare biosphere". Proc Natl Acad Sci USA 103:12115–12120. doi:10.1073/pnas.0605127103

Søreide JE, Hop H, Carroll ML, Falk-Petersen S, Hegseth EN (2006) Seasonal food web structures and sympagic–pelagic coupling in the European Arctic revealed by stable isotopes and a two-source food web model. Prog Oceanogr 71:59–87. doi:10.1016/j.pocean.2006.06.001

Sorhannus U, Ortiz JD, Wolf M, Fox MG (2010) Microevolution and speciation in *Thalassiosira weissflogii* (Bacillariophyta). Protist 161:237–249. doi:10.1016/j.protis.2009.10.003

Srinivas TN et al (2009) Bacterial diversity and bioprospecting for cold-active lipases, amylases and proteases, from culturable bacteria of kongsfjorden and Ny-alesund, Svalbard, Arctic. Curr Microbiol 59:537–547. doi:10.1007/s00284-009-9473-0

Stabeno P, Napp J, Mordy C, Whitledge T (2010) Factors influencing physical structure and lower trophic levels of the eastern Bering Sea shelf in 2005: sea ice, tides and winds. Prog Oceanogr 85:180–196

Stewart FJ, Fritsen CH (2004) Bacteria-algae relationships in Antarctic sea ice. Antarct Sci 16:143–156

Stewart R, Possingham H (2005) Efficiency, costs and trade-offs in marine reserve system design. Environ Model Assess 10:203–213. doi:10.1007/s10666-005-9001-y

Stirling I (1997) The importance of polynyas, ice edges, and leads to marine mammals and birds. J Mar Syst 10:9–21. doi:10.1016/S0924-7963(96)00054-1

Stoeck T, Taylor GT, Epstein SS (2003) Novel eukaryotes from the permanently anoxic Cariaco Basin (Caribbean Sea). Appl Environ Microbiol 69:5656–5663

Stoeck T, Hayward B, Taylor GT, Varela R, Epstein SS (2006) A multiple PCR-primer approach to access the microeukaryotic diversity in environmental samples. Protist 157:31–43. doi:10.1016/j.protis.2005.10.004

Stoeck T, Zuendorf A, Breiner HW, Behnke A (2007) A molecular approach to identify active microbes in environmental eukaryote clone libraries. Microb Ecol 53:328–339. doi:10.1007/s00248-006-9166-1

Stoeck T, Bass D, Nebel M, Christen R, Jones MDM, Breiner H-W, Richards TA (2010) Multiple marker parallel tag environmental DNA sequencing reveals a highly complex eukaryotic community in marine anoxic water. Mol Ecol 19:21–31. doi:10.1111/j.1365-294X.2009.04480.x

Stoecker DK, Gustafson DE, Black MMD, Baier CT (1998) Population dynamics of microalgae in the upper land-fast sea ice at a snow-free location. J Phycol 34:60–69. doi:10.1046/j.1529-8817.1998.340060.x

Stringer WJ, Groves JE (1991) Location and areal extent of Polynyas in the Bering and Chukchi Seas. Arctic 44(5):1–171

Stroeve J, Holland MM, Meier W, Scambos T, Serreze M (2007) Arctic sea ice decline: faster than forecast. Geophys Res Lett 34. doi:10.1029/2007GL029703

Suttle CA (2005) Viruses in the sea. Nature 437:356–361

Suttle CA (2007) Marine viruses — major players in the global ecosystem. Nat Rev Microbiol 5:801–812

Suzuki MT, Taylor LT, DeLong EF (2000) Quantitative analysis of small-subunit rRNA genes in mixed microbial populations via $5'$-nuclease assays. Appl Environ Microbiol 66:4605–4614

Tedersoo L et al (2010) 454 Pyrosequencing and Sanger sequencing of tropical mycorrhizal fungi provide similar results but reveal substantial methodological biases. New Phytol 188:291–301. doi:10.1111/j.1469-8137.2010.03373.x

Terrado R, Lovejoy C, Massana R, Vincent WF (2008) Microbial food web responses to light and nutrients beneath the coastal Arctic Ocean sea ice during the winter–spring transition. J Mar Syst 74:964–977. doi:10.1016/j.jmarsys.2007.11.001

Terrado R, Vincent WF, Lovejoy C (2009) Mesopelagic protists: diversity and succession in a coastal Arctic ecosystem. Aquat Microb Ecol 56:25–39

Terrado R, Medrinal E, Dasilva C, Thaler M, Vincent W, Lovejoy C (2011) Protist community composition during spring in an Arctic flaw lead polynya. Polar Biol 34:1901–1914. doi:10.1007/s00300-011-1039-5

Thiemann GW, Iverson SJ, Stirling I (2008) Polar bear diets and Arctic marine food webs: insights from fatty acid analysis. Ecol Monogr 78:591–613. doi:10.1890/07-1050.1

Thomas DN, Dieckmann GS (2002) Antarctic Sea ice--a habitat for extremophiles. Science 295:641–644. doi:10.1126/science.1063391

Thomas DN, Dieckmann GS (2010) Sea ice. Wiley-Blackwell, Oxford, 621 pp

Thomas DN, Kattner G, Engbrodt R, Giannelli V, Kennedy H, Haas C, Dieckmann GS (2001) Dissolved organic matter in Antarctic sea ice. Ann Glaciol 33:297–303. doi:10.3189/172756401781818338

Thomsen HA (1980) Two species of *Trigonaspis* gen. nov. (Prymnesiophyceae) from West Greenland. Phycologia 19:218–229. doi:10.2216/i0031-8884-19-3-218.1

Throndsen J (1978) Productivity and abundance of ultra- and nanoplankton in Oslofjorden. Sarsia 63:273–284. doi:10.1080/00364827.1978.10411349

Throndsen J, Kristiansen S (1991) *Micromonas pusilla* (Prasinophyceae) as part of pico-and nanoplankton communities of the Barents Sea. Polar Res 10(1) (Special Issue: Proceedings of the Pro Mare Symposium on Polar Marine Ecology, Part 1)

Tong S, Vørs N, Patterson DJ (1997) Heterotrophic flagellates, centrohelid heliozoa and filose amoebae from marine and freshwater sites in the Antarctic. Polar Biol 18:91–106. doi:10.1007/s003000050163

Tynan CT, DeMaster DP (1997) Observations and predictions of Arctic climatic change: potential effects on marine mammals. Arctic 50(4):289–399

Van den Hoek C, Mann DG, Jahns HM (1996) Algae: an introduction to phycology. Cambridge University Press, Cambridge

Vancoppenolle M, Goosse H, de Montety A, Fichefet T, Tremblay B, Tison J-L (2010) Modeling brine and nutrient dynamics in Antarctic sea ice: the case of dissolved silica. J Geophys Res Oceans 115. doi:10.1029/2009JC005369

Vaulot D, Eikrem W, Viprey M, Moreau H (2008) The diversity of small eukaryotic phytoplankton ($\leq$3 μm) in marine ecosystems. FEMS Microbiol Rev 32:795–820. doi:10.1111/j.1574-6976.2008.00121.x

Venter JC et al (2004) Environmental genome shotgun sequencing of the Sargasso Sea. Science 304:66–74. doi:10.1126/science.1093857

Vørs N (1992) Heterotrophic amoebae, flagellates and heliozoa from the Tvärminne area, Gulf of Finland, in 1988–1990. Ophelia 36:1–109. doi:10.1080/00785326.1992.10429930

Vørs N (1993) Marine heterotrophic amoebae, flagellates and heliozoa from Belize (Central America) and Tenerife (Canary Islands), with descriptions of new species, *Luffisphaera bulbochaete* N. Sp., *L. longihastis* N. Sp., *L. turriformis* N. Sp. and *Paulinella intermedia* N. sp. J Eukaryot Microbiol 40:272–287. doi:10.1111/j.1550-7408.1993.tb04917.x

Vyverman W et al (2010) Evidence for widespread endemism among Antarctic micro-organisms. Polar Sci 4:103–113. doi:10.1016/j.polar.2010.03.006

Walker G, Dorrell RG, Schlacht A, Dacks JB (2011) Eukaryotic systematics: a user's guide for cell biologists and parasitologists. Parasitology 138:1638–1663. doi:10.1017/s0031182010001708

Walsh JE, Chapman WL (2001) 20th-century sea-ice variations from observational data. Ann Glaciol 33:444–448. doi:10.3189/172756401781818671

Wang Q et al. (2014) Cloning, expression, purification, and characterization of glutaredoxin from Antarctic sea-ice bacterium *Pseudoalteromonas* sp. AN178. BioMed Res Int 2014:246871. doi:10.1155/2014/246871

Wells LE, Deming JW (2006) Modelled and measured dynamics of viruses in Arctic winter sea-ice brines. Environ Microbiol 8:1115–1121. doi:10.1111/j.1462-2920.2006.00984.x

Whitman WB, Coleman DC, Wiebe WJ (1998) Prokaryotes: the unseen majority. Proc Natl Acad Sci USA 95:6578–6583

Wright S, Ishikawa A, Marchant H, Davidson A, van den Enden R, Nash G (2009) Composition and significance of picophytoplankton in Antarctic waters. Polar Biol 32:797–808. doi:10.1007/s00300-009-0582-9

Zhu F, Massana R, Not F, Marie D, Vaulot D (2005) Mapping of picoeucaryotes in marine ecosystems with quantitative PCR of the 18S rRNA gene. FEMS Microbiol Ecol 52:79–92. doi:10.1016/j.femsec.2004.10.006